Environmental Science and Engineering

The ultimate goal of this series is to contribute to the protection of our environment, which calls for both profound research and the ongoing development of solutions and measurements by experts in the field. Accordingly, the series promotes not only a deeper understanding of environmental processes and the evaluation of management strategies, but also design and technology aimed at improving environmental quality. Books focusing on the former are published in the subseries Environmental Science, those focusing on the latter in the subseries Environmental Engineering.

Maulin P. Shah · Navneeta Bharadvaja ·
Lakhan Kumar
Editors

Biogenic Nanomaterials for Environmental Sustainability: Principles, Practices, and Opportunities

Springer

Editors
Maulin P. Shah
Applied and Environmental
Microbiology Lab
Industrial Wastewater Research Lab
Ankleshwar, Gujarat, India

Navneeta Bharadvaja
Department of Biotechnology
Delhi Technological University
Delhi, India

Lakhan Kumar
Department of Biotechnology
Delhi Technological University
Delhi, India

ISSN 1863-5520 ISSN 1863-5539 (electronic)
Environmental Science and Engineering
ISBN 978-3-031-45955-9 ISBN 978-3-031-45956-6 (eBook)
https://doi.org/10.1007/978-3-031-45956-6

This Springer imprint is published by the registered company Springer Nature Switzerland AG
The registered company address is: Gewerbestrasse 11, 6330 Cham, Switzerland

Contents

Chapter 1
Introduction to Bio-Nanotechnology

Raksha Anand, Kundan Kumar Mishra, and Navneeta Bharadvaja

Abstract Nanobiotechnology is the application of nanosized materials in various operational fields of biotechnology. The study of nanotubes, nanoparticles, nanofibers, nanowires, and other such nanometre-ranging structures is called nanoscience. This one billionth fine of a unit possesses properties for applications in various fields like bioremediation, nano-therapy, drug delivery, and so on. This book chapter describes the concept, key prospects, and features of bio-nanotechnology; the preparation, analysis, characterization, and application of nanomaterials for biological advancements. The major biological fields that benefit from nanotechnology have been briefly discussed.

Keywords Nanoscience · Bionanotechnology · Nanomaterials · Nano-therapy

1.1 Bionanotechnology

Nanotechnology and biology can have two approaches in the bigger picture: First, the use of nanomaterial tools and inspired processes in and on biological systems, and second, to use biological systems for nano-product development templates. Nanomaterials offer advantages like high stability, plasticity, and target selectivity overcoming the limitations of traditional biotechnological methods. Their wide range of applications in various majors qualifies nanotechnology to contribute towards a sustainable future and global economy (Jafarizadeh-Malmiri et al. 2019).

As declared in the National Nanotechnology Initiatives in the year 1999, the advanced area of Nanotechnology included the fabrication of an extensive range of nano-scaled materials, with properties dependent on their size and structure (Shahcheraghi et al. 2022). The early 21st century has witnessed the merging of

R. Anand · N. Bharadvaja (✉)
Department of Biotechnology, Delhi Technological University, New Delhi, India
e-mail: navneetab@dce.ac.in

K. K. Mishra
Department of Biomedical Engineering, Indian Institute of Technology, Ropar, India

M. P. Shah et al. (eds.), *Biogenic Nanomaterials for Environmental Sustainability: Principles, Practices, and Opportunities*, Environmental Science and Engineering,
https://doi.org/10.1007/978-3-031-45956-6_1

the massive advanced operational fields: Bio-technology and Nanotechnology. Nanomaterials are supposed to be 1–100 nm at least in one of their dimensions (Saleh 2020). Some of the important steps involved in nanotechnology can be hierarchically arranged as observing the bulk matter, measuring and manipulating, assembling, manufacturing, and controlling of biofunctionalization of nano-scale matter (National Nanotechnology Initiative 2023). The general procedure of nanomaterial synthesis can be categorised as either the top-down approach, where the bulk material is generally broken down to the nanoscale, or the bottom-up approach, where a monitored atom-by-atom or molecule-by-molecule addition is done (Abid et al. 2022). The dependence on non-renewable sources for persistent supply led to the emergence of bio-factories for nanomaterials, leading to the field of green nanotechnology (Kumar et al. 2022a; Kumar et al. 2022b). This is the biological mode of synthesis, while the other methods involve chemical (including sol–gel, vapor synthesis, and thermal decomposition) and mechanical routes (like milling) (Chinecherem Nkele and Ezema 2021).

Since the correlation ofthe structure–function relationship of nanomaterials is important for an efficient and successful application, nanomaterials are characterized for their functional, elemental, structural, morphological, and stability features. Hence relevant techniques are applied for determining the specific properties of nanoparticles (Chinecherem Nkele and Ezema 2021). For example, the morphology of nanomaterials is determined by scanning and transmission electron microscopy (SEM and TEM); optical microscopy finds applications in the determination of transmittance, luminescence, and reflectance features; X-ray diffractometry (XRD) and Fourier transform infrared (FTIR) being the most employed method for structural and elemental analysis respectively; and zeta potential stating its stability, to list the most common methods (Arakha and Jha 2018).

In order to understand the applications and advancements of nanotechnology and the journey of its amalgamation with biological sciences, it becomes important to trace the discovery timelines (Bayda et al. 2019). The integration of nanotechnology and different biological systems has led to the designing of efficient protocols and biological molecules of specificity and stability, helping in solving several critical problems. This chapter provides insight into the interdisciplinary researched applications of nanotechnology in enhancing biotechnological and industrial efficiencies.

1.2 Major Applications of Nanotechnology in Biotechnology

Nanomaterials find applications in diverse fields of medicine and healthcare, such as in drug delivery, for therapeutic and diagnostic processes. Some of these advanced applications are mentioned below.

1.2.1 Diagnostic Applications

Currently, diagnosing patients with most diseases requires the manifestation of visible symptoms before medical professionals can determine the patient's condition. However, by the time those symptoms are exhibited, treatment may not be as effective as it might have been. As a result, the earlier a disease is diagnosed, the greater the chance of a cure. In the current generation of medical devices, such as nanomaterials-based devices, there is greater sensitivity, as well as greater efficiency and economy. Some such applications of nanotechnology in diagnosis applications have been briefly mentioned.

1.2.1.1 Detection

In conventional clinical tests, specific antibodies bind to the disease-related targets in order to identify a molecule or disease-causing organism. In such tests, antibodies are usually conjugated with organic or inorganic dyes, and the signals are visualized using fluorescence microscopy. However, detection methods are often limited by the dyes, which limits their specificity and practicality. Nanomaterials are capable of withstanding substantially more cycles of excitations and light emissions than typical organic molecules, which decompose more easily.

1.2.1.2 Target-Specific Probes

Medical professionals prefer optical and colorimetric detection despite the advantages of magnetic detections. The company Nanosphere Inc (Northbrook, USA) developed a technique that allows doctors to visualize biological specimens' genetic makeup. Gold nanoparticles incorporated with DNA segments make the easy-to-read test for genetic sequences possible. In multiple nanospheres, sequences of interest form dense webs of gold balls when they attach to complementary DNA tentacles. The technology was shown to be highly sensitive in detecting anthrax, showing promising results when compared to the then-present tests.

1.2.1.3 Virus Detection

A multiplex colorimetric paper-based analytical device based on gold nanoparticles (Au-NPs) has recently been developed for the detection of DNA associated with viral infections, such as Coronavirus. Different nanomaterials such as AgNPs, MoS_2 nanosheets, QD-MP NPs, and Zr NPs have been applied in the detection of a range of Coronaviruses. To detect Coronaviruses, nanomaterials can be coupled with colorimetric sensing, electrochemiluminescence, immunosensing, photoluminescence, and chiro-immunosensing. Electrochemical devices can also be useful in the future for detecting new kinds of viruses due to their good ability to couple with nanomaterials. Nanomaterials also have applications in rapid RNA extraction from viruses like COVID-19 (Kailasa et al. 2021). In this aspect, nanomaterials can reduce analysis time and increase sensitivity, which can open the way for new approaches and higher performance in the future (Nikaeen et al. 2020).

1.2.1.4 Cancer Cell Detection

A cancer cell is both rare and physiologically distinct from its surrounding cells. Their significance lies in the detection and diagnosis of genetic defects. However, identifying and isolating these cancerous cells is challenging. Advancements in this field can be facilitated by nanomaterials. It has been demonstrated that 'nanosystems' can be used to sort cancerous cells from tissues in a highly efficient manner. Technology that employs cancer cells' unique characteristics, including differences in deformation, surface charge, receptor affinity, and ligand affinity, is used to treat cancer. Nano-based optical and electrochemical detection techniques can be useful in the development of point-of-care diagnostics for cancer cell detection. However, the nanoparticle-based sensor is regarded as an innovative research breakthrough for the early detection of cancer cells.

The extension of diagnosis application of nanomaterials has been applied in the fabrication of advanced forms of biosensors categorised as 'nano-biosensors' with diverse application fields (Sadana et al. 2018).

1.2.2 Therapeutic Applications of Nanomaterials and Drug Delivery

The use of nanoparticles as a therapeutic allows them to be delivered to targeted sites that are often difficult to reach by standard drugs. In this case, radio or magnetic signals can guide the therapeutic cargo to the site of the disease or infection if it is chemically attached to a nanoparticle. Additionally, nanodrugs can be programmed to release only when specific molecules are present or when external triggers are applied. Moreover, effective dosages can be reduced to avoid harmful side effects from potent medications. It is now possible to control drug release much more precisely than ever before by encapsulating drugs in nanomaterials and to the target site. The design of drugs enables them to carry therapeutic payloads (radiation, chemotherapy, or gene therapy) as well as imaging capabilities. Nanotechnology will improve the bioavailability of several agents that cannot be administered orally due to their poor bioavailability. In addition to protecting agents susceptible to degradation in extreme pH environments, nano-formulations can also extend the half-life of drugs by increasing their retention in the body through bio-adhesion. The delivery of antigens for vaccination is another broad application of nanotechnology. Nanotechnology can also contribute in the development of modern classes of vaccines. A recent study has demonstrated that microparticles and nanoparticles can enhance the immune response when encapsulated and developed into suitable animal models (Biju 2014).

1.2.3 Nanomaterials in Bioreactors and Energy Storage

Nanostructured materials are advantageous in offering huge surface-to-volume ratios, favorable transport properties, altered physical properties, and confinement effects resulting from the nanoscale dimensions. They have been extensively studied for energy-related applications such as solar cells, catalysts, thermoelectrics, lithium-ion batteries, supercapacitors, and hydrogen storage systems (Wang et al. 2020).

1.2.3.1 Energy Storage

Energy storage devices made from renewable sources are one of the best ways to address global environmental concerns and the current energy crisis. To construct highly efficient, stable, low cost and environment-friendly energy storage devices, suitable active materials must be explored. A large specific surface area nanomaterial was proposed as a guide to closing the gap between the achieved and theoretical capacitance without limiting the mass of the load. Batteries and supercapacitors are essential for storing the energy produced by renewable yet intermittent energy sources, such as solar and wind. Supercapacitors, which do not suffer from diffusion processes, store charge only at the surfaces, allowing them to achieve high power levels. Furthermore, supercapacitors are much more reversible and have a longer cycle life since they do not undergo a bulk phase change when charged and discharged. The bulk phase of electrode materials in batteries stores charge through redox reactions, which provides higher energy density but lower power performance than supercapacitors. Over the past few years, high-performance rechargeable batteries have become so ubiquitous and tangible that they have almost become common knowledge. Nanomaterials have contributed to improved performances of such cells (Savla et al. 2020).

1.2.3.2 Supercapacitors

A supercapacitor offers high energy density, fast charge–discharge rates, and a long lifespan, making it a great alternative to traditional capacitors and secondary batteries. Energy is stored in supercapacitors either through ion adsorption or redox reactions, where most of the charge is transferred near the electrode surface. Nanotechnology plays a role here. A supercapacitor is ideal for applications requiring a high-power density greater than lithium-ion batteries, requiring at least 10 $kWkg^{-1}$. Due to CDs' large specific surface area, hybridizing carbon dots (CDs) with symmetrical supercapacitors can significantly improve their electrochemical performance (Pang et al. 2020).

1.2.3.3 Batteries

Energy storage is a crucial function of batteries, which convert electrical energy into chemical energy via electrochemical reactions and release this energy through reverse reactions. The battery industry has enjoyed remarkable success in both research and commercialization over the past few decades. Battery demand is increasing in today's society as a result of common problems and societal advancement. In order to improve the performance of advanced batteries, they must have a high energy density, a high-power density, a long cycle life, and good safety characteristics. Battery development is also focusing on convenience and comfort, which require flexible, lightweight, and wearable batteries. A battery consists of three key components: electrodes, electrolyte, and separator. In order to create batteries that operate efficiently, electrode materials need to be electrochemically reactive, reversible, chemically stable, and non-toxic. The additional functionality of the battery is achieved using materials with special features. For example, flexible batteries require electrodes that are flexible enough to maintain high performance even under severe bending. Solid-state batteries require an electrolyte material with high ionic conductivity to ensure the ion transport rate inside the batteries, which cannot be achieved with general solid materials. In various energy applications, thermoelectric, piezoelectric, triboelectric, photovoltaic, and catalytic materials have played an important role at the nanoscale. Multifunctional inorganic nanomaterials play a crucial role in the development of advanced energy applications with enhanced performance due to their unique properties, such as excellent electrical and thermal conductivity, large surface areas, and chemical stability.

1.2.4 Nanomaterials for Environmental Remediation

Industrialization and its by-products have resulted in hazardous wastes, poisonous gasses, and smoke, which have contaminated the environment. Conventional techniques have been used to treat all types of organic and toxic waste by adsorption, biological oxidation, chemical oxidation, and incineration. In recent years, nanomaterials have gained much attention due to their ability to destroy demilitarization munitions wastes and complex industrial chemical wastes. Nanomaterials have attracted a great deal of interest in the environmental application. In the treatment of pollution in water and air, nanomaterials play a critical role in environmental remediation. Their unique properties (such as excellent adsorbents, catalysts, large specific surface areas, and high reactivities) can be used to degrade and scavenge pollutants in the air and water (Kumar and Bharadvaja 2022). The species adsorbed on nanomaterials can be removed by using a mild gravitational or magnetic force. Various nanomaterials play a significant role in influencing water and air quality in the natural environment based on their shapes. The magnetic properties of nano-adsorbents make them particularly attractive for water treatment as well as for water retention. Nanomaterials are also helpful in the detection as well as catalytic degradation of dyes, pesticides, and heavy metals (e.g., cadmium, copper, lead, mercury, arsenic, etc.) (Khin et al. 2012; Kumar et al. 2021).

1.2.5 Nanomaterials in Industries—Food, Oil, Cosmetics, Packaging

1.2.5.1 Nanomaterials in Food and Food Packaging

A nanomaterial can be used as an additive for polymers to enhance their performance due to its unique physical and chemical properties. A variety of applications can be achieved by using improved nanofillers like silicate and clay nanoplatelets, graphene, carbon nanotubes, and silica nanoparticles that keep their color, flavor, and texture, maintaining their stability during storage, and decreasing spoilage. The antimicrobial properties of nanomaterials have led to their application in food packaging. Among them, the control of the synthesis of metal nanomaterials, such as nano-Ag, is crucial to their antimicrobial applications. Nano-emulsions maintain the creamy texture of food alternately reducing the fat content of the food (Aswathanarayan and Vittal 2019). A number of studies have shown that nanoparticles can effectively inactivate microorganisms by disrupting their cell walls, interacting with thiol groups in DNA, phosphorus and sulfur, proteins, and enzymes, and through cell respiration. The generation of ROS caused by nanoparticles disrupts cell membranes, damages DNA and mitochondria, and disrupts electron transport across the cell membrane. Packaging materials can be made of a single polymer layer that prevents O_2 and H_2O from entering, or they can be reinforced by a nanocomposite layer to enhance barrier performance (Emamhadi et al. 2020).

1.2.5.2 Nanomaterials in Cosmetics

Nanomaterial-based cosmetics have some unique advantages over micro-scale cosmetics. Cosmetic manufacturers use nanomaterials to boost long-term results and stability. Because nanomaterials have a high surface area, the ingredients can be transported more efficiently through the skin. A key advantage of using nanomaterials in cosmetics is that they penetrate efficiently into the skin, facilitating the delivery of the product's ingredients, introducing new color elements (such as lipsticks and nail polishes), improving transparency (such as sunscreens), and extending the lasting effects of makeup. It is the ultimate goal of cosmetic manufacturers to deliver the right amount of ingredients to the desired areas of the body using nanomaterials and to achieve long-term stability with nanomaterials. Most commonly, nanomaterials are used in cosmetics as UV filters in skin care products (Fytianos et al. 2020).

1.2.6 Nanozymes, Nanomaterials as Nano-Bio-Catalysts/ Enzymes and Additives

Nanomaterials having inherent enzymatic properties or regular enzymes incorporated with nanomaterials are classified as Nanozymes (Xiong et al. 2021).

In recent years, nanomaterials (like Graphene, and graphene oxide (GO)) are increasingly used as immobilization matrices for enzymes in bio-catalytic applications. A variety of functional groups (such as hydroxyl, carboxyl, epoxy, carbonyl, phenol, lactone, and quinone) that can be manipulated on graphene surfaces, coupled with its unique electrochemical and mechanical properties, make graphene materials ideal for the attachment of enzymes. The technique of enzyme immobilization makes it possible to separate enzymes, modify their catalytic properties, and reuse them. However, GO-based carbon nanomaterials offer good immobilization options for enzymes because of their layered structure, large surface area, high functionalization potential, fascinating electronic properties, and excellent thermal and mechanical stability. In addition, these unique properties of GO have made it an effective candidate in a number of biotechnological, bioelectronic, bio-imaging, tissue engineering, and biosensing applications (Adeel et al. 2018).

1.2.7 Highly Advantageous Metal–Organic Frameworks (MOFs)

Nanomaterials can be used to exploit the metal-organic frameworks (MOF) of nanozymes. This not only adds to the porosity of the structure to allow an eased entry of molecules and exclusivity but also makes an overall improvement in the catalytic feature of the enzyme by adding to available active sites. These find applications in designing efficient biocatalysts, and novel biosensors as well as upgrading biomedical imaging systems (Cai et al. 2022).

1.2.8 Nanomaterials in Agriculture Industry

In the 21st century, nanomaterials have been applied to agriculture for the purpose of increasing the productivity of lands and crops, especially under suboptimal conditions. The application of nanotechnology to agricultural science remains relatively under-explored. It has been discovered that there are a number of nanomaterials with the potential to revolutionize agriculture, with both advantages and disadvantages. In addition to improving food quality, crop growth, and monitoring environmental conditions, they often solve many agricultural problems (e.g., problems in soil structure, detection of pollutants, plant disease, pests, and pathogens, delivery of pesticides, fertilizers, and nutrients, and genetic materials). The agricultural industry utilizes a wide range of nanomaterials, including single-walled carbon nanotubes (SWCNT), multi-walled carbon nanotubes (MWCNTs), or those from graphene oxide (GO), silver (Ag), iron (Fe), silicon (Si), zinc (Zn) and zinc oxide (ZnO). The promising aspects of this methodology include facilitating the delivery of numerous essential compounds, protecting against pathogens and diseases, improving nutrient absorption, and increasing pesticide and fungicide efficiency, thereby enhancing plant growth and ensuring that fertilizers are released in a site-specific and controlled manner. Due to the fact that nanomaterials can be applied to almost every part of agriculture, such as production, processing, storage, and transportation, they drastically improve efficiency, productivity, and agricultural protection (Khot et al. 2012).

1.3 Challenges and Way Forward

The risk of nanomaterials and their application in biological systems has not yet been adequately analysed. This demands a standardised method for the measurement of toxicity or hazardous effects on humans and the environment to be identified. Some of the key challenges that nanobiotechnology might face include public acceptance, safety aspects, risk assessment, and regulation. It is extremely important to draw the structure–function relationship of nanomaterials to attain a successful application. Along with the identification of the perfect nanomaterial for a particular function, sustainable, scalable, and economic production of such nanomaterials will be a challenge (Zhang 2017).

There are scopes for introducing biodegradable nanoparticles to deliver water-soluble or insoluble therapeutics with improved biocompatibility and retention time. The industrial sector can expand the horizons for nanobiotechnology in the production of ingredients like nano-additives and nanozymes. Medical devices and technology can advance their design to provide a molecular or sub-cellular interaction having high specificity (Fakruddin et al. 2012).

References

Abid N et al (2022) Synthesis of nanomaterials using various top-down and bottom-up approaches, influencing factors, advantages, and disadvantages: a review. Adv Colloid Interface Sci 300:102597. https://doi.org/10.1016/j.cis.2021.102597

Adeel M, Bilal M, Rasheed T, Sharma A, Iqbal HMN (2018) Graphene and graphene oxide: functionalization and nano-bio-catalytic system for enzyme immobilization and biotechnological perspective. Int J Biol Macromol 120:1430–1440. https://doi.org/10.1016/j.ijbiomac.2018.09.144

Arakha M, Jha S (2018) Synthesis and characterization of nanoparticles, pp 37–59. https://doi.org/10.1007/978-3-319-73326-5_2

Aswathanarayan JB, Vittal RR (2019) Nanoemulsions and their potential applications in food industry. Front Sustain Food Syst 3. https://doi.org/10.3389/fsufs.2019.00095

Bayda S, Adeel M, Tuccinardi T, Cordani M, Rizzolio F (2019) The history of nanoscience and nanotechnology: from chemical-physical applications to nanomedicine. Molecules 25(1):112. https://doi.org/10.3390/molecules25010112

Biju V (2014) Chemical modifications and bioconjugate reactions of nanomaterials for sensing, imaging, drug delivery and therapy. Chem Soc Rev 43(3):744–764. https://doi.org/10.1039/C3CS60273G

Cai X, Bao X, Wu Y (2022) Metal–organic frameworks as intelligent drug nanocarriers for cancer therapy. Pharmaceutics 14(12):2641. https://doi.org/10.3390/pharmaceutics14122641

Chinecherem Nkele A, Ezema FI (2021) Diverse synthesis and characterization techniques of nanoparticles. In: Thin films. IntechOpen. https://doi.org/10.5772/intechopen.94453

Emamhadi MA et al (2020) Nanomaterials for food packaging applications: a systematic review. Food Chem Toxicol 146:111825. https://doi.org/10.1016/j.fct.2020.111825

Fakruddin M, Hossain Z, Afroz H (2012) Prospects and applications of nanobiotechnology: a medical perspective. J Nanobiotechnology 10(1):31. https://doi.org/10.1186/1477-3155-10-31

Fytianos G, Rahdar A, Kyzas GZ (2020) Nanomaterials in cosmetics: recent updates. Nanomaterials 10(5):979. https://doi.org/10.3390/nano10050979

Jafarizadeh-Malmiri H, Sayyar Z, Anarjan N, Berenjian A (2019) Nanobiotechnology in food: concepts, applications and perspectives. Springer International Publishing, Cham. https://doi.org/10.1007/978-3-030-05846-3

Kailasa SK et al (2021) An overview of molecular biology and nanotechnology based analytical methods for the detection of SARS-CoV-2: promising biotools for the rapid diagnosis of COVID-19. Analyst 146(5):1489–1513. https://doi.org/10.1039/D0AN01528H

Khin MM, Nair AS, Babu VJ, Murugan R, Ramakrishna S (2012) A review on nanomaterials for environmental remediation. Energy Environ Sci 5(8):8075. https://doi.org/10.1039/c2ee21818f

Khot LR, Sankaran S, Maja JM, Ehsani R, Schuster EW (2012) Applications of nanomaterials in agricultural production and crop protection: a review. Crop Prot 35:64–70. https://doi.org/10.1016/j.cropro.2012.01.007

Kumar L, Bharadvaja N (2022) Biosynthesis, characterization, and evaluation of antibacterial and photocatalytic dye degradation activities of silver nanoparticles biosynthesized by *Chlorella sorokiniana*. Biomass Convers Biorefin. https://doi.org/10.1007/s13399-022-03433-w

Kumar L, Ragunathan V, Chugh M, Bharadvaja N (2021) Nanomaterials for remediation of contaminants: a review. Environ Chem Lett 19(4):3139–3163. https://doi.org/10.1007/s10311-021-01212-z

Kumar L, Mohan L, Anand S, Bhardwaj D, Bharadvaja N (2022a) Phyconanoremediation: a sustainable approach to deal with environmental pollutants heavy metals and dyes. Vegetos. https://doi.org/10.1007/s42535-022-00399-y

Kumar L, Kamal P, Soni K, Bharadvaja N (2022b) Phytonanotechnology for remediation of heavy metals and dyes. In: Nanotechnology for environmental remediation. Wiley, pp 139–162. https://doi.org/10.1002/9783527834143.ch9

National Nanotechnology Initiative (NNI) (2023) https://www.nano.gov/

Nikaeen G, Abbaszadeh S, Yousefinejad S (2020) Application of nanomaterials in treatment, anti-infection and detection of coronaviruses. Nanomedicine 15(15):1501–1512. https://doi.org/10.2217/nnm-2020-0117
Pang H, Cao X, Zhu L, Zheng M (2020) Nanomaterials for supercapacitors. In: Synthesis of functional nanomaterials for electrochemical energy storage. Springer Singapore, Singapore, pp 195–220. https://doi.org/10.1007/978-981-13-7372-5_7
Sadana A, Sadana N, Sadana R (2018) Nanobiosensors. In: A fractal analysis of chemical kinetics with applications to biological and biosensor interfaces, pp 69–96. https://doi.org/10.1016/B978-0-444-63872-4.00004-X
Saleh TA (2020) Nanomaterials: classification, properties, and environmental toxicities. Environ Technol Innov 20:101067. https://doi.org/10.1016/j.eti.2020.101067
Savla N, Anand R, Pandit S, Prasad R (2020) Utilization of nanomaterials as anode modifiers for improving microbial fuel cells performance. J Renew Mater 8(12):1581–1605. https://doi.org/10.32604/jrm.2020.011803
Shahcheraghi N, Golchin H, Sadri Z, Tabari Y, Borhanifar F, Makani S (2022) Nano-biotechnology, an applicable approach for sustainable future. 3 Biotech 12(3):65. https://doi.org/10.1007/s13205-021-03108-9
Wang H, Liang X, Wang J, Jiao S, Xue D (2020) Multifunctional inorganic nanomaterials for energy applications. Nanoscale 12(1):14–42. https://doi.org/10.1039/C9NR07008G
Xiong Y, Su L, Ye F, Zhao S (2021) Porous oxyhydroxide derived from metal-organic frameworks as efficient triphosphatase-like nanozyme for chromium(III) ion colorimetric sensing. ACS Appl Bio Mater 4(9):6962–6973. https://doi.org/10.1021/acsabm.1c00628
Zhang F (2017) Grand challenges for nanoscience and nanotechnology in energy and health. Front Chem 5. https://doi.org/10.3389/fchem.2017.00080

Chapter 2
Biogenic Nanomaterials: Synthesis, Characterization, and Applications

Salem S. Salem and Alsayed E. Mekky

Abstract From the healthcare to the energy sectors, nanotechnology plays a significant part in our modern lives. Nanomaterials (NMs) can be synthesized using a variety of physical, chemical, and biological processes for a variety of purposes. The employment of microorganisms in the biological manufacturing of NMs has substantial benefits over other methods and is rapidly being investigated. The production of NMs by microbes such as bacteria, fungus, and algae, as well as their uses, are discussed in this chapter. Microorganisms create a diverse environment in which NMs can be synthesized. The issues that must be addressed include optimum production and shortest time to get the appropriate size and shape, enhancing nanoparticle stability, and optimizing certain microorganisms for specific applications. There have been several discussions on the use of bio-nanoparticles in medicine, the environment, drug delivery, agriculture, and biochemical sensors.

Keywords Green synthesis · Nanomaterials · Characterization · Application

2.1 Introduction

Nanoscience is the study of the unique properties of materials between 1 and 100 nm, and nanotechnology is the application of such research to create or modify novel objects. The ability to manipulate structures at the atomic scale allows for the creation of nanomaterials (Salem et al. 2023). Nanomaterials have unique optical, electrical and/or magnetic properties at the nanoscale, and these can be used in the fields of electronics and medicine, amongst other scenarios (Soliman et al. 2023; Abdelghany et al. 2023; Shehabeldine et al. 2023, 2022; Hashem et al. 2022a). Nanomaterials are

---

Salem S. Salem—Scopus Author ID 57202162965.

S. S. Salem (✉) · A. E. Mekky
Botany and Microbiology Department, Faculty of Science, Al-Azhar University, Cairo 11884, Egypt
e-mail: salemsalahsalem@azhar.edu.eg

M. P. Shah et al. (eds.), *Biogenic Nanomaterials for Environmental Sustainability: Principles, Practices, and Opportunities*, Environmental Science and Engineering,
https://doi.org/10.1007/978-3-031-45956-6_2

unique as they provide a large surface area to volume ratio. Unlike other large-scaled engineered objects and systems, nanomaterials are governed by the laws of quantum mechanics instead of the classical laws of physics and chemistry. In short, nanotechnology is the engineering of useful objects and functional systems at the molecular or atomic scale (Ariga and Yamauchi 2020). Nanotechnologies have had a significant impact in almost all industries and areas of society as it offers (i) better built, (ii) safer and cleaner, (iii) longer-lasting and (iv) smarter products for medicine, communications, everyday life, agriculture and other industries (Doghish et al. 2022; Elakraa et al. 2022; Abdelaziz et al. 2022; Salem and Husen 2022). The use of nanomaterials in everyday products can be generally divided into two types. First, nanomaterials can be merged or added to a pre-existing product and improve the composite objects' overall performance by lending some of its unique properties (Elfadel et al. 2023). Otherwise, nanomaterials such as nanocrystals and nanoparticles can be used directly to create advanced and powerful devices attributed to their distinctive properties (Hammad et al. 2022a; Al-Zahrani et al. 2022a). The benefits of nanomaterials could potentially affect the future of nearly all industrial sectors (Salem 2022a; Saied et al. 2022). The beneficial use of nanomaterials can be found in sunscreens, cosmetics, sporting goods, electronics and several other everyday items (Salem et al. 2020, 2022a; Sharaf et al. 2022; Hashem et al. 2022b; Asmatulu et al. 2012).

2.2 Types of Nanoparticles

To date, several nanoparticles and nanomaterials have been investigated and approved for clinical use. Some common types of nanoparticles are discussed below.

2.2.1 Metallic Nanoparticles

The specific properties of metallic nanoparticles are it exhibits prospective optoelectronic and dimensional characteristics superior to their bulk metals (Srinoi et al. 2018). These particular traits render an increase in the surface to volume ratio, reactivity, efficiency, and functional modifications that can tap their potential in diverse applications as multifunctional technical tools (Seetharaman et al. 2018). Metallic nanoparticles include zinc, silver, copper, magnesium, selenium, titanium, iron, and gold nanoparticles (Mohamed et al. 2021; Saied et al. 2021; Abdelmoneim et al. 2022; Salem 2022b; Al-Zahrani et al. 2022b; Soliman et al. 2022). Iron oxide nanoparticles consist of a magnetic core (4–5 nm) and hydrophilic polymers, such as dextran or PEG (Shi Kam et al. 2004). Conversely, gold nanoparticles are composed of a gold atom core surrounded by negative reactive groups on the surface that can be functionalized by adding a monolayer of surface moieties as ligands for active targeting (Shi Kam et al. 2004). Metallic nanoparticles have been used as imaging contrast agents (Acharya and Sahoo 2011), in laser-based treatment, as optical biosensors, and drug delivery vehicles (Han et al. 2022; Chandrakala et al. 2022).

2.2.2 Liposomes

Liposomes are spherical vesicles with particle sizes ranging from 30 nm to several microns that consist of lipid bilayers (Kraft et al. 2014). Liposomes can be used to incorporate hydrophilic therapeutic agents inside the aqueous phase and hydrophobic agents in the liposomal membrane layer. Liposomes are versatile; their surface characteristics can be modified with polymers, antibodies and/or proteins, enabling macromolecular drugs, including nucleic acids and crystalline metals, to be integrated into liposomes (Lombardo et al. 2019; Katsuki et al. 2017). Poly(ethylene glycol) (PEG)ylated liposomal doxorubicin (Doxil®) is the first FDA-approved nanomedicine, which has been used for treatment of breast cancer, and it enhances the effective drug concentration in malignant effusions without the need to increase the overall dose (Lombardo et al. 2019; Katsuki et al. 2017).

2.2.3 Micelles

Micelles are amphiphilic surfactant molecules that consist of lipids and amphiphilic molecules (Letchford and Burt 2007). Micelles spontaneously aggregate and self-assemble into spherical vesicles under aqueous conditions with a hydrophilic outer monolayer and a hydrophobic core, and thus can be used to incorporate hydrophobic therapeutic agents. The unique properties of micelles allow for the enhancement of the solubility of hydrophobic drugs, thus improving bioavailability (Imran and Shah 2018). The diameter of micelles ranges from 10 to 100 nm. Micelles have various applications, such as drug delivery agents, imaging agents, contrast agents and therapeutic agents (Katsuki et al. 2017).

2.2.4 Carbon Nanotubes

Carbon nanotubes are cylindrical molecules that consist of rolled-up sheets of a single-layer of carbon atoms (graphene). They can be single-walled or multi-walled, or composed of several concentrically interlinked nanotubes (Zihan et al. 2022). Due to their high external surface area, carbon nanotubes can achieve considerably high loading capacities as drug carriers (Guo et al. 2017). Additionally, their unique optical, mechanical and electronic properties have made carbon tubes appealing as imaging contrast agents and biological sensors (Kurbanoglu and Ozkan 2018; Kaur et al. 2019).

2.2.5 *Dendrimers*

Dendrimers are macromolecules with branched repeating units expanding from a central core and consists of exterior functional groups (Singh et al. 2016). These functional groups can be anionic, neutral or cationic terminals, and they can be used to modify the entire structure, and/or the chemical and physical properties. Therapeutic agents can be encapsulated within the interior space of dendrimers, or attached to the surface groups, making dendrimers highly bioavailable and biodegradable (Sherje et al. 2018). Conjugates of dendrimers with saccharides or peptides have been shown to exhibit enhanced antimicrobial, antiprion and antiviral properties with improved solubility and stability upon absorption of therapeutic drugs (Ortega et al. 2020; Sorokina and Shifrina 2022). Polyamidoamine dendrimer-DNA complexes (called dendriplexes) have been investigated as gene delivery vectors and hold promise in facilitating successive gene expression, targeted drug delivery and improve drug efficacy (Palmerston Mendes et al. 2017). Dendrimers are promising particulate systems for biomedical applications, such as in imaging and drug delivery, due to their transformable properties (Kesharwani et al. 2018; Kokaz et al. 2022).

2.2.6 *Quantum Dots*

Quantum dots (QDs) are fluorescent semiconductor nanocrystals (1–100 nm) and have shown potential use for several biomedical applications, such as drug delivery and cellular imaging (Díaz-González et al. 2020; Liu et al. 2019). Quantum dots possess a shell-core structure, in which the core structure is typically composed of II-VI or III-V group elements of the periodic table. Due to their distinctive optical properties and size, with high brightness and stability, quantum dots have been employed in the field of medical imaging (Singh et al. 2022; Wan et al. 2021; Zhou et al. 2016).

2.2.7 *Green Nanotechnology*

Green nanotechnology is the application of green chemistry ideas to nanotechnology for the manufacture and processing of nanomaterials that is sustainable, safe, cost-effective, and environmentally friendly (Salem 2023; Said et al. 2023). The production of chemically based materials by biological materials (biomass, phytochemicals, microbial) and uses clean energy storage materials (energy efficient fuel/solar cells, green catalysis, etc.). The utmost goal of green nanotechnology is to decrease the hazardous effects of nanomaterials (Dutta and Das 2021). Green chemistry in nanomaterial manufacturing has a variety of health consequences, beginning with solvent selection, which involves choosing a renewable, biocompatible, and environmentally friendly solvent in the nanofabrication process. This new approach facilitates the development of safer, alternative materials for the synthesis of non-hazardous, commercial grade nanomaterials (Fouda et al. 2020). A nanomaterial is defined as a substance with external dimensions in the nanoscale. A nanoparticle is defined as a particle with at least one external dimension in the nanoscale. Nanoscale is defined as size range from about 1 nm to 100 nm. Therefore, we propose nanoscale to be defined as: a size ranges from about 1 nm to 100–1000 nm (Trotta and Mele 2019).

2.2.8 *Green Nanomaterial Preparation*

Nanomaterial preparations occurs through two different approaches bottom-up or top-down. The bottom-up approach involves preparing material from smaller building blocks. Top-down microfabrication methods are used to create nanomaterials by milling or grinding a bulk material into the appropriate form and structure. The popular green approaches used for synthesis of nanomaterials are explained as follows: 1. Using green solvents, such as ionic liquids, to replace toxic solvents. 2. Greener approaches such as sonochemical synthesis, microwave synthesis, hydrothermal synthesis, solvothermal synthesis, electrochemical synthesis, and biosynthesis are being used to replace traditional chemical synthesis processes (Arora 2020). Biosynthesis, a sophisticated multistep synthesis strategy regulated by catalytic enzymes, is one of the finest ways of green nanomaterial synthesis, in which natural ingredients are employed to create a desired material. Green biosynthesis creates nanomaterials using a number of processes, including bacteria, fungus, actinomycetes, algae, and plants. Bacteria are well recognized for their capacity to synthesis inorganic compounds both intracellularly and extracellularly; they have also exhibited this ability. Biosynthesis is used to make metal nanoparticles (Kapoor et al. 2021). *Bacillus sp.*, have been found to be effective in the production of Pd and Au nanoparticles, respectively (Zhang and Hu 2018). A fungal-mediated green method employing *Aspergillus flavus* and *Fusarium oxysporum* was also reported for the biological production of silver and gold nanoparticles (Balakumaran et al. 2016). Actinomycetes are biogenic microorganisms that merge prokaryotic bacteria and fungal features. Similarly, algae-based nanomaterial biosynthesis has been accomplished with success (Khanna et al. 2019). There are three different methods for the synthesis of nanomaterials: physical, chemical, and biological.

2.2.9 *Synthesis of Nanoparticles Using Fungi*

Fungi are the largest group among microbes, where are used in multiple applications in different sciences as bioremediation, enzyme production, nanotechnology, etc. (Salem et al. 2019, 2021; Selim et al. 2021; Fouda et al. 2018). Fungi have sparked a lot of interest in manufacturing metallic nanoparticles since they have several benefits over bacteria when it comes to nanoparticle synthesis. The simplicity of scaling up and downstream processing, as well as the economic feasibility and the existence of mycelia, which provides a larger surface area, are all significant benefit. A biomineralization mechanism is used in fungal-based NP production, which involves internal and extracellular enzymes and biomolecules reducing various metal ions. Silver has been the metal of choice for the manufacture and research of NPs. In addition, Au, Ti, Se, Cu and Zn have been identified as the next most important metal ions employed by fungus in the production of NPs (Mohamed et al. 2019; Hashem et al. 2021; Mohmed et al. 2017; Hammad et al. 2022b; Abu-Elghait et al. 2021). More research on NP biosynthesis has been done on *Fusarium*, *Aspergillus*, *Trichoderma*, *Verticillium*, *Rhizopus*, and *Penicillium* species (Salem and Fouda 2021; Shaheen et al. 2021a). Fungi-produced nanoparticles have been employed in a variety of applications, including anticancer drugs, medicine, antimicrobials, antivirals, diagnostics, antibiotics, antifungals, agriculture, bio-imaging, and industry. Agricultural and medicinal applications have been identified as the most common uses of NPs. When compared to bacteria cells, fungi cells produce a huge number of NPs. Fungi secrete more proteins, resulting in increased NPs output (Shaheen et al. 2021b).

2.2.9.1 Synthesis of Nanoparticles Using Yeast

The extracellular synthesis of nanoparticles in huge quantities, with straightforward downstream processing. Different processes used by yeast strains of different genera for nanoparticle formation result in significant differences in size, mono dispersity, particle position, and characteristics (Salem 2022a, b). These molecules determine the mechanism for the formation of nanoparticles and stabilize the complexes in the majority of the yeast species studied (Abdelfattah et al. 2023; El-Khawaga et al. 2023). Resistance is defined as the ability of a yeast cell to convert absorbed metal ions into complex polymer compounds that are not toxic to the cell. In the mass production of metal nanoparticles, yeast production is easy to manage in laboratory settings, and the rapid growth of yeast strains and the use of basic nutrients have various advantages. *Candida glabrata* and *Saccharomyces pombe* yeast strains have been described for the production of intracellular synthetized silver, titanium, cadmium sulphide, selenium, and gold nanoparticles for this purpose (Boroumand Moghaddam et al. 2015).

2.2.9.2 Synthesis of Nanoparticles Using Bacteria

Research has concentrated primarily on prokaryotes as a technique of synthesizing metallic nanoparticles (Salem et al. 2022b; Alsharif et al. 2020). Due to their ubiquity in the environment and their capacity to adapt to harsh situations, bacteria are a suitable candidate for study. They are also quick growing, affordable to cultivate and easy to manage. Growth parameters such as temperature, oxygenation and incubation time can be easily regulated. Bacteria are known to synthesis inorganic materials either intra or extra. *Pseudomonas stutzeri* was used to produce Ag-NPs outside the cells.

In addition, several bacterial strains (Gram negative as well as Gram positive) namely *A. calcoaceticus, B. amyloliquefaciens, B. flexus, B. megaterium* and *S. aureus* have been used for both extra and intracellular biosynthesis of Ag-NPs. These Ag-NPs are spherical, disk, cuboidal, hexagonal and triangular in shape. They have been fabricated using culture supernatant, aqueous cell-free extract or cells. *Rhodopseudomonas capsulata* was shown to be capable of producing Au-NPs of various sizes, with the form of the Au-NPs being regulated by pH. Bacteria are thought to be a possible biofactory for the production of NPs such as silver, gold, palladium, platinum, magnetite, titanium, titanium dioxide, cadmium sulphide, selenium, and other metal NPs (Salem and Fouda 2021).

2.2.9.3 Synthesis of Nanoparticles Using Actinomycetes

These actinomycetes have a good ability to make antibiotics as secondary metabolites. Actinomycetes have been found to have a significant role in the creation of metal nanoparticles. Actinomycetes are one of the less well-known microorganisms employed in the production of metal nanoparticles (Eid et al. 2020; Hassan et al. 2019). Actinomycetes produce nanoparticles with good poly-dispersity and stability, as well as high biocidal activity against a variety of diseases. *Thermoactinomycete* sp., *Rhodococcus* sp., *Streptomyces viridogens, Nocardia farcinica, Streptomyces hygroscopicus*, and *Thermomonospora* sp. have all effectively manufactured Au-NPs. *Streptomyces* spp., on the other hand, were used to successfully produce Cu, Ag, and Zn-NPs (Salem and Fouda 2021; Manimaran and Kannabiran 2017).

2.2.9.4 Synthesis of Nanoparticles Using Plant

Plant parts such as leaves, stems, roots, shoots, flowers, barks, seeds, and their metabolites have all been used to successfully synthesize nanoparticles (Al-Rajhi et al. 2022; Hashem and Salem 2022; Salem et al. 2022c). Plants with minimal costs and a high level of eco-friendliness are extremely sophisticated and advantageous to human uses. Using plant extracts such as *Cinnamom zeylanicum, Pinus resinosa, Ocimun sanctum, Curcuma longa, Anogeissus latifolia, Glycine max, Musa paradisica, Pulicaria glutinosa, Cinnamomum camphora, Doipyros kaki,* and *Gardenia jasminoides*, green production of NPs has been described (Salem and Fouda 2021; Aref and Salem 2020).

2.3 Application of Nanomaterials

2.3.1 Diseases Diagnosis and Imaging

Diagnosis of a disease is one of the most crucial steps in the healthcare process (Kute et al. 2022). All diagnoses are desired to be quick, accurate and specific to prevent 'false negative' cases. In vivo imaging is a non-invasive technique that identifies signs or symptoms within a patient's live tissues, without the need to undergo surgery (Medina et al. 2007). A previous improvement in diagnostic imaging techniques is the use of biological markers that can detect changes in the tissues at the cellular level. The aim of using a biological marker is to detect illnesses or symptoms, thereby serving as an early detection tool (Frank and Hargreaves 2003). Notably, some of these high precision molecular imaging agents have been developed through the use of nanotechnologies. In addition to diagnosis, imaging is also vital for detecting potential toxic reactions, in controlled drug release research, evaluating drug distribution within the body and closely monitoring the progress of a therapy. Potential drug toxicity can be reduced with the possibility of monitoring the distribution of drugs around the body and by releasing the drug as required (Sakiyama-Elbert and Hubbell 2001).

Imaging techniques such as X-ray, ultrasound, computed tomography, nuclear medicine and magnetic resonance imaging are well established, and are widely used in biochemical and medical research (Østergaard et al. 2008). However, these techniques can only examine changes on the tissue surface relatively late in disease progression, although they can be improved through the use of contrast and targeting agents based on nanotechnologies, to improve resolution and specificity, by indicating the diseased site at the tissue level (Mousa and Bharali 2011). Currently used medical imaging contrast agents are primarily small molecules that exhibit fast metabolism and a non-specific distribution, and can thus potentially result in undesirable toxic side effects (Lombardo et al. 2019). This particular area is where nanotechnologies make their most significant contribution in the field of medicine, by developing more powerful contrast agents for almost all imaging techniques, as nanomaterials exhibit lower toxicity, and enhanced permeability and retention effects in tissues. The size of the nanoparticles significantly influences its biodistribution, blood circulation half-life, cellular uptake, tissue penetration and targeting (Wang and Zhang 2022). The use of nanoparticles in X-rays has some limitations. In order to enhance the contrast, a number of heavy atoms must be delivered into the target site without causing any toxic reactions. This can be achieved using stable and inert surface atoms, such as gold (Bonačić-Koutecký and Guével n.d.). Hence, gold nanoshells have garnered significant attention, due to its low toxicity, gold nanoshells are heavy metal nanoparticles (dielectric core) encapsulated in gold shells and have been proposed to be one of the most promising materials in optical imaging of cancers (Pal et al. 2022). Gold nanoshells are cost-effective, safe due to its non-invasive property and may provide high resolution imaging. Gold nanoshells have similar physical characteristics to gold colloids, as they both possess a unified electronic response of the metal to light resulting in active optical absorption (Abbasi et al. 2017). Gold nanoshells are widely employed by researchers as contrast agents in the Optical Coherence Tomography of cancer cells, as the optical resonance of gold nanoshells can be adjusted accurately over a wide range, including near-infrared, where tissue transmissivity is higher (Abbasi et al. 2017; Hu et al. 2018; Kavalaraki et al. 2023).

2.3.2 *Drug Delivery*

Therapy typically involves delivering drugs to a specific target site. If an internal route for drug delivery is not available, external therapeutic methods, such as radiotherapy and surgical procedures are employed. These methods are often used interchangeably or in combination to combat diseases. The goal of therapy is to always selectively remove the tumours or the source of illness in a long-lasting manner (Liang et al. 2021). Nanotechnologies are making a compelling contribution in this area through the development of novel modes for drug delivery, and some of these methods have proven effective in a clinical setting and are clinically used (Saeedi et al. 2019). For example, doxorubicin a drug which exhibits high toxicity, can be delivered directly to tumour cells using liposomes (Doxil®) without affecting the heart or kidneys. Additionally, paclitaxel incorporated with polymeric mPEG-PLA micelles (Genexol-PM®) are used in chemotherapeutic treatment of metastatic breast cancers (Lombardo et al. 2019). The success of nanotechnologies in drug delivery can be attributed to the improved in vivo distribution, evasion of the reticuloendothelial system and the favorable pharmacokinetics (Ferreira et al. 2021).

A perfect drug delivery system encompasses two elements: Control over drug release and the targeting ability. Side effects can be reduced significantly, and drug efficiency can be ensured by specifically targeting and killing harmful or cancerous cells. Additionally, controlled drug release can also reduce the side effects of drugs (Singh et al. 2019). Benefits of nanoparticle drug delivery systems include minimized irritant reactions and improved penetration within the body due to their small size, allowing for intravenous and other delivery routes. The specificity of nanoparticle drug delivery systems is made possible by attaching nano-scaled radioactive antibodies that are complementary to antigens on the cancer cells with drugs, and these approaches have produced desirable results (Suri et al. 2007), exhibiting improved (i) drug bioavailability, (ii) delivery of drugs specifically to the target site, and (iii) uptake of low solubility drugs (Mazayen et al. 2022).

2.3.3 *Cancer Treatment*

Staggering numbers of individuals suffer from cancer worldwide, highlighting the need for an accurate detection method and novel drug delivery system that is more specific, efficient and exhibits minimal side effects (Ferlay et al. 2021). Anticancer treatments are often regarded as superior if the therapeutic agent can reach the specific target site without resulting in any side effects. Chemical modifications of the surface of nanoparticle carriers may improve this required targeted delivery. One of the best examples of modifications at the surface of nanoparticles is the incorporation of PEG or polyethylene oxide (Dezfuli et al. 2023). These modifications enhance not only the specificity of drug uptake, but also the tumor-targeting ability. Incorporating PEG avoids the detection of nanoparticles as foreign objects by the body's immune system, thus allowing them to circulate in the bloodstream until they reach the tumor. Additionally, the application of hydrogel in breast cancer is a prime example of this innovative technology. Herceptin is a type of monoclonal antibody used in breast cancer treatment by targeting human epidermal growth factor receptor 2 (HER2) on cancer cells. A vitamin E-based hydrogel has thus been developed that can deliver Herceptin to the target site for several weeks with just a single dose. Due to the improved retention of Herceptin within the tumor, the hydrogel-based drug delivery is more efficient than conventional subcutaneous and intravenous delivery modes, thus making it a better anti-tumor agent (Vlerken et al. 2007; Biswas et al. 2014). Nanoparticles can be modified in several ways to prolong circulation, enhance drug localisation, increase drug efficacy and potentially decrease the development of multidrug resistance through the use of nanotechnologies (Majidinia et al. 2020).

There are several studies using FDA-approved nano drugs, such as Abraxane®, Doxil® or Genexol-PM® as adjuvants in combinatory cancer treatment. Abraxane®, a paclitaxel albumin-stabilised nanoparticle formulation (nab-paclitaxel) has been approved for the treatment of metastatic breast cancer (Montero et al. 2011). There are >900 ongoing clinical trials involving nab-paclitaxel as an anticancer agent, based on Clinicaltrials.gov as of August 2020. Moreover, nab-paclitaxel, in combination with 5-chloro-2.4-dihydrooxypyridine, tegafur and oteracil potassium exhibited promising results when used for the treatment of HER2-negative breast cancer patients (Tsurutani et al. 2015). Doxorubicin, daunorubicin, paclitaxel and vincristine are among the most extensively investigated anticancer agents in liposome-based drug formulations (Lombardo et al. 2019).

2.3.4 Cardiovascular Diseases Treatment

Cardiovascular diseases are another field where the properties of nanoparticles may be leveraged (Park et al. 2020). Cardiovascular diseases are the leading cause of death globally, and the rates are increasing alarmingly, due to an increase in sedentary lifestyles. Common examples of cardiovascular diseases that affect several individuals includes stroke, hypertension and restriction or blockage of blood circulation in a specific area. These diseases are the most common causes of prolonged disability and death (McGill et al. 2008). Nanotechnologies offer novel avenues for therapeutic and diagnostic strategies for management of cardiovascular diseases. Most cardiovascular risk factors (for example, for hypertension, smoking, hypercholesterolemia, homocystinuria and diabetes mellitus) are associated with impaired nitric oxide (NO) endothelial production. Impaired endothelial function is established to be the first step in atherosclerosis. Gold and silica nanoparticles have been developed to improve NO supply for possible application in cardiovascular diseases, where low NO bioavailability occurs (Das et al. 2010). Systemic administration of the 17-βE loaded CREKA-peptide-modified-nanoemulsion system has been shown to reduce the levels of pathological contributors to early atherosclerosis by reducing lesion size, lowering the levels of circulating plasma lipids and decreasing the gene expression of inflammatory markers associated with the disease (Deshpande et al. 2016). Moreover, novel formulations of block copolymer micelles constructed using PEG and poly(propylene sulphide) have been demonstrated to suppress the levels of pro-inflammatory cytokines, and exhibited excellent potential for management of atherosclerosis (Wu et al. 2018).

Drug delivery via liposomes has been proven to be effective for prevention of platelet aggregation, atherosclerosis and thrombosis. Prostaglandin E-1 (PGE-1) exhibits a wide range of pharmacological properties, including vasodilation, inhibition of platelet aggregation, leukocyte adhesion, as well as exhibiting an anti-inflammatory effect. Liposomal drug delivery of PGE-1 (Liprostin™), is currently undergoing phase III clinical trials for the treatment of various cardiovascular diseases, such as restenosis following angioplasty (Bulbake et al. 2017). Additionally, the use of liposomes carrying the thrombolytic drug urokinase has also been assessed; cyclic arginyl-glycyl-aspartic acid (cRGD) peptide liposomes encapsulated with urokinase can selectively bind to the GPIIb/IIIa receptors, and this improves the thrombolytic efficacy of urokinase by almost 4-fold over free urokinase (Bulbake et al. 2017).

Efficacy and effectiveness of the conventional thrombolytic drugs can also be advanced via novel nano-therapeutic approaches. Drugs can be selectively targeted to vascular blockage sites through mechanical activation within blood vessels based on the high-fluid shear strains present within them. In vivo and in vitro studies have been encouraging, thus validating this approach for use in lysis of blood clots, using a significantly lower amount of thrombolytic drug (Katsuki et al. 2017; Zhang et al. 2018). One example of this technology is the use of dendrimers. Dendrimers have been used in several diseases as a means of delivering therapeutic agents. Plasminogen activator (rtPA) has been successfully attached to dendrimers producing an alternative drug delivery system, allowing for refinement of the rtPA-dendrimer complex concentration throughout the duration of treatment using different dilution proportions of each part of the complex (Najlah et al. 2007). Another potential role of nanoparticles is to decrease haemorrhaging, which is a severe side effect of thrombolytic agents. Targeted thrombolysis via rtPA bound to polyacrylic acid coated nanoparticles minimises the intracerebral haemorrhage, and enhances retention at the target site (Katsuki et al. 2017).

2.3.5 Nanotechnology and Agricultural

Nanotechnology has emerged as one of the most imperative tools in the recent agriculture sector (Acharya and Pal 2020). In this modern era of science and technology, there are no such fields left untouched by nanotechnology. Agriculture is no exception. This technology has broad applications in the agriculture and food sectors, telecommunication, the food industry, cosmetics, etc. Through nano biotechnology, we can understand the biology of different crops, which will eventually help enhance the yield and nutritional value of those crops through breeding (Elemike et al. 2019). The need for nanotechnology in the agriculture sector is felt to improve crops, diagnose plant diseases, and monitor plant health and soil quality (Salem and Husen 2023). These will ultimately contribute to enhanced plant performance, which is the main focus of every breeding program (Lowry et al. 2019).

2.3.6 Nano Fertilizers

One of the significant commitments of nanotechnology in agribusiness is the detailing of nano-based composts, i.e., nano fertilizers (Chhipa 2019). Nutrients can either be encased within nanomaterials like nanotubes or nonporous materials laminated with thin protective polymer film or rendered as particles or emulsions of nanoscale dimension (Yata et al. 2018). Nano fertilizers have interesting attributes like super high assimilation, an enormous surface region, pore-volume, and exceptionally requested pores. These properties enable slow delivery and advanced, proficient supplement take-up by harvests, making them more powerful than the majority of the most recent polymeric-sorted ordinary composts that have essentially been performed over the previous decades (Elemike et al. 2019; Kumar et al. 2019; Cheng et al. 2016). The primary delegates are nitrogen composts, potash manures, nanoporous zeolite, zinc nano fertilizers, zeolites, and nano clays that discharge supplements gradually for the duration of the existing pattern of the crops, there by lessening the dangers of filtering, adsorption, surface spillover, and deterioration (Cheng et al. 2016; Liu et al. 2021). Unlike traditional fertilizers, nano fertilizers can be made to release nutrients in a controlled way in a gradual manner into the soil, which may help prevent purification and contamination of bodies and the environment. Similarly, foliar application of Nano fertilizer has been reported to increase crop yields significantly (Chand Mali et al. 2020).

2.3.7 Seed Science

To improve the germination of the rain fed crop, researchers are working on metal oxide nanoparticles and carbon nanotubes. It is also found that carbo nano tubes can enhance the germination of tomato seeds through the better conveyance of moisture. The data shows that carbon nanotubes (CNTs) act as anew pore and facilitate the passage of water by penetrating the seed coat and acting as a way to channelize the water from the substrate into the seeds. Hence, it can enhance germination in the rain fed agricultural system (Salem and Husen 2022). It was found that TiNPs at 20 g/L concentration significantly increased the ear mass, seed number, biomass, stem elongation, flowering, yield, starch, and gluten content in wheat. Similarly, carbon nanotubes (CNTs) were reported to enhance the germination percentage of tomato seeds by penetrating their hard outer coat. Moreover, the germination percentage of other crops, like soybean, barley, corn, etc., was also increased when sprayed or encapsulated with multiwall carbon nanotubes (MWCNTs) (Acharya and Pal 2020). The use of nanoparticles (manganese, molybdenum) and multi walled carbon nanotubes (MWCNTs) has increased the germination percentage of *Brassica juncea*, *Phaseolus mungo*, etc. Silver nanoparticles are also used for seed dressing and surface sterilization of seed crops (Chhipa 2019). Germination rate, bud length, diameter, fresh bud weight, etc. were found to be higher in the seeds of cowpea, cabbage, and cucumber treated for 2 h, 4 h, and 12 h respectively withnano-863 than soaked in pure water. When soaked innano-863 treated water, garlic produced about 5 cm longer heights of bolts after about 20 days of treatment than the ones in the control group (Faridvand et al. 2021).

2.3.8 Seed Storage

Stored seeds emit several volatile aldehydes, which decide the degree of aging. These emitted gases affect other seeds too. So, to detect these volatile aldehydes and seeds showing signs of degradation, biosensors can be used, and they can be separated from the healthy ones before they come into use (Huang et al. 2015). In a like manner, the nano sensors can be used to identify the presence of insects or fungus precisely inside the stored grains in storage houses (Shukla et al. 2019).

2.3.9 Genetic Manipulation and Crop Improvement

Nanotechnology can be used to reform the genetic constituents of plants and can act as a defensive shield for pathogens, therefore helping in crop improvement (Badawy et al. 2021). With this technology, breeders can develop potentially enhanced plants that improve resistance to various environmental stresses such as drought, cold, disease, salinity, etc. For example, zinc nanoparticles are widely used to enhance the productivity of Pennisetum americium. The use of TiO2 NMs increases the productivity of mung beans (Pramanik et al. 2020) he advancement in nanotechnology empowered by gene sequencing is used to effectively identify and utilize plant genetic resources (Iavicoli et al. 2017). The application of nano-genomics based technology in plant breeding can be used to deploy nanomaterials as transporters of DNA or RNA in plant cells that divert the genes to the target location at a cellular level for gene expression (Sangeetha et al. 2017; Shandilya and Tarwadi 2021). The use of nano fertilizers improves crop yield and quality with higher nutrient efficiency while simultaneously reducing spillage in the environment and the cost of production, thus contributing to sustainable agriculture. For example, the application of phosphatic nano fertilizers has been found to increase the growth rate by (32%) and seed yield by (20%) of soybean (Glycine max) as compared to those treated with conventional fertilizers. Also, carbon nanoparticle nanotubes of Au, SiO_2, ZnO, and TiO_2 ameliorate the development of plants by enhancing elemental uptake and use of nutrients (Wani and Kothari 2018). A recent study on different crops has also shown increased germination, seedling growth, physiological activity like photosynthetic activity and nitrogen metabolism, m-RNA expression, and some positive changes in gene expression, fostering their potential use in crop improvement (Salem and Husen 2022; Cheng et al. 2016). Nutrient deficiency in crops can be managed by the use of Nano carriers that help in fulfilling the nutrient requirements of plants. Nanotechnology can also be used to recognize superior genes to improve crops for disease resistance or higher productivity (Fraceto et al. 2016). Just as genetically modified agriculture has given rise to a new level along the food chain, likewise, nanotechnology deployed from seeds to stomach, genome to gluten, will strengthen the grasp of agribusiness over global food farming at every stage (Kerry et al. 2017).

2.3.9.1 Resistance Against Abiotic Stress

Nano-barcode particles are usually manufactured through semi-automatic lamination of inactive metals such as Au, Ag, or Pt. Those nano barcodes are applied in the analysis of gene expression to develop resistance against ecological stresses and to induce resistance against corrosiveness and ultraviolet rays' effects on the chloroplast (Bhau et al. 2016). The use of various nanoparticles, GPS (Global Positioning System), and remote sensing technologies by farm managers helps widely detect multiple crop pests or evidence of different environmental stresses such as drought, pollution, nutrient deficiency, etc. to make a plan and protect our crops accordingly (Ndlovu et al. 2020).

2.3.9.2 Smart Delivery of Agrochemicals

The Nano biosensors that use carbon nanotubes or nano cantilevers are very small enough to trap and deliver even the smallest molecules and other individual proteins (Joshi et al. 2019). Smart sensors and smart delivery systems assist the horticultural business with combatting infections and other harvest microbes. The nano-based impetuses improve the effectiveness of pesticides and herbicides, permitting lower portions to be utilized (Abd-Elrahman and Mostafa 2015). For example, nano encapsulated pesticides increase the solubility of poorly soluble active ingredients and release them slowly. This activity of nano encapsulated agrochemicals allows for slow and effective release to a specific host, reducing toxicity to non-targeted organisms. Such controlled nano particulate delivery systems will require a designated conveyance approach centered on utilizing the information on the life-cycle and the conduct of the microbe or irritation (Rai and Ingle 2012). As per the research carried out by various workers, it has been found that nanoparticles of different metals are also cost-effective and the most reliable alternative for controlling insects, and pests. In this way, this technology has helped in the smart delivery of agrochemicals and is thus gaining popularity these days (Jatav and Nirmal 2013).

2.3.9.3 Hormonal Effects in Plants

Numerous plant regulators are being manufactured and utilized throughout the world to amplify the hormonal effects, including nano-5 and nano gro. A leading agricultural company, i.e., Syngenta has been fabricating Primo MAXX, a plant growth regulator, for the past couple of years (Bhau et al. 2016).

2.3.9.4 Disease and Pest Control

In 2011, a statement was fortified by Pimentel and Burgess that almost 545 million kg of pesticides is utilized on plants in the United States each year, but <0.15% of this reaches the preferred target and eliminates the pests (Kumar 2020). This statement reflects the overuse and uncontrolled application of pesticides and unsuccessful measures of pest control. Generally, plant diseases are generated by pathogenic microorganisms such as bacteria, fungi, nematodes, and viruses, which promptly reduce the crop quality and yield. Many practices are carried out to control diseases and pests, including cultural, physical, chemical, legislative, genetics, breeding, and integrated disease and pest management approaches. However, they can't control those (Kumar et al. 2019). The issue might lie in recognizing economic threshold levels outside of which they can't be controlled (Fraceto et al. 2016). Nano-based diagnostics can identify diseases and agrochemical residues by measuring differential protein in diseased and healthy crops, which leads to the identification of specific microorganism species and the stage of therapeutic drug application to prevent and stop disease (Vega-Vásquez et al. 2020). Moreover, nanotech applies to nanoparticle-mediated gene transfer in crops to create efficacious pest resistant and tolerant lines (Vega-Vásquez et al. 2020). The utilization of biomarkers unequivocally indicates disease stage (Chung et al. 2017), and nano silica-silver composite Silicon (Si) is known to be absorbed into plants to expand disease obstruction and stress resistance. Nanostructured frame-works have a broad scope of use in determining and treating diseases in plants, giving better management of insects and diseases. Allowing their lower portions to be used, which will also ensure environmental protection (Dhewa 2015). Utilizing nano sensors and carbon nano tubes further develops monitoring and protection of plants from plant diseases through prompting plant infection obstruction (Arora 2018).

2.3.9.5 Monitoring of Plant Growth Stages

Nano-sensors are immobilized bio-receptor probes that are designed for earlier response to alteration in the environment. Ultimately, it can be linked with GIS, and monitoring is possible to vast crop cultivation areas. With the help of Nano sensors, all possible growth stages of the plant, including the sensitive stages such as photosynthesis and seed germination, can be precisely monitored for obtaining higher and maximal yields (Alghuthaymi et al. 2021).

2.3.9.6 Secondary Metabolites Exploitation

Secondary metabolites (alkaloids, phenolics, and terpenoids) emitted by the plant as self-safe guarding tools of nature provide protective and defensive capacity against insects (Ghasemnezhad et al. 2019). The nanostructures can be used to discover precious metabolites; for instance, nano-convectors can detect terpenoids, anthocyanins, and other beneficial secondary metabolites in therapeutic herbs and shrubs. Furthermore, the nano-polymers can augment plant cell catalytic activity, thus producing more protein (Alghuthaymi et al. 2021). Mycotoxins are harmful auxiliary metabolites produced by fungi that are fat alto living beings. So, the pores of nanostructured polymer layers could be utilized as receptors to acknowledge comparable poisonous metabolites (Faridvand et al. 2021).

2.3.9.7 Opportunities

The utilization of nanotechnology and other fundamental science disciplines can produce optimized beneficiaries that can rectify the various challenges of agriculture, industry, health, communication, energy, and the environment (Cicek and Nadaroglu 2015). The world has now realized that nanoscale science and nanotechnologies can revolutionize the agriculture and food system and have given birth to 'agro-nanotechnology'. Although the first green revolution promised food to everyone, agriculture production is now facing a plateau, compelling the second green revolution, and this nanotechnology can grab this opportunity. Considering the current scenario, the most incredible opportunity of nanotechnology is 'crop improvement'. This technology helps develop resistant and improved plant genotypes through plant breeding or genetic engineering (Wieliczko and Floriańczyk 2022). Nanotech-derived devices (nano-sensors, nanoparticles) are widely used in plant breeding and genetic transformation to develop improved crops (Jatav and Nirmal 2013). Nano-biosensors possess unique physio-chemical properties such as being independent of pH, temperature, stirring, stability under normal storage conditions, etc. In addition to this, they are very tiny, cheap, portable, non-toxic, and non-antigenic, and hence can be operated easily. There is a vast opportunity in tissue culture, molecular breeding, genetic engineering, transgenic approach, and so on, through various approaches that include the following principles: using nanomaterial coated 'smart seeds', targeted delivery of agrochemicals and drugs, monitoring soil and plant health using biosensors, and decreasing losses through appropriate disease management practices. Besides all these opportunities, nanomaterials also have a chance to functionalize membranes and eliminate and or recover nutrients from a source of contaminated water and a waste stream (Lowry et al. 2019). It opens a broad range of opportunities in technical and mechanical fields, and its application in those fields can indeed bring prodigious positive changes in the coming decades (Jatav and Nirmal 2013). Due to the uncontrolled and unmanaged administration of chemicals on crops, adverse effects are visible in the ecosystem. These resistant and tardily degrading compounds like pesticides, insecticides, weedicides, and rodenticides enter the food chain of organisms and show detrimental effects. Thus, nanoparticles can play a nurturing role in the bioremediation of those contaminants by assisting microbial action and degrading them into innocuous. Furthermore, the powerful nanoparticles can be combined with botanical sides that are harmless to plants and the environment (Alghuthaymi et al. 2021). As well, nanotechnology can be permitted in agricultural research and applied zones for its advancement. Through nanotech, plant diseases can be prevented, early stresses can be detected and minimized, and diversified lines can be produced that are uninfluenced by harsh environmental fluctuations. Additionally, the amplification of nanotech in plant breeding can lead us to make our agricultural systems futuristic and quick-witted. Various advancing engineered tools can be accoutered with Nano devices that have the potential to supplant many cellular levels of machinery (Chhipa 2019).

2.3.9.8 Potential

Nanotechnology has come forward as a futuristic weapon to mitigate those challenges (Elsharkawy et al. 2022). The entrance of nanotechnology into plant breeding can bring colossal and drastic changes that will directly enhance the world economy. This technology shows great potential in embracing today's technologies despite competing with them. The versatile capacity of this tech will seize and up lift the growth of several fields, including agriculture, that will help to achieve the UN millennium goal by influencing food security and food safety in up-coming years (Kumar et al. 2019; Alghuthaymi et al. 2021). The applications of nano-fertilizers and nanopesticides enhance agricultural productivity with low cost and energy input with-out deteriorating functional bodies like soil, water, turf, and other vegetation (Tiwari et al. 2022; Nair 2021). Nanotechnology can further participate in the monitoring of water quality, crop disease incidences, growth rates, nutrient use efficiency, and pesticides through nano networks and wireless nano sensors for their efficient usage in field conditions to attain high-tech agriculture farming and sustainable development in agriculture (Nair 2021). In the last two decades, along with technological advancement, nanotechnology has emerged as a crucial and promising tool to revolutionize the agriculture and food system. The unique physicochemical property of this technology has huge potential in almost every sector, including agriculture, particularly in the crop improvement system (Jakhar et al. 2022). Nanotechnology has substantial potential in the agricultural sector. However, a few issues are still to be addressed, like risk assessment issues, sufficient study on nano toxicity, increasing the production scale, and lowering the cost (Jatav and Nirmal 2013). Another essential potential of nanotechnology is the development of an insect, pest, and disease resistant varieties through appropriate breeding programs that encompass nano-based materials. This can be achieved by incorporating nanoparticle-mediated genes or DNA of our interest into a targeted plant. Another essential potential of nanotechnology is the development of genetically modified (GM) crops with improved performance and increased yield, which is the most urgent need to meet the increasing demand for produce. And this can be achieved through various principles of nanobiotechnology, such as transgenic breeding, induced or natural mutation, etc. (Ahmar et al. 2021). Current agriculture is facing a broad spectrum of local as well as global challenges, which has led to various problems, including decreased production, environmental issues, etc. So, now is the time to integrate this beautiful and yet powerful technology into agriculture as it has the potential to transform agriculture (Acharya and Pal 2020). Nanotechnology has the phenomenal potential to facilitate and frame the next stage of the precision farming technique. Nevertheless, the primary focus of nanotechnology has always been placed on crop genomes to develop potentially enhanced plant varieties that are more tolerant to various biotic and abiotic stresses. Therefore, the latest and most revolutionary technology is nanotech, under nanoparticles, which acquire pioneering and unique properties that can revive and aggravate the world's agricultural and plant breeding sectors, among many more (Carmona et al. 2022).

2.3.9.9 Challenges

Despite these potential advantages, the agricultural sector is still comparably marginal and has not yet been able to make progress in the market to any greater extent relative to other fields of nanotech application (Jatav and Nirmal 2013). Nanotechnology has successfully been implemented in the sectors of genetics and plant breeding. However, nanotechnology remains insignificant and has not yet taken the market because of its poor cargo loading capacity (Kumar 2020). Despite those aforesaid opportunities and potentials of nanotechnology, some studies indicate the negative effects of nanoparticles on the human body, which might be trapped by the imputation of nanotechnology in the agricultural field. The inhalation of even low concentrations of nanosized titanium dioxide in rats induces microvascular dysfunction. Nanoparticles smaller than 1 m have the ability to pierce the healthy human skin and may affect the vascularized organs, including the brain and blood vessels.

Different governmental and non-governmental organizations and scientific authorities, as well as environmental, health, and safety councils, are giving their suggestions, opinions, and guidance for minimizing the risk of nanotechnology to human health, the environment, animals, and plants (Faridvand et al. 2021). The major challenge in adopting this technology on a large scales overcoming the risk issue. And as we know that technologies but safety-must', before the adoption of this technology, it is very important to assess the possible risks and consequences of using nanoparticles (Acharya and Pal 2020).

2.3.9.10 Positive Aspects

Nanotechnology intends to be an innovative technology that is implanted in plant breeding to develop precision and meet the intense demand for food in the world through its criterion (Elemike et al. 2019). The miniature materials manufactured through the utilization of nanotech, i.e., nanomaterials, ameliorate the efficacy of microorganisms to degenerate noxious and unwanted materials. The breakdown of those harmful substances in agricultural factors, including soil water and soil, subsequently induces safety by potentially suppressing toxic residue deposition (Fincheira et al. 2021). Furthermore, green nanomaterials, based on environmentally friendly concepts, are employed in the present agricultural system to enhance yield, input use efficiency, and reduce fertilization costs. These convenient materials play an outstanding role in repressing greenhouse gas emissions. In particular, the agricultural field minimizes the production of significant amounts of nitrous oxide, methane, and carbon dioxide (Iavicoli et al. 2017). Furthermore, the use of nanotechnology in plant breeding reduces chemical hazards, nutrient losses, pest outbreaks, and crop yields, while also helping to stabilize changing climatic conditions and levels of food security (Elemike et al. 2019). Nanotechnology helps in minimizing crop loss to a greater extent. Typically, the remediation process for stress caused by biotic and abiotic factors, disease, and insect-pest infestation starts only after the development of symptoms. Since this technology operates at the same nano scale s that of viruses or diseases, it potentially helps in their early detection and eradication (Shang et al. 2019). This technology will provide plants with augmented functions, helping them endure various biotic and abiotic stresses in a rapidly changing climate. Also, treating plants with low doses of antimicrobial micronutrients coated with nanoparticles increases the plant's ability to manage biotic stress from fungal root infection (Lowry et al. 2019). According to Hashem et al. 2022c nanomaterials synthesized from biopolymers (cellulose and starch) are safe for humans and have gained popularity around the world because they are non-toxic to humans and are generally regarded as safe.

2.3.9.11 Negative Aspects

Besides these ample benefits, nanotechnology also comes up with several negative impacts. Similar to that of all chemical processes, it may also have an unwanted and undesirable effect on non-targeted plants and plant-associated organisms. Therefore, for the large-scale adoption of this technology, there must be a clear understanding of its agro-ecological ramifications, potential, and nano toxicity (Baranwal et al. 2022). As a thumb rule, anything that has some benefits may also have some limitations, and so is the case with nanotechnology. The accumulation of nanomaterials in food products may give rise to various human health concerns and environmental issues (Mitra et al. 2023). In recent days, the use of nanotechnology in plant breeding and other prospective agricultural fields has prompted concerned authorities, including breeders, agriculturists, environmentalists, researchers, and even the general public, to express concerns about the initial and long-term safety of beings, the environment, and environmental parameters (Gupta 2021). Many studies have also reported the negative impact of using metal oxide nanoparticles on plant growth and seed germination, hence causing phyto toxicity, cytotoxicity, and genotoxicity (Agrawal et al. 2022; Bhattacharjee et al. 2022; Plaksenkova et al. 2020). Improper use of this technology can pose a greater threat to living organisms. The negative impact of nanomaterials on plants and soil microbes has been broadly identified. From the current studies, it is clear that engineered nanoparticles can be a prohibitive risk for humans, animals, and the ecosystem (Youssef and Elamawi 2020). To a certain extent, nanotechnologies have been found to impact the growth, enzyme activity, chlorophyll, and protein content of algae, though the effect is more dependent on the type of algae and the character of nanomaterials (Ritu et al. 2023). The use of nanotechnology has been banned in "organic food production in Canada", where an amendment was added to its national organic rules banning nanotechnology as a 'Prohibited Substance or Method' (Yin et al. 2020). An International Federation of Organic Agriculture Movements Position Paper on the Use of Nanotechnologies and Nanomaterials in Organic Agriculture rejected the use of nanotechnology in organic agriculture, indicating that nanoparticles behave in an unpredictable manner, which may be harmful to life (Rajput et al. 2021). A major concern while using nanomaterials is that they may be taken up by cells and cell nuclei of individuals where they may cause cell death or DNA mutation (Yin et al. 2020). Though carbon nanotubes facilitated better root elongation in onions and cucumbers, it was found that they significantly decreased the root length in tomatoes (Pramanik and Pramanik 2016). An international policy debate has now emerged in the last 5 years concerning an appropriate mechanism for the regulation and governance of nanotechnology (Baranwal et al. 2022). The nanoparticles incorporated into plants as nano pesticides, nano-herbicides, or nano fertilizers may block the vascular bundle in plants and prevent the transference of nutrients, minerals, water, and products of photosynthesis. They may also reduce pollination by creating physical obstruction between pollen and stigma (Ashraf et al. 2022).

References

Abbasi A, Park K, Bose A, Bothun GD (2017) Near-infrared responsive gold-layer some nanoshells. Langmuir 33(21):5321–5327. https://doi.org/10.1021/acs.langmuir.7b01273

Abdelfattah NAH, Yousef, MA, Badawy AA, Salem SS (2023) Influence of biosynthesized magnesium oxide nanoparticles on growth and physiological aspects of cowpea (*Vigna unguiculata* L.) plant cowpea beetle and cytotoxicity. Biotechnol J. https://doi.org/10.1002/biot.202300301

Abdelaziz AM, Salem SS, Khalil AMA, El-Wakil DA, Fouda HM, Hashem AH (2022) Potential of biosynthesized zinc oxide nanoparticles to control Fusarium wilt disease in eggplant (*Solanum melongena*) and promote plant growth. Biometals 35(3):601–616. https://doi.org/10.1007/s10534-022-00391-8

Abdelghany TM, Al-Rajhi AMH, Yahya R, Bakri MM, Al Abboud MA, Yahya R, Qanash H, Bazaid AS, Salem SS (2023) Phyto fabrication of zinc oxide nanoparticles with advanced characterization and its antioxidant, anticancer, and antimicrobial activity against pathogenic microorganisms. Biomass Convers Biorefin 13(1):417–430. https://doi.org/10.1007/s13399-022-03412-1

Abdelmoneim HEM, Wassel MA, Elfeky AS, Bendary SH, Awad MA, Salem SS, Mahmoud SA (2022) Multiple applications of CdS/TiO_2 nanocomposites synthesized via microwave-assisted sol-gel. J Cluster Sci 33(3):1119–1128. https://doi.org/10.1007/s10876-021-02041-4

Abd-Elrahman SH, Mostafa M (2015) Applications of nanotechnology in agriculture: an overview. Egypt J Soil Sci 55(2):197–214

Abu-Elghait M, Hasanin M, Hashem AH, Salem SS (2021) Ecofriendly novel synthesis of tertiary composite based on cellulose and myco-synthesized selenium nanoparticles: characterization, antibiofilm and biocompatibility. Int J Biol Macromol 175:294–303. https://doi.org/10.1016/j.ijbiomac.2021.02.040

Acharya A, Pal PK (2020) Agriculture nanotechnology: translating research outcome to field applications by influencing environmental sustainability. NanoImpact 19:100232. https://doi.org/10.1016/j.impact.2020.100232

Acharya S, Sahoo SK (2011) PLGA nanoparticles containing various anticancer agents and tumour delivery by EPR effect. Adv Drug Deliv Rev 63(3):170–183

Agrawal S, Kumar V, Kumar S, Shahi SK (2022) Plant development and crop protection using phytonanotechnology: a new window for sustainable agriculture. Chemosphere 299:134465. https://doi.org/10.1016/j.chemosphere.2022.134465

Ahmar S, Mahmood T, Fiaz S, Mora-Poblete F, Shafique MS, Chattha MS, Jung K-H (2021) Advantage of nanotechnology-based genome editing system and its application in crop improvement. Front Plant Sci 12. https://doi.org/10.3389/fpls.2021.663849

Alghuthaymi MA, Rajkuberan C, Rajiv P, Kalia A, Bhardwaj K, Bhardwaj P, Abd-Elsalam KA, Valis M, Kuca K (2021) Nanohybrid antifungals for control of plant diseases: current status and future perspectives. J Fungi 7(1):48

Al-Rajhi AMH, Salem SS, Alharbi AA, Abdelghany TM (2022) Ecofriendly synthesis of silver nanoparticles using Kei-apple (*Dovyalis caffra*) fruit and their efficacy against cancer cells and clinical pathogenic microorganisms. Arab J Chem 15(7). https://doi.org/10.1016/j.arabjc.2022.103927

Alsharif SM, Salem SS, Abdel-Rahman MA, Fouda A, Eid AM, El-Din Hassan S, Awad MA, Mohamed AA (2020) Multifunctional properties of spherical silver nanoparticles fabricated by different microbial taxa. Heliyon 6(5). https://doi.org/10.1016/j.heliyon.2020.e03943

Al-Zahrani FAM, Al-Zahrani NA, Al-Ghamdi SN, Lin L, Salem SS, El-Shishtawy RM (2022b) Synthesis of Ag/Fe_2O_3 nanocomposite from essential oil of ginger via green method and its bactericidal activity. Biomass Convers Biorefin. https://doi.org/10.1007/s13399-022-03248-9

Al-Zahrani FAM, Salem SS, Al-Ghamdi HA, Nhari LM, Lin L, El-Shishtawy RM (2022) Green synthesis and antibacterial activity of Ag/Fe_2O_3 nanocomposite using *Buddleja lindleyana* extract. Bioengineering 9(9). https://doi.org/10.3390/bioengineering9090452

Aref MS, Salem SS (2020) Bio-callus synthesis of silver nanoparticles, characterization, and antibacterial activities via *Cinnamomum camphora* callus culture. Biocatal Agric Biotechnol 27. https://doi.org/10.1016/j.bcab.2020.101689

Ariga K, Yamauchi Y (2020) Nanoarchitectonics from atom to life. Chem Asian J 15(6):718–728

Arora K (2018) Advances in nano based biosensors for food and agriculture. In: Gothandam KM, Ranjan S, Dasgupta N, Ramalingam C, Lichtfouse E (eds) Nanotechnology, food security and water treatment. Springer International Publishing, Cham, pp 1–52. https://doi.org/10.1007/978-3-319-70166-0_1

Arora A (2020) Nanomaterials: methods of generation. In: Nanotechnology. CRC Press, pp 11–22

Ashraf U, Zafar S, Ghaffar R, Sher A, Mahmood S, Noreen Z, Maqbool MM, Saddique M, Ashraf A (2022) Chapter 7 - Impact of nano chitosan-NPK fertilizer on field crops. In: Kumar DS, Madihally SV (eds) Role of chitosan and chitosan-based nanomaterials in plant sciences. Academic Press, pp 165–183. https://doi.org/10.1016/B978-0-323-85391-0.00008-3

Asmatulu E, Twomey J, Overcash M (2012) Life cycle and nano-products: end-of-life assessment. J Nanopart Res 14:1–8

Badawy AA, Abdelfattah NAH, Salem SS, Awad MF, Fouda A (2021) Efficacy assessment of biosynthesized copper oxide nanoparticles (CuO-NPs) on stored grain insects and their impacts on morphological and physiological traits of wheat (*Triticum aestivum L.*) plant. Biology 10(3). https://doi.org/10.3390/biology10030233

Balakumaran M, Ramachandran R, Balashanmugam P, Mukeshkumar D, Kalaichelvan P (2016) Mycosynthesis of silver and gold nanoparticles: optimization, characterization and antimicrobial activity against human pathogens. Microbiol Res 182:8–20

Baranwal J, Barse B, Fais A, Delogu GL, Kumar A (2022) Biopolymer: a sustainable material for food and medical applications. Polymers 14(5):983

Bhattacharjee R, Kumar L, Mukerjee N, Anand U, Dhasmana A, Preetam S, Bhaumik S, Sihi S, Pal S, Khare T, Chattopadhyay S, El-Zahaby SA, Alexiou A, Koshy EP, Kumar V, Malik S, Dey A, Proćków J (2022) The emergence of metal oxide nanoparticles (NPs) as a phytomedicine: a two-facet role in plant growth, nano-toxicity and anti-phyto-microbial activity. Biomed Pharmacother 155:113658. https://doi.org/10.1016/j.biopha.2022.113658

Bhau BS, Phukon P, Ahmed R, Gogoi B, Borah B, Baruah J, Sharma DK, Wann SB (2016) A novel tool of nanotechnology: nanoparticle mediated control of nematode infection in plants. In: Singh DP, Singh HB, Prabha R (eds) Microbial inoculants in sustainable agricultural productivity, vol 2: Functional Applications. Springer India, New Delhi, pp 253–269. https://doi.org/10.1007/978-81-322-2644-4_16

Biswas AK, Islam MR, Choudhury ZS, Mostafa A, Kadir MF (2014) Nanotechnology based approaches in cancer therapeutics. Adv Nat Sci: Nanosci Nanotechnol 5(4):043001. https://doi.org/10.1088/2043-6262/5/4/043001

Bonačić-Koutecký V, Le Guével X, Antoine R (n.d.) Engineering liganded gold nanoclusters as efficient theranostic agents for cancer applications. ChemBioChem n/a (n/a):e202200524. https://doi.org/10.1002/cbic.202200524

Boroumand Moghaddam A, Namvar F, Moniri M, Tahir P Md., Azizi S, Mohamad R (2015) Nanoparticles biosynthesized by fungi and yeast: a review of their preparation, properties, and medical applications. Molecules 20(9):16540–16565

Bulbake U, Doppalapudi S, Kommineni N, Khan W (2017) Liposomal formulations in clinical use: an updated review. Pharmaceutics 9(2):12

Carmona FJ, Guagliardi A, Masciocchi N (2022) Nanosized calcium phosphates as novel macronutrient nano-fertilizers. Nanomaterials 12(15):2709

Chand Mali S, Raj S, Trivedi R (2020) Nanotechnology a novel approach to enhance crop productivity. Biochem Biophys Rep 24:100821. https://doi.org/10.1016/j.bbrep.2020.100821

Chandrakala V, Aruna V, Angajala G (2022) Review on metal nanoparticles as nanocarriers: current challenges and perspectives in drug delivery systems. Emergent Mater:1–23

Cheng HN, Klasson KT, Asakura T, Wu Q (2016) Nanotechnology in agriculture. In: Nanotechnology: delivering on the promise volume 2, vol 1224. ACS symposium series, vol 1224. American Chemical Society, pp 233–242. https://doi.org/10.1021/bk-2016-1224.ch012

Chhipa H (2019) Chapter 6 - Applications of nanotechnology in agriculture. In: Gurtler V, Ball AS, Soni S (eds) Methods in microbiology, vol 46. Academic Press, pp 115–142. https://doi.org/10.1016/bs.mim.2019.01.002

Chung I-M, Rajakumar G, Gomathi T, Park S-K, Kim S-H, Thiruvengadam M (2017) Nanotechnology for human food: advances and perspective. Front Life Sci 10(1):63–72. https://doi.org/10.1080/21553769.2017.1365775

Cicek S, Nadaroglu H (2015) The use of nanotechnology in the agriculture. Adv Nano Res 3(4):207

Das A, Mukherjee P, Singla SK, Guturu P, Frost MC, Mukhopadhyay D, Shah VH, Patra CR (2010) Fabrication and characterization of an inorganic gold and silica nanoparticle mediated drug delivery system for nitric oxide. Nanotechnology 21(30):305102. https://doi.org/10.1088/0957-4484/21/30/305102

Deshpande D, Kethireddy S, Janero DR, Amiji MM (2016) Therapeutic efficacy of an ω-3-fatty acid-containing 17-β estradiol nano-delivery system against experimental atherosclerosis. PLoS One 11(2):e0147337. https://doi.org/10.1371/journal.pone.0147337

Dezfuli AAZ, Abu-Elghait M, Salem SS (2023) Recent Insights into nanotechnology in colorectal cancer. Appl Biochem Biotechnol. https://doi.org/10.1007/s12010-023-04696-3

Dhewa T (2015) Nanotechnology applications in agriculture: an update. Octa J Environ Res 3(2)

Díaz-González M, de la Escosura-Muñiz A, Fernandez-Argüelles MT, Alonso FJG, Costa-Fernandez JM (2020) Quantum dot bioconjugates for diagnostic applications. In: Surface-modified nanobiomaterials for electrochemical and biomedicine applications, pp 133–176

Doghish AS, Hashem AH, Shehabeldine AM, Sallam AAM, El-Sayyad GS, Salem SS (2022) Nanocomposite based on gold nanoparticles and carboxymethyl cellulose: synthesis, characterization, antimicrobial, and anticancer activities. J Drug Deliv Sci Technol 77. https://doi.org/10.1016/j.jddst.2022.103874

Dutta D, Das BM (2021) Scope of green nanotechnology towards amalgamation of green chemistry for cleaner environment: a review on synthesis and applications of green nanoparticles. Environ Nanotechnol Monit Manage 15:100418. https://doi.org/10.1016/j.enmm.2020.100418

Eid AM, Fouda A, Niedbała G, Hassan SED, Salem SS, Abdo AM, Hetta HF, Shaheen TI (2020) Endophytic *streptomyces laurentii* mediated green synthesis of Ag-NPs with antibacterial and anticancer properties for developing functional textile fabric properties. Antibiotics 9(10):1–18. https://doi.org/10.3390/antibiotics9100641

Elakraa AA, Salem SS, El-Sayyad GS, Attia MS (2022) Cefotaxime incorporated bimetallic silver-selenium nanoparticles: promising antimicrobial synergism, antibiofilm activity, and bacterial membrane leakage reaction mechanism. RSC Adv 12(41):26603–26619. https://doi.org/10.1039/d2ra04717a

Elemike EE, Uzoh IM, Onwudiwe DC, Babalola OO (2019) The role of nanotechnology in the fortification of plant nutrients and improvement of crop production. Appl Sci 9(3):499

Elfadel RG, Refat HM, Abd El-Wahab H, Salem SS, Owda ME, Abdel Reheim MAM (2023) Preparation of new surface coating based on modified oil-based polymers blended with ZnO and CuZnO NPs for steel protection. Sci Rep 13:7268. https://doi.org/10.1038/s41598-023-34085-z

El-Khawaga AM, Elsayed MA, Gobara M, Soliman AA, Hashem AH, Zaher AA, Mohsen M, Salem SS (2023) Green synthesized ZnO nanoparticles by *Saccharomyces cerevisiae* and their antibacterial activity and photocatalytic degradation. Biomass Conv Bioref. https://doi.org/10.1007/s13399-023-04827-0

Elsharkawy MM, Arafa RA, Omara RI, Kamel SM, Ismail W, Ismail S, Derbalah A (2022) Developing Ag_2O and Ag_2O/TiO_2 nanostructures as a new strategy for control late blight of potato caused by *Phytophthora infestans*. Physiol Mol Plant Pathol 120:101856. https://doi.org/10.1016/j.pmpp.2022.101856

Faridvand S, Amirnia R, Tajbakhsh M, El Enshasy HA, Sayyed RZ (2021) The effect of foliar application of magnetic water and nano-fertilizers on phytochemical and yield characteristics of fennel. Horticulturae 7(11):475

Ferlay J, Colombet M, Soerjomataram I, Parkin DM, Piñeros M, Znaor A, Bray F (2021) Cancer statistics for the year 2020: an overview. Int J Cancer 149(4):778–789. https://doi.org/10.1002/ijc.33588

Ferreira CA, Goel S, Ehlerding EB, Rosenkrans ZT, Jiang D, Sun T, Aluicio-Sarduy E, Engle JW, Ni D, Cai W (2021) Ultrasmall porous silica nanoparticles with enhanced pharmacokinetics for cancer theranostics. Nano Lett 21(11):4692–4699. https://doi.org/10.1021/acs.nanolett.1c00895

Fincheira P, Tortella G, Seabra AB, Quiroz A, Diez MC, Rubilar O (2021) Nanotechnology advances for sustainable agriculture: current knowledge and prospects in plant growth modulation and nutrition. Planta 254(4):66. https://doi.org/10.1007/s00425-021-03714-0

Fouda A, El-Din Hassan S, Salem SS, Shaheen TI (2018) In-Vitro cytotoxicity, antibacterial, and UV protection properties of the biosynthesized zinc oxide nanoparticles for medical textile applications. Microb Pathog 125:252–261. https://doi.org/10.1016/j.micpath.2018.09.030

Fouda A, Salem SS, Wassel AR, Hamza MF, Shaheen TI (2020) Optimization of green biosynthesized visible light active CuO/ZnO nano-photocatalysts for the degradation of organic methylene blue dye. Heliyon 6(9). https://doi.org/10.1016/j.heliyon.2020.e04896

Fraceto LF, Grillo R, de Medeiros GA, Scognamiglio V, Rea G, Bartolucci C (2016) Nanotechnology in agriculture: which innovation potential does it have? Front Environ Sci 4. https://doi.org/10.3389/fenvs.2016.00020

Frank R, Hargreaves R (2003) Clinical biomarkers in drug discovery and development. Nat Rev Drug Discov 2(7):566–580

Ghasemnezhad A, Ghorbanpour M, Sohrabi O, Ashnavar M (2019) Chapter Three - A general overview on application of nanoparticles in agriculture and plant science. In: Verma SK, Das AK (eds) Comprehensive analytical chemistry, vol 87. Elsevier, pp 85–110. https://doi.org/10.1016/bs.coac.2019.10.001

Guo Q, Shen X-t, Li Y-y, Xu S-q (2017) Carbon nanotubes-based drug delivery to cancer and brain. Curr Med Sci 37(5):635–641

Gupta N (2021) 2 Nanotechnology in agrifood sector: ethical, regulatory, and governance landscape in EU. In: Ethics in nanotechnology: social sciences and philosophical aspects, p 25

Hammad EN, Salem SS, Mohamed AA, El-Dougdoug W (2022a) Environmental impacts of ecofriendly iron oxide nanoparticles on dyes removal and antibacterial activity. Appl Biochem Biotechnol 194(12):6053–6067. https://doi.org/10.1007/s12010-022-04105-1

Hammad EN, Salem SS, Zohair MM, Mohamed AA, El-Dougdoug W (2022b) *Purpureocillium lilacinum* mediated biosynthesis copper oxide nanoparticles with promising removal of dyes. Biointerface Res Appl Chem 12(2):1397–1404. https://doi.org/10.33263/BRIAC122.13971404

Han YD, Kim KR, Lee KW, Yoon HC (2022) Retroreflection-based optical biosensing: from concept to applications. Biosens Bioelectron:114202

Hashem AH, Khalil AMA, Reyad AM, Salem SS (2021) Biomedical applications of mycosynthesized selenium nanoparticles using *Penicillium expansum* ATTC 36200. Biol Trace Elem Res 199(10):3998–4008. https://doi.org/10.1007/s12011-020-02506-z

Hashem AH, Salem SS (2022) Green and ecofriendly biosynthesis of selenium nanoparticles using *Urtica dioica* (stinging nettle) leaf extract: antimicrobial and anticancer activity. Biotechnol J 17(2). https://doi.org/10.1002/biot.202100432

Hashem AH, Shehabeldine AM, Ali OM, Salem SS (2022) Synthesis of chitosan-based gold nanoparticles: antimicrobial and wound-healing activities. Polymers 14(11). https://doi.org/10.3390/polym14112293

Hashem AH, Selim TA, Alruhaili MH, Selim S, Alkhalifah DHM, Al Jaouni SK, Salem SS (2022) Unveiling antimicrobial and insecticidal activities of biosynthesized selenium nanoparticles using prickly pear peel waste. J Funct Biomater 13(3). https://doi.org/10.3390/jfb13030112

Hashem AH, Abdelaziz AM, Attia MS, Salem SS (2022) Selenium and nano-selenium-mediated biotic stress tolerance in plants. In: Hossain MA, Ahammed GJ, Kolbert Z, El-Ramady H, Islam T, Schiavon M (eds) Selenium and nano-selenium in environmental stress management and crop quality improvement. Springer International Publishing, Cham, pp 209–226. https://doi.org/10.1007/978-3-031-07063-1_11

Hassan SED, Fouda A, Radwan AA, Salem SS, Barghoth MG, Awad MA, Abdo AM, El-Gamal MS (2019) Endophytic actinomycetes Streptomyces spp mediated biosynthesis of copper oxide nanoparticles as a promising tool for biotechnological applications. J Biol Inorg Chem. https://doi.org/10.1007/s00775-019-01654-5

Hu J, Sanz-Rodríguez F, Rivero F, Rodríguez EM, Torres RA, Ortgies DH, Solé JG, Alfonso F, Jaque D (2018) Gold nanoshells: contrast agents for cell imaging by cardiovascular optical coherence tomography. Nano Res 11(2):676–685. https://doi.org/10.1007/s12274-017-1674-4

Huang S, Wang L, Liu L, Hou Y, Li L (2015) Nanotechnology in agriculture, livestock, and aquaculture in China. A review. Agron Sustain Dev 35(2):369–400. https://doi.org/10.1007/s13593-014-0274-x

Iavicoli I, Leso V, Beezhold DH, Shvedova AA (2017) Nanotechnology in agriculture: opportunities, toxicological implications, and occupational risks. Toxicol Appl Pharmacol 329:96–111. https://doi.org/10.1016/j.taap.2017.05.025

Imran M, Shah MR (2018) Amphiphilic block copolymers–based micelles for drug delivery. In: Design and development of new nanocarriers. Elsevier, pp 365–400

Jakhar AM, Aziz I, Kaleri AR, Hasnain M, Haider G, Ma J, Abideen Z (2022) Nano-fertilizers: a sustainable technology for improving crop nutrition and food security. NanoImpact 27:100411. https://doi.org/10.1016/j.impact.2022.100411

Jatav G, Nirmal D (2013) Application of nanotechnology in soil-plant system. An Asian J Soil Sci 8(1):176–184

Joshi H, Choudhary P, Mundra S (2019) Future prospects of nanotechnology in agriculture. Int J Chem Stud 7(2):957–963

Kapoor RT, Salvadori MR, Rafatullah M, Siddiqui MR, Khan MA, Alshareef SA (2021) Exploration of microbial factories for synthesis of nanoparticles – a sustainable approach for bioremediation of environmental contaminants. Front Microbiol 12. https://doi.org/10.3389/fmicb.2021.658294

Katsuki S, Matoba T, Koga J-i, Nakano K, Egashira K (2017) Anti-inflammatory nanomedicine for cardiovascular disease. Front Cardiovasc Med 4:87

Kaur J, Gill GS, Jeet K (2019) Applications of carbon nanotubes in drug delivery: a comprehensive review. In: Characterization and biology of nanomaterials for drug delivery, pp 113–135

Kavalaraki A, Spyratou E, Kouri MA, Efstathopoulos EP (2023) Gold nanoparticles as contrast agents in ophthalmic imaging. Optics 4(1):74–99

Kerry RG, Gouda S, Das G, Vishnuprasad CN, Patra JK (2017) Agricultural nanotechnologies: Current applications and future prospects. In: Patra JK, Vishnuprasad CN, Das G (eds) Microbial biotechnology: Volume 1. Applications in agriculture and environment. Springer Singapore, Singapore, pp 3–28. https://doi.org/10.1007/978-981-10-6847-8_1

Kesharwani P, Gothwal A, Iyer AK, Jain K, Chourasia MK, Gupta U (2018) Dendrimer nanohybrid carrier systems: an expanding horizon for targeted drug and gene delivery. Drug Discov Today 23(2):300–314

Khanna P, Kaur A, Goyal D (2019) Algae-based metallic nanoparticles: synthesis, characterization and applications. J Microbiol Methods 163:105656

Kokaz SF, Deb PK, Borah P, Bania R, Venugopala KN, Nair AB, Singh V, Al-Shar'i NA, Hourani W, Gupta G (2022) Dendrimers: properties and applications in biomedical field. In: Nanoengineering of Biomaterials, pp 215–243

Kraft JC, Freeling JP, Wang Z, Ho RJ (2014) Emerging research and clinical development trends of liposome and lipid nanoparticle drug delivery systems. J Pharm Sci 103(1):29–52

Kumar A, Gupta K, Dixit S, Mishra K, Srivastava S (2019) A review on positive and negative impacts of nanotechnology in agriculture. Int J Environ Sci Technol 16(4):2175–2184. https://doi.org/10.1007/s13762-018-2119-7

Kumar R (2020) Application of nanotechnology in crop disease management. Universe Int J Interdisc Res 1(2)

Kurbanoglu S, Ozkan SA (2018) Electrochemical carbon based nanosensors: a promising tool in pharmaceutical and biomedical analysis. J Pharm Biomed Anal 147:439–457

Kute SS, Shreyas Madhav A, Kumari S, Aswathy S (2022) Machine learning–based disease diagnosis and prediction for E-healthcare system. In: Advanced analytics and deep learning models, pp 127–147

Letchford K, Burt H (2007) A review of the formation and classification of amphiphilic block copolymer nanoparticulate structures: micelles, nanospheres, nanocapsules and polymersomes. Eur J Pharm Biopharm 65(3):259–269

Liang J-L, Luo G-F, Chen W-H, Zhang X-Z (2021) Recent advances in engineered materials for immunotherapy-involved combination cancer therapy. Adv Mater 33(31):2007630. https://doi.org/10.1002/adma.202007630

Liu B, Jiang B, Zheng Z, Liu T (2019) Semiconductor quantum dots in tumor research. J Lumin 209:61–68

Liu C, Zhou H, Zhou J (2021) The applications of nanotechnology in crop production. Molecules 26(23):7070

Lombardo D, Kiselev MA, Caccamo MT (2019) Smart nanoparticles for drug delivery application: development of versatile nanocarrier platforms in biotechnology and nanomedicine. J Nanomater 2019

Lowry GV, Avellan A, Gilbertson LM (2019) Opportunities and challenges for nanotechnology in the agri-tech revolution. Nat Nanotechnol 14(6):517–522. https://doi.org/10.1038/s41565-019-0461-7

Majidinia M, Mirza-Aghazadeh-Attari M, Rahimi M, Mihanfar A, Karimian A, Safa A, Yousefi B (2020) Overcoming multidrug resistance in cancer: recent progress in nanotechnology and new horizons. IUBMB Life 72(5):855–871. https://doi.org/10.1002/iub.2215

Manimaran M, Kannabiran K (2017) Actinomycetes-mediated biogenic synthesis of metal and metal oxide nanoparticles: progress and challenges. Lett Appl Microbiol 64(6):401–408. https://doi.org/10.1111/lam.12730

Mazayen ZM, Ghoneim AM, Elbatanony RS, Basalious EB, Bendas ER (2022) Pharmaceutical nanotechnology: from the bench to the market. Future J Pharm Sci 8(1):12. https://doi.org/10.1186/s43094-022-00400-0

McGill HC, McMahan CA, Gidding SS (2008) Preventing heart disease in the 21st century. Circulation 117(9):1216–1227. https://doi.org/10.1161/CIRCULATIONAHA.107.717033

Medina C, Santos-Martinez MJ, Radomski A, Corrigan OI, Radomski MW (2007) Nanoparticles: pharmacological and toxicological significance. Br J Pharmacol 150(5):552–558. https://doi.org/10.1038/sj.bjp.0707130

Mitra D, Adhikari P, Djebaili R, Thathola P, Joshi K, Pellegrini M, Adeyemi NO, Khoshru B, Kaur K, Priyadarshini A, Senapati A, Del Gallo M, Das Mohapatra PK, Nayak AK, Shanmugam V, Panneerselvam P (2023) Biosynthesis and characterization of nanoparticles, its advantages, various aspects and risk assessment to maintain the sustainable agriculture: emerging technology in modern era science. Plant Physiol Biochem 196:103–120. https://doi.org/10.1016/j.plaphy.2023.01.017

Mohamed AA, Fouda A, Abdel-Rahman MA, Hassan SED, El-Gamal MS, Salem SS, Shaheen TI (2019) Fungal strain impacts the shape, bioactivity and multifunctional properties of green synthesized zinc oxide nanoparticles. Biocatal Agric Biotechnol 19. https://doi.org/10.1016/j.bcab.2019.101103

Mohamed AA, Abu-Elghait M, Ahmed NE, Salem SS (2021) Correction to: eco-friendly mycogenic synthesis of ZnO and CuO nanoparticles for in vitro antibacterial, antibiofilm and antifungal applications (Biol Trace Elem Res, (2021), 199, 7, (2788–2799), https://doi.org/10.1007/s12011-020-02369-4). Biol Trace Elem Res 199(7):2800–2801. https://doi.org/10.1007/s12011-020-02391-6

Mohmed AA, Fouda A, Elgamal MS, El-Din Hassan S, Shaheen TI, Salem SS (2017) Enhancing of cotton fabric antibacterial properties by silver nanoparticles synthesized by new egyptian strain fusarium keratoplasticum A1–3. Egypt J Chem 60:63–71. https://doi.org/10.21608/ejchem.2017.1626.1137

Montero AJ, Adams B, Diaz-Montero CM, Glück S (2011) Nab-paclitaxel in the treatment of metastatic breast cancer: a comprehensive review. Expert Rev Clin Pharmacol 4(3):329–334. https://doi.org/10.1586/ecp.11.7

Mousa SA, Bharali DJ (2011) Nanotechnology-based detection and targeted therapy in cancer: nano-bio paradigms and applications. Cancers 3(3):2888–2903

Nair PP (2021) A unique perspective in precision of nano-biotechnology for sustainable agricultural fields. In: Pal K (ed) Bio-manufactured nanomaterials: perspectives and promotion. Springer International Publishing, Cham, pp 299–320. https://doi.org/10.1007/978-3-030-67223-2_14

Najlah M, Freeman S, Attwood D, D'Emanuele A (2007) In vitro evaluation of dendrimer prodrugs for oral drug delivery. Int J Pharm 336(1):183–190. https://doi.org/10.1016/j.ijpharm.2006.11.047

Ndlovu N, Mayaya T, Muitire C, Munyengwa N (2020) Nanotechnology applications in crop production and food systems. Int J Plant Breed 7(1):624–634

Ortega MÁ, Guzmán Merino A, Fraile-Martínez O, Recio-Ruiz J, Pekarek L, G. Guijarro L, García-Honduvilla N, Álvarez-Mon M, Buján J, García-Gallego S (2020) Dendrimers and dendritic materials: from laboratory to medical practice in infectious diseases. Pharmaceutics 12(9):874

Østergaard M, Pedersen SJ, Døhn UM (2008) Imaging in rheumatoid arthritis–status and recent advances for magnetic resonance imaging, ultrasonography, computed tomography and conventional radiography. Best Pract Res Clin Rheumatol 22(6):1019–1044

Pal S, Koneru JK, Andreou C, Rakshit T, Rajasekhar VK, Wlodarczyk M, Healey JH, Kircher MF, Mondal J (2022) DNA-functionalized gold nanorods for perioperative optical imaging and photothermal therapy of triple-negative breast cancer. ACS Appl Nano Mater 5(7):9159–9169. https://doi.org/10.1021/acsanm.2c01502

Palmerston Mendes L, Pan J, Torchilin VP (2017) Dendrimers as nanocarriers for nucleic acid and drug delivery in cancer therapy. Molecules 22(9):1401

Park JH, Dehaini D, Zhou J, Holay M, Fang RH, Zhang L (2020) Biomimetic nanoparticle technology for cardiovascular disease detection and treatment. Nanoscale Horiz 5(1):25–42. https://doi.org/10.1039/C9NH00291J

Plaksenkova I, Kokina I, Petrova A, Jermaļonoka M, Gerbreders V, Krasovska M (2020) The impact of zinc oxide nanoparticles on cytotoxicity, genotoxicity, and miRNA expression in barley (*Hordeum vulgare* L.) seedlings. Sci World J 2020:6649746. https://doi.org/10.1155/2020/6649746

Pramanik S, Pramanik G (2016) Nanotechnology for sustainable agriculture in India. In: Ranjan S, Dasgupta N, Lichtfouse E (eds) Nanoscience in food and agriculture 3. Springer International Publishing, Cham, pp 243–280. https://doi.org/10.1007/978-3-319-48009-1_10

Pramanik P, Krishnan P, Maity A, Mridha N, Mukherjee A, Rai V (2020) Application of nanotechnology in agriculture. In: Dasgupta N, Ranjan S, Lichtfouse E (eds) Environmental nanotechnology volume 4. Springer International Publishing, Cham, pp 317–348. https://doi.org/10.1007/978-3-030-26668-4_9

Rai M, Ingle A (2012) Role of nanotechnology in agriculture with special reference to management of insect pests. Appl Microbiol Biotechnol 94(2):287–293. https://doi.org/10.1007/s00253-012-3969-4

Rajput VD, Singh A, Minkina T, Rawat S, Mandzhieva S, Sushkova S, Shuvaeva V, Nazarenko O, Rajput P, Komariah VKK, Singh AK, Rao M, Upadhyay SK (2021) Nano-enabled products: challenges and opportunities for sustainable agriculture. Plants 10(12):2727

Ritu, Sounkaria S, Sachdeva G, Das A, Verma SR, Saxena SC, Singh BP, Rahman S, Chandra P (2023) Chapter 6 - Potentialities of nanobiotechnology in nutrient management in the livestock products. In: Pratap Singh R, Adetunji CO, Singh RL, Singh J, Solanki PR, Singh KRB (eds)

Nanobiotechnology for the livestock industry. Elsevier, pp 111–137. https://doi.org/10.1016/B978-0-323-98387-7.00016-1
Saeedi M, Eslamifar M, Khezri K, Dizaj SM (2019) Applications of nanotechnology in drug delivery to the central nervous system. Biomed Pharmacother 111:666–675. https://doi.org/10.1016/j.biopha.2018.12.133
Said A, Abu-Elghait M, Atta HM, Salem SS (2023) Antibacterial activity of green synthesized silver nanoparticles using lawsonia inermis against common pathogens from urinary tract infection. Appl Biochem Biotechnol. https://doi.org/10.1007/s12010-023-04482-1
Saied E, Eid AM, Hassan SED, Salem SS, Radwan AA, Halawa M, Saleh FM, Saad HA, Saied EM, Fouda A (2021) The catalytic activity of biosynthesized magnesium oxide nanoparticles (MgO-NPs) for inhibiting the growth of pathogenic microbes, tanning effluent treatment, and chromium ion removal. Catalysts 11(7). https://doi.org/10.3390/catal11070821
Saied E, Salem SS, Al-Askar AA, Elkady FM, Arishi AA, Hashem AH (2022) Mycosynthesis of hematite (α-Fe_2O_3) nanoparticles using *Aspergillus niger* and their antimicrobial and photocatalytic activities. Bioengineering 9(8). https://doi.org/10.3390/bioengineering9080397
Salem SS (2022a) Baker's yeast-mediated silver nanoparticles: characterisation and antimicrobial biogenic tool for suppressing pathogenic microbes. BioNanoScience 12(4):1220–1229. https://doi.org/10.1007/s12668-022-01026-5
Salem SS (2022b) Bio-fabrication of selenium nanoparticles using baker's yeast extract and its antimicrobial efficacy on food borne pathogens. Appl Biochem Biotechnol 194(5):1898–1910. https://doi.org/10.1007/s12010-022-03809-8
Salem SS (2023) Application of Nano-materials. In: Raja R, Hemaiswarya S, Narayanan M, Kandasamy S, Jayappriyan K (eds) Haematococcus. Springer, Singapore. https://doi.org/10.1007/978-981-99-2901-6_10
Salem SS, Fouda A (2021) Green synthesis of metallic nanoparticles and their prospective biotechnological applications: an overview. Biol Trace Elem Res 199(1):344–370. https://doi.org/10.1007/s12011-020-02138-3
Salem SS, Husen A (2023) Zinc oxide nanoparticles: Plant response, interaction, phytotoxicity, and defence. In: Husen A (eds) nanomaterials and nanocomposites exposures to plants. Smart Nanomat Technol. Springer, Singapore. https://doi.org/10.1007/978-981-99-2419-6_12
Salem SS, Mohamed AA, Gl-Gamal MS, Talat M, Fouda A (2019) Biological decolorization and degradation of azo dyes from textile wastewater effluent by *Aspergillus niger*. Egypt J Chem 62(10):1799–1813. https://doi.org/10.21608/EJCHEM.2019.11720.1747
Salem SS, El-Belely EF, Niedbała G, Alnoman MM, Hassan SED, Eid AM, Shaheen TI, Elkelish A, Fouda A (2020) Bactericidal and in-vitro cytotoxic efficacy of silver nanoparticles (Ag-NPs) fabricated by endophytic actinomycetes and their use as coating for the textile fabrics. Nanomaterials 10(10):1–20. https://doi.org/10.3390/nano10102082
Salem SS, Fouda MMG, Fouda A, Awad MA, Al-Olayan EM, Allam AA, Shaheen TI (2021) Antibacterial, cytotoxicity and larvicidal activity of green synthesized selenium nanoparticles using *Penicillium corylophilum*. J Cluster Sci 32(2):351–361. https://doi.org/10.1007/s10876-020-01794-8
Salem SS, Husen A (2022) Effect of engineered nanomaterials on soil microbiomes and their association with crop growth and production. In: Engineered nanomaterials for sustainable agricultural production, soil improvement and stress management. pp 311–336. https://doi.org/10.1016/B978-0-323-91933-3.00010-6
Salem SS, Hashem AH, Sallam AAM, Doghish AS, Al-Askar AA, Arishi AA, Shehabeldine AM (2022) Synthesis of silver nanocomposite based on carboxymethyl cellulose: antibacterial, antifungal and anticancer activities. Polymers 14(16). https://doi.org/10.3390/polym14163352
Salem SS, Ali OM, Reyad AM, Abd-Elsalam KA, Hashem AH (2022) *Pseudomonas indica*-mediated silver nanoparticles: antifungal and antioxidant biogenic tool for suppressing mucormycosis fungi. J Fungi 8(2). https://doi.org/10.3390/jof8020126

Salem SS, Badawy MSEM, Al-Askar AA, Arishi AA, Elkady FM, Hashem AH (2022) Green biosynthesis of selenium nanoparticles using orange peel waste: characterization, antibacterial and antibiofilm activities against multidrug-resistant bacteria. Life 12(6). https://doi.org/10.3390/life12060893

Salem SS, Hammad EN, Mohamed AA, El-Dougdoug W (2023) A comprehensive review of nanomaterials: types, synthesis, characterization, and applications. Biointerface Res Appl Chem 13(1). https://doi.org/10.33263/BRIAC131.041

Sangeetha J, Thangadurai D, Hospet R, Purushotham P, Karekalammanavar G, Mundaragi AC, David M, Shinge MR, Thimmappa SC, Prasad R, Harish ER (2017) Agricultural nanotechnology: concepts, benefits, and risks. In: Prasad R, Kumar M, Kumar V (eds) Nanotechnology: an agricultural paradigm. Springer Singapore, Singapore, pp 1–17. https://doi.org/10.1007/978-981-10-4573-8_1

Sakiyama-Elbert SE, Hubbell JA (2001) Functional biomaterials: design of novel biomaterials. Ann Rev Mater Res 31(1):183–201. https://doi.org/10.1146/annurev.matsci.31.1.183

Seetharaman PK, Chandrasekaran R, Gnanasekar S, Chandrakasan G, Gupta M, Manikandan DB, Sivaperumal S (2018) Antimicrobial and larvicidal activity of eco-friendly silver nanoparticles synthesized from endophytic fungi *Phomopsis liquidambaris*. Biocatal Agric Biotechnol 16:22–30. https://doi.org/10.1016/j.bcab.2018.07.006

Selim MT, Salem SS, Mohamed AA, El-Gamal MS, Awad MF, Fouda A (2021) Biological treatment of real textile effluent using *Aspergillus flavus* and *Fusarium oxysporium* and their consortium along with the evaluation of their phytotoxicity. J Fungi 7(3). https://doi.org/10.3390/jof7030193

Shaheen TI, Fouda A, Salem SS (2021a) Integration of cotton fabrics with biosynthesized CuO nanoparticles for bactericidal activity in the terms of their cytotoxicity assessment. Ind Eng Chem Res 60(4):1553–1563. https://doi.org/10.1021/acs.iecr.0c04880

Shaheen TI, Salem SS, Fouda A (2021) Current advances in fungal nanobiotechnology: mycofabrication and applications. In: Lateef A, Gueguim-Kana EB, Dasgupta N, Ranjan S (eds) Microbial nanobiotechnology: principles and applications. Springer Singapore, Singapore, pp 113–143. https://doi.org/10.1007/978-981-33-4777-9_4

Shandilya G, Tarwadi K (2021) Immunity in early life. In: Nanotechnology in Plant growth promotion and protection. pp 93–112. https://doi.org/10.1002/9781119745884.ch6

Shang Y, Hasan MK, Ahammed GJ, Li M, Yin H, Zhou J (2019) Applications of nanotechnology in plant growth and crop protection: a review. Molecules 24(14):2558

Sharaf MH, Nagiub AM, Salem SS, Kalaba MH, El Fakharany EM, Abd El-Wahab H (2022) A new strategy to integrate silver nanowires with waterborne coating to improve their antimicrobial and antiviral properties. Pigm Resin Technol. https://doi.org/10.1108/PRT-12-2021-0146

Shehabeldine AM, Amin BH, Hagras FA, Ramadan AA, Kamel MR, Ahmed MA, Atia KH, Salem SS (2023) Potential antimicrobial and antibiofilm properties of copper oxide nanoparticles: time-kill kinetic essay and ultrastructure of pathogenic bacterial cells. Appl Biochem Biotechnol 195(1):467–485. https://doi.org/10.1007/s12010-022-04120-2

Shehabeldine AM, Salem SS, Ali OM, Abd-Elsalam KA, Elkady FM, Hashem AH (2022) Multifunctional silver nanoparticles based on chitosan: antibacterial, antibiofilm, antifungal, antioxidant, and wound-healing activities. J Fungi 8(6). https://doi.org/10.3390/jof8060612

Sherje AP, Jadhav M, Dravyakar BR, Kadam D (2018) Dendrimers: a versatile nanocarrier for drug delivery and targeting. Int J Pharm 548(1):707–720

Shi Kam NW, Jessop TC, Wender PA, Dai H (2004) Nanotube molecular transporters: internalization of carbon nanotube– protein conjugates into mammalian cells. J Am Chem Soc 126(22):6850–6851

Shukla P, Chaurasia P, Younis K, Qadri OS, Faridi SA, Srivastava G (2019) Nanotechnology in sustainable agriculture: studies from seed priming to post-harvest management. Nanotechnol Environ Eng 4(1):11. https://doi.org/10.1007/s41204-019-0058-2

Singh J, Jain K, Mehra NK, Jain N (2016) Dendrimers in anticancer drug delivery: mechanism of interaction of drug and dendrimers. Artif Cells Nanomed Biotechnol 44(7):1626–1634

Singh AP, Biswas A, Shukla A, Maiti P (2019) Targeted therapy in chronic diseases using nanomaterial-based drug delivery vehicles. Signal Transduct Target Ther 4(1):33. https://doi.org/10.1038/s41392-019-0068-3

Singh KR, Nayak V, Sabui P, Mallick S, Singh J, Singh RP (2022) Bioinspired quantum dots: promising nanosystems for biomedical application. Nano-Struct Nano-Objects 32:100921

Soliman MKY, Abu-Elghait M, Salem SS, Azab MS (2022) Multifunctional properties of silver and gold nanoparticles synthesis by *Fusarium pseudonygamai*. Biomass Convers Biorefin. https://doi.org/10.1007/s13399-022-03507-9

Soliman MKY, Salem SS, Abu-Elghait M, Azab MS (2023) Biosynthesis of silver and gold nanoparticles and their efficacy towards antibacterial, antibiofilm, cytotoxicity, and antioxidant activities. Appl Biochem Biotechnol 195(2):1158–1183. https://doi.org/10.1007/s12010-022-04199-7

Sorokina SA, Shifrina ZB (2022) Dendrimers as antiamyloid agents. Pharmaceutics 14(4):760

Srinoi P, Chen Y-T, Vittur V, Marquez MD, Lee TR (2018) Bimetallic nanoparticles: enhanced magnetic and optical properties for emerging biological applications. Appl Sci 8(7):1106

Suri SS, Fenniri H, Singh B (2007) Nanotechnology-based drug delivery systems. J Occup Med Toxicol 2(1):16. https://doi.org/10.1186/1745-6673-2-16

Tiwari AK, Mishra A, Pandey G, Gupta MK, Pandey PC (2022) Nanotechnology: a potential weapon to fight against COVID-19. Part Part Syst Charact 39(1):2100159. https://doi.org/10.1002/ppsc.202100159

Trotta F, Mele A (2019) Nanomaterials: classification and properties. Nanosponges: synthesis and applications: 1st edn. Wiley, London

Tsurutani J, Kuroi K, Iwasa T, Miyazaki M, Nishina S, Makimura C, Tanizaki J, Okamoto K, Yamashita T, Aruga T, Shigekawa T, Komoike Y, Saeki T, Nakagawa K (2015) Phase I study of weekly nab-paclitaxel combined with S-1 in patients with human epidermal growth factor receptor type 2-negative metastatic breast cancer. Cancer Sci 106(6):734–739. https://doi.org/10.1111/cas.12658

van Vlerken LE, Vyas TK, Amiji MM (2007) Poly(ethylene glycol)-modified nanocarriers for tumor-targeted and intracellular delivery. Pharm Res 24(8):1405–1414. https://doi.org/10.1007/s11095-007-9284-6

Vega-Vásquez P, Mosier NS, Irudayaraj J (2020) Nanoscale drug delivery systems: from medicine to agriculture. Front Bioeng Biotechnol 8. https://doi.org/10.3389/fbioe.2020.00079

Wan J, Zhang X, Fu K, Zhang X, Shang L, Su Z (2021) Highly fluorescent carbon dots as novel theranostic agents for biomedical applications. Nanoscale 13(41):17236–17253

Wang X, Zhang W (2022) The Janus of Protein Corona on nanoparticles for tumor targeting, immunotherapy and diagnosis. J Controlled Release

Wani KA, Kothari R (2018) Agricultural nanotechnology: applications and challenges. Ann Plant Sci 7:2146–2148

Wieliczko B, Floriańczyk Z (2022) Priorities for research on sustainable agriculture: the case of Poland. Energies 15(1):257

Wu T, Chen X, Wang Y, Xiao H, Peng Y, Lin L, Xia W, Long M, Tao J, Shuai X (2018) Aortic plaque-targeted andrographolide delivery with oxidation-sensitive micelle effectively treats atherosclerosis via simultaneous ROS capture and anti-inflammation. Nanomed Nanotechnol Biol Med 14(7):2215–2226. https://doi.org/10.1016/j.nano.2018.06.010

Yata VK, Tiwari BC, Ahmad I (2018) Nanoscience in food and agriculture: research, industries and patents. Environ Chem Lett 16(1):79–84. https://doi.org/10.1007/s10311-017-0666-7

Yin J, Dong Z, Liu Y, Wang H, Li A, Zhuo Z, Feng W, Fan W (2020) Toxicity of reduced graphene oxide modified by metals in microalgae: effect of the surface properties of algal cells and nanomaterials. Carbon 169:182–192. https://doi.org/10.1016/j.carbon.2020.07.057

Youssef MS, Elamawi RM (2020) Evaluation of phytotoxicity, cytotoxicity, and genotoxicity of ZnO nanoparticles in *Vicia faba*. Environ Sci Pollut Res 27(16):18972–18984. https://doi.org/10.1007/s11356-018-3250-1

Zhang H, Hu X (2018) Biosynthesis of Pd and Au as nanoparticles by a marine bacterium Bacillus sp. GP and their enhanced catalytic performance using metal oxides for 4-nitrophenol reduction. Enzyme Microb Technol 113:59–66. https://doi.org/10.1016/j.enzmictec.2018.03.002

Zhang N, Li C, Zhou D, Ding C, Jin Y, Tian Q, Meng X, Pu K, Zhu Y (2018) Cyclic RGD functionalized liposomes encapsulating urokinase for thrombolysis. Acta Biomater 70:227–236. https://doi.org/10.1016/j.actbio.2018.01.038

Zhou J, Deng W, Wang Y, Cao X, Chen J, Wang Q, Xu W, Du P, Yu Q, Chen J (2016) Cationic carbon quantum dots derived from alginate for gene delivery: one-step synthesis and cellular uptake. Acta Biomater 42:209–219

Zihan Y, Chang Q, Yaxin Z, Haoting L, Yufei L (2022) A Review of the extremely structural properties of carbon nanotubes, synthesis and purification methods. Am J Mater Synth Process 7(2):29–35

Chapter 3
Synthesis of Biogenic Nanomaterials, Their Characterization, and Applications

Simran Kaur, Gunjit Setia, Mudra Sikenis, and Saroj Kumar

Abstract Nanotechnology has advanced over the last ten years, which has sparked new biotechnological, imaging, and diagnostic device developments. Nanomaterials offer several scientific benefits and have exciting industrial potential. One such category of nanomaterials receiving attention is biogenic nanomaterials. Biogenic nanomaterials are characterized as materials that are derived from biological organisms. The production of nanomaterials using biosynthetic pathways may replace current chemical processes, opening a wide range of potential to develop more environmentally friendly products that can be applied to industrial operations. Due to its uses, researchers have directed their focus to nanomaterials, resulting in the development of more comprehensive platforms in fields including biomedicine and nanobiotechnology. Synthesizing biogenic nanomaterials is a critical step in research that can provide significant benefits in the future as we attempt to meet the growing demand for nanomaterials for biological needs. This chapter provides various methods of biogenic synthesis along with physical and chemical attributes. Further, this chapter expands into the various techniques employed for the characterization of nanoparticles along with the applications of synthetic biological nanomaterials in the environmental, healthcare, and industrial world.

Keywords Nanomaterials · Biogenic · Synthesis · Extracellular · Intracellular · Environment · Healthcare

Simran Kaur, Gunjit Setia, Mudra Sikenis—Authors contributed equally to this work.

S. Kaur · G. Setia · M. Sikenis · S. Kumar (✉)
School of Biosciences, Apeejay Stya University, Gurugram, Haryana 122103, India
e-mail: sarojbiotech@gmail.com

M. P. Shah et al. (eds.), *Biogenic Nanomaterials for Environmental Sustainability: Principles, Practices, and Opportunities*, Environmental Science and Engineering,
https://doi.org/10.1007/978-3-031-45956-6_3

3.1 Introduction

Nanotechnology is the technology that provides the most effective methods used to fabricate nanomaterials (NMs) within the most extensive areas of NM research (Kumar et al. 2020). The main characteristic of NMs is that their dimensions are less than 100 nm, which helps them to bond between bulk material and molecular structure materials and makes them suitable for various applications (immune modulator for tumor treatment, antibacterial, antifungal agents, drug delivery, etc.). Nanobiotechnology is a developing area of nanotechnology and biotechnology that integrates research from many other disciplines, including chemistry, biology, medicine, and material science, to build the necessary therapeutic agents. Among the various nanomaterials, metal and metal oxide nanoparticles are most effective due to their significant biomedical applications and increased area-to-volume ratio. Furthermore, synthesis of nanomaterial from plant species and herbal extracts has shown reducing and capping properties in the novel field of nanoscience. Initially, the natural plant extract of the pure compound was utilized in the green synthesis of nanomaterials. Furthermore, isolated plant-based products such as cellulose, glucose, starch, or extracts have been used in the synthesis of nanomaterials (Alle et al. 2020).

The approaches adapted for nanoparticle synthesis are the top-down and the bottom-up approaches. The top-down approach involves cutting and breaking bulk material into finite particles at the nanoscale. The bottom-up method involves joining atom by atom, protein cluster by cluster, or molecule by molecule to create or build up the nanoscale material (Fig. 3.1). Various methods have been developed for synthesizing nanomaterials, including chemical reduction methods, electrochemical reduction, and different other routes. Generally, nanomaterials are synthesized chemically, including toxic chemicals, and releasing these chemicals into the environment may be a significant concern (Villaseñor-Basulto et al. 2019). Due to environmental concerns, chemical synthesis is not preferred. Biogenic synthetic protocols are being used to produce nanomaterials due to their non-toxic, eco-friendly nature, which provides a way to replace toxic chemicals with natural products. Biogenic synthesis utilizes the products such as vitamins, amino acids, and carbohydrates from plants, bacteria, algae, or fungi. Biogenic synthesis follows bottom-up approaches, which help produce medium to tiny nanomaterials. Various biogenic sources have been used to synthesize nanomaterials, showing advantages such as higher production rate, cost-effectiveness, and low energy requirements (Sawalha et al. 2021).

Metal nanoparticles consist of metallic cores like gold, platinum, or their compounds. These nanoparticles have become the main focus of interest due to their characteristic properties compared to their natural bulk material (Elizabeth et al. 2019; Khan 2020). Several methods like microemulsion, sputtering and mechanical milling are used to prepare these nanoparticles. Recent advancements in the nanobiotechnology field have found that the use of biodegradable food waste is a significant and feasible source for extraction of metallic nanoparticles as they contain different organic compounds like polyphenols, flavonoids, carotenoids, and vitamins, which as a templating agent (Sharma et al. 2019). Metal oxide nanoparticles are the

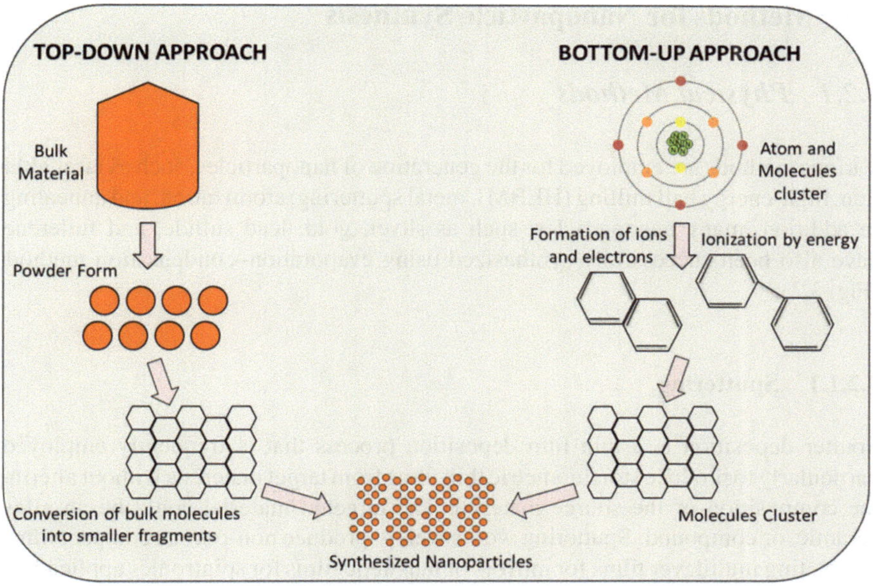

Fig. 3.1 Represents the precise elucidation of the top-down and bottom-up approaches

most commonly used nanoparticles as they show optical properties like UV absorption and color absorption within the visible region (Denizli et al. 2021). Wide varieties of nanoparticles have been synthesized under this class using metal as targets like aluminum, titanium, nickel, and many more. These nanoparticles have several key advantages, including structural modifications that can change cell parameters and lattice symmetry, changes in electrochemical properties brought on by the quantum confinement effect, and modifications to surface properties that result in a sharp increase in the band gap, which can affect conductivity (Baig et al. 2021). These various properties of metal oxide nanoparticles have unearthed potential which is yet there to discover.

This chapter covers the synthesis and characterization of biogenic nanomaterials. The biogenic mechanism of nanomaterials is represented in three aspects: physical, chemical, and biological (green nanomaterials). This chapter also describes the applications of biogenic nanomaterials in biotechnology, biomedical and pharmaceutical domains using bacteria, fungi, and plant extracts.

3.2 Methods for Nanoparticle Synthesis

3.2.1 *Physical Methods*

Various methods are employed for the generation of nanoparticles, such as laser ablation, high-energy ball milling (HEBM), metal sputtering, atomization, and annealing. In addition, many nanoparticles, such as silver, gold, lead sulfide, and fullerene, have also been successfully synthesized using evaporation–condensation methods (Fig. 3.2).

3.2.1.1 Sputtering

Sputter deposition is a thin film deposition process that is frequently employed, particularly to produce stoichiometric thin films from target materials without altering the composition of the source material. The targeted material could be an alloy, ceramic, or compound. Sputtering works well to produce non-porous compact films. Depositing multilayer films for mirrors or magnetic films for spintronics applications is a very effective process.

Some inert gas ions, such as ArC, are incident on a target at high energy during sputter deposition. The ion-target interaction can be a complicated process depending on the energy of the ions and the mass ratio of the ions to the target atoms. The incident ions become neutral at the surface, but due to their energy, incident ions may penetrate, bounce back, displace some of the target atoms, produce photons while losing energy to the target atoms, or even sputter out some targeted atoms/ molecules, clusters known as metal nanoparticles. Direct current (DC) sputtering, radio frequency (RF) sputtering, and magnetron sputtering are three methods that can be used for sputter deposition. Low base pressure ensures that the required purity is

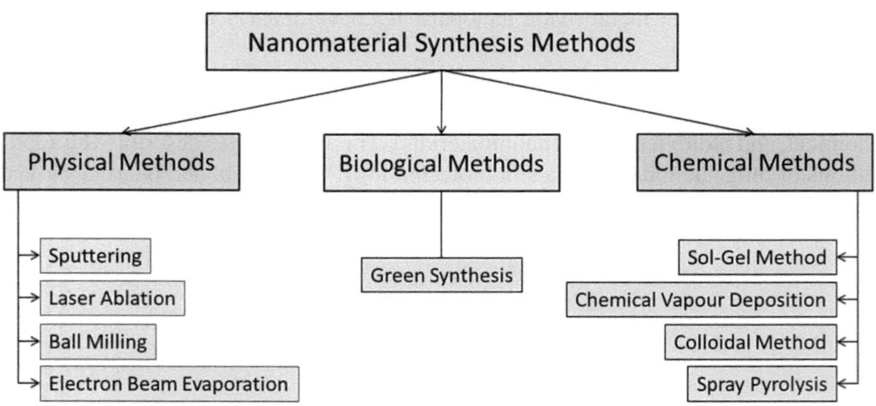

Fig. 3.2 Represents the various methods that are employed for the synthesis of nanoparticles

obtained even when the system is operating at high gas pressure during the deposition process (Rane et al. 2018).

3.2.1.2 Laser Vapourization (Ablation)

With this technique, high-power laser pulses are used, which results in the vaporization of the material. The setup of laser ablation consists of an ultra-high vacuum (UHV) or high vacuum system with a laser beam, solid target, cooled substrate, and a facility for introducing inert or reactive gases. Any material that can be formed into a solid object may be synthesized into clusters. Because some metal surfaces frequently reflect other wavelengths, such as IR or visible light, it is typically necessary to employ a laser operating in the UV range, such as an excimer (excited monomers) laser. Atoms from a solid source are vaporized by a strong laser beam, and as they hit with atoms of inert gas (or reactive gas) and cool on them, they form clusters. On the cooled substrate, they condense. In determining particle size and dispersion, gas pressure is a key factor. Alloys or compounds are created by the simultaneous evaporation of one substance, and the mixing of the two evaporated materials in an inert atmosphere (Bensebaa 2013).

3.2.1.3 Ball Milling

It is one of the simplest methods for producing powdered nanoparticles of several metals and alloys. Usually, a large amount of small particles are synthesized in a single period of time in large quantities. Balls of hardened steel or alloy and powder or flakes of interesting substance are placed in the container depending on the amount to be prepared. Any size and form can be used as an initial material. Generally, a ball-to-material mass ratio is 2:1. The impact energy is increased when heavy milling balls collide with each other, resulting in a reduction in grain size with continuous movement of the larger particle. Gas used to fill the container with balls is simple air or inert gas, but it is considered an extra source. During the collision, the temperature may rise from 100 to 1000 °C, so cryo-cooling may be used to dissipate the heat produced in the container.

Along with gases, liquids can also be employed in the milling process. The containers are rotated around their own axes at high speed (a few hundred revolutions per minute). They may also rotate around a central axis called "planetary ball mills" (Petridis et al. 2019).

3.2.2 Chemical Methods

The most commonly used chemical methods include the sol–gel method, the microemulsion technique, hydrothermal synthesis, polyols synthesis, and chemical vapor synthesis. Generally, these reactions are completed in solution, and the product has colloidal qualities. Consequently, the normal term utilized for the phenomenon is coprecipitation, which involves the concurrence of various phenomena: reduction, nucleation, growth, coarsening, or agglomeration.

3.2.2.1 Sol–gel Method

Sol–gel is a wet-compound cycle that includes the development of an inorganic colloidal suspension (sol) and gelation of the sol in a constant fluid state (gel) to shape a three-layered network structure. This process is known as a sol–gel method because the liquid precursor is changed into a sol during the production of the metal-oxide nanoparticles, and the sol is then further transformed into a network structure known as a gel (Danks et al. 2016). Metal alkoxides are the traditional precursors used in the sol–gel process for creating nanomaterials. Sol–gel synthesis of nanoparticles can be accomplished in several steps. Steps include the metal oxide being hydrolyzed in water or with the aid of alcohol to create a sol. The solvent viscosity increases due to condensation in the following stages, creating porous structures that are then allowed to oxidize. Metal-hydroxo- or metal-oxo-polymers are formed in the solution due to the creation of hydroxo- (M-OH-M) or oxo- (M–O-M) bridges during the condensation or polycondensation processes. Also, the material's structure, characteristics, and porosity change during the polycondensation process. Aging results in a reduction in porosity and an increase in the spacing between colloidal particles. After the aging process, the gel is dried, removing water and organic solvents. At last, calcination is done to produce nanoparticles (Parashar et al. 2020). The type of precursor, rate of hydrolysis, aging period, pH, and molar ratio of the precursor to water all directly impact the end product made using the sol–gel process (De Coelho Escobar and Dos Santos 2014).

3.2.2.2 Chemical Vapor Deposition

Chemical vapor deposition is a strong innovation for producing high-quality solid thin films and coatings. Although widely used in modern industries, it is continuously being developed as it is adapted to new materials (Sun et al. 2021).

Chemical vapor deposition plays a significant role in the production of nanomaterials based on carbon. In the vapor phase, thin solid films are deposited from the chemical precursors. Precursors are the thin coating on the substrate surface and are considered necessary for CVD as they exhibit sufficient volatility, high chemical purity, high evaporation stability, low cost, a non-hazardous nature, and long shelf life. Additionally, its decomposition shouldn't leave behind any contaminants. For instance, during the CVD process, a substrate is heated to high temperatures to produce carbon nanotubes, and precursor gas containing carbon (such as hydrocarbons) is introduced to the system, leading to the release of carbon atoms which are then cooled and joined again to create carbon nanotubes on the substrate (Shah and Tali 2016).

3.2.3 Biological Methods

The biological methods incorporate the synthesis of nanoparticles using bacteria, fungi, or plants.

Green nanoparticle synthesis involves using plant or plant parts to bio-reduct metal ions into their elemental form in size range from 1 to 100 nm. The course of green synthesis is more proficient, less difficult, and conservative and can effectively be increased to perform bigger tasks. For example, silver, gold, palladium, iron, and zinc oxide nanoparticles have handily been orchestrated through green synthesis. The bio-reduction of metallic particles is carried out by the phytocompounds present in the plant extracts, such as polyphenols, terpenoids, and polyols. The nanoparticles so created have noteworthy antimicrobial, cell reinforcement, and synergist properties derived from the phytocompounds that decrease them into nanoparticles. These properties have brought about a wide scope of nanoparticle applications, such as drug conveyance, the chemical industry, beauty care products, food, and drugs. In addition, other possible parts of nanoparticles are at present being investigated by the local research area across the globe (Pal et al. 2019).

3.3 Conventional and Biogenic Nanomaterials

Conventional synthesis includes the usage of chemicals, which, when released into the environment, cause major concern and have resulted in adverse effects on plants, soil, and agricultural products. Biogenic synthesis includes a large surface/volume ratio, and the use of biological components is identified to be ecologically beneficial. When released into the environment, they are known to cause minimal environmental pollution and, therefore, are considered non-toxic (Table 3.1) (Bandala et al. 2020).

3.4 Biosynthesis of Nanoparticles

The biological synthesis of nanomaterial is determined on the basis of types of biomaterial and biosystems. Figure 3.3 shows the synthesis mechanism of Au–Ag alloy nanoparticles from *Trichoderma harzianum*, fungal biomass which acts as a reducing and capping agent. The salts of the targeted alloy were separated into specific ions in the solution. However, the lower metal ion concentration is harmful to the biological system. Here, the fungal biomass releases NADH-dependent enzyme, which oxidizes

Table 3.1 Major advantages and drawbacks recognized for conventional and biogenic nanomaterials

Material	Advantages	Drawbacks
Conventional nanomaterials	–Smaller and chemically simple NMs –Reliable shape, size, and characteristics –Standardized synthetic procedure –Full-scale application feasible	–Unknown toxicity when released into the environment –Highly reactive or may be unstable –Unknown effects on human health –Lifecycle of the product is known
Biogenic nanomaterials	–Eco-friendly synthesis process –Avoid the use of toxic/organic solvents –Unstable but stabilized by capping agents –Recyclable biogenic sources	–Larger and chemically more complex due to plant-based –Unknown feasibility for full-scale applications –High variability of sizes, shapes, and surface characteristics –Life cycle unknown

into NAD+ and neutralizes the ions by reduction. These neutralized ions go through a nucleation reaction that leads to the synthesis of Au–Ag alloy nanoparticles. After the nucleation, fungal biomass release sulfur-containing proteins, which encapsulate the nanoparticle and thus provide stability. The analysis confirmation is done using Fourier Transform (Tripathi et al. 2015).

3.5 Biogenic Synthesis of Nanoparticles

Biogenic nanomaterials have various applications in different areas of the world. Similarly, there are various synthetic approaches used to produce such products, but the use of biological organisms as a resource for bio-nano factories is a new procedure for the synthesis of different nano-sized particles. Biological organism includes bacteria, plants, and fungi (Table 3.2). There are numerous biosynthetic processes involved in synthesizing nanoparticles in the research. Depending on whether these nanostructures are formed inside or outside of the microbial cells, these biosynthetic processes can be classified as intracellular or extracellular.

3.5.1 Extracellular Synthesis of Biogenic Nanoparticles

Extracellular synthesis is one of the most popular methods for synthesizing nanoparticles due to its potential in a variety of sectors, including biosensors, medicine, etc. It involves the reduction of metal ions for NPs synthesis by the use of microbial

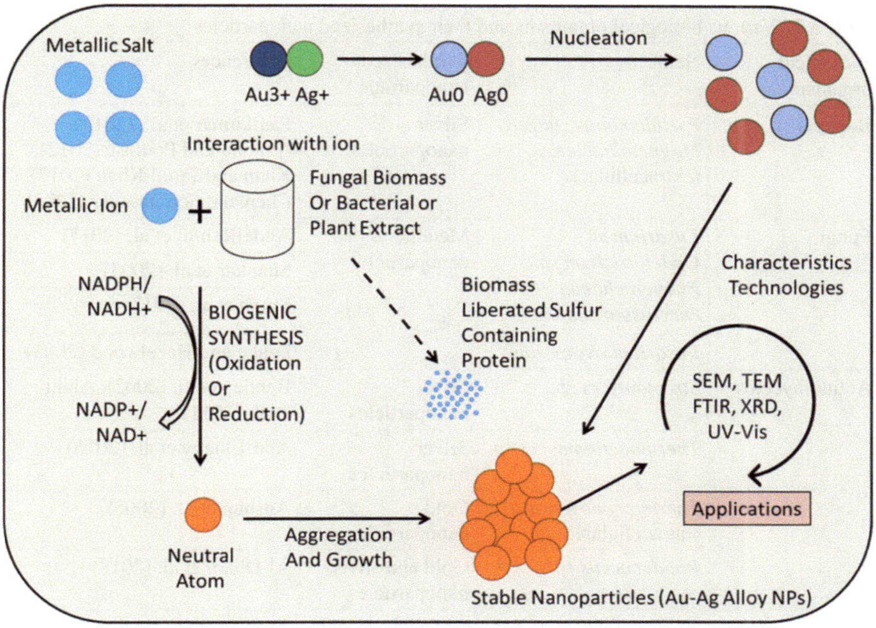

Fig. 3.3 Overview of biogenic synthesis of Au–Ag alloy nanoparticles using fungal biomass

enzymes and proteins or other organic molecules. Silver nanomaterial synthesis is the widely used method with huge applications among the metallic nanomaterials. For example, the bacteria *Pseudomonas stutzeri* produced silver nanoparticles with sizes ranging from 16 to 40 nm and a diameter of 27 nm, magnetite (Fe_3O_4) or greigite (Fe_3S_4) nanocrystals by magneto bacteria, and the production of siliceous material in diatoms (Khan and Jamil Khan 2017; Klaus et al. 1999). Various strains of *Fusarium oxysporum* were used in the extracellular synthesis of nanoparticles with the help of the hydrogenase enzyme present in the fungus broth. This extracellular enzyme exhibits evident characteristics and functions as an electron transporter in the bioreduction process, which makes it capable of converting metal ions to nanoparticles (Durán et al. 2005). These microorganisms releases hydroquinone, those function as a reducing agent or electron transporter by reducing metal ions to the appropriate nanoparticles.

3.5.2 *Intracellular Synthesis of Biogenic Nanoparticles*

Intracellular synthesis is another method used for the synthesis of nanomaterial. It involves the electrostatic attraction of metal ions by carboxyl groups present in the cell walls of microorganisms which allows the metal ions to pass through them and

Table 3.2 Various biological organisms and their synthesized nanoparticles

Biological organism	Name	Synthesized nanoparticle	References
Bacteria	*Pseudomonas stutzeri, Magneto bacteria* (extracellular)	Silver nanoparticles	Paulkumar et al., (2013), (Prabhu and Poulose (2012), Khan and Jamil Khan (2017), Chen and Schluesener (2008)
Fungi	*Fusarium sp., Colletotrichum sp., Phanerochaete chrysosporium*	Metallic nanoparticle	AbdelRahim et al. (2017)
			Shankar et al. (2003)
			Singh et al. (2018)
	Fusarium oxysporum		Drake and Hazelwood (2005)
Actinomycetes	*Streptomyces sp.*	Silver nanoparticles	Bentley et al. (2002), Alani et al. (2011)
	Thermoactinomyces sp.	Silver nanoparticles	Abd-Elnaby et al. (2016)
	Thermomonospora sp. (intracellular)	Gold nanoparticles	Anshup et al. (2005)
	Rhodococcus sp.	Gold and silver nanoparticles	Al-Dhabi et al. (2018)
Yeast	*Candida glabrata, and Schizosaccharomyces pombe*	CdS nanocrystals	Castillo-Henríquez et al. (2020)
	Torulopsis sp.	PBS nanocrystallites	Kowshik et al. (2002)
Algae	*Bacillus subtilis and Staphylococcus aureus*	Gold nanoparticles	Uma Suganya et al. (2015), Singaravelu et al. (2007)
	Sargassum wightii		
	Diatoms (extracellular)	Siliceous material	Khan and Jamil Khan (2017)
Plant	*P. foetida*	Silver nanoparticles	Lade and Shanware (2020)
	Azadirachta indica (neem) (intracellular)	Gold and silver nanoparticles	Shankar et al. (2004)

reduction in the ion concentration results in the synthesis of the metallic nanoparticle. For example, the bioreduction of ferric to ferrous is preceded by the precipitation of amorphous oxide, which leads to a sequence of changes that result in the formation of magnetite nanoparticles. Also, some microorganisms such as bacteria and fungi undergo intracellular synthesis in which the cell wall is used to transport metal ions, wherein the interaction between positive and negative ions takes place, and consequently, enzyme present in the cell of bacteria or fungi reduce these ions to metal ions. For example, silver and gold nanocrystals are synthesized by the fungi *Aspergillus fumigatus* and *Colletotrichum sp.* (Bhainsa and D'Souza 2006).

3.5.3 Synthesis of Bacteria-Mediated Nanoparticles

Bacteria are potential biofactories that synthesis metallic nanoparticles like silver and gold, as they are known to produce inorganic materials either internally or externally. However, some bacteria are resistant to silver and can accumulate as much as 25% of total dry-weight biomass on their cell wall (Paulkumar et al. 2013).

The first evidence of bacteria synthesizing silver nanoparticles was recorded using the *Pseudomonas stutzeri* AG259 strain isolated from a silver mine (Prabhu and Poulose 2012).

The most widely acknowledged component in the silver biosynthesis is the use of nitrate reductase enzyme, which is responsible for converting nitrate to nitrite. Alpha-nicotinamide adenine dinucleotide phosphate reduced form (NADPH) dependent nitrate reductase is present in this in vitro production of silver by bacteria, eliminating the down-streaming process step that is typically necessary for other situations (Chen and Schluesener 2008).

Evidence supports intracellular *Escherichia coli* DH5α mediated synthesis of gold nanoparticles from chloroauric acid with mostly spherical morphology and some triangular and quasihexagonal morphology deposited on the cell surface.

3.5.4 Synthesis of Fungi-Mediated Nanoparticles

Fungi are utilized to synthesize metallic nanoparticles because of their tolerance and ability for metal bioaccumulation with high binding capacity and intracellular intake, similar to bacteria (AbdelRahim et al. 2017). Many fungies such as *Fusarium sp.* (Drake and Hazelwood 2005) *Colletotrichum sp.* (Shankar et al. 2003), and *Phanerochaete chrysosporium* (Singh et al. 2018) has been used for nanoparticle generation. This generation is more advanced in fungi than other microorganisms because fungi grow more quickly and are simpler to fabricate and handle in a laboratory environment than bacteria.

The nanoparticle synthesis mechanism in fungi is different as fungus creates a large amount of enzyme used to reduce silver ions that induce the formation of the metallic nanoparticle. The extracellular enzyme like naphthoquinones and anthraquinones facilitate the reduction process. In the case of *F. oxysporum*, it is believed that NADPH-dependent nitrate reductase and shuttle quinine extracellular process are responsible for nanoparticle formation (Drake and Hazelwood 2005).

3.5.5 Synthesis of Actinomycetes-Mediated Nanoparticles

Actinomycetes are members of the phylum anti-bacteria and share similar characteristics to fungi. Due to their properties, they are responsible for two-thirds of the total antimicrobial compounds currently used (Bentley et al. 2002). Due to the lack of research, very little evidence of nanoparticle generation from actinomycetes is available, since its utilization is limited. However, scientists have reported that the formation of a monodisperse gold particle with an average size of 8 nm by using a new extremophilic actinomycete is possible in certain conditions.

The proteins responsible for nanoparticle stabilization are present on nanoparticle surfaces using FTIR analysis (AbdelRahim et al. 2017). Intracellular accumulation of gold nanoparticles is observed in a halotolerant actinomycete, i.e., *Rhodococcus sp.*, and particles are found to be accumulated on the cell wall and cell membranes as AG0 (AbdelRahim et al. 2017). Similarly, silver nanoparticles are observed to have been synthesized by many actinomycetes such as *Streptomyces sp.* (Alani et al. 2011), *Thermoactinomyces sp.* (Abd-Elnaby et al. 2016) and *Rhodococcus sp.* (Al-Dhabi et al. 2018). The nanoparticle synthesis is applied to many biomedical applications and has been successfully tested for its antimicrobial properties against pathogens (Castillo-Henríquez et al. 2020).

3.5.6 Synthesis of Yeast-Mediated Nanoparticles

Yeast, a unicellular eukaryotic organism that has been the most used, microbe synthesized nanoparticles to create semiconductors of various types. Some Monodispersed, spherical, peptide-bound CdS quantum crystallites 20 Å in size were synthesized using yeast *Candida glabrata*, and *Schizosaccharomyces pombe* produced wurtzite hexagonal CdS crystals in mid-log phase 1–1.5 nm in size. Similarly, PBS nanocrystallites with a size of 2–5 nm were successfully created intracellularly in the vacuoles of *Torulopsis sp.*, and the resulting nanoparticles were employed to create an ideal diode (Kowshik et al. 2002). The extracellular production of silver nanoparticles has been carried out using the yeast strain MKY3, which is silver-tolerant. In its log phase, the strain synthesized silver nanoparticles 2–5 nm in size when challenged with one mM soluble silver. The particles were separated based on the differential thawing of the sample. The formation of silver nanoparticles was confirmed by optical absorption, x-ray diffraction, transmission electron microscopy and x-ray photoelectron spectroscopy (Singaravelu et al. 2007; Uma Suganya et al. 2015).

3.5.7 Synthesis of Algae-Mediated Nanoparticles

Algae have been used in numerous experiments to create nanoparticles. According to reports, *Spirulina platensis*, a Bluegreen alga, was employed in the protein-mediated production of uniform-sized gold nanoparticles with an average size of about 5 nm. These were used for antibacterial assays against *Bacillus subtilis* and *Staphylococcus aureus* (Uma Suganya et al. 2015). *Sargassum wightii*, a brown seaweed, has been reported to synthesize stable, well-dispersed gold nanoparticles of 8–12 nm, as confirmed by UV–visible spectrum, transmission electron spectroscopy, and x-ray diffraction analysis (Singaravelu et al. 2007). Similarly, silver nanoparticles were synthesized using *S. wightii* and tested for their antibacterial potential (Govindaraju et al. 2009). In addition, gold nanoparticles, EPS-gold, and silica-gold bio nanocomposites have been synthesized with diatoms such as Navicula atomus and Diadesmis gallica (Schröfel et al. 2011).

3.5.8 Synthesis of Plant-Mediated Nanoparticles

The biological method is one of the methods for the production of nanomaterial using plant extract. The production of plant-mediated nanoparticles has the upper hand compared to the other class as the plants are cost-effective and can accumulate and detoxify heavy metals, thereby harboring the environment from harmful pollutants (Shahid et al. 2017). Many investigations have exhibited nano-formulation using plant parts, including callus, seeds, fruit, stem, and flower, and the formation of byproducts after the nanoparticle synthesis is identified to be both eco-friendly and biodegradable.

The nanoparticle synthesis using plant extract is initiated by weighing the specific amount of plant sample in grams and then washing and boiling the selected plant part with distilled water. The extract is then squeezed and filtered with a muslin cloth or syringe filter tube. The type of nanoparticle to be synthesized decides the salt solution be added after the filtration process. The solution displays the change in color, indicating the formation of nanoparticles which can later be separated (Fig. 3.4). This noticeable color change is the primary indicator for the synthesis of nanoparticles (Fig. 3.5). Sometimes the mixture does not change color, and it could be due to the high acidity of the medium. To overcome this obstacle, a solution of NaOH can be used to produce an alkaline solution. It is essential to apply salt extract gradually through a syringe at a moderate volume to properly synthesize nanoparticles. The produced nanoparticles may be in colloidal form. Nanoparticle production is confirmed by a peak at 420–460 nm or 350–525 nm with a reduction in the intensity of the plasmon absorbance. Therefore, nanoparticles, such as silver, gold, copper, titanium, etc., are synthesized using plant extract.

For instance, using plants to produce silver nanoparticles has drawn attention in recent years because its rapid, eco-friendly, non-pathogenic, and economical protocol provides a single-step technique for the biosynthetic process. The simplest and least expensive method to make silver nanoparticles is to reduce and stabilize silver ions using a mixture of biomolecules already present in plant extracts, such as proteins, amino acids, polysaccharides, terpenes, alkaloids, phenolics, saponins, and vitamins (Roy and Das 2015). The main benefits of employing plant extracts for the synthesis of silver nanoparticles include their accessibility, safety, and general lack of toxicity. They also have a wide range of metabolites that can help reduce silver ions and can be produced more quickly than bacteria. Because of photo chemicals, plant-assisted reduction is the primary mechanism considered for this process. In addition,

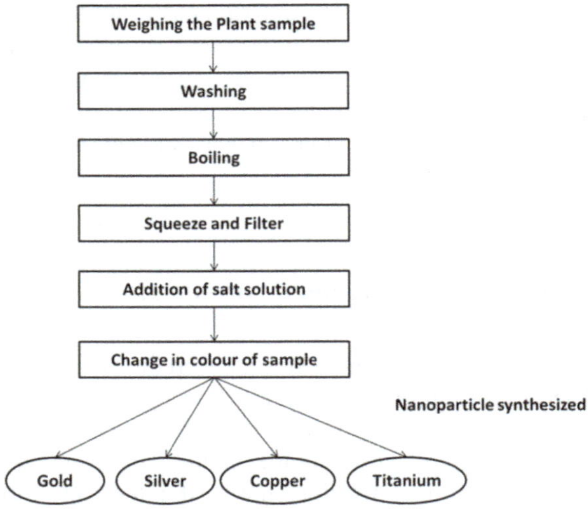

Fig. 3.4 Representation of the synthesis of nanoparticles from plant sample

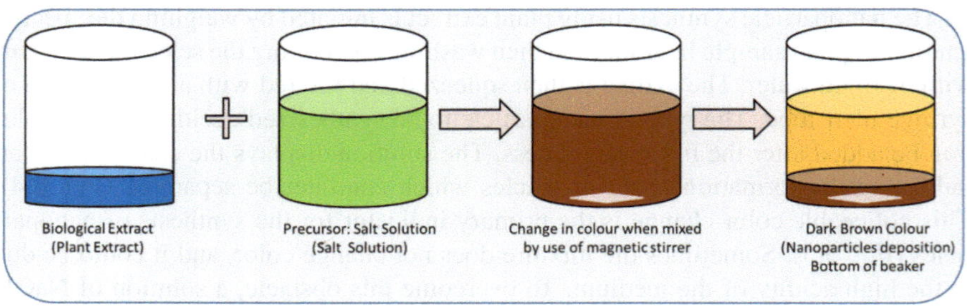

Fig. 3.5 Nanoparticles are readily synthesized by combining the correct concentration of both solutions plant extract and specific salt solution for synthesized nanomaterial. Synthesis of the nanoparticle is confirmed by a change in color of the solution from light green to dark brown, detected by the naked eye

the relatively high levels of steroids, saponins, carbohydrates, and flavonoids act as reducing agents and phytoconstituents as the capping agents, which provide stability to the silver nanoparticles.

3.6 Challenges Associated with the Biogenic Synthesis of Nanomaterials

Important functional properties like size and shape, mono-dispersity, surface charge, plasmonic response, medical diagnostics response, and bio-functional or catalytic activity are key factors in the potential application, fate, and behavior of nanoparticles in the environment and towards living systems. One of the most difficult and frequent problems to be resolved for developing and implementing novel green synthesis protocols is the controlled synthesis of the designed greener products using safer techniques while retaining nanoparticle function and efficiency (Korbekandi et al. 2009). The application of green, sustainable synthesis routes for the production of metal nanoparticles still needs extensive research and innovative solutions to establish a promising and sustainable trend because the organisms used in nanoparticle synthesis can range from simple prokaryotic bacterial cells to complex eukaryotes.

It is possible to generate NPs with a specific structure, size, and distribution by changing the synthesis methods, reducing agents, and stabilizers (Iravani 2011). Various plant extracts are used with specific synthesis methods to obtain the correct size, shape, and surface molecules. Similar variations in character were seen in NPs synthesized from the same plant material due to different synthesis techniques. The researchers use varied methods to prepare extracts and keep them in proper precursor concentration, temperature, and pH. These technologies produce NPs with a variety of size and shape characteristics. The leaves of plants are among the materials that can be used widely for large-scale synthesis. The leaves are the only plant material that can be effectively used due to their production on a large scale and without harm to the environment. Furthermore, Consuming leaves will not impact plants, but using other resources such as flowers, fruit, seed, root, or latex may affect agricultural yield. Notably, plant-based nanoparticles are reusable, stable, and safe for the environment.

Another crucial factor that needs to be considered is the synthesis parameters' optimization. The yield and the production rate are critical factors in the decision to use microbes or plants for the industrial-scale synthesis of metal nanoparticles. Therefore, it is essential to correctly manage and optimize the concentrations and ratios of the extract and salt, the experimental circumstances such as synthesis time, pH, and temperature, the buffer to be utilized, and the stirring velocity throughout production. By maximizing the concentration of the green extract, the size, shape, and concentration of the nanoparticles may all be controlled (Yang and Li 2013).

3.6.1 Effect of pH

The synthesis of nanoparticles is mainly affected by pH as it disrupts their shape and size. It has been shown that the product's shape was less consistent and likely to cluster when it was prepared at a lower pH. The appropriate size and shape of the particles can be evenly created while synthesizing NPs under various pH conditions. The pH affects the surrounding surface of nanoparticles by protonating and deprotonating molecular atoms during the nucleation and development stages of NPs (Barisik et al. 2014). The pH change from normal to alkaline results in the development of colloidal form nanoparticles thereby avoiding aggregation of nanoparticles (Patil and Chandrasekaran 2020). At acidic pH, the nanoparticles exhibit tetrahedral, decahedral, hexagonal, icosahedron, irregular, and rod shapes, indicating the importance of pH in synthesizing gold nanoparticles (Armendariz et al. 2004).

3.6.2 Effect of Precursor and Reducing Agents Concentration

It is seen that the concentration of reactants like precursors and reducing agents show an effect on the size of NPs being formed. This occurrence might be caused by the interaction between various reducing agents and the surface of prepared nuclei, which in turn enhances the secondary reduction of silver ions on the surface of the nuclei. This results in the elevated growth rate of the NPs, thereby positively affecting the size of the NPs. On the other hand, using an excessive amount of reducing agents may increase the bridging effect between the NPs that are generated, leading to the aggregation of NPs. The aggregation may occur because the surface of the prepared nuclei took up too many metal ions. The uptake of metal ions results in the occurrence secondary reduction process on the surface, thereby leading to the effect in the formation of large-size NPs (Ghosh 2017; Shankar et al. 2005).

3.6.3 Effect of Temperature

Temperature is classified as another important factor affecting the synthesis of nanoparticles. The nucleation and development of more significant nanoparticles are typically favorably influenced by temperature (Kaviya et al. 2011). Low temperatures are favorable for growth, yet it has been found that when the reactive temperature rises, the rate of all reactions also rises. Due to its noticeably varied influence on the nucleation kinetics constant k_1 and growth kinetics constant k_2, temperature affects the size of NPs under sufficient and insufficient quantities of the precursors (Shaba et al. 2021). Since most metal ions are consumed in creating nuclei as the reaction temperature rises, the secondary reduction process on the surface of the formed nuclei is prevented. As a result, small, highly scattered NPs are produced at a higher yield (Liu et al. 2020). The reaction medium's incubation time significantly impacts the properties of NPs made utilizing biological techniques. The aggregation or shrinkage of particles may cause differences in character throughout protracted incubation, and the self-life of particles may influence the potential particles (Patra and Baek 2015).

3.6.4 Effect of Capping Agents

Capping agents are the ligands that bind and form a bond with the nanoparticles, thereby altering their properties (Javed et al. 2020). These agents are recognized for playing a vital role in the synthesis of nanoparticles. The properties of nanoparticles are influenced by the addition of capping agents, as they can modify the particles' morphology, size, and surface. Although the use of chemical capping agents has improved the morphology of the developed nanoparticles, studies indicate the difficulty associated with their degradation, thereby affecting the environment (Liu et al. 2005). Due to the dire need, the researchers are now focusing on developing environmentally friendly nanoparticles that are recognized as green capping agents. Biomolecules and polysaccharides fall under the category of green capping agents and are designed in laboratories and at the industrial levels. Biomolecules are suitable candidates as they are not toxic and are reducing in nature. They act as capping agents and provide a unique configuration to the subjected nanoparticle. Polysaccharides are an ideal candidate for capping agents because they are cheap, non-toxic, biodegradable, and safe to use. These green capping agents have proven to be a boon for nanoparticle synthesis as they have worked as an alternative to the toxic chemical, thereby functioning as an environment-friendly agent.

3.7 Characterization of Biogenic Nanoparticles

Biogenic nanoparticles possess significant advantages in various domains, including healthcare and industry. Since each nanoparticle has unique characteristics, it is crucial to characterize them to get the most out of them. Characterization is essential because it allows for a more thorough evaluation of the particle and a better understanding of its nature. Dynamic light scattering (DLS), atomic force microscopy (AFM), Raman spectroscopy (RS), and scanning and transmission electron microscopy (SEM/TEM) are just a few of the methods used to characterize nanoparticles. These methods shed light on the nanoparticle's many features, including its size, shape, geometry, surface properties, and interactions with its environment (Fig. 3.6).

3.7.1 Topography or Surface Morphology of Nanoparticles

Topography is the understanding of the surface features of the nanoparticles. Particle size and distribution are other essential characteristics of the nanoparticle. The size can significantly affect the properties of the NPs, especially in the medical domain. Therefore, it is crucial to analyze the particle size and morphology, as the aggregation of the nanoparticles can affect their stability and functioning. Therefore, electron

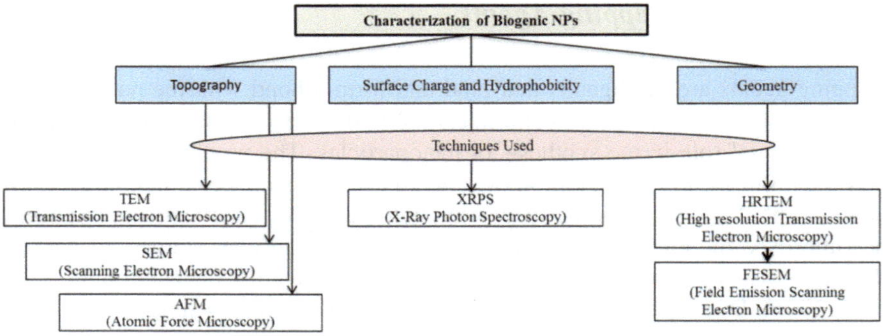

Fig. 3.6 Characterization of biogenic nanoparticles

microscopy is used to analyze the particle size of the NPs to designate them to their respective application. The techniques employed for the topography characterization are AFM, SEM, and TEM. In SEM for analysis, knocked-off electrons are used, whereas, in TEM, transmitted electrons are used for analysis. SEM can detect the aggregation and morphology of the nanoparticle and provides an image of higher resolution than TEM (Mughal et al. 2021). AFM is another technique for extracting information such as topography, size, and distribution of nanoparticles (Lin et al. 2014).

3.7.1.1 Transmission Electron Microscopy

Transmission Electron Microscopy or TEM is the technique widely employed for understanding the shape and size of the NPs as it gives a direct image along with the atomic composition of the NP (KRUMEICH 2022; Montes et al. 2011). It works by analyzing the interaction between the projection electron beam and the subjected sample, provided the sample should be ultra-thin for the electrons to pass. The electron beam strikes the sample, and during this time, some electrons are transmitted through the sample while others are scattered (Borchert et al. 2005). The transmitted electrons are analyzed as the interaction between the transmitted electrons, and the sample gives us information on the nanoparticle's topography, morphology, shape, and size (Vilela et al. 2012). TEM also provides information on the aggregation of the NPs, thereby holding applications in the medical domain, such as sensing and diagnostics (Saupe et al. 2006).

3.7.1.2 Scanning Electron Microscopy

SEM, or scanning electron microscopy, is used to obtain the high-resolution images necessary to characterize the NPs. It also works on the same principle as TEM, provided TEM analyzes the transmitted electrons, whereas SEM analyzes the knocked-off electrons. In the case of SEM, the nanoparticle solution is converted into dry powder and is coated with the conductive metal (Zur Mühlen et al. 1996). After coating, the electron beam is projected, and the electrons being knocked off from the sample surface are obtained and analyzed. SEM provides information about the morphology and sample size but fails to deliver exact information about the nanoparticle size distribution.

3.7.1.3 Atomic Force Microscopy

The method utilized to obtain a high-resolution three-dimensional image of the nanoparticle is called atomic force microscopy, or AFM (Vilela et al. 2012). AFM offers topographical information about the sample. It functions by measuring the forces that are either attracted to or repellent from the sample and the probe (Zur Mühlen et al. 1996). The probe has a pointed tip that is attached to a cantilever. Upon interaction with the sample surface, forces cause the cantilever to deviate. The deviation is quantified and examined by the laser beam that reflects at the back of the cantilever. The advantage of AFM is that no specific surface modification is needed, making it appropriate for even non-conducting samples.

3.7.2 Geometry of Nanoparticles

Each nanoparticle has a specific shape and size, and these characteristics are responsible for imparting unique properties to nanoparticles, which are exploited for the particular application. It is well known that the methods used to characterize the geometry of nanoparticles have exceptional magnification and resolution, making it possible to detect the size and shape. Field-emission scanning electron microscopy (FeSEM) and high-resolution transmission electron microscopy (HRTEM) and are the principal techniques employed (Singh et al. 2016). They are renowned for identifying the atomic composition and providing details on the aggregation's state, structure, and morphology.

3.7.2.1 High-Resolution Transmission Electron Microscopy

High-resolution transmission electron microscopy, or HRTEM, is a technique employed to analyze the geometry of nanoparticles. It is an imaging mode of TEM that utilizes phase-contrast imaging to deliver results. In phase-contrast imaging, scattered as well as transmitted electrons are used to produce the final image (Williams and Carter 2009). In the case of HRTEM, a larger objective aperture is required so as to collect scattered electrons. The high resolution of HRTEM helps in the identification of arrays of atoms aligned in the crystalline structure, along with the analysis of the internal structure of nanoparticles.

3.7.2.2 Field-Emission Scanning Electron Microscopy

Field-emission scanning electron microscopy, or FeSEM, is a technique used to obtain a high-resolution image to understand the morphology and structure of nanoparticles. In FeSEM, the sample is subjected to a high-energy beam of electrons. In order to produce a high-energy beam, the electron transmitters require high vacuum conditions. Upon striking the beam, the electrons that are knocked off are detected by the detectors and are analyzed so as to produce the final image, which helps to analyze the structure, aggregation, and morphology of the subjected nanoparticles.

3.7.3 Surface Charge and Hydrophobicity of Nanoparticles

Surface charge and hydrophobicity are the characteristics of nanoparticles that determine the interaction of nanoparticles to the hydrophobic and hydrophilic environment. It is essential to analyze the intensity of the surface charge as it would evaluate the interaction between the NPs and the biological environment. The surface charge of NPs is determined indirectly through the zeta potential. Zeta potential provides information about the stability of the nanoparticle, which in turn is influenced by the surface charge. The X-ray photon spectroscopy technique, which identifies the chemical group on the surface of NPs, is one of many methods used for surface analysis.

3.7.3.1 X-ray Photon Spectroscopy

XRPS or X-ray photon spectroscopy is a quantitative technique widely employed for the surface characterization of nanoparticles. It works on the principle of the photoelectric effect, i.e., the sample is subjected to a monochromatic X-Ray (Lu 2011). The electrons of the sample absorb the photons, attain higher energy levels, and get ejected from the sample. The energy associated with the ejected electrons is analyzed to obtain the final spectra. The spectra obtained contain the plot of electrons ejected versus binding energy. The spectra help to analyze the structure, ligand exchange, composition, and functional groups associated with the nanoparticles.

3.7.4 Other Techniques Used for Characterization

Various other techniques are also employed for the characterization of nanoparticles. Some of the commonly used techniques are as follows.

3.7.4.1 Dynamic Light Scattering

Dynamic Light Scattering, or DLS, is widely utilized to identify the Brownian nanoparticles' particle size in the colloidal solution. The nanoparticles in the colloidal solution are in Brownian motion and are struck by monochromatic light. The Doppler shift that occurs when the incident light strikes the NPs causes a change in wavelength. Therefore, analysis of the wavelength shift provides data on the nanoparticle size. In DLS, a reduced concentration of NP is required to avoid multiple scattering, which might provide deviated results (Fultz and Howe 2013).

3.7.4.2 X-ray Diffraction

XRD or X-Ray Diffraction is an imaging technique that is employed to deduce the morphology of the nanoparticles. It helps researchers to understand the nature of phase, crystalline structure, and the various lattice parameters of the nanoparticles. It works on the principle of scattering of X-Rays. The x-ray is projected on the nanoparticles, and its interaction with the revolving electrons results in the scattering of X-rays which are then analyzed for the characterization of the nanoparticles (Mourdikoudis et al. 2018). XRD proves to have the upper hand when analyzing the sample in powder form; it provides statistically representative, volume-averaged values. The obtained results help to understand the composition of the subjected nanoparticles by comparing the intensity and the position of the obtained peaks with the reference peak patterns. The reference patterns can be obtained from the International Centre for Diffraction Data database (Abbas 2019). The XRD peaks are broad for particles below 3 nm, thereby not being a suitable technique for amorphous materials.

3.7.4.3 UV–Vis Spectroscopy

UV–Vis spectroscopy is another powerful and cost-effective technique that is used for the characterization of nanoparticles. It works by analyzing the intensity of light reflected by the nanoparticle upon passing and is compared with the reference material. The analysis of the optical properties helps researchers analyze the nanoparticles' morphological features, refractive index, and agglomeration. The LSPR of the nanoparticles influences the absorbance as it permits only the highly sensitive photons for absorption. The results are analyzed on the basis of the LSPR, as any change in LSPR would, in turn, affect the absorbance of light.

3.7.4.4 Fourier Transform Infrared Spectroscopy

There are also several other techniques that help in the determination of the structure, composition, size, and other basic features of the NPs. Fourier transform infrared spectroscopy (FTIR) is a technique based on the measurement of the absorption of electromagnetic radiation with wavelengths within the mid-infrared region (4000–400 cm^{-1}) (Abbas 2019). If a molecule absorbs IR radiation, the dipole moment is somehow modified, and the molecule becomes IR active. A recorded spectrum gives the position of bands related to the strength and nature of bonds and specific functional groups, thus providing information concerning molecular structures and interactions (Dankovich and Smith 2014). FTIR spectroscopy analysis is a method based on the principle of infrared spectroscopy, and it has extended its area of application to the study of nano-scaled objects during the last decade.

3.8 Applications of Biogenic Nanoparticles

The characteristics of biogenic nanoparticles differ from those of the bulk substance; thus, the characteristics of both may not apply to each other. Nevertheless, researchers use these unique characteristics of nanoparticles because they may tailor their smaller size and geometry to meet various industries' expanding needs and demands. Due to its applicability, researchers have shifted their focus to nanomaterials, opening up greater platforms in industries like biomedicine, textiles, food production, and pharmaceuticals.

3.8.1 Nanoparticle and Drug Delivery

The use of nanoparticles for the delivery of the drug is garnering attention in the healthcare domain as the nanoparticles prove to be beneficial compared to the conventional drug delivery techniques due to their specific targeting, biocompatibility, and reduced toxicity. Nanoparticles ranging from 10 to 100 nm are an ideal candidate for targeted delivery due to their small size, thereby improving the existing drug delivery system. The features of nanoparticles are the reason which makes them an ideal candidate for emerging drug delivery systems. Surface to mass ratio of the nanoparticles is much more as compared to the other compounds. Its large functional surface has the upper hand as it helps carry compounds like drugs and proteins.

The use of nanoparticles enables the delivery of drugs at hard-to-reach sites and also reduces the chances of toxicity without affecting the therapeutic effects (Groneberg et al. 2006). The drugs are entrapped in the nanoparticle, which is then targeted to the site of action (Fig. 3.7). In order to utilize nanoparticles for therapeutic intervention, the drug to be transported is either attached to the particle's surface or integrated into the particle's matrix. The small size of the nanoparticles allows the penetration of the cell membranes, which results in the targeted delivery, thereby elevating the therapeutic index. In order to develop effective delivery strategies, it is important to have a thorough understanding of the interactions between the nanoparticles and the various components, such as the biological environment, cell surface

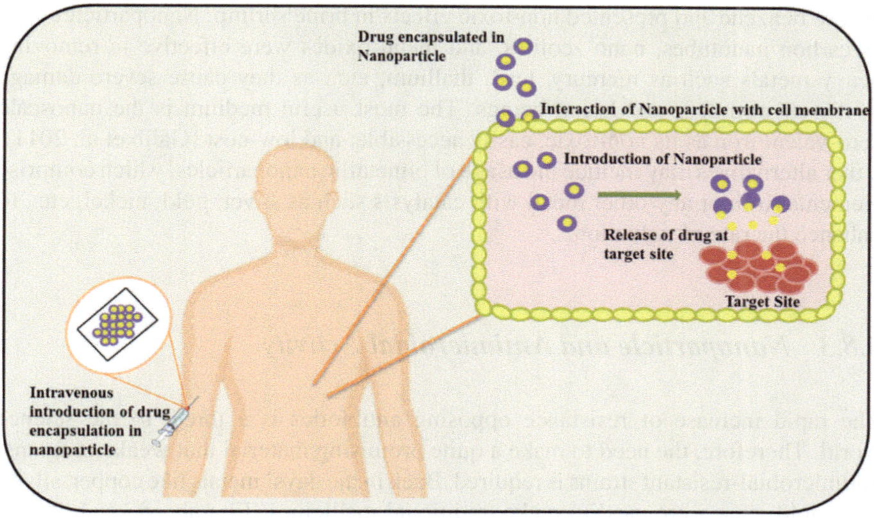

Fig. 3.7 Overview of nanoparticle-based drug delivery

receptors, and the targeted cell. It is also important to analyze the molecular mechanisms involved, the progression of disease in the body, and the factors involved with the drug, such as its retention in the body, its site of action, and involvement of multiple drug administration (Williams and Carter 2009).

For instance, the researchers, in vivo, were able to overcome drug resistance in the case of human colorectal cancer. The anticancer drug paclitaxel was entrapped in emulsifying wax nanoparticles and was administered to a tumor that was resistant to the drug. The results demonstrated that drug resistance prevailed in the human colon adenocarcinoma cell line (HCT-15), indicating the endless possibilities of the use of nanoparticles as effective drug carriers (Williams and Carter 2009).

3.8.2 Nanoparticle and Environment

The advancements and developments in new technology have profoundly impacted the environment. But nanotechnology development has assured us that we can retain the once-damaged environment through proper utilization. Nanoparticles can be used to fulfill the complaints of having clean, accessible water by providing quick and fundable assessment and management to remove impurities from water. For example, the filter paper made from Cu NPs during water purification helped reduce the bacteria *E. coli* and resulted in a low level of copper in drinking water (Rodd et al. 2014). Nanoparticles have also proved to be beneficial in cleaning up oil spills. With the help of engineered surface chemistry, the carbon particles were able to give stabilized oil in water emulsions (Fu et al. 2014). In addition, the carbon black [CB] NPs could absorb benzene and presented non-toxic effects in brine shrimp. Nanoparticles such as carbon nanotubes, nano zeolites, and metal oxides were effective in removing heavy metals such as mercury, lead, thallium, etc., as they cause severe damage to the environment and living beings. The most useful medium is the nanoscale zero-valent iron as its non-toxic, easily accessible, and low-cost (Galib et al. 2011). Other alternatives may include the usage of bimetallic nanoparticles, which comprise elemental iron or any other metal with catalysts such as silver, gold, nickel, etc., to enhance the rate of reduction.

3.8.3 Nanoparticle and Antimicrobial Activity

The rapid increase of resistance opposing antibiotics is a threat to the science world. Therefore, the need to make a quite promising material that weakens against antimicrobial-resistant strains is required. Back in the days' metals like copper, silver, iron, gold, etc., were used to make traditional medicines (Eltarahony et al. 2018). Using this past knowledge, the researchers decided to make nano-based metallic oxide NPs as they in-heretically possess antimicrobial activity.

The plant-derived metallic NPs displayed various biocidal activities opposing Gram-positive and Gram-negative bacteria and eukaryotes (Baptista et al. 2018). It also showed that metallic NPs gave effective inhibition against resistant restraints such as ampicillin-resistant *E. coli*, methicillin-resistant *S. aureus* [MSRA], etc. (Slavin et al. 2017). Presently the antimicrobial activity of metallic NPs is incorporated from plants against bacterium and fungi. Furthermore, it is assumed that NPs have a higher surface-to-volume ratio than their bulk counterparts, which easily enhances their interaction with the cell membrane (Xia et al. 2008).

Fungi cell structure is composed of the cell membrane and cell wall. The cell membrane consists of phospholipids, and the cell wall comprises mannoproteins, β-1,3-D-glucan and β-1,6-D-glucan proteins, chitin, proteins, lipids, and polysaccharides [chitin, glucan, and mannan or galactomannan]. The metallic NPs work by interacting with the cell membrane and cell wall. The diffusion is followed by inhibiting a vital cell wall component, i.e., β-glucan synthase (Arciniegas-Grijalba et al. 2017). ROS is followed by oxidative stress that interacts will macromolecules and cause cell lysis (Zeng and Dong 2021).

3.8.4 Nanoparticle and Anticancer Property

Cancer is a leading cause of death due to its high mortality rate. The existing chemotherapeutics become inefficient, i.e., after some cycles, the patient becomes resistant to the administered drugs. Metallic nanoparticles are emerging as a ray of hope for cancer treatment due to their small size, larger surface area, tumor specificity, and apoptotic activity, i.e., they can induce cell death and possess cytotoxic properties. There are different mechanisms adopted by NPs for anticancer activity, the first one being apoptosis, wherein the elevated levels of ROS result in oxidative stress leading to DNA fragmentation and cell lysis (Lim et al. 2011). Another one is the interaction of NPs with cell membranes and the induction of changes in cell permeability (Castro-Aceituno et al. 2016). Panax ginseng fresh leaves were used to synthesize the silver NPs (PgAgNPs). These nanoparticles demonstrated cytotoxic effects, which in turn cause oxidative stress in A549, MCF7, and HepG2 cancer cell lines (Nethi et al. 2019). It inhibited the EGF, and an increase in phosphorylation of EGF receptors was found in A549 cells. In addition, PgAgNPs interfered with the cell morphology and elevated the apoptosis process. This interlinkage might be the mechanism followed by PgAgNPs for their anticancer activity. The utilization of NPs for cancer treatment is an emerging domain, and in-depth in vivo studies are required to understand better the mechanisms followed by NPs.

3.8.5 Nanoparticle and Wound Healing

A wound is an injury to the skin tissue in response to stimuli or trauma. Various treatment therapies are adopted, such as chemotherapeutics, plant-based therapy, dressing, and vascular surgery, but every technique has its side effects. Nanoparticles are emerging as an approach to wound healing. It includes two groups, one of which acts as a drug and promotes wound healing, and the other acts as a delivery agent for repair (Hamdan et al. 2017a). For instance, Garg et al. (2014) demonstrated the antibacterial and healing activity of silver NPs in excision wounds in albino Wistar male rats (Naraginti et al. 2016). The root extract of *Arnebia nobilis* was used to synthesize the silver NPs. It was formulated with hydrogel and was applied to rats. NPs showed no toxic signs on the rats, and cells were re-epithelialized. The use of NPs for wound healing is in the budding stage. It requires more extensive research for a better understanding of the molecular mechanism and the response of the body to the NPs so as use them to their full therapeutic potential.

3.9 Conclusion and Future Perspective

The green or biogenic synthesis of the nanoparticles has opened doors to the eco-friendly way as compared to the existing conventional methods. Green synthesis has the upper hand in terms of negligible toxicity and reduced expense. Apart from the benefits, some challenges are also encountered while dealing with the biogenic nanoparticles, such as uniformity and homogeneity of nanomaterials and the risk of infection with the biogenic nanomaterials.

The growing surge of nanomaterials and their applications in various domains indicate the research interest of the researchers in the biogenic synthesis of NPs. The use of biosynthesis pathways to produce nanomaterials could replace the existing chemical methods, leading to a wide array of opportunities to produce more environmentally friendly products that can be used in industrial work. Nanomaterials present a wide range of applications in various domains, but it's still in their early stages.

The biogenic synthesis of Nanomaterials is an emerging and promising domain, but there is much to explore as we are yet at the initial pages. Nevertheless, better technologies and extensive research on nanomaterials would yield quality results that would help uplift and improve the environmental, healthcare, and industrial world.

References

Abbas Q (2019) Understanding the UV-Vis spectroscopy for nanoparticles. https://doi.org/10.4172/2324-8777.1000268

Abd-Elnaby HM, Abo-Elala GM, Abdel-Raouf UM, Hamed MM (2016) Antibacterial and anticancer activity of extracellular synthesized silver nanoparticles from marine Streptomyces rochei MHM13. Egypt J Aquat Res 42(3):301–312. https://doi.org/10.1016/J.EJAR.2016.05.004

AbdelRahim K, Mahmoud SY, Ali AM, Almaary KS, Mustafa AEZMA, Husseiny SM (2017) Extracellular biosynthesis of silver nanoparticles using *Rhizopus stolonifer*. Saudi J Biol Sci 24(1):208–216. https://doi.org/10.1016/J.SJBS.2016.02.025

Alani F, Moo-Young M, Anderson W (2011) Biosynthesis of silver nanoparticles by a new strain of Streptomyces sp. compared with *Aspergillus fumigatus*. World J Microbiol Biotechnol 28(3):1081–1086. https://doi.org/10.1007/S11274-011-0906-0

Al-Dhabi NA, Ghilan AKM, Arasu MV, Duraipandiyan V (2018) Green biosynthesis of silver nanoparticles produced from marine Streptomyces sp. Al-Dhabi-89 and their potential applications against wound infection and drug resistant clinical pathogens. J Photochem Photobiol B 189:176–184. https://doi.org/10.1016/J.JPHOTOBIOL.2018.09.012

Alle M, Lee SH, Kim JC (2020) Ultrafast synthesis of gold nanoparticles on cellulose nanocrystals via microwave irradiation and their dyes-degradation catalytic activity. J Mater Sci Technol 41:168–177. https://doi.org/10.1016/J.JMST.2019.11.003

Anshup, Sai Venkataraman J, Subramaniam C, Rajeev Kumar R, Priya S, Santhosh Kumar TR, Omkumar RV, John A, Pradeep T (2005) Growth of gold nanoparticles in human cells. Langmuir 21(25):11562–11567. https://doi.org/10.1021/LA0519249/ASSET/IMAGES/MEDIUM/LA0519249N00001.GIF

Arciniegas-Grijalba PA, Patiño-Portela MC, Mosquera-Sánchez LP, Guerrero-Vargas JA, Rodríguez-Páez JE (2017) ZnO nanoparticles (ZnO-NPs) and their antifungal activity against coffee fungus Erythricium salmonicolor. Appl Nanosci (Switzerland) 7(5):225–241. https://doi.org/10.1007/S13204-017-0561-3/FIGURES/11

Armendariz V, Herrera I, Peralta-Videa JR, Jose-Yacaman M, Troiani H, Santiago P, Gardea-Torresdey JL (2004) Size controlled gold nanoparticle formation by Avena sativa biomass: use of plants in nanobiotechnology. J Nanopart Res 6(4):377–382. https://doi.org/10.1007/S11051-004-0741-4

Baig N, Kammakakam I, Falath W, Kammakakam I (2021) Nanomaterials: a review of synthesis methods, properties, recent progress, and challenges. Mater Adv 2(6):1821–1871. https://doi.org/10.1039/D0MA00807A

Bandala ER, Stanisic D, Tasic L (2020) Biogenic nanomaterials for photocatalytic degradation and water disinfection: a review. Environ Sci: Water Res Technol 6(12):3195–3213. https://doi.org/10.1039/D0EW00705F

Baptista PV, McCusker MP, Carvalho A, Ferreira DA, Mohan NM, Martins M, Fernandes AR (2018) Nano-strategies to fight multidrug resistant bacteria-"A battle of the titans". Front Microbiol 9(JUL):1441. https://doi.org/10.3389/FMICB.2018.01441/XML/NLM

Barisik M, Atalay S, Beskok A, Qian S (2014) Size dependent surface charge properties of silica nanoparticles. J Phys Chem C 118(4):1836–1842. https://doi.org/10.1021/JP410536N/ASSET/IMAGES/MEDIUM/JP-2013-10536N_0011.GIF

Bensebaa F (2013) Dry production methods. Interface Sci Technol 19:147–184. https://doi.org/10.1016/B978-0-12-369550-5.00003-3

Bentley SD, Chater KF, Cerdeño-Tárraga AM, Challis GL, Thomson NR, James KD, Harris DE, Quail MA, Kieser H, Harper D, Bateman A, Brown S, Chandra G, Chen CW, Collins M, Cronin A, Fraser A, Goble A, Hidalgo J, et al (2002) Complete genome sequence of the model actinomycete *Streptomyces coelicolor* A3(2). Nature 417(6885):141–147. https://doi.org/10.1038/417141a

Bhainsa KC, D'Souza SF (2006) Extracellular biosynthesis of silver nanoparticles using the fungus *Aspergillus fumigatus*. Colloids Surf, B 47(2):160–164. https://doi.org/10.1016/J.COLSURFB.2005.11.026

Borchert H, Shevchenko EV, Robert A, Mekis I, Kornowski A, Grübel G, Weller H (2005) Determination of nanocrystal sizes: a comparison of TEM, SAXS, and XRD studies of highly monodisperse CoPt 3 particles. Langmuir 21(5):1931–1936. https://doi.org/10.1021/LA0477183/ASSET/IMAGES/MEDIUM/LA0477183N00001.GIF

Castillo-Henríquez L, Alfaro-Aguilar K, Ugalde-álvarez J, Vega-Fernández L, de Oca-Vásquez GM, Vega-Baudrit JR (2020) Green synthesis of gold and silver nanoparticles from plant extracts and their possible applications as antimicrobial agents in the agricultural area. Nanomaterials 10(9):1763. https://doi.org/10.3390/NANO10091763

Castro-Aceituno V, Ahn S, Simu SY, Singh P, Mathiyalagan R, Lee HA, Yang DC (2016) Anticancer activity of silver nanoparticles from Panax ginseng fresh leaves in human cancer cells. Biomed Pharmacother 84:158–165. https://doi.org/10.1016/J.BIOPHA.2016.09.016

Chen X, Schluesener HJ (2008) Nanosilver: a nanoproduct in medical application. Toxicol Lett 176(1):1–12. https://doi.org/10.1016/J.TOXLET.2007.10.004

Dankovich TA, Smith JA (2014) Incorporation of copper nanoparticles into paper for point-of-use water purification. Water Res 63:245–251. https://doi.org/10.1016/J.WATRES.2014.06.022

Danks AE, Hall SR, Schnepp Z (2016) The evolution of 'sol–gel' chemistry as a technique for materials synthesis. Mater Horiz 3(2):91–112. https://doi.org/10.1039/C5MH00260E

De Coelho Escobar C, Dos Santos JHZ (2014) Effect of the sol–gel route on the textural characteristics of silica imprinted with Rhodamine B. J Sep Sci 37(7):868–875. https://doi.org/10.1002/JSSC.201301143

Denizli A, Nguyen TA, Rajendran S, Yasin G, Nadda AK (2021) Nanosensors for smart agriculture. Nanosens Smart Agric:1–797. https://doi.org/10.1016/C2019-0-05461-1

Drake PL, Hazelwood KJ (2005) Exposure-related health effects of silver and silver compounds: a review. Ann Occup Hyg 49(7):575–585. https://doi.org/10.1093/ANNHYG/MEI019

Durán N, Marcato PD, Alves OL, De Souza GIH, Esposito E (2005) Mechanistic aspects of biosynthesis of silver nanoparticles by several *Fusarium oxysporum* strains. J Nanobiotechnol 3(1):1–7. https://doi.org/10.1186/1477-3155-3-8/FIGURES/9

Elizabeth PS, Néstor MM, David QG (2019) Nanoparticles as dental drug-delivery systems. Nanobiomater Clin Dent:567–593. https://doi.org/10.1016/B978-0-12-815886-9.00023-1

Eltarahony M, Zaki S, Elkady M, Abd-El-Haleem D (2018) Biosynthesis, characterization of some combined nanoparticles, and its biocide potency against a broad spectrum of pathogens. J Nanomater2018. https://doi.org/10.1155/2018/5263814

Fu F, Dionysiou DD, Liu H (2014) The use of zero-valent iron for groundwater remediation and wastewater treatment: a review. J Hazard Mater 267:194–205. https://doi.org/10.1016/J.JHAZMAT.2013.12.062

Fultz B, Howe J (2013) High-resolution STEM and related imaging techniques, pp 587–615. https://doi.org/10.1007/978-3-642-29761-8_12

Galib, BarveM, Mashru M, Jagtap C, Patgiri BJ, Prajapati PK (2011) Therapeutic potentials of metals in ancient India: a review through Charaka Samhita. J Ayurveda Integr Med 2(2):55. https://doi.org/10.4103/0975-9476.82523

Garg S, Chandra A, Mazumder A, Mazumder R (2014) Green synthesis of silver nanoparticles using *Arnebia nobilis* root extract and wound healing potential of its hydrogel. Asian J Pharm. https://doi.org/10.4103/0973-8398.134925

Ghosh S (2017) Novel anticancer platinum and palladium nanoparticles from Barleria prionitis. Global J Nanomed 2(5). https://doi.org/10.19080/gjn.2017.02.555600

Govindaraju K, Kiruthiga V, Kumar VG, Singaravelu G (2009) Extracellular synthesis of silver nanoparticles by a marine alga, *Sargassum wightii* grevilli and their antibacterial effects. J Nanosci Nanotechnol 9(9):5497–5501. https://doi.org/10.1166/JNN.2009.1199

Groneberg DA, Rabe KF, Fischer A (2006) Novel concepts of neuropeptide-based drug therapy: vasoactive intestinal polypeptide and its receptors. Eur J Pharmacol 533(1–3):182–194. https://doi.org/10.1016/J.EJPHAR.2005.12.055

Hamdan S, Pastar I, Drakulich S, Dikici E, Tomic-Canic M, Deo S, Daunert S (2017a) Nanotechnology-driven therapeutic interventions in wound healing: potential uses and applications. ACS Cent Sci 3(3):163. https://doi.org/10.1021/ACSCENTSCI.6B00371

Iravani S (2011) Green synthesis of metal nanoparticles using plants. Green Chem 13(10):2638–2650. https://doi.org/10.1039/C1GC15386B

Javed R, Zia M, Naz S, Aisida SO, Ain N ul, Ao Q (2020) Role of capping agents in the application of nanoparticles in biomedicine and environmental remediation: recent trends and future prospects. J Nanobiotechnol 18(1):1–15. https://doi.org/10.1186/S12951-020-00704-4

Kaviya S, Santhanalakshmi J, Viswanathan B (2011) Green synthesis of silver nanoparticles using *Polyalthia longifolia* leaf extract along with D-sorbitol: study of antibacterial activity. J Nanotechnol. https://doi.org/10.1155/2011/152970

Khan NT, Jamil Khan M (2017) Biogenic nanoparticles: an introduction to what they are and how they are produced. Int J Biotechnol Bioeng 3(3):66–70. https://doi.org/10.25141/2475-3432-2017-3.0066

Khan SA (2020) Metal nanoparticles toxicity: role of physicochemical aspects. In: Metal nanoparticles for drug delivery and diagnostic applications, pp 1–11. https://doi.org/10.1016/B978-0-12-816960-5.00001-X

Klaus T, Joerger R, Olsson E, Granqvist CG (1999) Silver-based crystalline nanoparticles, microbially fabricated. Proc Natl Acad Sci USA 96(24):13611–13614. https://doi.org/10.1073/PNAS.96.24.13611/ASSET/B0642F07-61EA-4201-83F5-85F29CFC1A09/ASSETS/GRAPHIC/PQ2392063003.JPEG

Korbekandi H, Iravani S, Abbasi S (2009) Production of nanoparticles using organisms. 29(4):279–306. https://doi.org/10.3109/07388550903062462

Kowshik M, Ashtaputre S, Kharrazi S, Vogel W, Urban J, Kulkarni SK, Paknikar KM (2002) Extracellular synthesis of silver nanoparticles by a silver-tolerant yeast strain MKY3. Nanotechnology 14(1):95. https://doi.org/10.1088/0957-4484/14/1/321

Krumeich F (2022) introduction into transmission and scanning transmission. March

Kumar H, Bhardwaj K, Dhanjal DS, Nepovimova E., Şen F, Regassa H, Singh R, Verma R, Kumar V, Kumar D, Bhatia SK, Kuča K (2020) Fruit extract mediated green synthesis of ṃetallic nanoparticles: a new avenue in pomology applications. Int J Mol Sci 21(22):8458. https://doi.org/10.3390/IJMS21228458

Lade BD, Shanware AS (2020) Phytonanofabrication: methodology and factors affecting biosynthesis of nanoparticles. In: Smart nanosystems for biomedicine, optoelectronics and catalysis. https://doi.org/10.5772/INTECHOPEN.90918

Lim ZZJ, Li JEJ, Ng CT, Yung LYL, Bay BH (2011) Gold nanoparticles in cancer therapy. Acta Pharmacol Sin 32(8):983. https://doi.org/10.1038/APS.2011.82

Lin PC, Lin S, Wang PC, Sridhar R (2014) Techniques for physicochemical characterization of nanomaterials. Biotechnol Adv 32(4):711–726. https://doi.org/10.1016/J.BIOTECHADV.2013.11.006

Liu FK, Ko FH, Huang PW, Wu CH, Chu TC (2005) Studying the size/shape separation and optical properties of silver nanoparticles by capillary electrophoresis. J Chromatogr A 1062(1):139–145. https://doi.org/10.1016/J.CHROMA.2004.11.010

Liu H, Zhang H, Wang J, Wei J (2020) Effect of temperature on the size of biosynthesized silver nanoparticle: deep insight into microscopic kinetics analysis. Arab J Chem 13(1):1011–1019. https://doi.org/10.1016/J.ARABJC.2017.09.004

Lu LT (2011) Water-dispersible magnetic nanoparticles for biomedical applications: synthesis and characterisation. https://doi.org/10.17638/03062809

Montes MO, Mayoral A, Deepak FL, Parsons JG, Jose-Yacamán M, Peralta-Videa JR, Gardea-Torresdey JL (2011) Anisotropic gold nanoparticles and gold plates biosynthesis using alfalfa extracts. JNR 13(8):3113–3121. https://doi.org/10.1007/S11051-011-0230-5

Mourdikoudis S, Pallares RM, Thanh NTK (2018) Characterization techniques for nanoparticles: comparison and complementarity upon studying nanoparticle properties. Nanoscale 10(27):12871–12934. https://doi.org/10.1039/C8NR02278J

Mughal B, Zaidi SZJ, Zhang X, Hassan SU (2021) Biogenic nanoparticles: synthesis, characterisation and applications. Appl Sci 11(6):2598. https://doi.org/10.3390/APP11062598

Naraginti S, Kumari PL, Das RK, Sivakumar A, Patil SH, Andhalkar VV (2016) Amelioration of excision wounds by topical application of green synthesized, formulated silver and gold nanoparticles in albino Wistar rats. Mater Sci Eng, C 62:293–300. https://doi.org/10.1016/J.MSEC.2016.01.069

Nethi SK, Das S, Patra CR, Mukherjee S (2019) Recent advances in inorganic nanomaterials for wound-healing applications. Biomater Sci 7(7):2652–2674. https://doi.org/10.1039/C9BM00423H

Pal G, Rai P, Pandey A (2019) Green synthesis of nanoparticles: a greener approach for a cleaner future. In: Green synthesis, characterization and applications of nanoparticles, pp 1–26. https://doi.org/10.1016/B978-0-08-102579-6.00001-0

Parashar M, Shukla VK, Singh R (2020) Metal oxides nanoparticles via sol–gel method: a review on synthesis, characterization and applications. J Mater Sci: Mater Electron 31(5):3729–3749. https://doi.org/10.1007/S10854-020-02994-8

Patil S, Chandrasekaran R (2020) Biogenic nanoparticles: a comprehensive perspective in synthesis, characterization, application and its challenges. J Genet Eng Biotechnol 18(1):1–23. https://doi.org/10.1186/S43141-020-00081-3

Patra JK, Baek KH (2015) Green nanobiotechnology. J Nanomater2014. https://doi.org/10.1155/2014/417305

Paulkumar K, Rajeshkumar S, Gnanajobitha G, Vanaja M, Malarkodi C, Annadurai G (2013) Biosynthesis of silver chloride nanoparticles using *Bacillus subtilis* MTCC 3053 and assessment of its antifungal activity. ISRN Nanomater 2013:1–8. https://doi.org/10.1155/2013/317963

Petridis LV, Kokkinos NC, Mitropoulos AC, Kyzas GZ (2019) Graphene aerogels for oil absorption. Interface Sci Technol 30:173–197. https://doi.org/10.1016/B978-0-12-814178-6.00008-X

Prabhu S, Poulose EK (2012) Silver nanoparticles: mechanism of antimicrobial action, synthesis, medical applications, and toxicity effects. Int Nano Lett 2(1):1–10. https://doi.org/10.1186/2228-5326-2-32

Rane AV, Kanny K, Abitha VK, Thomas S, Thomas S (2018) Methods for synthesis of nanoparticles and fabrication of nanocomposites. In: Synthesis of inorganic nanomaterials: advances and key technologies. Elsevier Ltd. https://doi.org/10.1016/B978-0-08-101975-7.00005-1

Rodd AL, Creighton MA, Vaslet CA, Rangel-Mendez JR, Hurt RH, Kane AB (2014) Effects of surface-engineered nanoparticle-based dispersants for marine oil spills on the model organism Artemia franciscana. Environ Sci Technol 48(11):6419–6427. https://doi.org/10.1021/ES500892M/SUPPL_FILE/ES500892M_SI_001.PDF

Roy S, Das TK (2015) Plant mediated green synthesis of silver nanoparticles-a review. Int J Plant Biol 3(3):1044

Saupe A, Gordon KC, Rades T (2006) Structural investigations on nanoemulsions, solid lipid nanoparticles and nanostructured lipid carriers by cryo-field emission scanning electron

microscopy and Raman spectroscopy. Int J Pharm 314(1):56–62. https://doi.org/10.1016/J.IJPHARM.2006.01.022
Sawalha H, Abiri R, Sanusi R, Shaharuddin NA, Noor AAM, Shukor NAA, Abdul-Hamid H, Ahmad SA (2021) Toward a better understanding of metal nanoparticles, a novel strategy from eucalyptus plants. Plants 10(5):929. https://doi.org/10.3390/PLANTS10050929
Schröfel A, Kratošová G, Bohunická M, Dobročka E, Vávra I (2011) Biosynthesis of gold nanoparticles using diatoms—silica-gold and EPS-gold bionanocomposite formation. J Nanopart Res 13(8):3207–3216. https://doi.org/10.1007/S11051-011-0221-6
Shaba EY, Jacob JO, Tijani JO, Suleiman MAT (2021) A critical review of synthesis parameters affecting the properties of zinc oxide nanoparticle and its application in wastewater treatment. Appl Water Sci 11(2):1–41. https://doi.org/10.1007/S13201-021-01370-Z
Shah KA, Tali BA (2016) Synthesis of carbon nanotubes by catalytic chemical vapour deposition: a review on carbon sources, catalysts and substrates. Mater Sci Semicond Process 41:67–82. https://doi.org/10.1016/J.MSSP.2015.08.013
Shahid M, Dumat C, Khalid S, Schreck E, Xiong T, Niazi NK (2017) Foliar heavy metal uptake, toxicity and detoxification in plants: a comparison of foliar and root metal uptake. J Hazard Mater 325:36–58. https://doi.org/10.1016/J.JHAZMAT.2016.11.063
Shankar SS, Ahmad A, Pasricha R, Sastry M (2003) Bioreduction of chloroaurate ions by geranium leaves and its endophytic fungus yields gold nanoparticles of different shapes. J Mater Chem 13(7):1822–1826. https://doi.org/10.1039/B303808B
Shankar SS, Rai A, Ahmad A, Sastry M (2004) Rapid synthesis of Au, Ag, and bimetallic Au core–Ag shell nanoparticles using Neem (Azadirachta indica) leaf broth. J Colloid Interface Sci 275(2):496–502. https://doi.org/10.1016/J.JCIS.2004.03.003
Shankar SS, Rai A, Ahmad A, Sastry M (2005) Controlling the optical properties of lemongrass extract synthesized gold nanotriangles and potential application in infrared-absorbing optical coatings. Chem Mater 17(3):566–572. https://doi.org/10.1021/CM048292G/ASSET/IMAGES/MEDIUM/CM048292GN00001.GIF
Sharma D, Kanchi S, Bisetty K (2019) Biogenic synthesis of nanoparticles: a review. Arab J Chem 12(8):3576–3600. https://doi.org/10.1016/J.ARABJC.2015.11.002
Singaravelu G, Arockiamary JS, Kumar VG, Govindaraju K (2007) A novel extracellular synthesis of monodisperse gold nanoparticles using marine alga, Sargassum wightii Greville. Colloids Surf, B 57(1):97–101. https://doi.org/10.1016/J.COLSURFB.2007.01.010
Singh P, Kim YJ, Zhang D, Yang DC (2016) Biological synthesis of nanoparticles from plants and microorganisms. Trends Biotechnol 34(7):588–599. https://doi.org/10.1016/J.TIBTECH.2016.02.006
Singh J, Dutta T, Kim KH, Rawat M, Samddar P, Kumar P (2018) 'Green' synthesis of metals and their oxide nanoparticles: applications for environmental remediation. J Nanobiotechnol 16(1):1–24. https://doi.org/10.1186/S12951-018-0408-4
Slavin YN, Asnis J, Häfeli UO, Bach H (2017) Metal nanoparticles: understanding the mechanisms behind antibacterial activity. J Nanobiotechnol 15(1):1–20. https://doi.org/10.1186/S12951-017-0308-Z/FIGURES/4
Sun L, Yuan G, Gao L, Yang J, Chhowalla M, Gharahcheshmeh MH, Gleason KK, Choi YS, Hong BH, Liu Z (2021) Chemical vapour deposition. Nat Rev Methods Primers 1(1):5. https://doi.org/10.1038/S43586-020-00005-Y
Tripathi RM, Gupta RK, Bhadwal AS, Singh P, Shrivastav A, Shrivastav BR (2015) Fungal biomolecules assisted biosynthesis of Au–Ag alloy nanoparticles and evaluation of their catalytic property. IET Nanobiotechnol 9(4):178–183. https://doi.org/10.1049/IET-NBT.2014.0043
Uma Suganya KS, Govindaraju K, Ganesh Kumar V, Stalin Dhas T, Karthick V, Singaravelu G, Elanchezhiyan M (2015) Blue green alga mediated synthesis of gold nanoparticles and its antibacterial efficacy against Gram positive organisms. Mater Sci Eng, C 47:351–356. https://doi.org/10.1016/J.MSEC.2014.11.043

Vilela D, González MC, Escarpa A (2012) Sensing colorimetric approaches based on gold and silver nanoparticles aggregation: Chemical creativity behind the assay. A review. Analytica Chimica Acta 751:24–43. https://doi.org/10.1016/J.ACA.2012.08.043

Villaseñor-Basulto DL, Pedavoah MM, Bandala ER (2019) Plant materials for the synthesis of nanomaterials: greener sources. Handbook Ecomater 1:105–121. https://doi.org/10.1007/978-3-319-68255-6_88/COVER

Williams DB, Carter CB (2009) High-resolution TEM. In: Transmission electron microscopy, pp 483–509. https://doi.org/10.1007/978-0-387-76501-3_28

Xia T, Kovochich M, Liong M, Mädler L, Gilbert B, Shi H, Yeh JI, Zink JI, Nel AE (2008) Comparison of the mechanism of toxicity of zinc oxide and cerium oxide nanoparticles based on dissolution and oxidative stress properties. ACS Nano 2(10):2121–2134. https://doi.org/10.1021/NN800511K/SUPPL_FILE/NN800511K_SI_001.PDF

Yang N, Li WH (2013) Mango peel extract mediated novel route for synthesis of silver nanoparticles and antibacterial application of silver nanoparticles loaded onto non-woven fabrics. Ind Crops Prod 48:81–88. https://doi.org/10.1016/J.INDCROP.2013.04.001

Zeng R, Dong J (2021) the hippo signaling pathway in drug resistance in cancer. Cancers 13(2):318. https://doi.org/10.3390/CANCERS13020318

Zur Mühlen A, Zur Mühlen E, Niehus H, Mehnert W (1996) Atomic force microscopy studies of solid lipid nanoparticles. Pharm Res 13(9):1411–1416. https://doi.org/10.1023/A:1016042504830

Chapter 4
Biogenic Synthesis of Nanomaterials Using Diverse Microbial Nano-Factories

Kanchanlata Tungare, Priyamvada Arte, Aqdas Khan, Khadija Turabi, Renitta Jobby, and Mustansir Bhori

Abstract In recent years, nanotechnology has made integral contributions to the field of biomedicine, agriculture and environmental sciences. Innumerable multi-functional nanomaterials have been developed using physical and chemical techniques. However, there has emerged a need for the development of sustainable and eco-friendly synthesis processes to improve their biological applications. "Green synthesis" or biogenic synthesis is a bottom-up approach where biogenic nanomaterials are synthesized by utilizing matter of biological origin. One such approach employs the use of microbial organisms which encompass a broad range of living organisms including a variety of bacteria, algae, yeast, fungi, as well as virus which can be utilized to fabricate highly biocompatible, dynamic and well-defined nanostructures with multifaceted applications. This chapter discusses the mechanisms which enable microorganisms to behave like nano-factories and explores the role of various classes of microbes in the production of nano-materials with emphasis on recent investigations and advancements. It also aims to provide a comprehensive insight into the developmental challenges and future prospects presented by this novel approach.

Keywords Biogenic nanomaterials · Bottom-up approach · Biosynthesis · Microorganisms · Novel nano-factories

K. Tungare · P. Arte · A. Khan · K. Turabi · M. Bhori (✉)
School of Biotechnology and Bioinformatics, D Y Patil Deemed to Be University, CBD Belapur, Plot No 50, Sector 15, Navi Mumbai 400614, Maharashtra, India
e-mail: mustansir.bhori@inveniolife.com

K. Tungare
e-mail: kanchanlata.singh@dypatil.edu

P. Arte
e-mail: pri.art.bt18@dypatil.edu

A. Khan
e-mail: aqdas.khan@ruhr-uni-bochum.de

M. P. Shah et al. (eds.), *Biogenic Nanomaterials for Environmental Sustainability: Principles, Practices, and Opportunities*, Environmental Science and Engineering,
https://doi.org/10.1007/978-3-031-45956-6_4

4.1 Introduction

Nanotechnology is an exceptionally dynamic field of research involved in developing nanomaterials of varying dimensions, having applicability in a multitude of fields including agriculture, energy, environmental sciences and biomedicine. For years, nanomaterials have been synthesized via the "bottom-up" or "top-down" approaches, often employing chemical or physical means of material synthesis (Abid et al. 2021). Chemical-based methodologies often necessitate the use of substances that hinder their biocompatibility and can potentially be hazardous to the environment. Furthermore, physical methods provide lower yield while demanding a high consumption of energy and pressure thus becoming an accessory in the ever-increasing energy crisis (Rahimi and Doostmohammadi 2019). Therefore, there is an exigency for the development of reliable, sustainable and non-toxic methods of nanomaterial synthesis which could potentially ameliorate their biomedical applications as well.

Lately, researchers have refocused their attention to strategies that utilize materials of biological origin like plant extracts and microorganisms which can subjugate the issues associated with chemical and physical approaches (Shah et al. 2015; Ijaz et al. 2020). Production methods employing microscopic organisms are especially superior since they encompass a plethora of living entities including bacteria, algae, yeast, fungi as well as viruses which through biogenic enzymatic processes can facilitate the fabrication of diverse "biogenic" nanostructures with defined sizes, shapes, mono-dispersity and properties in a relatively brief timeframe (Zhang et al. 2011). This is further corroborated by the fact that microbes can be grown rapidly in diverse environmental conditions with varying pH, temperature and pressure and can be

K. Turabi
e-mail: kturabi@unmc.edu

R. Jobby
Amity Institute of Biotechnology, Amity University Maharashtra, Mumbai-Pune Expressway, Bhatan, Panvel, Navi Mumbai 410206, Maharashtra, India

Amity Centre of Excellence in Astrobiology, Amity University Maharashtra, Mumbai-Pune Expressway, Bhatan, Panvel, Navi Mumbai 410206, Maharashtra, India

R. Jobby
e-mail: rjobby@mum.amity.edu

P. Arte
Anatek Services PVT LTD, Santacruz Railway Bridge, Sen Nagar, Santacruz East, Mumbai 400055, Maharashtra, India

A. Khan
Molecular and Developmental Stem Cell Biology, Ruhr Universität Bochum, Bochum 44799, Germany

K. Turabi
Eppley Institute for Research in Cancer and Allied Diseases, Fred & Pamela Buffet Cancer Center, University of Nebraska Medical Center, 986805 Nebraska Medical Center, Omaha, NE 68198-6805, United States of America

easily maintained. Furthermore, this "Green synthesis" route avoids the involvement of toxic chemicals or intensive energy consumption making the process entirely sustainable and improving its applications in fields like food industry, pharmaceuticals, medicine and wastewater treatment where the biocompatibility of nanomaterials is pivotal (Li et al. 2017; Ranjha et al. 2022).

The purpose of this chapter is to briefly discuss the mechanisms which enable microorganisms to behave like functional nano-factories and explore the role of various types of microbial organisms in the production of nano-materials with an emphasis on recent investigations and advancements in the fabrication of metallic, metal oxide, non-metallic nanomaterials. It also provides a comprehensive insight within the developmental challenges and future prospects presented by this novel approach.

4.2 Mechanism

Biogenic synthesis of nanomaterials can be achieved using unicellular to multicellular microbes including diverse bacteria, fungi, actinomycetes, algae and viruses. Their ability to fabricate nanomaterials can be attributed to the cellular detoxification mechanism which allows them to capture and reduce metallic and non-metallic ions into nanomaterials by enzymatic reduction (Li et al. 2011). Biomolecules such as amino acids or proteins, lipids, exopolysaccharides, etc. secreted by the microbes and their constituents can constitute "natural" stabilizing and capping agents thus preventing agglomeration without the incorporation of other stabilizing agents (Chugh et al. 2021). These microbial components also form biocompatible coatings on the final nanomaterials which enables them to efficiently interact with other biological systems due to the presence of certain functional groups. This feature of biogenic synthesis is particularly important since capping agents have a considerable impact on the nanomaterial size, shape, stability, surface energy, agglomeration, dispersion, and electrostatic and steric hindrance (Sidhu et al. 2022).

Furthermore, the process of biogenic nanomaterial production may be distinguished into extracellular and intracellular processes based on the location and environment in which the nanomaterials are generated. Bacteria and fungi are two main classes of microbes who biosynthesize using this dualistic approach.

4.2.1 Extracellular Biosynthesis of Nanomaterials

The extracellular pathway entails capturing and accumulating ions on the surface of microbial cells, as well as reducing ions in the presence of produced enzymes. In such processes, the surface proteins and enzymes are involved as well (Mughal et al. 2021). For instance, extracellular enzymes like nitrate reductase and NADH-dependent oxidoreductases have been demonstrated to aid electron transfer from donor groups to metal ions to form silver and gold nanoparticles (Lahiri et al. 2021a, b). Koul et al. and colleagues reported another method of extracellular synthesis which involved the centrifugation of microbial cells containing broth where the supernatant consisting of reductase enzymes would be collected and separately allowed to react with metal ions leading to nanoparticle production (Koul et al. 2021). A similar method is often employed in fungi-mediated synthesis methods where addition of metal precursors to an aqueous filtrate containing fungal biomolecules can lead to the formation of nanomaterials in the dispersion. The dispersions are then purified using filtration, centrifugation and dialysis techniques to collect the nanomaterials. These methods are widely used due to their simplicity and reproducibility since they do not require the cells to release nanomaterials formed within their external surfaces (Guilger-Casagrande and Lima 2019; Singh et al. 2018). Furthermore, the remaining biomass pellet can be used for intracellular synthesis as well, making the process highly sustainable (Das et al. 2014). Finally, certain metal-reducing bacterial strains like those belonging to *Geobactor* and *Shewanella* species possess a unique extracellular electron transfer (EET) process which enables them to exchange electrons with the external environment and mediate extracellular reduction reactions leading to controllable nanomaterial fabrication. Though these methods focus mainly on metallic nanostructures, non-metallic nanoparticles such as copper and chalcogen nanoparticles have been produced using the EET pathway (Zou et al. 2021).

4.2.2 Intracellular Biosynthesis of Nanomaterials

Several theories have been put forth to determine the intricacies behind the intracellular synthesis of nanomaterials though the exact mode of action employed by several microorganisms is still unclear. But the most well accepted theory is the cellular uptake of ions propelled by the electrostatic interactions between the negatively charged cell membranes or cell walls and the positively charged metal ions. These cationic ions diffuse across the outer microbial membranes and travel through the cytoplasmic membrane through a variety of ion channels and transport proteins where they are reduced by cellular enzymes into nanomaterials (Pandit et al. 2022). Metal and non-metal ions are primarily reduced to provide intracellular inorganic NMs through the action of enzymes like NADPH-dependent or NADH-dependent reductase, nitrite and nitrate reductases, as well as non-enzyme proteins and peptides like phytochelatin (PC) and metallothionein (MT). For example, NADPH-dependent nitrate reductases produced by *Penicillium fellutanum* were able to reduce ions into nanoparticles in a short proportion of time (Kathiresan et al. 2009). Moreover, microorganisms like *Bacillus licheniformis, Fusarium oxysporum* and *Rhodobacter sphaeroides* have biosynthesized nanostructures via the intracellular NAD(P)H-dependent oxidoreductases, sulfate and sulfite reductases produced by the microbial cells (Choi and Lee 2020). Furthermore, intracellular enzymes such as cytochrome oxidases have also been discovered to aid in the reduction of metal ions to NPs by electron transfer between cytoplasm components such as vitamins, and organic acids (Lahiri et al. 2021a, b).

4.3 Strategies for Nanomaterial Synthesis Using Microorganisms

In contrast to physically and chemically derived approaches, utilizing various microorganisms as a suitable means for synthesizing nanomaterials has expanded quickly and is considered as a greener nanomaterial synthesis approach. Among multiple microorganisms, fungi and bacteria are some of the prominent biogenic sources as they can produce a higher concentration of the reducing enzyme needed to produce nanomaterials (Lahiri et al. 2021a, b). Further they are also preferred due to their ease of cultivation, and the morphological characteristics of the synthesized nanomaterials can be controlled; this can significantly lower the expenses of large scale implementation (Lahiri et al. 2021a, b).

This section highlights the various microorganism-based biogenic sources for efficiently synthesizing nanomaterials.

4.3.1 Bacteria-Mediated Synthesis of Nanomaterials-with Cyanobacteria

Bacterial synthesis of nanomaterials has become a prominent global strategy in green nanotechnology. Various bacterial species have been implemented fro synthesisng numerous nanomaterials as a means to replace the common chemical and physical production strategies. Recent research has focused on developing a regulated and scalable procedure for producing mono-dispersed, relatively stable nanomaterials (Iravani 2014). Therefore, bacterial biomass and cellular extracts have been employed in nanoparticle synthesis as a greener alternative strategy.

4.3.1.1 Metallic Nanomaterials

Bacterial species have proven to be a promising agent for manufacturing nanomaterials due to their efficient heavy metal reducing capabilities (Iravani 2014). Due to their capacity to accumulate and attach to metallic ions as well as convert metals hazardous to them to nontoxic nanomaterials, they provide a novel strategy for nanomaterial production (Tsekhmistrenko et al. 2020).

Bacteria produce metallic nanomaterials both intracellularly and extracellularly. However, the extracellular production and extraction of the nanomaterials are more effective and simpler in comparison. Further, research has also indicated that dead bacteria can also be employed along with viable bacterial cells for synthesizing nanomaterials (Tsekhmistrenko et al. 2020; Salunke et al. 2015).

Nitrate reductase is a frequently employed bacterial enzyme for synthesizing nanomaterials, including for Ag and Au nanoparticles (Tsekhmistrenko et al. 2020). The bacterial species, *Bacillus clausii* was utilized to synthesise silver nanoparticles intracellularly (Mukherjee et al. 2018). The study found the synthesis due to two bacterial intracellular reductase enzymes, nitrate reductase and sulphite reductase (Mukherjee et al. 2018). Similarly, multiple researches conducted explored the biosynthesis of silver nanomaterials using the bacterial species *E. coli* (Iravani 2014). A common factor among all these studies was the use of bacterial supernatant for the synthesis. The supernatant enzymes aided the reduction of silver cations to silver particles, which accumulate and form silver nanoparticles. Further, many other bacterial species have been utilized to synthesize silver nanomaterials. *B. subtilis* was employed by Saifuddin et al. to synthesize spherical and triangular silver nanostructures extracellularly (Saifuddin et al. 2009) by utilizing supernatant culture and microwave irradiation in water.

E. coli is also commonly used for manufacturing gold nanomaterials. In *E. coli*, gold nanomaterials were synthesized with varying morphology, including spherical, triangular, and rod shaped (Deplanche and Macaskie 2008). Varying the pH lead to the production of Ag nanoparticles heterogeneous in morphology and size (Deplanche and Macaskie 2008). Crystalline gold nanoclusters were observed when synthesized in strains of *Lactobacillus*, which were also capable of forming Au and Ag-Au alloy nanocrystals (Nair and Pradeep 2002). Additionally, *Rhodopseudomonas capsulata* was found to extracellularly synthesize stable gold nanomaterials of varying morphology, based on the pH of the precursor ion solution utilized (Iravani 2014). The gold ions underwent reduction in the presence of *R. capsulata*. Upon pH reduction of the mixture, gold nanoplates were produced along with nanoparticles, and when it was increased, nanowire production was seen (Iravani 2014).

Shewanella oneidensis is a bacterial species capable of metal ion reduction both aerobically and anaerobically. *S. oneidensis* has the ability to synthesise multiple metal nanomaterials, including palladium nanoparticles, and tellurium nano rods, by reducing hazardous soluble tellurite and silver nanoparticles exhibiting antibacterial activity (Iravani 2014). Another *Shewanella* species was found to reduce hexachlorideplatinate (IV) ions resulting in the formation of platinum nanomaterials (Iravani 2014).

Cyanobacteria have been frequently employed for manufacturing metallic nanomaterials as they can absorb various heavy metallic ions; further, they are the chief atmospheric nitrogen fixing bacteria (Pathak et al. 2019). The nitrogenase enzyme in cyanobacteria catalyses the conversion of molecular dinitrogen gas to ammonia, which aids the reduction of metallic salts and regulates nanoparticle formation (Pathak et al. 2019). The utilization of cyanobacteria for the biogenesis of nanoparticles has proven to be an efficient strategy due to their rapid production rate and high biomass efficiency. They produce nanomaterials through intracellula as well as extracellular mechanisms.

The production of Cadmium Sulphide (CdS) nanomaterials is significantly aided by the photosynthetic pigment, phycobilin, found in cyanobacteria such as *Phormidium tenue* and *Nostoc carneum* (Hamida et al. 2020). Another cyanobacterial pigment, C-phycocyanin, found in *Spirulina* and *Limnothrix* species and *Nostoc linckia*, was found to play a role in producing silver nanoparticles with varying morphological characteristics (Hamida et al. 2020). Further, cyanobacterial proteins present in *Spirulina plantensis* were used to extracellularly produce gold, silver, and bimetallic gold-silver core–shell nanomaterials. These proteins aid the stability of produced nanomaterials. Additionally, cyanobacterial nitrogenase catalysed metallic ion reduction lead to Palladium, Platinum, Gold, as well as Silver nanoparticles formation from the strains *C. pulvinate*, *A. flos-aquae*, and *L. foveolaru*m (Hamida et al. 2020). Extracellular polysaccharides from *Nostoc commune* can also produce silver nanoparticles (Hamida et al. 2020).

4.3.1.2 Metal Oxide Nanomaterials

Metallic oxide-based nanomaterials can adopt a wide range of configurations that exhibit metallic, semiconductor, or insulator features and have distinctive physicochemical properties. The physicochemical characteristics of metallic oxide nanomaterials are distinct from those of heavier metal oxides due to their exceedingly reduced size (Jeevanandam et al. 2016). Thus, since their improved properties possess a wide array of applications, a need for a greener methodology to synthesize metal oxide metallic oxide-based nanomaterials is essential. Bacteria-mediated synthesis of these nanomaterials has been increasingly adopted over the years due to the reduced toxicity and the ability to regulate the morphological characteristics of the nanomaterials synthesized by bacterial species (Jeevanandam et al. 2016). Even though bacterial synthesis provides many advantages for synthesizing such nanomaterials, chemical methodologies are still more frequently adopted for their production. Thus limited research is present on bacterial-mediated metallic oxide nanomaterial formation..

Magnetococcus marinus, a magnetotactic bacteria, has been utilized for the intracellular biosynthesis of magnetite, iron (II, III) oxide, nanocrystals (Valverde-Tercedor et al. 2015). The bacterial species contains proteins including MamC, MamCnts, and Mms6 enclosed within the membrane, which are involved in the nanocrystal production process (Valverde-Tercedor et al. 2015). Based on the proteins, the magnetite nanocrystals were observed to possess differential properties, such as Mms6-derived nanocrystals were super paramagnetic, whereas nanocrystals produced by MamC and MamCnts were larger (Valverde-Tercedor et al. 2015). These proteins were seen to create the magnetite nanocrystals by creating micelles with a negative surface, to which the iron cations bound and resulted in the formation of nanomaterials. Similarly, *Magnetospirillum magnetotacticum*, another magnetotactic bacteria, has been used to produce intracellular membrane-bound magnetite nanocrystals, whereas the bacterial species *Bacillus subtilis* and *Proteus vulgaris* synthesized iron oxide nanoclusters extracellularly (Jeevanandam et al. 2016).

Another widely synthesized metal oxide nanomaterial, having a wide array of applications, is zinc oxide (ZnO). *Lactobacillus plantarum*, a lactic acid bacteria, has been found to synthesize sphere shaped ZnO nanocrystals intracellularly (Mohd Yusof et al. 2019). Another lactic acid bacteria species, *Lactobacillus sporogens*, was used to produce differentially sized ZnO nanomaterials in a hexagonal morphology (Mohd Yusof et al. 2019). Further, *Aeromonas hydrophila* produced spherical nanoparticles, *Bacillus megaterium* gave rise to rod and cube shaped nanomaterials, and *Sphingobacterium thalpophilum* synthesized triangular ZnO nanoparticles (Mohd Yusof et al. 2019). Cupric oxide (CuO) nanomaterials may also be produced by bacterial means. Various bacteria, including *Pseudomonas*, *Rhodococcus*, *Brevundimonas*, *Marinomonas*, and *Bacillus* species were found to produce small, uniformly sized CuO nanoparticles by reducing $CuSO_4$ at low temperatures (John et al. 2021).

Titanium dioxide (TiO_2) nanomaterials have also been synthesized using multiple bacterial species. *Bacillus mycoides* and *Bacillus subtilis* have been used to produce spherical TiO_2 nanoclusters from titanyl hydroxide and metatitanic acid, respectively (Jeevanandam et al. 2016). Additionally, iron reducing bacterial species are associated with uranium oxide nanoparticle production. Bacterial species such as *Desulfosporosinus desulfuricans* and *Geobacter metallireducens* have been found to detect trace amounts of uranium in their environment and, following the reduction of iron (III), consequently, reduce uranium (IV) and accumulate it to form uranium oxide nanoparticles (Jeevanandam et al. 2016). Further, multiple species of *Shewanella* are adopted to synthesize small, crystalline manganese oxide nanomaterials from manganese aerobically (Jeevanandam et al. 2016).

Several cyanobacterial strains also have the ability to synthesise metal oxide nanomaterials. Cell extracts of *Nostoc* and *Desertifilum* strains were used to synthesize star, and rod shaped ZnO nanomaterials, respectively (Ebadi et al. 2022). Furthermore, *Cylindrospermum stagnale* and *Spirulina platensis* have been used to synthesize sphere shaped crystalline CuO nanomaterials that possess prominent antibacterial activity.

4.3.1.3 Non Metallic Nanomaterials

Bacteria-mediated nanomaterial synthesis has proven to provide a more efficient and greener methods for the biogenesis of silica-based nanomaterials. Widely used conventional methods adopted for the synthesis of such silica-based particles comprise of flame synthesis, sol–gel, and reverse micro emulsion. However, these methodologies pose various limitations such as large nanomaterial size, being costly, and being non-ecofriendly (Show et al. 2015). These challenges may be overcome by substituting the current methods with bacteria.

Show et al. synthesized silicon dioxide nanoparticles from organic and inorganic substrates, tetraethyl orthosilicate and magnesium trisilicate, respectively; using the biomass of a thermophilic bacterial strain of *Thermoanaerobacter* species, BKH1 (Show et al. 2015). The bacterial enzyme bioremediase was also associated with formation of uniformly distributed, irregular in shape SiO_2 nanoparticles deposited on the cellular surface of BKH1 strains (Show et al. 2015). These nanoparticles were found to be stable and negatively charged, which was responsible for preventing cluster formation.

Thus, bacteria-mediated synthesis through the reduction process by interaction with various bacterial proteins and enzymes, via numerous pathways provides a greener approach for the production of nanomaterials. One of the key advantages of employing this technique involving bacterial species is the cost-effective large-scale adaptation of the process, without the utilization of toxic chemicals. However, this method also has a few limitations, such as the overall procedure being relatively slow and the regulation of the morphological characteristics of the nanoparticle may be challenging.

4.4 Actinomycetes-Mediated Synthesis of Nanomaterials

Actinomycetes are a group of gram-positive bacilli, which also share common properties with fungi. Thus, they are also regarded as ray fungi since they exhibit mycelial growth (Kumari et al. 2020). Further, actinomycetes can be successfully employed for the biogenesis of various nanomaterials. Nanoparticle synthesis by actinobacteria is a more efficient and ecofriendly approach than the conventional methods of production, which rely on high pressures and temperatures and toxic chemicals (Kumari et al. 2020). Genetic makeup of actinomycetes may also be manipulated for regulating morphological characteristics of nanomaterials (Kumari et al. 2020).

Actinobacteria can produce nanomaterials through intracellular and extracellular modes. The intracellular route involves the attachment of precursor ions to carboxyl groups electrostatically on the enzyme located on the mycelial cell wall surface (Manimaran and Kannabiran 2017). Further, they result in ion reduction, which then accumulate, giving rise to nanomaterials. Whereas the extracellular method entails the actinobacterial electron-shuttling enzymes commonly taking part in the N_2 fixation. The enzymes are responsible for reducing precursors and result in the formation of nanomaterials (Manimaran and Kannabiran 2017). Nanomaterials extracellularly produced by actinomycetes are more commonly employed due to its efficiency and wide range of commercial applications.

4.4.1 Metallic Nanomaterials

Due to their metal, detoxifying capabilities actinobacteria can reduce soluble toxic metal ions into their non toxic metal nanomaterials by biomineralization, intracellular bioaccumulation, or extracellular precipitation (Kumari et al. 2020). Various actinobacterial components including extracellular enzymes, intracellular proteins, co-factors, cell wall constituents, etc. are involved in the formation of metallic nanomaterials. Since metallic ions carry a negative charge, when exposed to a metallic salt solution, they bind to the cell wall, after which they are reduced and form their corresponding nanomaterials (Kumari et al. 2020). The characteristics of the nanomaterials produced by actinomycetes depend on several factors, such as the amount of reducing enzymes present in the extract and the temperature and pH, which determines the production rate, morphology, and stability of nanomaterials (Kumari et al. 2020).

A commonly synthesized metallic nanomaterial from actinobacteria is Ag nanomaterial. The actinobacterial species of *Rhodococcus* can synthesize silver nanoparticles intracellularly with the help of membrane bound enzymes. The nanoparticles obtained were found to sphere shaped and 10 nm in size (Gahlawat and Choudhury 2019). Similarly, various species of *Streptomyces* can produce Ag nanoparticles extracellularly. Nitrate reductase, found in *Streptomyces* species, synthesize extracellular silver nanoparticles by reducing Ag ions to Ag nuclei, which accumulate to result in the formation of the nanomaterials. The silver nanomaterials synthesized through this mechanism by *Streptomyces* species LK3 were found to have strong pesticidal activity (Gahlawat and Choudhury 2019). Another actinobacterial species, *Streptacidiphilus durhamensis*, was employed to synthesize antibacterial Ag nanoparticles having a sphere morphology (Gahlawat and Choudhury 2019).

A range of actinomycetes have been used to also synthesized gold nanoparticles. The *Rhodococcus* species produced Au nanomaterials intracellularly when reacted with a chloroauric solution; the enzymes located between the mycelial wall and the membrane were involved in the process (Ovais et al. 2018). Other species of actinomycetes such as *Thermomonospora* and *Actinobacter* have also been found to synthesize gold nanoparticles having a hexagon shape and are up to 50 nm in size. Furthermore, copper nanoparticles have also been synthesized using *Streptomyces capillispiralis* Ca-1 and marine endophytic actinomycetes (Gahlawat and Choudhury 2019). Other metallic nanomaterials such as zinc, manganese, and selenium nanomaterials have also been synthesized by multiple species of Streptomyces (Kumari et al. 2020).

4.4.2 Metal Oxide Nanomaterials

Even though actinomycetes have been considerably exploited for synthesizing metal nanomaterials, they have not yet been explored extensively for biosynthesis of metal oxide nanomaterials. Copper oxide nanoparticles are one type of nanomaterials synthesized by actinomycetes. When actinomycetes extract was reacted with copper sulphate solution, spherical CuO nanoparticles were observed (Waris et al. 2021). The secondary metabolites present in the actinobacteria were involved in the formation, as the functional group on the nanomaterials was also present on the secondary metabolites in the extract. Further, the biomass extract from actinomycetes isolated from *Oxalis corniculata* produced sphere-shaped copper oxide nanoparticles, 80 nm in size upon reaction with copper sulphate solution (Waris et al. 2021).

Rajivgandhi et al. synthesized pure ZnO nanosheets using actinomycetes species *Nocardiopsis* species GRG1 biomass (Rajivgandhi et al. 2018). The ZnO nanosheets produced were seen to have prominent antibacterial activity as they prevented the formation of biofilms (Rajivgandhi et al. 2018). The production of zinc oxide nanoparticles from actinomycetes strains isolated from a soil sample was also reported (Rajivgandhi et al. 2022). These nanomaterials were synthesized using actinobacterial isolates when treated with a zinc sulphate solution by reducing the zinc ions. The nanoparticles obtained were determined to be crystalline and had high stability, which prevented them from agglomerating (Rajivgandhi et al. 2022).

4.4.3 Non Metallic Nanomaterials

The species *Actinobacter* was found to have the ability to synthesize silicon-silica nanocomposites using potassium hexafluorosilicate as a substrate (Singh et al. 2008). The bacterium was found to produce the Si/SiO_2 nanocomposites extracellularly using oxidase and reductase enzymes. The nanocomposites were quasi-sphered, crystalline, and 10 nm in size (Singh et al. 2008). Similarly, silicon dioxide nanocomposites were produced from wheat bran and rice husk by utilizing actinomycetes. The nanocomposites obtained were crystalline and approximately 240 nm in size (Karande et al. 2021).

Thus, the amount of data available on the biosynthesis of other types of nanomaterials by actinomycetes, apart from metallic-based, remains limited. Further, investigation and research on actinomycetes are required to identify prominent species capable of synthesizing nanomaterials having important commercial applications.

4.5 Yeast-Mediated Synthesis of Nanomaterials

Yeast strains have proven to be prominent bionanofactories for a wide array of nanomaterials. They pose multiple advantages as a mode of nanomaterial synthesis in contrast to other conventional strategies and biogenic sources such as bacteria. Some of these include a more straightforward culture method and greater biomass, a more significant extent of metabolite accumulation within cells, greater endurance and capacity for metal ion uptake, and better metal attachment ability to the cell walls (Roychoudhury 2020).

Further, biomineralization, the primary method for nanomaterial production, can be induced by yeast cells serving as a blueprint (Roychoudhury 2020). Their cell membrane aids the envelopment of the nanomaterial synthesized within the yeast cells. Similar to the other biogenic sources, yeast is also capable of extracellular and intracellular synthesis of nanomaterials. Extracellular synthesis involves the enzymes in yeast filtrate that reduce the precursor ions and lead to the formation of various types of nanomaterials. Similarly, yeast-mediated intracellular synthesis of nanomaterials is also reduction-based, where reducing enzymes within the cells convert precursor ions to their nanomaterial form (Roychoudhury 2020).

4.5.1 Metallic Nanomaterials

The primary goal of yeast-mediated nanomaterial biogenesis is to reduce nanomaterial cytotoxic activity through the cells' stress responses using substances like phytochelatins, metallothioneins, and glutathione, that possess the ability to attach metal ions and also exhibit unique redox and nucleophilic characteristics necessary for the bioreduction of metallic ions (Roychoudhury 2020). From aqueous metal salt solutions, the metal ions penetrate the cells through passive diffusion, after which oxidoreductase enzymes present on the plasma membrane reduce the metallic ions, forming metallic nanomaterials. Based on the mechanism of synthesis adopted by the yeast cells, metallic nanomaterials of varying morphology and properties.

Cadmium-based nanomaterials are most commonly produced among the various types of metallic nanomaterials synthesized using yeast (Hulkoti and Taranath 2014). Due to their semiconductor characteristics, cadmium-based nanomaterials are found to possess a multitude of applications. *Schizosaccharomyces pombe*, *Trichosporon jirovecii*, and *Candida glabrata* are some of the yeast species utilized for CdS nanomaterials' biogenesis. These species were found to synthesize CdS nanomaterials upon culturing with cadmium salts. The presence of the cadmium salts leads to the generation of γ-glutamyl peptide resulting in the cadmium-peptide complex formation (Hulkoti and Taranath 2014). Thus, when these yeast strains are treated with solutions contain cadmium salts, they result in intracellular synthesis of CdS nanomaterials. Further, the yeast species, *Saccharomyces cerevisiae*, can extracellularly synthesise of fluorescent Cadmium Telluride quantum dots. Upon culturing with $CdCl_2$ and Na_2TeO_3 solutions, the yeast cells synthesized monodispersed, crystalline CdTe nanomaterials (Bao et al. 2010).

Multiple yeast species including, *Saccharomyces cerevisiae*, *Candida albicans*, R*hodoturula glutinis*, and *Geotrichum candidum* have been employed for synthesizing Ag nanomaterials (Roychoudhury 2020). These species synthesize the Ag nanomaterials upon the addition of silver salt solutions. The nanomaterials were produced with the help of sulphate and nitrate reductase enzymes. Ag nanomaterials were also found to be produced as a result of complex formation with phytochelatins (Roychoudhury 2020).

Further, Au nanomaterials of different morphology and size were produced by *Pichia jadinii* species (Hulkoti and Taranath 2014). The synthesis involved Au ion reduction by enzymes present within the cells. The cell wall-embedded proteins were also played an integral role in their synthesis; they lead to the formation of a peptide coating on the nanomaterials, preventing agglomeration (Hulkoti and Taranath 2014). *Yarrowia lipolytica* also has the ability to produce gold nanomaterials using a protease enzyme attached to the yeast cell wall (Roychoudhury 2020). The reductase enzymes extracted from the yeast species *Hansenula anomala* were also used to produce Au nanoparticles (Menon et al. 2017).

Additionally, the *Toluropsis* yeast species, when introduced to Pb salt solutions, synthesized PbS nanocrystals intracellularly (Kowshik et al. 2002). The marine yeast *Rhodosporidium diobovatum* also intracellularly produced uniform PbS nanocrystals having great stability (Seshadri et al. 2011). Other metal nanomaterials, including selenium nanoparticles, were formed by culturing with sodium selenite solution and palladium nanoparticles using palladium acetate and hydrazine solution (Roychoudhury 2020).

4.5.2 Metal Oxide Nanomaterials

Along with metallic nanomaterials, yeasts are also widely employed for synthesizing numerous metal oxide-based nanomaterials with various applications. Antimony (III) oxide (Sb_2O_3) nanoparticles have been biosynthesized by the species *Saccharomyces cerevisiae* intracellularly (Hulkoti and Taranath 2014). Oxidoreductase enzymes present in membrane and cytoplasm, and quinones aided their production (Jha et al. 2009). The resulting nanomaterials were found to have a face-centered cubic structure.

Similarly, *Saccharomyces cerevisiae* was also used to synthesize titanium dioxide (TiO_2) nanomaterials having a diameter less than 12 nm (Roychoudhury 2020). Titanium ions in titanium (III) chloride solution underwent nucleation and formed lamellar TiO_2 nanoparticles at high temperatures in the presence of reducing enzymes urease and amylase (Roychoudhury 2020). The nanoparticles were spherical and 6.7 nm in diameter. Further, *S. cerevisiae* have also synthesized monodispersed manganese dioxide nanomaterials having hexagon and sphere shapes and approximately 34 nm in size (Salunke et al. 2015).

Additionally, *Pichia kudriavzevi* and *Pichia fermentans* JA2 were reported to synthesize ZnO nanomaterials extracellularly with reaction-dependent morphology (Mohd Yusof et al. 2019). Further, *S. cerevisiae* and *Rhodotorula mucilaginos*a yeast species has been used to synthesize copper oxide nanoparticles (Seabra and Durán 2015). Apart from various metal oxide nanomaterials, *S. cerevisiae* has also been found to biosynthesize barium carbonate nanoparticles intracellularly, having uniform distribution with diameter of 5 nm (Chang et al. 2021).

4.5.3 Non Metallic Nanomaterials

Similar to bacteria and actinomycete-mediated green synthesis of nanomaterials, limited yeast species have been studied to explore the ability of these microorganisms to synthesize nanomaterials other than metal-based nanomaterials. However, *S. cerevisiae*, which, as mentioned previously, has been used for synthesizing multiple metal nanomaterials, can also produce silica nanomaterials (Karande et al. 2021). The yeast species synthesized the silica nanoparticles, 25 nm in size, in the presence of sodium silicate solution. Along with silica nanoparticles, one study reported the production of carbon dots using beer yeast exhibiting fluorescence and antibacterial properties (Gao et al. 2019).

4.6 Algae-Mediated Synthesis of Nanomaterials

Algae are commonly used in biosynthesis of different metallic and metallic oxide nanoparticles due to the quick growth, ease of handling, and tenfold faster biomass growth than higher plants on average. Presently, many strains of algae are studied for the sustainable development of various kinds of nanoparticles (Chaudhary et al. 2020).

4.6.1 Metallic Nanomaterials

Metallic NPs consisting of copper, gold and silver nanoparticles are among the majority of produced nanoparticles from brown algae. More than fifty percent of the published data on metallic NPs are to the production of AgNPs from various algae strains (Azizi et al. 2014). This is due to the fact that AgNPs offer superior physicochemical properties compared to their heavier forms, which leads to them being particularly valuable in several fields, including the jewellery, paint, fabric, dentistry, pharmaceuticals, and wound repair (Maneerung et al. 2008; Mohanpuria et al. 2008).

In a study, spherical silver nanoparticles with size 96 nm were produced extracellularly from *T. conoides* and displayed potent antimicrobial action against *C albicans, S aureus, Aspergillus niger,* and *P aeruginosa* (Rajeshkumar et al. 2012). Similar form of nanoparticles that are widely produced from brown algae strains is gold nanoparticles, they show a variety of bioactivities with major clinical uses (Khanna et al. 2019). Sphere-like gold nanoparticles with size 15–20 nm were produced extracellularly from *L japonica* using chemical compounds that acted as capping as well as reducing agents (Ghodake and Lee 2011). Due to a potent reducing agent, *Porphyra vietnamensis* is among the most prevalent red algae species described for the production of several types of nanoparticles (NPs) (Rao et al. 2007).

A variety of red algae strains, including *K alvarezii, G dura, G acerosa, Palmaria decipiens* and many more, have been described in the literature as capable of biosynthesized AgNPs. In biological applications, the size and form of NPs are key parameters that play a significant role (Pugazhendhi et al. 2018). Reportedly, the gold nanoparticles generated from several red algae strains are mostly 20–60 nm in size and shaped sphere like. *Lemanea fluviatilis* is a type of red marine algae being studied for the manufacture of AuNPs. It produced 5.9 nm-sized face-centered cubic crystalline silver nanoparticles (Subbiah et al. 2017).

Blue-green algae are extensively utilized for the production of different nanoparticles. From S. platensis, spherical AgNPs (2–8 nm) were produced for efficient usage in the health, and feed sectors. Apart from *S. platensis*, gold nanoparticles of various dimensions have been produced from a number of additional blue-green algae, including *Oscillato riawillei, Cylindrospermum stagn, Plectonema boryanum* and *Microchaete diplosiphon* (Mukherjee et al. 2002).

Similar to AgNPs, S. platensis also contributed to the production of AuNPs. Proteins acting as reducing agents have been implicated in the *S. platensis* based extracellular production of cubic octahedral and spherical silver nanoparticles, as reported by many studies (Husain et al. 2015). Another major freshwater green algae, *Chlamydomonas reinhardtii*, was discovered to be engaged in the regulation of cadmium sulphide bimetallic nanoparticles (Rao and Pennathur 2017). Along with these metallic nanoparticles, algae have also been utlised for the biogenic synthesis of Palladium Nanoparticles (PdNPs). The extract of the algal species *Sargassum bovinum* has been utilised to synthesize octahedral PdNPs having a 5–10 nm diameter (Momeni and Nabipour 2015). These monodispersed synthesised metallic nanoparticles were found to be highly sensitive, stable and selective, and were used as sensors for detecting hydrogen peroxide in solutions (Momeni and Nabipour 2015).

4.6.2 Metal-Oxide Nanomaterials

In addition to metallic NPs, it has been reported that brown algae may biosynthesize other metal oxide nanoparticles, such as zinc and titanium oxide nanoparticles (Sirelkhatim et al. 2015). According to one research, Zinc oxide nanoparticles are produced by combining powdered form of *S. muticum* with water and heating it till it is fully dissolved. Next, a zinc acetate salt mixture was prepared, and the mixture was stirred continuously for hours until NPs were produced. The hexagonal ZnONPs produced ranged in size from 35 to 57 nm and some were capped by bioactive molecules such as sulphate, amines, hydroxyl, and carbonyl (Azizi et al. 2014). Research has also shown that amide, carboxylic, and nitro compounds-rich *Gracilaria edulis* was employed to synthesise spherical AgNPs and octahedral ZnONPs (Priyadharshini et al. 2014). Another algal species, *Sargassum myriocystum*, was found to extracellularly synthesise ZnONPs having a wide range of morphology (Khanna et al. 2019).

The algal seaweed, *Sargassum muticum*, was also utilised for the biogenic synthesis of magnetic iron oxide nanoparticles having a cubic structure and 14–18 nm size (Mahdavi et al. 2013). The research conducted highlighted that the sulphated polysacharides present in the seaweed extract played a key role in the nanoparticle synthesis as stabilising and reducing agents (Mahdavi et al. 2013). The algal species *Bifurcaria bifurcata* have also been studied to evaluate their potential to synthesise Copper Oxide Nanoparticles (CuONPs) (Khanna et al. 2019). The

CuONPs synthesised extracellularly from the algae in this study were found to be spherical with a varying diameter between 5 and 45 nm (Khanna et al. 2019). These crystallin CuONPs synthesised from brown algae extract were found to have prominent antibacterial activity (Khanna et al. 2019). Another biogenically synthesised metallic oxide nanoparticles exhibiting antibacterial potential are Zirconia (ZrO_2) nanoparticles (Kumaresan et al. 2018). The Zirconia nanoparticles were synthesised from the marine brown algae, *Sargassum wightii* (Kumaresan et al. 2018). A green combustion method was utilised for their synthesis and resulted in 4.8 nm tetragonal nanoparticles (Kumaresan et al. 2018).

4.6.3 Non-metal Nanomaterials

Silica nanoparticles were produced by the breakdown and poly-condensation of silicon alkaloids by proteins and peptides in *C. vulgaris* extract. Using varied molar ratios of $AgNO_3$ and $HAuCl_4$, *Gracilaria edulis* synthesized effectively bimetallic Ag-Au nanoparticles. These bimetallic nanoparticles demonstrate strong anticancer activity towards breast cancer cell types. Algae has also been utilised for the green synthesis of carbon-based nanoparticles. Fluorescent carbon nanoparticles were synthesised by a hydrothermal process from the abundant algal species *Cladophora vagabunda* (Calangian et al. 2018). The synthesised nanoparticles were synthesised through an extracellular process using the extract of the algal species and resulted in the formation of Carbon nanoparticles having a 42.78 nm diameter and exhibiting fluorescence. The algal bloom obtained from fresh water, which contained various algal species of Cyanophyceae, Chlorophyceae, Bacillariophyceae and Eugenophyceae, have also been utilised to synthesise photoluminescent carbon dots (Ramanan et al. 2016). These spherical nanodots had an 8 nm diameter and showed high water solubility, high stability, remarkable cell permeability and low cytoxicity (Ramanan et al. 2016). Thus, these biogenically synthesised nanoparticles possess potential as biomarkers for cancer detection. In addition to silica-NPs, the biosynthesis of various metallic, bimetallic, oxide—based, and semiconductor nanoparticles is now underway, and a large amount of studies and trials are in the early stages (Chaudhary et al. 2020).

4.7 Fungi-Mediated Synthesis of Nanomaterials

Fungi are capable of producing a vast array of chemicals that may be utilized in a range of situations. Filamentous fungus as well as other fungi are believed to generate around 6,400 bioactive compounds (Bérdy 2005). Because of their resistance to heavy metals and potential to internalize and bioaccumulate metals, these organisms are commonly employed as reducing and stabilizing agents. Fungi may also be easily grown in huge quantities and create nanoparticles with regulated dimensions

(Guilger-Casagrande and Lima 2019). Fungi have an advantage that they generate vast amounts of proteins, a few could be employed for quick nanoparticle manufacturing (Vahabi et al. 2011). Fungal nanoparticles may be biosynthesized intracellularly or extracellularly. The metallic precursor is introduced to the fungal culture and therefore is absorbed in the event of intracellular synthesis. Extracellular production involves the addition of a metallic precursor to an aqueous filtrate which solely has fungal macromolecules, this leads to the creation of free nanomaterials in the dispersion (Guilger-Casagrande and Lima 2019).

4.7.1 Metallic Nanomaterials

Many researchers have talked about the production of metallic nanoparticles (gold and silver) on the surface and membrane of fungal mycelium (Singh et al. 2013). Mukherjee et al. used the fungus *Verticillium* to study the formation of intracellular AgNPs. The authors discovered that exposing fungal biomass to water Ag + ions cause intracellular metal ion reduction and the creation of 25 12 nm AgNPs. Electron microscopy analyses of fungal cells revealed that AgNPs were formed underneath the cell wall surface, likely as a result of metal ion reduction by enzymes buried in the cell wall membrane (Mukherjee et al. 2001). The majority of fungi that are used for extracellular biomass independent production of AgNPs are harmful to plants and/or humans. This makes processing and removal of the biomass a significant impediment to the process's commercialisation. As a result, an unique strategy of assessing a non-pathogenic fungus for the effective production and capping of nanoparticles is required (Thakkar et al. 2010). Gold nanoparticles were created utilizing a variety of fungal species, including *Aspergillus oryzae, Fusarium oxysporum, Verticillium luteoalbum Collitotrichum sp, and Trichothecium sp.*

4.7.2 Metal-Oxide Nanomaterials

Aspergillus flavus TFR7 fungal spores were employed to make TiO_2 nanoparticles, and *Aspergillus terreus* was used to make zinc oxide nanoparticles *Fusarium oxysporum* fungal spores were employed to create unique tertiary oxide structures of barium titanate and bismuth oxide. Various fungal cells, including *Fusarium oxysporum* are used to effectively produce silica, titanium, magnetite, cadmium selenide and zirconia nanoparticles, as well as cadmium sulphide (Jeevanandam et al. 2016). In 2013, it was reported for the first time that large cerium oxide nanoparticles were produced using the halophilic Humicola sp. (Khan and Ahmad 2013). When the fungus *Humicola sp.* is exposed to aqueous solutions of the oxide precursor cerium (III) nitrate hexahydrate ($CeN_3O_96H_2O$), it produces extracellular CeO_2 nanoparticles containing Ce(III) as well as Ce(IV). *Lactobacillus sp.* and *S. cerevisiae* were used to synthesize Titanium oxide nanoparticles. TiO_2 nanomaterials were also produced using A. flavus using a simple aqueous reduction method and characterized using FTIR, XRD, SEM, and AFM (Rajakumar et al. 2012).

4.8 Virus-Mediated Synthesis of Nanomaterials

Producing inorganic nanoparticles with the necessary arrays along the length is challenging for bacteria and fungus. However, viruses do not face this issue. The use of viruses for synthesizing nanomaterials is a very novel technique and so far and in particular two viruses—Tobacco Mosaic Virus and M13 Bacteriophage are majorly used (Das et al. 2017). Tobacco mosaic virus (TMV) is the most prominent example because of its favourable characteristics, such as its high aspect ratio, limited size distribution, different biochemical capabilities shown on the surface, and tolerance with a variety of chemical conjugations. These characteristics are also amenable to genetic engineering, which opens the door to the synthesis of various metallic and non-metallic nanomaterials with wide-ranging uses (Lee et al. 2021).

Tobacco mosaic virus (TMV) was genetically modified by researchers to surface display a peptide with effective metal ion binding and reducing capabilities, and the results showed that, when combined to 3 mM potassium tetrachloroaurate, this construct could result in the development of discrete 10 and 40 nm gold nanoparticles, which had previously been impossible with wild type TMV. Different analytical physicochemical methods confirmed the crystalline and stable nature of these nanoparticles (Love et al. 2015). The silica based nanoparticles have a core of thick silica and are encircled by a radial array of TMV pieces that are each 50 nm in length, making them less than 150 nm in size. The success of the crystals indicates that a similar technique may be used to manufacture a broad variety of inorganic oxides, semiconductors, and metal-based nanoparticles with interiors. Scientists have developed silica nanoparticles coated with tobacco mosaic virus (TMV). As a delivery mechanism, the nanoparticles' porous shape allows for a high medical payload capacity, while TMV acts as a biocompatible covering to improve cell contacts. The generated TMV nanohybrid nanoparticles resemble little wool balls and exhibit improved cell absorption. Drug delivery, contrast agent imaging, and immunotherapy are just a few of the medicinal uses that might be developed for the nanoparticles (Marín-Caba et al. 2019). TMV was used to provide a template for the synthesis of CdS nanowires, which were then deposited in the virus's inner center channel. The diameter of TMV/CdS was measured to be 4.0 nm, confirming the accuracy of the previous measurements (Yang et al. 2021).

4.9 Developmental Challenges and Future Prospects

In the past decade, researchers have worked with a plethora of microorganisms to develop nanostructures with a focus on improving their biocompatibility for biomedical and agricultural applications and a multitude of nanomaterials have been developed and studied so far. However, despite rapid development, the biogenic synthesis approach presents some gaps and limitations in successful nanomaterial production. It can be observed that the majority of these research works have only been carried out at a laboratory scale as this methodology presents certain challenges for industrial scale-ups. The primary issue with microbe-mediated synthesis is the selection of suitable microbial strains with the right biochemical attributes required for nanomaterial development. Moreover, utilization of toxic microbial strain might lead to products with adverse effects. The synthesis process itself requires highly controlled conditions so that the microorganisms can grow and carry out enzymatic reactions to gain a substantial yield (Ovais et al. 2018). Additionally, several mechanisms involved in biosynthesis are still unknown and the details of the pathways in these methodologies are important to ensure reproducibility of the production process at an industrial scale. By understanding the intricacies behind the synthesis of these nanostructures, scientists will also be able to gain a better perspective on modulating the shape, size, morphology and mono-dispersity of the final product which are essential parameters that influence the therapeutic potency of nanomaterials. Furthermore, though biogenic nanomaterials are proposed and proven to be biocompatible to a certain extent, their interactions with living systems and potential toxicities need to be assessed to determine their applications in food, agriculture and medicine (Vecchio et al. 2012; Chen 2018). Finally, difficulties related to drug release mechanisms, pharmacokinetics, bio-distribution, stability and development of non-aggregating nanoparticles are yet to be addressed in detail (Gahlawat and Roy Choudhury 2019). Thus, future investigations are a necessity for biogenic nanomaterials to move on to translational and clinical research and be practically utilized on an industrial scale. With modern interventions such as in silico techniques, computational design, and artificial intelligence, discovery of more effective biogenic nanomaterials can be made easier. Hence, if these developmental challenges are overcome, biogenic nanomaterials have the potential to make revolutionary contributions in the future.

4.10 Conclusion

Biogenic nanomaterial synthesis methods are highly sought after due to the ever-increasing need for novel biomaterials with extensive applicability in biomedical, pharmaceutical, food as well as agriculture industries. Over the course of the last decade, significant progress has been achieved in the development and production of nanomaterials fabricated using microorganisms. This green chemistry strategy provides an eco-friendly and highly sustainable alternative for conventional nanomaterial synthesis methodologies. Furthermore, the nanomaterials produced using these methods exhibit high biocompatibility and safety due to the absence of hazardous chemicals and utilization of naturally occurring microbes whose abundance allows the process to be cost-effective as well. In this chapter, we discuss the diverse range of microbial populations that can be utilized in nanomaterial synthesis of metallic, metal-oxide and non-metallic nanomaterials. The microorganisms include various strains of bacteria, actinomycetes, yeast, algae, fungi and viruses. We observed that a significant number of investigations utilizing this approach employed bacteria and yeast mediated synthesis methodologies and produced mainly metallic and metal-oxide nanomaterials. Additionally, a brief overview of the mechanisms (intracellular and extracellular modes) that enable microbes to act as functional "bio-nano-factories" and produce nanostructures was also provided. Finally, we also highlighted the developmental challenges that these methods pose which range from the selection of correct microbial strains, optimal growth environments, enzymatic reactions and synthesis mechanisms to modulating the shape, size, morphology and mono-dispersity of the final nanostructures. We conclude that microbial nanomaterials have tremendous potential and with extensive fine-tuned investigations using modern *in-silico* tools, they can play an integral role in the future of modern bionanotechnology (Table 4.1).

Table 4.1 Biogenic synthesis of nanomaterials by diverse microorganisms

Sr. No	Origin	Nano—materials	Shape	Size (nm)	Mechanism of synthesis	References
	Bacteria					
1	Bacillus clausii	Ag	Spherical	30–80	Intracellular	Mukherjee et al. (2018)
2	Bacillus megaterium	ZnO	Rod, Cube	45–95	Extracellular	Yusof et al. (2019)
3	Bacillus mycoides	TiO_2	Spherical	40–60	Extracellular	Jeevanandam et al. (2016)
4	Bacillus subtilis	Ag	Spherical, Triangular	5–60	Extracellular	Saifuddin et al. (2009)
5	Desertifilum sp. EAZ03	ZnO	Rod	88	Extracellular	Ebadi et al. (2022)
6	Escherichia coli	Ag	Spherical	10–100	Extracellular	Iravani (2014)
7	Escherichia coli	Au	Triangular, Hexagonal, Rod	10–50	Intracellular	Deplanche and Macaskie (2008)
8	Lactobacillus	Au	Hexagonal, Triangular, Crystalline	20–50, >100	Extracellular, Intracellular	Nair and Pradeep (2002)
9	Lactobacillus	Ag-Au alloys	Crystalline, Clusters	100–300	Extracellular, Intracellular	Nair and Pradeep (2002)
10	Lactobacillus plantarum	ZnO	Spherical	7–19	Intracellular	Yusof et al. (2019)
11	Limnothrix sp.	Ag	Spherical, Elongated	23–27	Extracellular	Hamida et al. (2020)
12	Magnetococcus marinus	Fe_3O_4	Crystalline	20–40	Intracellular	Valverde-Tercedor et al. (2015)
13	Magnetospirillum magnetotacticum	Fe_3O_4	Cubo-octahedrons	47	Intracellular	Jeevanandam et al. (2016)
14	Nostoc carneum	Ag	Spherical	7–27	Extracellular	Hamida et al. (2020)

(continued)

Table 4.1 (continued)

Sr. No	Origin	Nano—materials	Shape	Size (nm)	Mechanism of synthesis	References
15	Nostoc commune	Ag	Spherical	15–45	Extracellular	Hamida et al. (2020)
16	Phormidium tenue	CdS	Spherical	5	Extracellular	Hamida et al. (2020)
17	Proteus vulgaris	Fe_2O_3	Spherical	20–30	Extracellular	Jeevanandam et al. (2016)
18	Rhodopseudomonas capsulata	Au	Spherical, Nanoplate	10–20	Extracellular	Iravani (2014)
19	Spirulina sp.	Ag	Spherical	11–15	Extracellular	Hamida et al. (2020)
20	Thermoanaerobacter sp. BKH1	SiO_2	Spherical	10–20	Extracellular	Show et al. (2015)
	Actinomycetes					
21	Actinobacter sp.	Si/SiO_2 nanocoposites	Quasi-sphered	10	Extracellular	Singh et al. (2008)
22	Actinomycete extract from soil	CuO	Spherical	60	Extracellular	Waris et al. (2021)
23	Nocardiopsis sp. GRG1	ZnO	Sheets	>500	Extracellular	Rajivgandhi et al. (2018)
24	Oxalis corniculata	CuO	Spherical	80	Extracellular	Waris et al. (2021)
25	Rhodococcus sp.	Ag	Spherical	10–15	Intracellular	Gahlawat and Choudhury (2019)
26	Rhodococcus sp.	Au	Spherical	8–12	Intracellular	Ovais et al. (2018)
27	Streptacidiphilus durhamensis	Ag	Spherical	8–48	Extracellular	Gahlawat and Choudhury (2019)
28	Streptomyces capillispiralis	Cu	Spherical	3–59	Extracelluler	Gahlawat and Choudhury (2019)

(continued)

Table 4.1 (continued)

Sr. No	Origin	Nano—materials	Shape	Size (nm)	Mechanism of synthesis	References
29	Streptomyces sp. LK3	Ag	Spherical	5	Extracellular	Gahlawat and Choudhury (2019)
30	Thermanospora	Au	Hexagonal	50	Extracellular	Ovais et al. (2018)
	Yeast					
31	Beer yeast	Carbon dots	Spherical	< 5	–	Gao et al. (2009)
32	Candida glabrata	CdS	Crystalline	2	Intracellular	Hulkoti and Taranath (2014)
33	Candida albicans	Ag	Spherical	10–20	Extracellular	Gahlawat and Choudhury (2019)
34	Geotrichum candidum	Ag	–	2–20	Extracellular	Roychoudhury (2020)
35	Pichia fermentans JA2	ZnO	Smooth, Elongated	–	Extracellular	Yusof et al. (2019)
36	Pichia jadinii	Au	Spherical, Hexagonal, Triangular	<100	Intracellular	Hulkoti and Taranath (2014)
37	Pichia kudriavzevi	ZnO	Hexagonal	10–61	Extracellular	Yusof et al. (2019)
38	Rhodosporidium diobovatum	PbS	Spherical	2–5	Intracellular	Seshadri et al. (2011)
39	Rhodoturula glutinis	Ag	Spherical	15–35	Extracellular	Gahlawat and Choudhury (2019)
40	Rhodotorula mucilaginos	Cu_2O	Spherical	52–111	Extracellular	Seabra and Durán (2015)
41	Saccharomyces cerevisiae	Ag	Spherical	2–20	Extracellular	Gahlawat and Choudhury (2019)

(continued)

Table 4.1 (continued)

Sr. No	Origin	Nano—materials	Shape	Size (nm)	Mechanism of synthesis	References
42	Saccharomyces cerevisiae	$BaCO_3$	Crystalline	5	Intracellular	Chang et al. (2021)
43	Saccharomyces cerevisiae	CdTe	Cubic	2–4	Extracellular	Bao et al. (2010)
47	Saccharomyces cerevisiae	MnO_2	Hexagonal, Spherical	34	Extracellular	Salunke et al. (2015)
44	Saccharomyces cerevisiae	Sb_2O_3	Face-centered cubic	2–10	Intracellular	Jha et al. (2009)
45	Saccharomyces cerevisiae	Si	Spherical	25	Extracellular	Karande et al. (2021)
46	Saccharomyces cerevisiae	TiO_2	Spherical	<12	Extracellular	Roychoudhury (2020)
48	Schizosaccharomyces pombe	CdS	Crystalline	1–2	Intracellular	Hulkoti and Taranath (2014)
49	Toluropsis sp.	PbS	Cubic, Hexagonal	4–8	Intracellular	Kowshik et al. (2002)
50	Trichosporon jirovecii	CdS	Spherical	6–15	Intracellular	Hulkoti and Taranath (2014)
51	Yarrowia lipolytica	Au	Hexagonal, Crystalline	15	Intracellular	Roychoudhury (2020)
	Algae					
52	Turbinaria conoides	Ag	Spherical	96	Extracellular	Rajeshkumar et al. (2012)
53	L. japonica	Au	Sphere-like	15–20	Extracellular	Ghodake and Lee (2011)

(continued)

Table 4.1 (continued)

Sr. No	Origin	Nano—materials	Shape	Size (nm)	Mechanism of synthesis	References
54	Lemanea fluviatilis	Au	Face-centered cubic	5.9	Extracellular	Subbiah et al. (2017)
55	Rhizoclonium fontinale	Au	Spherical, nanotriangular, nanohexagonal, rod-shaped	5–88	Intracellular	Khanna et al. (2019)
56	S. platensis	Ag	Sphere	2–8	Extracellular	Mukherjee et al. (2002)
57	Oscillato riawillei	Ag	Sphere	10–25	Intracellular	Mukherjee et al. (2002)
58	Plectonema boryanum	Ag	Octahedral	200	Intracellular	Mukherjee et al. (2002)
59	Cylindrospermum stagn	Ag	Pentagonal	38–88	Extracellular	Mukherjee et al. (2002)
60	Sargassum bovinum	Pd	Octahedral	5–10	Extracellular	Momeni and Nabipour (2015)
60	S. muticum	Zn	Hexagonal	35–57	Extracellular	Azizi et al. (2014)
61	Sargassum myriocystum	ZnO	Spherical, triangular, rod-shaped, hexagonal, rectangular	18–42	Extracellular	Khanna et al. (2019)
62	Sargassum wightii	ZrO_2	Tetragonal	4.8	–	Kumaresan et al. (2018)
63	Sargassum muticum	Fe_3O_4	Cubic	14–18	Extracellular	Mahdavi et al. (2013)
64	Bifurcaria bifurcata	CuO	Spherical	5–45	Extracellular	Khanna et al. (2019)
65	Sargassum plagiophyllum	AgCl	Spherical	18–42	Extracellular	Dhas et al. (2014)

(continued)

Table 4.1 (continued)

Sr. No	Origin	Nano—materials	Shape	Size (nm)	Mechanism of synthesis	References
66	Scenedesmus-24	CdS	Oval	120–175	Intracellular	(Jena et al., 2014)
67	Cladophora vagabunda	Fluorescent Carbon NPs	–	42.78	Extracellular	Calangian et al. (2018)
68	Fresh water algal bloom	Carbon dots	Spherical	8	Extracellular	Ramanan et al. (2016)
	Fungi					
69	Verticillium	Ag	Hexagonal	12–25	Intracellular	Mukherjee et al. (2001)
70	Agaricus bisporus	Ag	Spherical	2–5	Extracellular	Loshchinina et al. (2018)
71	Amylomyces rouxii	Au	Spherical	12–17	Extracellular	Sarkar et al. (2012)
72	Arthroderma fulvum	Au	Spherical	15	Extracellular	Xue et al. (2016)
73	Aspergillus versicolor	Cu	Spherical	23–82	Extracellular	Ammar et al. (2019)
74	Candida albicans	Au	Spherical	20–80	Extracellular	(Rahimi et al., 2016)
75	Emericella nidulans	Au	Spherical	10–20	Extracellular	Silva et al. (2017)
76	Neurospora crassa	Pt	Spherical	4–35	Intracellular	Castro-Longoria et al. (2012)
77	Trichosporon jirovecii	CdS	Spherical	6–15	Extracellular	El-Baz et al. (2016)
78	Agaricus arvensis	Se	Spherical	150–550	Extracellular	Loshchinina et al. (2018)
79	Macrophomina phaseolina	Au	Spherical	5–40	Extracellular	Chowdhury et al. (2014)
	Virus					

(continued)

Table 4.1 (continued)

Sr. No	Origin	Nano—materials	Shape	Size (nm)	Mechanism of synthesis	References
80	Tobacco mosaic virus	Si	Spherical	150		Marín-Caba et al. (2019)
81	Tobacco mosaic virus	Ag	Spherical	10–40		Love et al. (2015)
82	Tobacco mosaic virus	CdS	Spherical	4		Yang et al. (2021)

References

Abid N, Khan AM, Shujait S, Chaudhary K, Ikram M, Imran M, Haider J, Khan M, Khan Q, Maqbool M (2021) Synthesis of nanomaterials using various top-down and bottom-up approaches, influencing factors, advantages, and disadvantages: a review. Adv Colloid Interface Sci 102597

Ammar HA, Rabie GH, Mohamed E (2019) Novel fabrication of gelatin-encapsulated copper nanoparticles using Aspergillus versicolor and their application in controlling of rotting plant pathogens. Bioprocess Biosyst Eng 42(12):1947–1961. https://doi.org/10.1007/s00449-019-02188-5

Azizi S, Ahmad MB, Namvar F, Mohamad R (2014) Green biosynthesis and characterization of zinc oxide nanoparticles using brown marine macroalga Sargassum muticum aqueous extract. Mater Lett 116:275–277. https://doi.org/10.1016/j.matlet.2013.11.038

Bao H, Hao N, Yang Y, Zhao D (2010) Biosynthesis of biocompatible cadmium telluride quantum dots using yeast cells. Nano Res 3(7):481–489

Bérdy J (2005) Bioactive microbial metabolites. J Antibiot 58(1), Article 1. https://doi.org/10.1038/ja.2005.1

Calangian MTF, Ildefonzo AB, Manzano VKS, Agcaoili GJT, Ganado RJJ, Yago ACC, & Vasquez RD, Franco Jr CF (2018) Facile synthesis of biologically derived fluorescent carbon nanoparticles (Fcnps) from an abundant marine alga and its biological activities. Oriental J Chem 34(2):791–799. https://doi.org/10.13005/ojc/340224

Castro-Longoria E, Moreno-Velasquez SD, Vilchis-Nestor AR, Avalos-Borja EA, Borja MA (2012) Production of platinum nanoparticles and nanoaggregates using neurospora crassa. 22(7):1000–1004. https://doi.org/10.4014/jmb.1110.10085

Chang Y, Chen S, Liu T, Liu P, Guo Y, Yang L, Ma X (2021) Yeast cell route: a green and facile strategy for biosynthesis of carbonate nanoparticles. Cryst Eng Commun 23(26):4674–4679

Chaudhary R, Nawaz K, Khan AK, Hano C, Abbasi BH, Anjum S (2020) An overview of the algae-mediated biosynthesis of nanoparticles and their biomedical applications. Biomolecules 10(11):1498. https://doi.org/10.3390/biom10111498

Chen H (2018) Metal based nanoparticles in agricultural system: Behavior, transport, and interaction with plants. Chem Speciat Bioavailab 30(1):123–134

Choi Y, Lee SY (2020) Biosynthesis of inorganic nanomaterials using microbial cells and bacteriophages. Nat Rev Chem 4(12):638–656

Chowdhury S, Basu A, Kundu S (2014) Green synthesis of protein capped silver nanoparticles from phytopathogenic fungus Macrophomina phaseolina (Tassi) Goid with antimicrobial properties against multidrug-resistant bacteria. Nanoscale Res Lett 9(1):365. https://doi.org/10.1186/1556-276X-9-365

Chugh D, Viswamalya VS, Das B (2021) Green synthesis of silver nanoparticles with algae and the importance of capping agents in the process. J Genet Eng Biotechnol 19(1):126. https://doi.org/10.1186/s43141-021-00228-w

Das RK, Pachapur VL, Lonappan L, Naghdi M, Pulicharla R, Maiti S, Cledon M, Dalila LMA, Sarma SJ, Brar SK (2017) Biological synthesis of metallic nanoparticles: Plants, animals and microbial aspects. Nanotechnol Environ Eng 2(1):18. https://doi.org/10.1007/s41204-017-0029-4

Das VL, Thomas R, Varghese RT, Soniya EV, Mathew J, Radhakrishnan EK (2014) Extracellular synthesis of silver nanoparticles by the Bacillus strain CS 11 isolated from industrialized area. 3 Biotech, 4(2):121–126. https://doi.org/10.1007/s13205-013-0130-8

Deplanche K, Macaskie LE (2008) Biorecovery of gold by Escherichia coli and Desulfovibrio desulfuricans. Biotechnol Bioeng 99(5):1055–1064

Dhas TS, Kumar VG, Karthick V, Angel KJ, Govindaraju K (2014) Facile synthesis of silver chloride nanoparticles using marine alga and its antibacterial efficacy. Spectrochim Acta Part A, Mol Biomol Spectrosc 120:416–420. https://doi.org/10.1016/j.saa.2013.10.044

Ebadi M, Zolfaghari MR, Aghaei SS, Zargar M, Noghabi KA (2022) Desertifilum sp. EAZ03 cell extract as a novel natural source for the biosynthesis of zinc oxide nanoparticles and antibacterial, anticancer and antibiofilm characteristics of synthesized zinc oxide nanoparticles. J Appl Microbiol 132(1):221–236.

El-Baz AF, Sorour NM, Shetaia YM (2016) Trichosporon jirovecii-mediated synthesis of cadmium sulfide nanoparticles. J Basic Microbiol 56(5):520–530. https://doi.org/10.1002/jobm.201500275

Gahlawat G, Choudhury AR (2019) A review on the biosynthesis of metal and metal salt nanoparticles by microbes. RSC Adv 9(23):12944–12967

Gahlawat G, Roy Choudhury A (2019) A review on the biosynthesis of metal and metal salt nanoparticles by microbes. RSC Adv 9(23):12944–12967. https://doi.org/10.1039/C8RA10483B

Gao Z, Zhao C-X, Li Y-Y, Yang Y-L (2019) Beer yeast-derived fluorescent carbon dots for photoinduced bactericidal functions and multicolor imaging of bacteria. Appl Microbiol Biotechnol 103(11):4585–4593

Ghodake G, Lee DS (2011) Biological synthesis of gold nanoparticles using the aqueous extract of the brown algae laminaria Japonica. J Nanoelectron Optoelectron 6(3):268–271. https://doi.org/10.1166/jno.2011.1166

Guilger-Casagrande M, Lima RD (2019) Synthesis of silver nanoparticles mediated by fungi: a review. Front Bioeng Biotechnol 7. https://doi.org/10.3389/fbioe.2019.00287

Hamida RS, Ali MA, Redhwan AMO, Bin-Meferij MM (2020) Cyanobacteria–a promising platform in green nanotechnology: a review on nanoparticles fabrication and their prospective applications. Int J Nanomed 15:6033–6066

Hulkoti NI, Taranath TC (2014) Biosynthesis of nanoparticles using microbes—a review. Colloids Surf, B 121:474–483

Husain S, Sardar M, Fatma T (2015) Screening of cyanobacterial extracts for synthesis of silver nanoparticles. World J Microbiol Biotechnol 31(8):1279–1283. https://doi.org/10.1007/s11274-015-1869-3

Ijaz I, Gilani E, Nazir A, Bukhari A (2020) Detail review on chemical, physical and green synthesis, classification, characterizations and applications of nanoparticles. Green Chem Lett Rev 13(3):223–245. https://doi.org/10.1080/17518253.2020.1802517

Silva LPC, Oliveira JP, Keijok WJ, Silva ARD, Aguiar AR, Guimarães MCC, Ferraz CM, Araújo JV, Tobias FL, Braga FR (2017) Extracellular biosynthesis of silver nanoparticles using the cell-free filtrate of nematophagous fungus *Duddingtoniaflagrans*. Int J Nanomed 12:6373–6381. https://doi.org/10.2147/IJN.S137703

Iravani S (2014) Bacteria in nanoparticle synthesis: current status and future prospects. Int Sch Res Not 2014:1–18

Jeevanandam J, Chan YS, Danquah MK (2016) Biosynthesis of metal and metal oxide nanoparticles. ChemBioEng Rev 3(2):55–67

Jena J, Pradhan N, Aishvarya V, Nayak RR, Dash BP, Sukla LB, Panda PK, Mishra BK (2014) Biological sequestration and retention of cadmium as cds nanoparticles by the microalga scenedesmus-24. J Appl Phycol 27(6):2251–2260. https://doi.org/10.1007/s10811-014-0499-8

Jha AK, Prasad K, Prasad K (2009) A green low-cost biosynthesis of Sb_2O_3 nanoparticles. Biochem Eng J 43(3):303–306

John MS, Nagoth JA, Zannotti M, Giovannetti R, Mancini A, Ramasamy KP, Miceli C, Pucciarelli S (2021) Biogenic synthesis of copper nanoparticles using bacterial strains isolated from an antarctic consortium associated to a psychrophilic marine ciliate: characterization and potential application as antimicrobial agents. Mar Drugs 19(5):263

Karande SD, Jadhav SA, Garud HB, Kalantre VA, Burungale SH, Patil PS (2021) Green and sustainable synthesis of silica nanoparticles. Nanotechnol Environ Eng 6(2):29

Kathiresan K, Manivannan S, Nabeel MA, Dhivya B (2009) Studies on silver nanoparticles synthesized by a marine fungus, Penicillium fellutanum isolated from coastal mangrove sediment. Colloids Surf, B 71(1):133–137

Khan SA, Ahmad A (2013) Fungus mediated synthesis of biomedically important cerium oxide nanoparticles. Mater Res Bull 48(10):4134–4138. https://doi.org/10.1016/j.materresbull.2013.06.038

Khanna P, Kaur A, Goyal D (2019) Algae-based metallic nanoparticles: Synthesis, characterization and applications. J Microbiol Methods 163:105656. https://doi.org/10.1016/j.mimet.2019.105656

Koul B, Poonia AK, Yadav D, Jin J-O (2021) Microbe-mediated biosynthesis of nanoparticles: applications and future prospects. Biomolecules 11(6):886. https://doi.org/10.3390/biom11060886

Kowshik M, Vogel W, Urban J, Kulkarni SK, Paknikar KM (2002) Microbial Synthesis of Semiconductor PbS Nanocrystallites. 14(11):815–818

Kumaresan M, Vijai Anand K, Govindaraju K, Tamilselvan S, Ganesh Kumar V (2018) Seaweed sargassum wightii mediated preparation of zirconia (zro_2) nanoparticles and their antibacterial activity against gram positive and gram negative bacteria. Microb Pathog 124:311–315. https://doi.org/10.1016/j.micpath.2018.08.060

Kumari S, Tehri N, Gahlaut A, Hooda V (2020) Actinomycetes mediated synthesis, characterization, and applications of metallic nanoparticles. Inorg Nano Met Chem 1–10

Lahiri D, Nag M, Sheikh HI, Sarkar T, Edinur HA, Pati S, Ray RR (2021b) Microbiologically-synthesized nanoparticles and their role in silencing the biofilm signaling cascade. Front Microbiol 12:636588

Lahiri D, Nag M, Sheikh HI, Sarkar T, Edinur HA, Pati S, Ray RR (2021) Microbiologically-Synthesized nanoparticles and their role in silencing the biofilm signaling cascade. Front Microbiology 12. https://doi.org/10.3389/fmicb.2021.636588

Lee KZ, Basnayake Pussepitiyalage V, Lee Y-H, Loesch-Fries LS, Harris MT, Hemmati S, Solomon KV (2021) Engineering tobacco mosaic virus and its virus-like-particles for synthesis of biotemplated nanomaterials. Biotechnol J 16(4):2000311. https://doi.org/10.1002/biot.202000311

Li C, Zhou L, Yang H, Lv R, Tian P, Li X, Zhang Y, Chen Z, Lin F (2017) Self-assembled exopolysaccharide nanoparticles for bioremediation and green synthesis of noble metal nanoparticles. ACS Appl Mater Interfaces 9(27):22808–22818. https://doi.org/10.1021/acsami.7b02908

Li X, Xu H, Chen Z-S, Chen G (2011) Biosynthesis of nanoparticles by microorganisms and their applications. J Nanomater

Loshchinina EA, Vetchinkina EP, Kupryashina MA, Kursky VF, Nikitina VE (2018) Nanoparticles synthesis by Agaricus soil basidiomycetes. J Biosci Bioeng 126(1):44–52. https://doi.org/10.1016/j.jbiosc.2018.02.002

Love A, Makarov V, Sinitsyna O, Shaw J, Yaminsky I, Kalinina N, Taliansky M (2015) A genetically modified tobacco mosaic virus that can produce gold nanoparticles from a metal salt precursor. Front Plant Sci 6. https://www.frontiersin.org/articles/10.3389/fpls.2015.00984

Mahdavi M, Namvar F, Ahmad M, Mohamad R (2013) Green biosynthesis and characterization of magnetic iron oxide (Fe_3O_4) nanoparticles using seaweed (Sargassum muticum) aqueous extract. Molecules 18(5):5954–5964. https://doi.org/10.3390/molecules18055954

Maneerung T, Tokura S, Rujiravanit R (2008) Impregnation of silver nanoparticles into bacterial cellulose for antimicrobial wound dressing. Carbohyd Polym 72(1):43–51. https://doi.org/10.1016/j.carbpol.2007.07.025

Manimaran M, Kannabiran K (2017) Actinomycetes-mediated biogenic synthesis of metal and metal oxide nanoparticles: progress and challenges. Lett Appl Microbiol 64(6):401–408

Marín-Caba L, Chariou PL, Pesquera C, Correa-Duarte MA, Steinmetz NF (2019) Tobacco mosaic virus-functionalized mesoporous silica nanoparticles, a wool-ball-like nanostructure for drug delivery. Langmuir 35(1):203–211. https://doi.org/10.1021/acs.langmuir.8b03337

Menon S, Rajeshkumar S, Kumar V (2017) A review on biogenic synthesis of gold nanoparticles, characterization, and its applications. Resour-Effic Technol 3(4):516–527

Mohanpuria P, Rana NK, Yadav SK (2008) Biosynthesis of nanoparticles: technological concepts and future applications. J Nanopart Res 10(3):507–517. https://doi.org/10.1007/s11051-007-9275-x

Mohd Yusof H, Mohamad R, Zaidan UH, Abdul Rahman NA (2019) Microbial synthesis of zinc oxide nanoparticles and their potential application as an antimicrobial agent and a feed supplement in animal industry: a review. J Anim Sci Biotechnol 10(1):57

Momeni S, Nabipour I (2015) A simple green synthesis of palladium nanoparticles with sargassum alga and their electrocatalytic activities towards hydrogen peroxide. Appl Biochem Biotechnol 176(7):1937–1949. https://doi.org/10.1007/s12010-015-1690-3

Mughal B, Zaidi SZJ, Zhang X, Hassan SU (2021) Biogenic nanoparticles: synthesis, characterisation and applications. Appl Sci 11(6):2598

Mukherjee P, Ahmad A, Mandal D, Senapati S, Sainkar SR, Khan MI, Parishcha R, Ajaykumar PV, Alam M, Kumar R, Sastry M (2001) Fungus-mediated synthesis of silver nanoparticles and their immobilization in the mycelial matrix: a novel biological approach to nanoparticle synthesis. Nano Lett 1(10):515–519. https://doi.org/10.1021/nl0155274

Mukherjee P, Senapati S, Mandal D, Ahmad A, Khan MI, Kumar R, Sastry M (2002) Extracellular synthesis of gold nanoparticles by the fungus fusarium oxysporum. ChemBioChem 3(5):461–463. https://doi.org/10.1002/1439-7633(20020503)3:5%3c461::AID-CBIC461%3e3.0.CO;2-X

Mukherjee K, Gupta R, Kumar G, Kumari S, Biswas S, Padmanabhan P (2018) Synthesis of silver nanoparticles by Bacillus clausii and computational profiling of nitrate reductase enzyme involved in production. J Genet Eng Biotechnol 16(2):527–536

Nair B, Pradeep T (2002) Coalescence of nanoclusters and formation of submicron crystallites assisted by *Lactobacillus* Strains. Cryst Growth Des 2(4):293–298

Ovais M, Khalil AT, Ayaz M, Ahmad I, Nethi SK, Mukherjee S (2018a) Biosynthesis of metal nanoparticles via microbial enzymes: a mechanistic approach. Int J Mol Sci 19(12):4100. https://doi.org/10.3390/ijms19124100

Pandit C, Roy A, Ghotekar S, Khusro A, Islam MN, Emran TB, Lam SE, Khandaker MU, Bradley DA (2022) Biological agents for synthesis of nanoparticles and their applications. J King Saud Univ-Sci 34(3):101869. https://doi.org/10.1016/j.jksus.2022.101869

Pathak J, Rajneesh Ahmed H, Singh DK, Pandey A, Singh SP, Sinha RP (2019) Recent developments in green synthesis of metal nanoparticles utilizing cyanobacterial cell factories. In: Nanomaterials in plants, algae and microorganisms. Elsevier, pp 237–265

Priyadharshini RI, Prasannaraj G, Geetha N, Venkatachalam P (2014) Microwave-mediated extracellular synthesis of metallic silver and zinc oxide nanoparticles using macro-algae (Gracilaria edulis) extracts and its anticancer activity against human PC3 cell lines. Appl Biochem Biotechnol 174(8):2777–2790. https://doi.org/10.1007/s12010-014-1225-3

Pugazhendhi A, Prabakar D, Jacob JM, Karuppusamy I, Saratale RG (2018) Synthesis and characterization of silver nanoparticles using Gelidium amansii and its antimicrobial property against various pathogenic bacteria. Microb Pathog 114:41–45. https://doi.org/10.1016/j.micpath.2017.11.013

Rahimi H-R, Doostmohammadi M (2019) Nanoparticle synthesis, applications, and toxicity. In: Applications of nanobiotechnology. IntechOpen. https://doi.org/10.5772/intechopen.87973

Rahimi G, Alizadeh F, Khodavandi A (2016) Mycosynthesis of silver nanoparticles from candida albicans and its antibacterial activity against escherichia coli and staphylococcus aureus. Trop J Pharm Res 15(2), Article 2. https://doi.org/10.4314/tjpr.v15i2.21

Rajakumar G, Rahuman AA, Roopan SM, Khanna VG, Elango G, Kamaraj C, Zahir AA, Velayutham K (2012) Fungus-mediated biosynthesis and characterization of TiO_2 nanoparticles and their activity against pathogenic bacteria. Spectrochim Acta Part A, Mol Biomol Spectrosc 91:23–29. https://doi.org/10.1016/j.saa.2012.01.011

Rajeshkumar S, Kannan C, Annadurai G (2012) Green synthesis of silver nanoparticles using marine brown algae turbinaria conoides and its antibacterial activity. 9

Rajivgandhi G, Maruthupandy M, Muneeswaran T, Anand M, Manoharan N (2018) Antibiofilm activity of zinc oxide nanosheets (ZnO NSs) using Nocardiopsis sp. GRG1 (KT235640) against MDR strains of gram negative Proteus mirabilis and Escherichia coli. Process Biochem 67:8–18

Rajivgandhi G, Mythili Gnanamangai B, Heela Prabha T, Poornima S, Maruthupandy M, Alharbi NS, Kadaikunnan S, Li W-J (2022) Biosynthesized zinc oxide nanoparticles (ZnO NPs) using actinomycetes enhance the anti-bacterial efficacy against K. Pneumoniae. J King Saud Univ-Sci 34(1):101731

Ramanan V, Thiyagarajan SK, Raji K, Suresh R, Sekar R, Ramamurthy P (2016) Outright green synthesis of fluorescent carbon dots from eutrophic algal blooms for in vitro imaging. ACS Sustain Chem Eng 4(9):4724–4731. https://doi.org/10.1021/acssuschemeng.6b00935

Ranjha MMAN, Shafique B, Rehman A, Mehmood A, Ali A, Zahra SM, Roobab U, Singh A, Ibrahim SA, Siddiqui SA (2022) Biocompatible nanomaterials in food science, technology, and nutrient drug delivery: recent developments and applications. Front Nutr 8:778155. https://doi.org/10.3389/fnut.2021.778155

Rao MD, Pennathur G (2017) Green synthesis and characterization of cadmium sulphide nanoparticles from Chlamydomonas reinhardtii and their application as photocatalysts. Mater Res Bull 85:64–73. https://doi.org/10.1016/j.materresbull.2016.08.049

Rao PVS, Mantri VA, Ganesan K (2007) Mineral composition of edible seaweed Porphyra vietnamensis. Food Chem 102(1):215–218. https://doi.org/10.1016/j.foodchem.2006.05.009

Roychoudhury A (2020) Yeast-mediated green synthesis of nanoparticles for biological applications. Indian J Pharm Biol Res (IJPBR) 8(3):26–31

Saifuddin N, Wong CW, Yasumira AAN (2009) Rapid biosynthesis of silver nanoparticles using culture supernatant of bacteria with microwave irradiation. E-J Chem 6(1):61–70

Salunke BK, Sawant SS, Lee S-I, Kim BS (2015) Comparative study of MnO_2 nanoparticle synthesis by marine bacterium Saccharophagus degradans and yeast Saccharomyces cerevisiae. Appl Microbiol Biotechnol 99(13):5419–5427

Sarkar J, Ray S, Chattopadhyay D, Laskar A, Acharya K (2012) Mycogenesis of gold nanoparticles using a phytopathogen Alternaria alternata. Bioprocess Biosyst Eng 35(4):637–643. https://doi.org/10.1007/s00449-011-0646-4

Seabra A, Durán N (2015) Nanotoxicology of metal oxide nanoparticles. Metals 5(2):934–975

Seshadri S, Saranya K, Kowshik M (2011) Green synthesis of lead sulfide nanoparticles by the lead resistant marine yeast, Rhodosporidium diobovatum. Biotechnol Prog 27(5):1464–1469

Shah M, Fawcett D, Sharma S, Tripathy SK, Poinern GEJ (2015) Green synthesis of metallic nanoparticles via biological entities. Materials 8(11):7278–7308. https://doi.org/10.3390/ma8115377

Show S, Tamang A, Chowdhury T, Mandal D, Chattopadhyay B (2015) Bacterial (BKH1) assisted silica nanoparticles from silica rich substrates: a facile and green approach for biotechnological applications. Colloids Surf, B 126:245–250

Sidhu AK, Verma N, Kaushal P (2022) Role of biogenic capping agents in the synthesis of metallic nanoparticles and evaluation of their therapeutic potential. Front Nanotechnology 3. https://doi.org/10.3389/fnano.2021.801620

Singh S, Bhatta UM, Satyam PV, Dhawan A, Sastry M, Prasad BLV (2008) Bacterial synthesis of silicon/silica nanocomposites. J Mater Chem 18(22):2601

Singh H, Du J, Singh P, Yi TH (2018) Extracellular synthesis of silver nanoparticles by Pseudomonas sp. THG-LS1.4 and their antimicrobial application. J Pharm Anal 8(4):258–264. https://doi.org/10.1016/j.jpha.2018.04.004

Singh D, Rathod V, Ninganagouda S, Herimath J, Kulkarni P (2013) Biosynthesis of silver nanoparticle by endophytic fungi Pencillium sp. Isolated from Curcuma longa (turmeric) and its antibacterial activity against pathogenic gram negative bacteria. J Pharmacy Res 7(5):448–453. https://doi.org/10.1016/j.jopr.2013.06.0035

Sirelkhatim A, Mahmud S, Seeni A, Kaus NHM, Ann LC, Bakhori SKM, Hasan H, Mohamad D (2015) Review on zinc oxide nanoparticles: antibacterial activity and toxicity mechanism. Nano-Micro Lett 7(3):219–242. https://doi.org/10.1007/s40820-015-0040-x

Subbiah M, Bhuvaneswari S, Vajiravelu S (2017) Green synthesis, characterization of silver nanoparticles of a marine red alga spyridia fusiformis and their antibacterial activity. Int J Pharm Pharm Sci 9:192. https://doi.org/10.22159/ijpps.2017v9i5.17105

Thakkar KN, Mhatre SS, Parikh RY (2010) Biological synthesis of metallic nanoparticles. Nanomed: Nanotechnol Biol Med 6(2):257–262. https://doi.org/10.1016/j.nano.2009.07.002

Tsekhmistrenko SI, Bityutskyy VS, Tsekhmistrenko OS, Horalskyi LP, Tymoshok NO, Spivak MY (2020) Bacterial synthesis of nanoparticles: a green approach. Biosyst Divers 28(1):9–17

Vahabi K, Mansoori GA, Karimi Dorcheh S (2011) Biosynthesis of silver nanoparticles by fungus trichoderma. Reesei

Valverde-Tercedor C, Montalbán-López M, Perez-Gonzalez T, Sanchez-Quesada MS, Prozorov T, Pineda-Molina E, Fernandez-Vivas MA, Rodriguez-Navarro AB, Trubitsyn D, Bazylinski DA, Jimenez-Lopez C (2015) Size control of in vitro synthesized magnetite crystals by the MamC protein of Magnetococcus marinus strain MC-1. Appl Microbiol Biotechnol 99(12):5109–5121

Vecchio G, Galeone A, Brunetti V, Maiorano G, Rizzello L, Sabella S, Cingolani R, Pompa PP (2012) Mutagenic effects of gold nanoparticles induce aberrant phenotypes in Drosophila melanogaster. Nanomedicine: Nanotechnol, Biol Med 8(1):1–7

Waris A, Din M, Ali A, Ali M, Afridi S, Baset A, Ullah Khan A (2021) A comprehensive review of green synthesis of copper oxide nanoparticles and their diverse biomedical applications. Inorg Chem Commun 123:108369

Xue B, He D, Gao S, Wang D, Yokoyama K, Wang L (2016) Biosynthesis of silver nanoparticles by the fungus Arthroderma fulvum and its antifungal activity against genera of Candida, Aspergillus and Fusarium. Int J Nanomed 11:1899–1906. https://doi.org/10.2147/IJN.S98339

Yang Y, Yang K, Wang J, Cui D, Zhao M (2021) Fabrication and characterization of CdS nanowires templated in tobacco mosaic virus with improved photocatalytic ability. Appl Microbiol Biotechnol 105(21):8255–8264. https://doi.org/10.1007/s00253-021-11596-1

Zhang X, Yan S, Tyagi RD, Surampalli RY (2011) Synthesis of nanoparticles by microorganisms and their application in enhancing microbiological reaction rates. Chemosphere 82(4):489–494. https://doi.org/10.1016/j.chemosphere.2010.10.023

Zou L, Zhu F, Long Z, Huang Y (2021) Bacterial extracellular electron transfer: a powerful route to the green biosynthesis of inorganic nanomaterials for multifunctional applications. J Nanobiotechnology 19(1):1–27

Chapter 5
Environmental Pollutants Remediation Using Phyto-Nanoparticles: An Overview on Synthesis, Characterization, and Remediation Potential

Roopal Pal, Lakhan Kumar, Shaubhik Anand, and Navneeta Bharadvaja

Abstract Accumulation of environmental pollution has created a misbalance between the environment and human health. Industrialization, incomplete fuel combustion, use of fertilizers, unsafe disposal of pollutants directly into the atmosphere contributes to their buildup in the environment. Progress in the development of remediation techniques has led the way to employ green synthesized nanoparticles. Phyto-nanoparticles are an improvement in the field of bioremediation in eco-friendly, non-toxic, and cost-effective ways. Nanoparticles adhered to phytocompounds are explored for their potential to remediate pollutants like heavy metals, dyes, pharmaceutical residues, polycyclic aromatic hydrocarbons, biocides, etc. Nanoparticle's size, morphology, and properties are optimized by regulating factors like pH, temperature, light exposure, agitation, etc. For example, Iron nanoparticles synthesized from plants like tea are used to biodegrade heavy metals like chromium, arsenic in contaminated water. This chapter aims to recapitulate the process of bioremediation via emerging Phyto-nanoparticles, their synthesis from easily available plants, techniques used to characterize synthesized nanoparticles, and to investigate remediation potential to act on environmental pollutants and control environmental pollutants matrices in the atmosphere.

Keywords Phyto-nanoparticles · Heavy metals · Dyes · Pharmaceutical products · Polycyclic aromatic hydrocarbons

R. Pal · L. Kumar · S. Anand · N. Bharadvaja (✉)
Plant Biotechnology Laboratory, Department of Biotechnology, Delhi Technological University, Delhi, India
e-mail: navneetab@dce.ac.in

M. P. Shah et al. (eds.), *Biogenic Nanomaterials for Environmental Sustainability: Principles, Practices, and Opportunities*, Environmental Science and Engineering,
https://doi.org/10.1007/978-3-031-45956-6_5

5.1 Introduction

Environmental pollutants are affecting human health at an alarming rate throughout the world. Ever-increasing anthropogenic activities and widespread industrial implementation, increasing world population have escalated the content of environmental pollutants (Rasheed et al. 2019). Researchers are constantly conducting studies on health risks associated with various pollutants like heavy metals, dyes, biocides like pesticides and insecticides, pharmaceutical products, polycyclic aromatic hydrocarbons. Every type of environmental pollution i.e., water, air, and pollution pose a great risk to the human body as well as the environment. India has accounted for 1·67 million deaths in the year 2019, 0.98 million deaths were due to particulate pollution while 0.61 M deaths were due to household air pollution (Balakrishna et al. 2017). Mostly, pollutants enter water bodies, soil, and air via careless product disposal and not treating the main source of environmental pollutants through conventional treatment plants (Norfazilah Wan Ismail and Umairah Mokhtar 2020). Release of pollutants like heavy metal even in minute quantity leaves damaging outcome in environment affecting all living beings (Chugh et al. 2022). Past few years, researchers have experimented with several techniques for efficient removal of environmental pollutants from water, air, and soil due to challenges faced with conventional technologies. Bioremediation is the efficient and controlled process to biodegrade toxic chemical environmental pollutants from water, air, and soil (Stroo 2015; Bourquin and Pedersen 2015). Nanotechnology is a key emerging technique with its unique properties for the remediation of environmental pollutions. Nanoparticles are synthesized from three methods i.e., physical, chemical, and biological. Physically and chemically synthesized nanoparticles are not cost-effective and due to the involvement of chemicals, they are toxic and non-eco-friendly (Kumar et al. 2021). Therefore, Biological methods, involving plants as the key source for nanoparticles are a better alternative for the biodegradation of pollutants.

Phytoremediation is an eco-friendly method of pollution remediation with the inclusion of nanotechnology; due to the smaller size of nanoparticles (1–100 nm), magnetic properties, and ease in scale-up make it easier for adsorption of pollutants from contaminated source (Kumar et al. 2021). Phyto-nanoremediation, which is a known combination of two techniques nanotechnology and phytoremediation, has proven to be efficient to degrade and remove environmental pollutants (Srivastav et al. 2019). In Phyto-nanoremediation, the toxicity of pollutants is altered with the interaction of nanoparticle (Roberto et al. 2020). The incorporation of green strategies like the use of plant capping agents and stabilizers to replace chemicals has become a vital step. Several sustainable ways are assessed for pollutant remediation from different sources like water, soil, and air. For example, the zero-valent iron nanoparticle is commonly used for remediation of a wide range of environmental pollutants like polycyclic aromatic hydrocarbons, biocides, heavy metals, etc. (Grieger et al. 2015). Fe_3O_4 iron oxide nanoparticles due to their magnetic properties have been used to assess their efficiency to remove heavy metals pollutants from contaminated soil and water (Vázquez-Núñez et al. 2020). Considering the vast annual industrial effluent dye emission, researchers have made numerous studies for managing dye pollutants. For example, green synthesized zinc oxide nanoparticles employed to degrade methylene blue dye (Ishwarya et al. 2018), photocatalytic degradation of azo dye by zirconium oxide nanoparticles (Sai Saraswathi and Santhakumar 2017), remediation of rhodamine B dye with the help of zero-valent iron nanoparticles (Khan and Al-Thabaiti 2018), and silver nanoparticles synthesized from *Morinda tinctoria* leaf extract used to degrade methylene blue dye shows 95% efficiency in 72 h contact time (Vanaja et al. 2014). Studies helped surface several challenges associated with Phyto-nanoremediation like these studies are conducted at a smaller scale, to overcome this challenge, experimenting at a larger scale is required (Srivastav et al. 2019). In this chapter, we have presented knowledge on plant-mediated synthesis of nanomaterials, methods available for their characterization and potential against several environmental pollutants including heavy metals, dyes, PAHs, biocides, and pharmaceutical products etc.

5.2 Phyto-Nanoparticle Synthesis

The bottom-up approach of nanoparticle synthesis from plant parts like roots, leaves, stems, flowers, fruit, peel, etc. is the greener method (Figure-1). Phytocompounds like flavonoids, polyphenolic compounds help in nanoparticle synthesis from metal salts (Husen and Siddiqi 2014). The method based on plant extract has recently gained importance because of non-toxicity, ease of availability of plants (Noruzi 2015).Nanoparticles green synthesis depends upon three conditions such as an ecofriendly solvent, a natural reducing agent, and a nonhazardous stabilizing agent (Singh et al. 2018). Initially, plant parts like leaves, fruit, peel, etc. are drenched under running tap water to remove any dirt particles and then later sterilized by rinsing with double distilled water. Dry cleaned and sterilized plant parts under direct sunlight or at room temperature (Jadoun et al. 2021) to macerate into fine powder. Powdered plant parts are heated in Milli-Q water as per the requirement with constant stirring. The heated solution is filtered using filter paper, the filtrate (plant extract) could be stored in a flask at 4 °C for a week (Ramadhan et al. 2019). Studies have found that for the extraction of certain phytocompounds like polyphenol, a specific time is necessary (Wang et al. 2019). The reaction mixture is prepared by mixing plant extract and metal salt in the desired ratio. Factors like pH, the different concentration ratios of the plant to metal salt, temperature, light, agitation play crucial role in determining the properties and morphology of synthesized nanoparticles (Noruzi 2015). Nanoparticle green synthesis from various plant sources and their properties are mentioned in Table 5.1 and Fig. 5.1.

Table 5.1 Various nanoparticles synthesized from different plants, plant & metal salt ratio, reaction temperature & agitation, size, and shape

Nanoparticles (size/shape)	Plant material	Plant extract: Metal salt ratio	Reaction: temp/agitation	Morphology size & shape	References
Tin oxide (SnO_2) nanoparticles	*Vitex agnus-castus* fruits extract	$SnCl_2$ (80 ml, 0.01 M)	60° C (6 h)/ 35,000 rpm	8 nm diameter, spherical	Ebrahimian et al. (2020)
Zero Valent Silver (Ag-NPs) nanoparticle	*Ficus Benjamina* leaves extract	Leaf's extract: $AgNO_3$ (1 Mm) 2: 25	25° C/stirred for 40 min	dendritic structure	Al-Qahtani (2017)
Zinc oxide (ZnO) nanoparticles	*Sphagneticola trilobata*	Leaf extract: Zinc acetate dehydrates (0.1 M) 25 ml: 1.3 ml	60° C/6000 rpm	65-80 nm, irregular & complex	Shaik et al. (2020)
Zero valent Iron nanoparticle	*Eucalyptus globulus* leaf extract	Leaf extract : $FeSO_4 7H_2O$ (0.1 M) 1:1	37° C/stirred continuously	NP in range of 50-80 nm, spherical shaped	Madhavi et al. (2013)
Zero-valent Iron nanoparticles	*Rosa damascene* (RD), *Thymus vulgaris* (TV), *Urtica dioica* (UD)	Leaf extract: $FeCl_2.4H_2O$ (0.1 M) 2.3	At room temp/mixing	100 nm, irregularly shaped	Fazlzadeh et al. (2017)
Silver nanoparticles (Ag-NP)	*Araucaria heterophylla* *Azadirachta indica* *Prosopis chilensis*	Plant gum extract: AgNO3(3 mM) 1:10	37° C/10,000 rpm for 5 min	AH–<30 nm AI–<35 nm PC–50 nm	Samrot et al. (2019)
Iron nanoparticle (Fe-NP)	*Eucalyptus* leaf extract	Eucalyptus leaf extract: FeSO4 (0.05 M) (1:2)	80° C for 2 h	size range between20 to 80 nm	Weng et al. (2016)

(continued)

Table 5.1 (continued)

Nanoparticles (size/shape)	Plant material	Plant extract: Metal salt ratio	Reaction: temp/agitation	Morphology size & shape	References
Cupric oxide nanoparticles (CuO-NPs)	*Calotropis procera*	Plant latex: $CuSO_4.5H_2O$ (0.2 M)	37 °C, 10,000 rpm for 10 min	15–20 nm	Dubey and Sharma (2017)
Iron nanoparticles (Fe-NPs)	*Nephrolepis auriculata*	Plant extract: $FeCl_3$(0.1 M) (2:1)	37 °C, Continuous stirring for 10 min	40–70 nm, spheroidal	Yi et al. (2019)

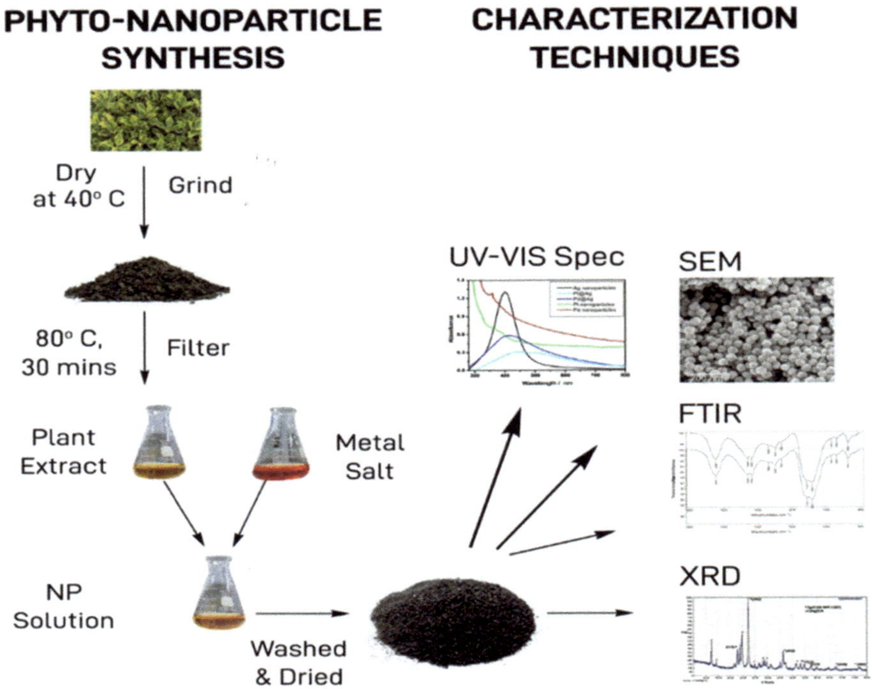

Fig. 5.1 Green synthesis process from plants and characterization of Phyto-nanoparticles

5.3 Phyto-Nanoparticle Characterization Methods

Characterization is a crucial step after nanoparticle synthesis for creating a better understanding of their properties like shape, structure, structural components, etc., and full potential. The basic requirement of characterization is the separation and cleaning of nanoparticles. Amongst many separations and washing methods, centrifugation is most suitable. Advancement in technology has widened the alternatives of characterization based on the target of analysis. Techniques like UV–Vis's spectrophotometer, FTIR, SEM, TEM, XRD, and AFM are used to characterize the particles sized as small as 10^{-9} nm. UV–Vis's spectrophotometer is a technique based on optical properties, measures the specified surface plasmon resonance of nanoparticles in the 300 nm–800 nm range (Rana et al. 2020). The presence of nanoparticles is confirmed with color change in the plant extract in additionto metal salt and UV–Vis measures the absorption peak of color change. Green synthesis of AuNP from leaf extract of *Saraca asoca* observed color change to ruby red and absorption peak at 532 nm confirming their presence (Patra et al. 2018). Fourier transform infrared (FT-IR) spectroscopy is another spectroscopy technique to give a chemical analysis of nanoparticles based on bond vibration and rotation of specific functional groups in nanoparticles. FTIR helps with identifying changes in structure upon the interaction of phytocompounds with metal ions (Faghihzadeh et al. 2016). Comparative study of FTIR scan of plant extract with nanoparticle can help with identifying capping and reducing agents involved in nanoparticle synthesis. Most common phytocompounds like flavonoids, tannins, terpenoids, etc. have been found to serve as reducing agents with absorption peaks in the range of 1,000–1,800 cm^{-1}(Noruzi 2015). FTIR of gold nanoparticles synthesized from *Mangifera indica* showed 1,624, 1,737, 1,444 cm^{-1} absorption bands for C = C, C = O, C–N bonds resp (Philip 2010).

Microscopy-based techniques like SEM and TEM determines morphology like the one shape and size structure of nanoparticles whereas TEM has an edge over SEM for identifying structural difference like crystalline to amorphous. In SEM, an electron beam is passed through the sample scans the surface ultimately providing the 3D structure of the particles (Rana et al. 2020). In TEM, an electron beam is transmitted through the sample at certain angles and is diffracted at other angles. Deflected electrons are blocked with the help of an aperture to let unscattered electrons pass through it and create a contract image also known as a Light field image. Dark-field images are obtained with deflected electrons (Titus et al. 2019). For example, TEM images of AuNP synthesized from leaf extract of *Chamaecostus Cuspidatus* displayed a spherical shape with 50 nm in size whereas SEM images of the same nanoparticles showed various shapes (Philip 2010). Characterization techniques for nanoparticle identification are mentioned below in Table 5.2.

5.4 Remediation Potential of Phyto-Nanoparticles

The environmental hazard crisis has taken a grip globally for decades. Management of environmental pollutants becomes the utmost priority for researchers. Reach of environmental pollutants like heavy metals, dyes, biocides, pharmaceutical products into the ground surface and main water sources are inevitable. Phyto-nanoremediation is intensely explored and recommended for managing the remediation of environmental pollutants (Govindappa et al. 2018). Phyto-nanoremediation utilizes nanoparticles synthesized from plants as the main resource to convert pollutants to a desirable state for biodegradation and then degrades pollutants to less toxic form (Ahmad et al. 2017). Based on structural properties like the high surface area to mass ratio of Phyto-nanoparticles, contaminants are adsorbed, oxidized, or co-precipitated on the treatment of contaminated soil, industrial wastewater, tannery wastewater (Kumar et al. 2022) and drinking water sources (Rana et al. 2020). Sorption is a key process in nanoremediation, adsorption, and absorption. During adsorption, the pollutant interacts with sorbent at surface level; next, the pollutant permeates sorbent at deeper layers to form a solution (Vázquez-Núñez et al. 2020). Along with adsorption, nanoparticles are capable of photocatalytic, electrostatic, and active surface interactions for acting on pollutants (Pandey 2018). In a study conducted by Ahmed Rather, Using *Bergenia ciliata* extract as reducing and capping agent for green synthesis of zinc nanoparticles in order to evaluate the remediation potential for textile dyes like methylene blue from water resources. Zinc Nanoparticles could degrade 82% of MB

Table 5.2 Various characterization techniques used for nanoparticles

Nanoparticles	Plant source	Metal salt: plant extract	Reaction (agitation/temp)	Observation (colour change) Initial: Final	Characterization method & result	References
Iron oxide NP	*Mentha pulegium L*	$Fecl_3$:leaf extract 20 ml:200 ml 1:10	Continuous stir at 75 °C for 1 h	Light brown to Dark black	UV–Vis spectra; (Abs. peak) 275-301 nm FT-IR spectrum: (Wavenumber range) 510 and 594 cm^{-1} SEM: cubic & irregular shaped	Bouafia and Laouini (2020)
Iron oxide NP	*Carica papaya*	$FeCl_3.6H_2O$ (50 mL of 0.1 M): leaves extract 1:1	Magnetic stir for 30 min	Pale yellow to intense black	FT-IR: 4000–400 cm^{-1} EDX: C-25% O-31% Fe-32% XRD: (crystallite size) 4.58 nm TGA: 25.41% (weight loss at 24.38 °C)	Hossen et al. (2020)

(continued)

Table 5.2 (continued)

Nanoparticles	Plant source	Metal salt: plant extract	Reaction (agitation/temp)	Observation (colour change) Initial: Final	Characterization method & result	References
Iron oxide NP	*Laurus nobilis L*	0.1 M FeCl3 $0.6H_2O$: leaves extract 30 ml/30 ml (1:1)	Continuous stirring at room temperature	Dark precipitate formation	UV–Vis: 285 nm FT-IR: 600, 450 cm^{-1} EDS: Fe-69% O-22% TEM: 8.03 ± 8.99 nm (average particle size)	Jamzad and Kamari 2020)
Iron oxide NP	*Borassus flabellifer*	Ferric chloride (0.2 M) and Ferrous sulphate (0.1 M) in 2:1 ratio:seed coat extract(25ML) (1:1)	Stirring at room temp	Brown to dark precipitate formation	XRD: (crystalline size) 35 nm UV–Vis: 350 nm TGA: 0 °C–700 °C SEM: hexagonal shaped iron	Sandhya and Kalaiselvam (2020)
Iron oxide NP	*Hibiscus rosa-sinensis*	Ferric and ferrous chloride: Extract (1:1)	200 rpm/room temp	Instant dark precipitate	UV–Vis: 229 nm FITR: 567.94 cm^{-1} suggesting Fe–O vibration of Fe_3O_4 NP SEM: 65 nm	Razack et al. (2020)

(continued)

Table 5.2 (continued)

Nanoparticles	Plant source	Metal salt: plant extract	Reaction (agitation/temp)	Observation (colour change) Initial: Final	Characterization method & result	References
Gold nanoparticles (Au-NPs)	*Capsicum annumL.* leaf extract	gold (III) chloride trihydrate ($HAuCl_4.3H_2O$): leaf extract	Room temp/ Shaking	Yellow to red	UV–Vis: 335.26 nm FTIR: 3324–3329, 2100 and 1636 cm^{-1} shift suggesting -OH, -CN and C-O groups SEM: rectangular, square shaped NP X-Ray Diffraction: (peak) 111°and 311°	Baran et al. (2020)
Silver nanoparticle (AuNP)	*Calophyllum tomentosum* Leaf extract	Leaf extract: $AgNO_3$ (5 mM) (2:1)	Room temp/ normal stirring	Pale yellow to brown	UV–Vis: 428 nm FTIR: 3401 cm^{-1} suggesting O–H stretch 1041 cm^{-1} suggesting Flavanones presence SEM: spherical	Govindappa et al. (2018)

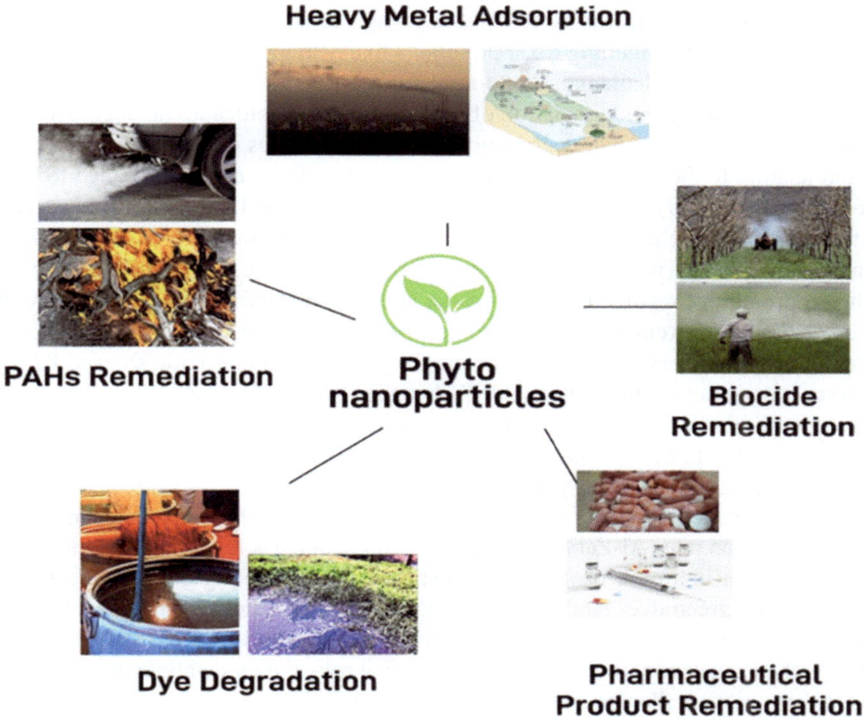

Fig. 5.2 Applications of phyto-nanoparticles in remediation of environmental pollutants like heavy metals, dyes, PAHs, Pharmaceutical products, and biocides

dye in the span of 52 min under the influence of sunlight for photocatalytic degradation (Ahmed Rather et al. 2021). When in fact, green synthesized Fe-NP from *Citrus paradisi* extract degraded 80% of MB in 6 h under room temperature (Kumar et al. 2020).

Figure 5.2 illustrates potential applications of Phyto-nanoparticles in abatement of several pollutants including heavy metals, dyes, biocides, pharmaceutical products, and PAHs etc.

5.4.1 Remediation Potential of Phyto-Nanoparticles for Dyes

With increasing industrialization and improper disposal of industrial waste like synthetic dyes as well as effluent water containing high amounts of dyes, health hazards are alarmingly escalating. Despite the environmental and health hazard, dyes are constantly in demand for food, plastic, pharmaceutical industries, and especially textile industries. In textile factories, after use, approximately 15% of the dye is left and discharged into nearby water bodies (Singh et al. 2018). Mismanagement of effluent treatment containing toxic pollutants like heavy metals in an open environment is a real-time challenge. Phyto-nanoremediation has been found eco-friendly and efficient for the biodegradation of industrial dyes. Photocatalytic degradation is a key phenomenon for dye degradation; irradiation excites electrons to conduction band leading to the generation of electron–hole pair. Dyes are biodegraded to non-toxic forms like CO_2, H_2O by the action of newly generated hydroxyl radical serving as strong oxidizing agents (Marimuthu et al. 2020; Rostami-Vartooni et al. 2016). In studies like Ag/TiO_2 acted on azo dyes for photocatalytic, degradation of dye has shown productive results. Light absorbing ability and stability of silver doped with TiO_2 helps keep excited electrons on the surface to prevent recombination of electron–hole pair (Seery et al. 2007). Doping Ag with Titanium oxide showed 82.3% photocatalytic degradation of MB dye (Whang et al. 2009). Incorporation of nanoparticles in phytoremediation has increased remediation efficiency such as Methylene blue removal with *Lemna minor* showed 80% removal efficiency (Imron et al. 2019) whereas silver nanoparticle synthesized from honey showed 92% removal efficiency in 72 h,under the influence of acidic pH (pH 2) exposure of reactive sites increases the degradation rate (Al-Zaban et al. 2021). Iron nanoparticles synthesized from tea with high concentration of polyphenols show approx. 190 mg/g removal capacity of malachite green dye. High reduction potential of Epicatechin and Catechin like polyphenols rich tea reduces iron ions effectively (Das and Eun 2018). The size of the synthesized nanoparticles also plays a crucial role in their catalytic performance, as the size gets smaller the performance get better (Akbari et al. 2020).

Zeta potential analysis of Fe-NP biosynthesized from tea extracts for degradation of cationic dyes shows interaction between negative charge on Fe-NP and positive charge on the surface of dye like rhodamine B, malachite green. Adsorption process initiates with adsorption of dye on surface reactive sites of Fe-NP. H_2O in the vicinity reacts with Fe^0 thus loses electrons taken up by H^+ and creates reducibility by producing active hydrogen. Ultimately, benzene rings of dye molecules are interrupted by accepting electrons from active hydrogen (Xiao et al. 2020).

Nanoparticles have been found exposed in the environment be it for treating environmental pollutants or for any other use. In the environment, nanoparticles can serve as a precursor for other particles to bind with and affect atmospheric chemistry, the state of microorganisms in the environment, and pollutants (Balakrishna et al. 2017). Several studies conducted on photocatalytic degradation of dyes are mentioned in Table 5.3.

5.4.2 *Remediation Potential of Phyto-Nanoparticles for Heavy Metals*

Amongst several environmental pollutants, Heavy metals are second most reported class of toxic pollutants (Frankenberger and Losi 2015). Exposure time and dose of

Table 5.3 Remediation potential of Phyto-nanoparticles for dyes degradation

Phyto-nanoparticle	Plants	Dye	Concentration range (NP: Dye)	Reaction time	Degradation efficiency or degradation rate constant	pH & Temp/ light	References
Fe-NP	*Trigonella foenum-graecum*	Methylene orange	100 ml of 25 mg/L: 30 mg	30 min	95%	UV irradiation	Radini et al. (2018)
Ag-NPs	*Gardenia jasminoides Ellis*	Coomassie brilliant blue	2 ml: 1 ml	4 h	92%	Sunlight irradiation	Saravanakumar et al. (2018)
Fe_3O_4 (polyaniline)Au-NP	*Allium Sp*	Methylene blue Methylene orange	1 mg: 5 mL	1 min 1.5 min	10^3–10^4 X more degradation with NPs		Yu et al. (2019)
ZnO-NP	*Calliandra haematocephala*	Methylene blue	50 mg/L: 20 mg/L	4.5 h	88%	Sunlight irradiation	Vinayagam et al. (2020)
$Ca(OH)^{2-}$NPs	*Androraphis echioides*	Methylene blue	50 mg: 50 ml (10 ppm dye)	90 min	97.5% 96.38%	UV Sunlight irradiation	Mathivanan et al. (2018)
FeO-NP	*Punica granatum*	Aniline blue	0.25 mg/ml: 1 mg/ml	150 min	90%	70 °C pH 10	Kaur et al. (2018)
Au-NP	*Elaeis guineensis*	Methylene blue		60 min	92.55%	visible light irradiation	Ahmad et al. (2020)
Ag-NP		Eosin yellow	1 mg/mL: 0.1 mg/mL	20 min	91.17% (sunlight) 75.41% (UV)	Room temp	Bhavya et al. (2021)
AgNP	*Prosopis farcta*	Methylene blue	10 mg: 20 ml(10 mg/L)	30 min	70.2%	Visible light irradiation	Miri et al. (2018)
ZnO-NP	Oakfruit	Basic violet 3	0.05 g: 50 mL	30 min	79%	Visible light irradiation	Sorbiun et al. (2018)

(continued)

Table 5.3 (continued)

Phyto-nanoparticle	Plants	Dye	Concentration range (NP: Dye)	Reaction time	Degradation efficiency or degradation rate constant	pH & Temp/ light	References
Ag-NPs	Dandelion Extract	Methyl orange, Rhodamine B	10 μL: 10 mL (25 mg/L)	90 secs 40 secs	$0.0393\ s^{-1}$ $0.1038\ s^{-1}$	Room temp	Lai et al. (2019)
SnO_2-NP	*Psidium guajava*	Reactive yellow 186	1 mg: 10 mL (40 ppm)	180 min	90%	Sunlight irradiation	Kumar et al. (2018)
Fe-NP	*L. speciosa*	Allura red, Brilliant blue, Green S	0.05 mg: 5 mL (50 mg/ L): 1 ml (105) H_2O_2	240 min	$0.0131\ min^{-1}$ –$0.0177\ min^{-1}$		Garole et al. (2018)
Ag-NP	*Zingiber officinale*	Methylene blue	0.5 ml Ag-NP: 1 0 μM MB dye	<10 min	99.9%	In the presence of $NaBH_4$	Mehata((2021)
CdS-NP	*Calotropis gigantea*	Methylene blue Eosin yellow	25 mg/L	60 min	85.53% 91.12%	Sunlight irradiation	Ayodhya and Veerabhadram (2017)
CeO_2-NP	*Calotropis procera*	Methylene orange	3.72 gm in 10 ml: 10 ml	50 min	98%	–	Muthuvel et al. (2020)

heavy metals influence their potential of toxicity to human health and environmental risk. Mismanaged discharge of heavy metals like Cadmium, Chromium, Copper, Arsenic, lead, and zinc contaminated industrial effluent into hydrosphere remains one of the key reasons for contamination (Barakat 2011). Anthropogenic activities like the use of fertilizers, fuel combustion, mining, etc. contribute to the heavy metal index (Ahmad et al. 2020). Heavy metal like Cr is found in toxic form like Cr^{6+} in wastewater whereas Cr^{3+} is non-hazardous but is often oxidized into hazardous form like Cr^{6+}. Therefore, adsorption and recovery of heavy metals from water sources is crucial task (Ahmed et al. 2013). Due to the high surface area to volume ratio of nanoparticles, the adsorption of pollutants to their surface helps with their co-precipitation. Along with adsorption, their magnetic property extensively carries out decontamination of pollutants (Bhavya et al. 2021). For example, Fe-NP synthesized from *Eucalyptus* adsorbs pentavalent arsenic through Intradiffusion model i.e., first, adsorption of arsenic on Fe-NP surface followed by deeper penetration until equilibrium is achieved. FeNP showed the adsorbent capacity of arsenic in wastewater by 27.7% (Rufus et al. 2019). Phyto-nanoparticles exhibit benefits over phytoremediation, since phytoremediation is a time-consuming process due to the slow growth of plants, lesser biomass, and the remediation process takes more time (Miri et al. 2018). With the use of nanoparticles in the phytoremediation process, nanoparticles stabilize heavy metals by adsorption and facilitate their easy remediation (Sorbiun et al. 2018). For example, Phytoremediation of arsenic from contaminated soil by *Vetiveria zizanoides* showed 71% removal of Arsenic from soil in 90 days on the other handiron nanoparticles synthesized from *Prangos ferulacea* could remove 93.8% of arsenic in 20 min at 2 g/L concentration on nanoparticles. Adsorption of Arsenic ions by iron nanoparticles was noticed to be maximum around neutral pH but in acidic pH with more protons, adsorption declined due to lesser availability of sorption surface for As (Karimi et al. 2019). The fact that nanoparticles like nZVI are intended to remediate heavy metal pollutants from soil, they may be sediment in soil and reach to groundwater and make drinking water toxic (Roberto et al. 2020). FTIR analysis of Ag-NP synthesized from *Convolvulus arvensis* for heavy metals (Cu^{2+}) adsorption shows involvement of –COOH and –OH functional groups in adsorption of copper ions through ion exchange and formation of coordination bonds between adsorbent and heavy metal ion (Al-Senani and Al-Kadhi 2020). A comparative study done on five different plants (black tea, green tea, eucalyptus, oak tree, pomegranate) for Fe-NP green synthesis for arsenic removal from contaminated water. Experiments revealed *eucalyptus* Fe-NP had highest adsorption efficiency of 39.9 mg/g. EDS study on Fe-NP result suggested a possible connection between the iron content and adsorption capacity. Highest Fe percentage of *Eucalyptus* Fe-NP amongst other nanoparticles showed highest adsorption capacity for arsenic (Kamath et al. 2020). Several studies employed Phyto-nanoparticles for remediation of heavy metals has been listed in Table 5.4.

Table 5.4 Remediation potential of Phyto-nanoparticles for heavy metal removal

Nanoparticles (NP)	Plant source	Heavy metal (HM)	Concentration range (NP: HM)	Reaction time	Removal efficiency or adsorption capacity	pH & temp/light	References
Fe_2O_3-NP		Pb^2 Cu^{2+}	0.4 g/L: 0.2 mg/L(Pb) 0.1 g/L(Cu)	1 min	74% (Pb) 28% (Cu) Adsorption capacity		Rajput et al. (2017)
Fe-Cu-NP	Green tea	Cr(VI)	0.01 g: 25 ml	60 min	94.7% removal efficiency	pH 5: 30 °C	Zhu et al. (2018)
Fe-NP		Arsenic	2 g/L: 0.1 g/L	20 min	93.9% removal efficiency		Karimi et al. (2019)
Zn-NP		Cr(VI)	100 μg/mL: 1 g/L	8 h	81.1% Reduction percentage		Shaik et al. (2020)
Fe-NP	*Pyrus sinkiangensis*	Cr(VI)	0.02 g: mg/L	120 min	99.5%	pH 5.0 55 °C	Rong et al. (2020)
MgO-NP	*Aspergillus niger*	Cr, Co, Pb, Cd, Ni	1.0 mg mL^{-1}: variant	180 min	94.2%, 63.4%, 72.7%, 74.1% 70.8%		Fouda et al. (2021)

(continued)

Table 5.4 (continued)

Nanoparticles (NP)	Plant source	Heavy metal (HM)	Concentration range (NP: HM)	Reaction time	Removal efficiency or adsorption capacity	pH & temp/light	References
FexO/FeSNP	*Vaccinium floribundum Kunth*	Cu^{2+}, Zn^{2+}, Mn^{2+} Ni^{2+}, Cd^{2+}, Cr^{6+}	0.01 g: 5 mg/L, 5 mg/L, 4 mg/L, 2.5 mg/L, 3 mg/L, 1 mg/L	5 min	99.8%, 99.5%, 71.4% 99.6% 99.6% 99%	pH 6	Abril et al. (2018)
Ag-NP	*Ficus benjamina*	Cd(II)	0.05 g: 30 mg/l	40 min	85%	pH 6: 500 rpm	Al-Qahtani (2017)
Fe-S-NP	*Phoenix dactylifera*	Cr^{6+}	25 mg: 50 ml (4.5 mg/l)	120 min	97%	pH 7	Bhattacharjee et al. (2021)
Fe-NP	Green tea	Cr(VI)			90.2%	pH 5	Hao et al. (2021)
$CuFe_2O$-NP	*Simarouba glauca*	Pb(II)	0.1 g: 100 ml (10 mg/L)	150 min	80%	pH 6	Sreekala et al. (2021)
Fe_3O_4-NP	*Moringa oleifera*	Pb(II)	0.2 g/L: 200 mg/L	60 min	93.70%	pH 5 120 rpm	Gautam et al. (2020)
Fe-NP	*yerba mate*	Cr(VI)	0.5 mg: 25 ml (15.6 mg/L)	20 min	80%	pH 3	García et al. (2019)

5.4.3 Remediation Potential of Phyto-Nanoparticles for Pharmaceutical Products

Pharmaceutical residues are emerging pollutants, due to their ability to assimilate in animals and humans as well as their stability to persist in the environment makes them hazardous for human health and the environment. Their connectivity from humans and animals to treatment plants through sewage water elevates their risk of exposure (Yan et al. 2017). The inefficiency of other techniques like chlorination and filtration etc., used for remediating pharmaceutical byproducts, encourages the utilization of cost-effective and eco-friendly processes such as Phyto-nanoremediation (Husein et al. 2019). Open release or exposure of free nanoparticles even for remediation purposes can cause inhalation by workers or farmers and can cause skin exposure. Nanoparticles tend to cross biological barriers and enter the human body resulting in the potential release of ROS hence oxidative stress (Guzmán et al. 2006).

Studies conducted in recent years have made suggestions for remediation of contaminated water resources with Phyto-nanoremediation (Malakootian et al. 2019). For example, Phyto-nanoremediation of Ibuprofen from the water via composite Fe-NP synthesized from black tea had shown 92% of adsorption capacity in half-hour considering polyphenol and caffeine as reducing as well as capping agent in the process of green synthesis (Ali et al. 2016). *Cicer arietinum* used to investigate phytoremediation potential to degrade ciprofloxacin was 60% in 7 days (Shikha 2016) although when employed magnetite nanoparticles synthesized from peel extract of lemon, grape, cucumber for degrading antibiotics showed more than 90% remediation of ampicillin (Stan et al. 2017). Experiments done over the last few years have concluded that the adsorption capacity of iron nanoparticles for antibiotics is not only decided by smaller size and higher surface area of nanoparticles but also by the molecular interaction between the drug particles and nanoparticles serving to be potential for antibiotic removal (Ivashchenko et al. 2015). Some studies attribute the adsorption capacity of nanoparticles for drug particles to pH. Study mentioned in the table below, removal of drugs like Ibuprofen, Naproxen, and Diclofenac by copper nanoparticles green synthesized from *Tilia.* In acidic pH, the electrostatic attraction of copper nanoparticles towards drug particles can be the cause of higher absorption because at a pH value of 4.5 the drug particles exist as cationic form. Other than pH, the adsorbent dose of nanoparticles also provides evident proof for their better adsorption capacity. Higher adsorbent dose provides larger surface area for the absorption of drug particles whereas Lesser nanoparticle concentration provides lesser surface area for adsorption of drug particles on copper nanoparticles therefore by increasing the dose of copper nanoparticles the possibility of adsorption of the drug on copper nanoparticle increases (Husein et al. 2019).Details of remediation of pharmaceuticals residues with the help of photo-nanoparticles are mentioned in the Table 5.5.

Table 5.5 Remediation potential of Phyto-nanoparticles for pharmaceutical productsremoval

Nanoparticles	Plant source	Pharmaceutical product	Concentration range (NP: PP)	Reaction time	Reduction rate	pH & temp	References
GO-Fe-NP	Green tea	Mitoxantrone	0.8 g/L: 10 mg/L	20 min	95%	30 °C: 250 rpm	Wu et al. (2020)
Fe_3O_4-NP	*Euphorbiacochinchinensis*	doxorubicin	0.01 g: 20 mg/L	5 min	80.2%	pH 6 30 °C	Weng et al. (2018)
Fe-NP	*Ceratonia silique*	Amoxicillin	0.4 g: 0.0287 g (250 mL)	200 min	99%	pH 2	Aksu Demirezen et al. (2019)
Fe_3O_4-NP	*Citrus limon*	Piperacillin Tetracycline ERY	50 mg: 5 mL	200 min	>90%	pH 5.5, 40 °C,	Stan et al.0 2017)
Cu-NP	*Tilia*	Ibuprofen, Naproxen Diclofenac	10 mg: 24.8 29.6 32.3 mg/g	30 min	74.4% 86.9% 91.4%	pH 4.5 24 °C	Husein et al. (2019)
Zn-Fe_2O_3-NP	*Monsonia burkeana*	Sulfisoxazole	25 mg	45 min	68%	pH 12	Makofane et al. 2021)
Fe_3O_4-NP	*Excoecaria cochinchinensis*	Rifampicin	10 mg: 84.8 mg/g (20 ml)	60 min	98.4%	30 °C	Cai et al. (2019)
Cu-NP Fe-NP Ag-NP	Green Tea	Aflatoxin B_1	5 µg/mL: 100 ng/ mL	120 min	90% >90% >70%	37 °C 150 rpm	Asghar et al. (2018)
ZnO-NP	*Punica granatum*	Flumequine	10 mg:15 mg/L (50 ml)	18 min	97.6%	pH 7.9	Essawy et al. (2020)
Fe-NP	*Urtica dioica* *Thymus vulgaris*	Cephalexin	0.1 g/L:25 mg/L	20 min	≈100 ≈100	pH 2	Leili et al. (2018)

(continued)

Table 5.5 (continued)

Nanoparticles	Plant source	Pharmaceutical product	Concentration range (NP: PP)	Reaction time	Reduction rate	pH & temp	References
Fe_3O_4-Au-NP	*Carum carvi*	Imatinib Imipenem	0.1 m: 10 ppm	20 min	92% 96%	UV irradiation	Mirsadeghi et al. (2020)
Zero valent Fe-NP	*Camelliasinensis* *Vitis vinifera*	Ibuprofen	14 mmol/L: 2.8 mg/kg	200 h	pH 7	95%	Machado et al. (2013)
Fe-Pd (bentonite coated) NP	*Punica granatum*	Tetracycline	1000 mg/L: 20 mg/L	180 mins	pH 7	80%	Gopal et al. (2020)
NiFe-NP	*Punica granatum*	Tetracycline	300 mg/L: 20 mg/L	90 mins		93%	Ravikumar et al. (2019)
Tripyamid-TiO_2-NP	*Aleovera*	Ciprofloxacin	5 mg: 50 ml	60 min	UV light irradiation	90%	Li et al. (2020)

5.4.4 Remediation Potential of Phyto-Nanoparticles for Polycyclic Aromatic Hydrocarbons (PAHs)

Organic pollutants like polycyclic aromatic hydrocarbons are actively assessed for toxicity and ability to accumulate in the environment hence for their degradation, environmentally friendly technology like Phyto-nanoremediation fulfills the demand via adsorption, degrading, photocatalytic (Rani and Shanker 2018). Incomplete combustion of fuel, petroleum refineries, volcanic eruptions, fuel seepage are some common anthropogenic and natural sources of generation of PAHs. PAHs released directly into the atmosphere and their accumulation in the atmosphere tends to find a way into the soil and human body. Mutagenic and carcinogenic properties of PAHs make them the pollutants of higher risk to human health (Abdel-Shafy and Mansour 2016). Process of bioremediation could get expensive due to involvement of certain techniques (Wilson and Jawson 2015) thus managing pollutants like polycyclic aromatic hydrocarbons with eco-friendly Phyto-nanoparticles technique is suitable for treating soil, water bodies, and sediments (Patel et al. 2020). SiO_2NP synthesized from the root extract of *Erigeron annuus* were used to remove PAHs from the soil, experiment showed promising results by protecting plants grown in PAHs contaminated soils thus indicating remediation potential for PAHs (Zuo et al. 2020). Efficient adsorbing and photocatalytic properties of nanoparticles help with the degradation of organic pollutants like PAHs while optimizing certain factors affecting their activity like pH, catalytic time, temperature, the concentration of pollutant, and photocatalyst (Hassan et al. 2015). Comparative studies on the remediation potential of phytoremediation to Phyto-nanoremediation demonstrate the benefits of using nanoparticles for bioremediation as a better alternative. Phytoremediation of pyrene showed 57% removal efficiency (Jeelani et al. 2017) on the other hand silver nanoparticle synthesized from *Allium sativum* could remediate more than 85% of pyrene under conditions like higher temperature around 65 °C as well as under acidic environment (Abbasi et al. 2014).

Catalytic activity of silver and gold nanoparticles synthesized from *Breynia rhamnoides* for 4-nitrophenol (vehicle emission (Perrone et al. 2014)) shows catalytic rate constant of $\sim 9.2 \times 10^{-3}$ s at minimal initial concentration. As a result of the experiment, increasing the concentration of the plant extract significantly decreased the catalytic efficacy of Ag-NP and AuNP and the catalytic efficacy increases with increasing the dose of Ag-NP and AuNP because of the higher surface area (Gangula et al. 2011). As mentioned in the table below, study conducted on Ag-NP synthesized from *Dodonaea viscosa* extract and $NaBH_4$ to reduce a toxic pollutant like 4-nitrophenol into a less toxic product like 4-aminophenol (Serrà 2020) and 2-nitrophenol into 2-aminophenol. The best result of reduction were obtained from the Ag-NP synthesized in the ratio of 30:25 for extract and metal salt and this particular group of Ag-NP served the most greatest surface area for the adsorption of organic pollutant to Ag-NP (Shah et al. 2021).Details of the remediation potential of Phyto-nanoparticles are compiled in the Table 5.6.

Table 5.6 Remediation potential of various Phyto-nanoparticles forpolycyclic aromatic hydrocarbons PAHs

Nano particles	Plant sources	PAHs	Concentrationrange (NP: PAHs)	Reactiontime	pH/temp/ light	Degradation rate/ efficiency	References
Iron hexacyanoferrates-NP	*Sapindus-mukorossi*	Anthracene, Phenanthrene Chrysene, Fluorene	25 mg: 50 mg L^{-1}	48 hours	pH 7/ Sunlight & UV	80–90% (An, Ph) 70–80% (Ch, Fl)	Shanker et al. (2017)
ZnS-NP	*Ficus johannis*	Naphthalene, Anthracene, Chlorophenol	10 mg: 50 mL	120 min	UV irradiation	≈100% ≈100% >80%	Rajabi et al. (2020)
ZnO/SiO_2NP	*Butea monosperma*	Acenaphthylene	1 g/l: 50 mg/L	4 hours	pH 9 30 °C	73%	Bharati and Suresh (2017)
Cu-NPs Ag-NP	*Azadirachta indica*	Naphthalene	5 mg/l: 7 mg/l	30 mins	150 rpm	98.07% 89.71%	Abbas et al. (2020)
$ZnFe_2O_4$ chitosan coated-NP	*Azadirachta-indica*	Anthracene Phenanthrene	20 mg: 2 mgL^{-1}	12 hours	pH 7/ sunlight	95% 92.2%	Rani and Rachna (2020)
Au-NP	*Artemisia capillaris*	4-nitrophenol	100 µL: 1 mL(0.4 mM)	20 mins	24 °C 18,000 rpm	1.46×10^{-3}	Lim et al. (2016)
Fe_3O_4-Pd-Graphene oxide-NP	*W. coagulans*	4-nitrophenol	5 mg: 25 ml (2.5 Mm)	1 min		0.051 s^{-1} (Kinetic rate constant)	Atarod et al. (2016)
Pd-Graphene oxide-NP	*Berberis vulgaris*	Nitroarenes	6.0 mg: 1.0 mmol	90 mins	50 °C	98%	Nasrollahzadeh et al. (2016)
Au-NP	*Delonix regia*	O-nitroaniline	0.005 g: 3 mL		50 °C8	35.38×10^{-2} Min^{-1} (Reductionrate)	Dauthal and Mukhopadhyay (2016)

(continued)

Table 5.6 (continued)

Nano particles	Plant sources	PAHs	Concentrationrange (NP: PAHs)	Reactiontime	pH/temp/ light	Degradation rate/ efficiency	References
Fe_3O_4-Cu-NP	*Ilybum marianum*	Nitroarene	2.0 mol%: 1.0 mmol		50 °C	95%	Sajadi et al. (2016)
Ag-NP	*Dodonaea viscosa*	4-Nitrophenol 2-Nitrophenol	10 mg: 0.1 mm	18 min		93.1% 96.8%	Shah et al. (2021)

5.4.5 Remediation Potential of Phyto-Nanoparticles for Biocides

Biocides are a group of chemical products, natural or synthetic origin, used to inhibit the growth of unwanted microbes. Accumulation of byproducts of biocides like 2,4-dichlrophenol in soil and water, holds for major environmental pollution (Hasanin et al. 2021). The release of these compounds in open surface water deteriorates the condition of drinking water. Biocides enter the environment through water running through agricultural fields, leaching from building materials (Durak et al. 2021). As the agricultural demands increases globally, fertilizers, pesticides, and herbicides are vigorously used as common practice in agriculture, which leads to defiling of surface soil, surface, and groundwater (Rasheed et al. 2019). Health concerns related to exposure to biocides are as serious as benign or malignant tumors, genetic disorders, and reproductive disorders (Lorenz 2017). Trifluralin, pendimethalin and Monsanto, Imazapyr are a few of the most widely used pesticides and herbicides respectively (Özkara et al. 2016). The intervention of Phyto-nanoremediation technology in remediation of toxic pollutants like pesticides and herbicides assist with low-cost management of environmental pollutant. A remediation experiment conducted on green synthesized Ag/Cu nanoparticles from papaya extract has shown degradation of pesticide like chlorpyrifos into its metabolite 3,5,6-trichloropyridinol (TCP) confirmed by the appearance of characteristic peak of TCP in UV–Vis spectroscopy and LC–MS (Rosbero and Camacho 2017) whereas chemically synthesized iron nanoparticle degraded 95% hexachlorocyclohexane pesticide in 2 days (Elliott et al. 2008). Phytoremediation of *Ricinus communis* for organochlorine Pesticide like chlorpyrifos had remediation potential of approx. 25%in 66 days (Rissato et al. 2015),this inefficiency of remediation could be overcome in a shorter period by utilizing nanoparticles like silver nanoparticle (100 mg) to remediate 3 mg/l chlorpyrifos in 5 h. Concentration ratio of both Ag-NP and Pesticides serves as a factor for effective degradation results (Manimegalai et al. 2014). Mineralization with the help of nanoparticles effectively degrades pesticides while avoiding the drawbacks for other methods used to degrade pesticides like expensiveness and time consumption (Zayed et al. 2001). These approaches based on green nanotechnology utilizing plants to replace the function of chemicals are turning to be promising for eco-friendly degradation of hazardous pollutant. Although participation of nanoparticles does leave behind a concern due to their smaller size, once they are released in an aqueous solution it is challenging to recover dispersed nanoparticles. Immobilizing of nanoparticles on a biological matrix can make their recovery easier than just release them in solutions (Rawtani et al. 2018). Details of Phyto-nanoparticles and their remediation potential for biocides are compiled in the Table 5.7.

Table 5.7 Remediation potential of Phyto-nanoparticles to degrade biocides (pesticides/herbicides)

Nano particle	Plant sources	Biocides (pesticides/ herbicides)	Concentration range (NP:biocide)	Reaction time	Factor affecting (Ph/light /temp)	Degradation rate/removal efficiency	References
Fe-NP	*Camellia sinensis*	Ametryn (Herbicide)	2.5 g/L: 30 μg/L	30 mins	pH 7/20 °C	88%	Ali et al. (2016)
Fe-NP	*Euphorbia cochinchensis*	2,4-Dichlorophenol	1 g/L: 30 mg/L	120 mins	pH 6.26	83.5%	Gan et al. (2018)
Fe-NP	*Eucalyptus*	Ametryn	2.83 mg/L:	135 mins	210 rpm pH 5	100%	Sangami and Manu (2017)
Fe-NP	*Euphorbia cochinchensis*	2,4-dichlorophenol	1 g/L: 50 mg/L	50 mins	pH 6.8 250 rpm 30 °C	64.0%	Guo et al. (2017)
Ag-NP	*Colocasia esculenta*	4-nitrophenol 2-nitrophenol	10 mg:	6 mins 16 mins		97.83% 92.97%	Ismail et al. (2018)
C-NP	*Punica granatum*	Glyphosate Trifluralin 2,4-Dichlorophenoxyacetic Acid	0.5 gs: 50 mg/L	20 mins	pH 9	88% 94% 92%	Yousefi (2017)
Ag-NP	*Camellia japonica*	Nitrobenzene	5 mg/mL: 50 μM		pH 7	R^2–0.990	Karthik et al. (2017)
Co_3O_4-MCM-41	*Prunus persica* Rice husk	Acephate	0.2 g: 50 mg/L	60 mins	pH 10	100%	AbuKhadra et al. (2020)

5.5 Conclusion

Deterioration of human health and quality of lifestyle has been majorly affected by the overwhelming increase in global population, Industrialization demand, fuel combustion, vehicle emission, etc. To revive the compromised status of the environment, development, and employment of transformative technology like Phyto-nanoparticles for bioremediation of environmental pollutants is desired. Phyto-nanoremediation is revolutionary from batch experiment to large scale-up, easy availability of plants, reusability of nanoparticles, natural reducing and stabilizing agents, application in soil, water, air resources, less energy demand to a sustainable approach. Through optimization of key factors associated with nanoparticles and their synthesis, like pH, light exposure, agitation, temperature, alkalinity, the concentration of metal ion precursor and plant, their working efficiency could be enhanced for better interaction and adsorption between nanoparticles and pollutants. Reviewed literature has proved supporting data for synergistic relation between bioremediation and nanotechnology for sustainable treatment of different pollutants. To completely accept phyto-nanoremediation as a stable and secure method to deal with different sorts of pollutants, more studies need to be done to completely understand their potential, working mechanism, consequences, and residual toxicity in water, soil, and air. Finally, Phyto-nanoremediation is a potent emerging technology whose global acceptance can alter the state of environmental pollution.

References

Abbas S, Nasreen S, Haroon A, Ashraf MA (2020) Synhesis of silver and copper nanoparticles from plants and application as adsorbents for naphthalene decontamination. Saudi J Biol Sci 27:1016–1023. https://doi.org/10.1016/j.sjbs.2020.02.011

Abbasi M, Saeed F, Rafique U (2014) Preparation of silver nanoparticles from synthetic and natural sources: remediation model for PAHs. In: IOP Conf Ser Mater Sci Eng., Institute of Physics Publishing, p 012061. https://doi.org/10.1088/1757-899X/60/1/012061.

Abdel-Shafy HI, Mansour MSM (2016) A review on polycyclic aromatic hydrocarbons: Source, environmental impact, effect on human health and remediation. Egypt J Pet 25:107–123. https://doi.org/10.1016/j.ejpe.2015.03.011

Abril M, Ruiz H, Cumbal LH (2018) Biosynthesis of multicomponent nanoparticles with extract of mortiño (vaccinium floribundum kunth) berry: application on heavy metals removal from water and immobilization in soils. J Nanotechnol 2018:10. https://doi.org/10.1155/2018/9504807

AbuKhadra MR, Mohamed AS, El-Sherbeeny AM, Elmeligy MA (2020) Enhanced photocatalytic degradation of acephate pesticide over MCM-41/Co3O4 nanocomposite synthesized from rice husk silica gel and Peach leaves. J Hazard Mater 389:122129. https://doi.org/10.1016/j.jhazmat.2020.122129

Ahmad N, Bhatnagar S, Saxena R, Iqbal D, Ghosh AK, Dutta R (2017) Biosynthesis and characterization of gold nanoparticles: Kinetics, in vitro and in vivo study. Mater Sci Eng C 78:553–564. https://doi.org/10.1016/j.msec.2017.03.282

Ahmad T, Bustam MA, Zulfiqar M, Moniruzzaman M, Idris A, Iqbal J, Asghar HMA, Ullah S (2020) Controllable phytosynthesis of gold nanoparticles and investigation of their size and morphology-dependent photocatalytic activity under visible light. J Photochem Photobiol A Chem 392:112429. https://doi.org/10.1016/j.jphotochem.2020.112429

Ahmed MA, Ali SM, El-Dek SI, Galal A (2013) Magnetite–hematite nanoparticles prepared by green methods for heavy metal ions removal from water. Mater Sci Eng B 178:744–751. https://doi.org/10.1016/J.MSEB.2013.03.011

Ahmed Rather G, Nanda A, Ahmad Pandit M, Yahya S, sofi MA, Barabadi H, Saravanan M (2021) Biosynthesis of Zinc oxide nanoparticles using Bergenia ciliate aqueous extract and evaluation of their photocatalytic and antioxidant potential. Inorg Chem Commun 134:109020. https://doi.org/10.1016/J.INOCHE.2021.109020

Akbari A, Sabouri Z, Hosseini HA, Hashemzadeh A, Khatami M, Darroudi M (2020) Effect of nickel oxide nanoparticles as a photocatalyst in dyes degradation and evaluation of effective parameters in their removal from aqueous environments. Inorg Chem Commun 115:107867. https://doi.org/10.1016/J.INOCHE.2020.107867

Aksu Demirezen D, Yıldız YS, Demirezen Yılmaz D (2019) Amoxicillin degradation using green synthesized iron oxide nanoparticles: Kinetics and mechanism analysis. Environ Nanotechnology, Monit Manag 11:100219. https://doi.org/10.1016/j.enmm.2019.100219

Ali I, Al-Othman ZA, Alwarthan A (2016) Synthesis of composite iron nano adsorbent and removal of ibuprofen drug residue from water. J Mol Liq 219:858–864. https://doi.org/10.1016/j.molliq.2016.04.031

Ali I, AL-Othman ZA, Alwarthan A (2016) Green synthesis of functionalized iron nano particles and molecular liquid phase adsorption of ametryn from water. J Mol Liq 221:1168–1174 (2016). https://doi.org/10.1016/j.molliq.2016.06.089

Al-Qahtani KM (2017) Cadmium removal from aqueous solution by green synthesis zero valent silver nanoparticles with Benjamina leaves extract. Egypt J Aquat Res 43:269–274. https://doi.org/10.1016/j.ejar.2017.10.003

Al-Senani GM, Al-Kadhi N (2020) The synthesis and effect of silver nanoparticles on the adsorption of Cu2+ from aqueous solutions. Appl Sci 10:4840. https://doi.org/10.3390/app10144840

Al-Zaban MI, Mahmoud MA, AlHarbi MA (2021) Catalytic degradation of methylene blue using silver nanoparticles synthesized by honey, Saudi. J Biol Sci 28:2007–2013. https://doi.org/10.1016/j.sjbs.2021.01.003

Asghar MA, Zahir E, Shahid SM, Khan MN, Asghar MA, Iqbal J, Walker G (2018) Iron, copper and silver nanoparticles: Green synthesis using green and black tea leaves extracts and evaluation of antibacterial, antifungal and aflatoxin B1 adsorption activity. LWT—Food Sci Technol 90:98–107. https://doi.org/10.1016/j.lwt.2017.12.009

Atarod M, Nasrollahzadeh M, Mohammad Sajadi S (2016) Green synthesis of Pd/RGO/Fe3O4 nanocomposite using Withania coagulans leaf extract and its application as magnetically separable and reusable catalyst for the reduction of 4-nitrophenol. J Colloid Interface Sci 465:249–258. https://doi.org/10.1016/j.jcis.2015.11.060

Ayodhya D, Veerabhadram G (2017) One-pot green synthesis, characterization, photocatalytic, sensing and antimicrobial studies of Calotropis gigantea leaf extract capped CdS NPs. Mater Sci Eng B 225:33–44. https://doi.org/10.1016/J.MSEB.2017.08.008

Barakat MA (2011) New trends in removing heavy metals from industrial wastewater. Arab J Chem 4:361–377. https://doi.org/10.1016/j.arabjc.2010.07.019

Baran MF, Acay H, Keskin C (2020) Determination of antimicrobial and toxic metal removal activities of plant-based synthesized (Capsicum annuum L. Leaves), ecofriendly, gold nanomaterials. Glob Challenges 4:1900104. https://doi.org/10.1002/gch2.201900104

Bharati R, Suresh S (2017) Biosynthesis of ZnO/SiO 2 nanocatalyst with palash leaves' powder for treatment of petroleum refinery effluent. Resour Technol 3:528–541. https://doi.org/10.1016/j.reffit.2017.08.004

Bhattacharjee S, Habib F, Darwish N, Shanableh A (2021) Iron sulfide nanoparticles prepared using date seed extract: Green synthesis, characterization and potential application for removal of ciprofloxacin and chromium. Powder Technol 380:219–228. https://doi.org/10.1016/j.powtec.2020.11.055

Bhavya G, Belorkar SA, Mythili R, Geetha N, Shetty HS, Udikeri SS, Jogaiah S (2021) Remediation of emerging environmental pollutants: A review based on advances in the uses of eco-friendly biofabricated nanomaterials. Chemosphere 275:129975. https://doi.org/10.1016/j.chemosphere.2021.129975

Bouafia A, Laouini SE (2020) Green synthesis of iron oxide nanoparticles by aqueous leaves extract of Mentha Pulegium L.: Effect of ferric chloride concentration on the type of product. Mater Lett 265:127364. https://doi.org/10.1016/j.matlet.2020.127364

Bourquin AW, Pedersen TA (2015) Bioremediation engineering. Bioremediation Sci Appl:259–272. https://doi.org/10.2136/SSSASPECPUB43.C15

Cai W, Weng X, Chen Z (2019) Highly efficient removal of antibiotic rifampicin from aqueous solution using green synthesis of recyclable nano-Fe3O4. Environ Pollut 247:839–846. https://doi.org/10.1016/j.envpol.2019.01.108

Chugh M, Kumar L, Bhardwaj D, Bharadvaja N (2022) Bioaccumulation and detoxification of heavy metals: an insight into the mechanism. Dev Wastewater Treat Res Process:243–264. https://doi.org/10.1016/B978-0-323-85657-7.00013-4

Das PR, Eun JB (2018) A comparative study of ultra-sonication and agitation extraction techniques on bioactive metabolites of green tea extract. Food Chem 253:22–29. https://doi.org/10.1016/J.FOODCHEM.2018.01.080

Dauthal P, Mukhopadhyay M (2016) Phyto-synthesis and structural characterization of catalytically active gold nanoparticles biosynthesized using Delonix regia leaf extract. 3 Biotech 6:1–9. https://doi.org/10.1007/s13205-016-0432-8

Dubey S, Sharma YC (2017) Calotropis procera mediated one pot green synthesis of Cupric oxide nanoparticles (CuO-NPs) for adsorptive removal of Cr(VI) from aqueous solutions. Appl Organomet Chem 31:1–15. https://doi.org/10.1002/aoc.3849

Durak J, Rokoszak T, Skiba A, Furman P, Styszko K (2021) Environmental risk assessment of priority biocidal substances on Polish surface water sample. Environ Sci Pollut Res 28:1254–1266. https://doi.org/10.1007/s11356-020-11581-7

Ebrahimian J, Mohsennia M, Khayatkashani M (2020) Photocatalytic-degradation of organic dye and removal of heavy metal ions using synthesized SnO2 nanoparticles by Vitex agnus-castus fruit via a green route. Mater Lett 263:127255. https://doi.org/10.1016/j.matlet.2019.127255

Elliott DW, Lien H-L, Zhang W (2008) Zerovalent iron nanoparticles for treatment of ground water contaminated by hexachlorocyclohexanes. J Environ Qual 37:2192–2201. https://doi.org/10.2134/JEQ2007.0545

Essawy AA, Alsohaimi IH, Alhumaimess MS, Hassan HMA, Kamel MM (2020) Green synthesis of spongy Nano-ZnO productive of hydroxyl radicals for unconventional solar-driven photocatalytic remediation of antibiotic enriched wastewater. J Environ Manage 271:110961. https://doi.org/10.1016/j.jenvman.2020.110961

Faghihzadeh F, Anaya NM, Schifman LA, Oyanedel-Craver V (2016) Fourier transform infrared spectroscopy to assess molecular-level changes in microorganisms exposed to nanoparticles, Nanotechnol. Environ Eng 1:1–16. https://doi.org/10.1007/s41204-016-0001-8

Fazlzadeh M, Rahmani K, Zarei A, Abdoallahzadeh H, Nasiri F, Khosravi R (2017) A novel green synthesis of zero valent iron nanoparticles (NZVI) using three plant extracts and their efficient application for removal of Cr(VI) from aqueous solutions. Adv Powder Technol 28:122–130. https://doi.org/10.1016/j.apt.2016.09.003

Fouda A, Hassan SED, Saied E, Hamza MF (2021) Photocatalytic degradation of real textile and tannery effluent using biosynthesized magnesium oxide nanoparticles (MgO-NPs), heavy metal adsorption, phytotoxicity, and antimicrobial activity. J Environ Chem Eng 9:105346. https://doi.org/10.1016/j.jece.2021.105346

Frankenberger WT, Losi ME (2015) Applications of bioremediation in the cleanup of heavy metals and metalloids. Bioremediation Sci Appl:173–210. https://doi.org/10.2136/SSSASPECPUB43.C11

Gan L, Li B, Guo M, Weng X, Wang T, Chen Z (2018) Mechanism for removing 2,4-dichlorophenol via adsorption and Fenton-like oxidation using iron-based nanoparticles. Chemosphere 206:168–174. https://doi.org/10.1016/j.chemosphere.2018.04.162

Gangula A, Podila R, R. M, Karanam L, Janardhana C, Rao AM (2011) Catalytic reduction of 4-nitrophenol using biogenic gold and silver nanoparticles derived from breynia rhamnoides. Langmuir 27:15268–15274. https://doi.org/10.1021/la2034559

García FE, Senn AM, Meichtry JM, Scott TB, Pullin H, Leyva AG, Halac EB, Ramos CP, Sacanell J, Mizrahi M, Requejo FG, Litter MI (2019) Iron-based nanoparticles prepared from yerba mate extract. Synthesis, characterization and use on chromium removal. J Environ Manage 235:1–8. https://doi.org/10.1016/j.jenvman.2019.01.002

Garole VJ, Choudhary BC, Tetgure SR, Garole DJ, Borse AU (2018) Detoxification of toxic dyes using biosynthesized iron nanoparticles by photo-Fenton processes. Int J Environ Sci Technol 15:1649–1656. https://doi.org/10.1007/s13762-017-1510-0

Gautam PK, Shivalkar S, Banerjee S (2020) Synthesis of M. oleifera leaf extract capped magnetic nanoparticles for effective lead [Pb (II)] removal from solution: Kinetics, isotherm and reusability study. J Mol Liq 305:112811. https://doi.org/10.1016/j.molliq.2020.112811

Gopal G, Kvg R, Salma M, Lavanya Agnes Angalene J, Chandrasekaran N, Mukherjee A (2020) Green synthesized Fe/Pd and in-situ Bentonite-Fe/Pd composite for efficient tetracycline removal. J Environ Chem Eng 8:104126. https://doi.org/10.1016/j.jece.2020.104126

Govindappa M, Hemashekhar B, Arthikala MK, Ravishankar Rai V, Ramachandra YL (2018) Characterization, antibacterial, antioxidant, antidiabetic, anti-inflammatory and antityrosinase activity of green synthesized silver nanoparticles using Calophyllum tomentosum leaves extract. Results Phys 9:400–408. https://doi.org/10.1016/j.rinp.2018.02.049

Grieger KD, Hjorth R, Rice J, Kumar N, Bang J (2015) Nano-remediation : tiny particles cleaning up big environmental problems, Iucn. pp 1–7

Guo M, Weng X, Wang T, Chen Z (2017) Biosynthesized iron-based nanoparticles used as a heterogeneous catalyst for the removal of 2,4-dichlorophenol. Sep Purif Technol 175:222–228. https://doi.org/10.1016/j.seppur.2016.11.042

Guzmán KAD, Taylor MR, Banfield JF (2006) Environmental risks of nanotechnology: National nanotechnology initiative funding, 2000–2004. Environ Sci Technol 40:1401–1407. https://doi.org/10.1021/es0515708

Hao R, Li D, Zhang J (2021) Insights into the removal of Cr(VI) from aqueous solution using plant-mediated biosynthesis of iron nanoparticles. Environ Technol Innov 23:101566. https://doi.org/10.1016/j.eti.2021.101566

Hasanin M, Abdelhameed RM, Dacrory S, Abou-Yousef H, Kamel S (2021) Photocatalytic degradation of pesticide intermediate using green eco-friendly amino functionalized cellulose nanocomposites. Mater Sci Eng B 270:115231. https://doi.org/10.1016/J.MSEB.2021.115231

Hassan SSM, Azab WIME, Ali HR, Mansour MSM (2015) Green synthesis and characterization of ZnO nanoparticles for photocatalytic degradation of anthracene. Adv Nat Sci Nanosci Nanotechnol 6:45012. https://doi.org/10.1088/2043-6262/6/4/045012

Hossen S, Yusuf M, Chandra S, Das T, Saha O, Rahaman M, Islam J (2020) Heliyon Green synthesis of iron oxide nanoparticle using Carica papaya leaf extract : application for photocatalytic degradation of remazol yellow RR dye and antibacterial activity. Heliyon. 6:e04603. https://doi.org/10.1016/j.heliyon.2020.e04603

Husein DZ, Hassanien R, Al-Hakkani MF (2019) Green-synthesized copper nano-adsorbent for the removal of pharmaceutical pollutants from real wastewater samples. Heliyon. 5:e02339. https://doi.org/10.1016/j.heliyon.2019.e02339

Husen A, Siddiqi KS (2014) Phytosynthesis of nanoparticles: Concept, controversy and application. Nanoscale Res Lett 9:1–24. https://doi.org/10.1186/1556-276X-9-229

Imron MF, Kurniawan SB, Soegianto A, Wahyudianto FE (2019) Phytoremediation of methylene blue using duckweed (Lemna minor). Heliyon. 5:e02206. https://doi.org/10.1016/j.heliyon.2019.e02206

Ishwarya R, Vaseeharan B, Kalyani S, Banumathi B, Govindarajan M, Alharbi NS, Kadaikunnan S, Al-anbr MN, Khaled JM, Benelli G (2018) Facile green synthesis of zinc oxide nanoparticles using Ulva lactuca seaweed extract and evaluation of their photocatalytic, antibiofilm and insecticidal activity. J Photochem Photobiol B Biol 178:249–258. https://doi.org/10.1016/j.jphotobiol.2017.11.006

Ismail M, Khan MI, Khan SB, Akhtar K, Khan MA, Asiri AM (2018) Catalytic reduction of picric acid, nitrophenols and organic azo dyes via green synthesized plant supported Ag nanoparticles. J Mol Liq 268:87–101. https://doi.org/10.1016/j.molliq.2018.07.030

Ivashchenko O, Lewandowski M, Peplińska B, Jarek M, Nowaczyk G, Wiesner M, Załęski K, Babutina T, Warowicka A, Jurga S (2015) Synthesis and characterization of magnetite/silver/ antibiotic nanocomposites for targeted antimicrobial therapy. Mater Sci Eng C 55:343–359. https://doi.org/10.1016/J.MSEC.2015.05.023

Jadoun S, Arif R, Jangid NK, Meena RK (2021) Green synthesis of nanoparticles using plant extracts: a review. Environ Chem Lett 19:355–374. https://doi.org/10.1007/s10311-020-01074-x

Jamzad M, Kamari M (2020) Green synthesis of iron oxide nanoparticles by the aqueous extract of Laurus nobilis L . leaves and evaluation of the antimicrobial activity. J Nanostructure Chem. https://doi.org/10.1007/s40097-020-00341-1

Jeelani N, Yang W, Xu L, Qiao Y, An S, Leng X (2017) Phytoremediation potential of Acorus calamus in soils co-contaminated with cadmium and polycyclic aromatic hydrocarbons. Sci Rep 7:1–9. https://doi.org/10.1038/s41598-017-07831-3

K. Balakrishnan, S. Dey, T. Gupta, R.S. Dhaliwal, M. Brauer, A.J. Cohen, J.D. Stanaway, G. Beig, T.K. Joshi, A.N. Aggarwal, Y. Sabde, H. Sadhu, J. Frostad, K. Causey, W. Godwin, D.K. Shukla, G.A. Kumar, C.M. Varghese, P. Muraleedharan, A. Agrawal, R.M. Anjana, A. Bhansali, D. Bhardwaj, K. Burkart, K. Cercy, J.K. Chakma, S. Chowdhury, D.J. Christopher, E. Dutta, M. Furtado, S. Ghosh, A.G. Ghoshal, S.D. Glenn, R. Guleria, R. Gupta, P. Jeemon, R. Kant, S. Kant, T. Kaur, P.A. Koul, V. Krish, B. Krishna, S.L. Larson, K. Madhipatla, P.A. Mahesh, V. Mohan, S. Mukhopadhyay, P. Mutreja, N. Naik, S. Nair, G. Nguyen, C.M. Odell, J.D. Pandian, D. Prabhakaran, P. Prabhakaran, A. Roy, S. Salvi, S. Sambandam, D. Saraf, M. Sharma, A. Shrivastava, V. Singh, N. Tandon, N.J. Thomas, A. Torre, D. Xavier, G. Yadav, S. Singh, C. Shekhar, T. Vos, R. Dandona, K.S. Reddy, S.S. Lim, C.J.L. Murray, S. Venkatesh, L. Dandona, The impact of air pollution on deaths, disease burden, and life expectancy across the states of India: the Global Burden of Disease Study (2017) Lancet Planet. Heal 3(2019):e26–e39. https://doi.org/10.1016/S2542-5196(18)30261-4

Kamath V, Chandra P, Jeppu GP (2020) Comparative study of using five different leaf extracts in the green synthesis of iron oxide nanoparticles for removal of arsenic from water. 22:1278–1294. https://doi.org/10.1080/15226514.2020.1765139

Karimi P, Javanshir S, Sayadi MH, Arabyarmohammadi H (2019) Arsenic removal from mining effluents using plant-mediated, green-synthesized iron nanoparticles. Processes 7:759. https://doi.org/10.3390/pr7100759

Karthik R, Govindasamy M, Chen SM, Cheng YH, Muthukrishnan P, Padmavathy S, Elangovan A (2017) Biosynthesis of silver nanoparticles by using Camellia japonica leaf extract for the electrocatalytic reduction of nitrobenzene and photocatalytic degradation of Eosin-Y. J Photochem Photobiol B Biol 170:164–172. https://doi.org/10.1016/j.jphotobiol.2017.03.018

Kaur P, Thakur R, Malwal H, Manuja A, Chaudhury A (2018) Biosynthesis of biocompatible and recyclable silver/iron and gold/iron core-shell nanoparticles for water purification technology, Biocatal. Agric. Biotechnol 14:189–197. https://doi.org/10.1016/j.bcab.2018.03.002

Khan Z, Al-Thabaiti SA (2018) Green synthesis of zero-valent Fe-nanoparticles: Catalytic degradation of rhodamine B, interactions with bovine serum albumin and their enhanced antimicrobial activities. J Photochem Photobiol B Biol 180:259–267. https://doi.org/10.1016/j.jphotobiol.2018.02.017

Kumar M, Mehta A, Mishra A, Singh J, Rawat M, Basu S (2018) Biosynthesis of tin oxide nanoparticles using Psidium Guajava leave extract for photocatalytic dye degradation under sunlight. Mater Lett 215:121–124. https://doi.org/10.1016/j.matlet.2017.12.074

Kumar B, Smita K, Galeas S, Sharma V, Guerrero VH, Debut A, Cumbal L (2020) Characterization and application of biosynthesized iron oxide nanoparticles using Citrus paradisi peel: A sustainable approach. Inorg Chem Commun 119:108116. https://doi.org/10.1016/J.INOCHE.2020.108116

Kumar L, Ragunathan V, Chugh M, Bharadvaja N (2021) Nanomaterials for remediation of contaminants: a review. Environ Chem Lett 1:3. https://doi.org/10.1007/s10311-021-01212-z

Kumar L, Khushbu, Chugh M, Bharadvaja N (2022) Microbial remediation of tannery wastewater. Dev Wastewater Treat Res Process:303–328. https://doi.org/10.1016/B978-0-323-85657-7.00011-0

Lai X, Guo R, Xiao H, Lan J, Jiang S, Cui C, Ren E (2019) Rapid microwave-assisted bio-synthesized silver/Dandelion catalyst with superior catalytic performance for dyes degradation. J Hazard Mater 371:506–512. https://doi.org/10.1016/j.jhazmat.2019.03.039

Leili M, Fazlzadeh M, Bhatnagar A (2018) Green synthesis of nano-zero-valent iron from Nettle and Thyme leaf extracts and their application for the removal of cephalexin antibiotic from aqueous solutions. Environ Technol (United Kingdom) 39:1158–1172. https://doi.org/10.1080/09593330.2017.1323956

Li Y, Fu Y, Zhu M (2020) Green synthesis of 3D tripyramid TiO2 architectures with assistance of aloe extracts for highly efficient photocatalytic degradation of antibiotic ciprofloxacin. Appl Catal B Environ 260:118149. https://doi.org/10.1016/j.apcatb.2019.118149

Lim SH, Ahn EY, Park Y (2016) Green synthesis and catalytic activity of gold nanoparticles synthesized by artemisia capillaris water extract. Nanoscale Res Lett 11:1–11. https://doi.org/10.1186/s11671-016-1694-0

Lorenz ES (2017) Potential health effects of pesticides. Pestic Educ Progr. https://extension.psu.edu/potential-health-effects-of-pesticides (Accessed 17 May 2021)

Machado S, Stawiński W, Slonina P, Pinto AR, Grosso JP, Nouws HPA, Albergaria JT, Delerue-Matos C (2013) Application of green zero-valent iron nanoparticles to the remediation of soils contaminated with ibuprofen. Sci Total Environ 461–462:323–329. https://doi.org/10.1016/j.scitotenv.2013.05.016

Madhavi V, Prasad TNVKV, Reddy AVB, Ravindra Reddy B, Madhavi G (2013) Application of phytogenic zerovalent iron nanoparticles in the adsorption of hexavalent chromium, Spectrochim. Acta—Part A Mol Biomol Spectrosc 116:17–25. https://doi.org/10.1016/j.saa.2013.06.045

Makofane A, Motaung DE, Hintsho-Mbita NC (2021) Photocatalytic degradation of methylene blue and sulfisoxazole from water using biosynthesized zinc ferrite nanoparticles. Ceram Int. https://doi.org/10.1016/j.ceramint.2021.04.274

Malakootian M, Yaseri M, Faraji M (2019) Removal of antibiotics from aqueous solutions by nanoparticles: a systematic review and meta-analysis. Environ Sci Pollut Res 26:8444–8458. https://doi.org/10.1007/s11356-019-04227-w

Manimegalai G, Shanthakumar S, Sharma C (2014) Silver nanoparticles: synthesis and application in mineralization of pesticides using membrane support. Int. Nano Lett. 4:1–5. https://doi.org/10.1007/s40089-014-0105-8

Marimuthu S, Antonisamy AJ, Malayandi S, Rajendran K, Tsai PC, Pugazhendhi A, Ponnusamy VK (2020) Silver nanoparticles in dye effluent treatment: A review on synthesis, treatment methods, mechanisms, photocatalytic degradation, toxic effects and mitigation of toxicity. J Photochem Photobiol B Biol 205:111823. https://doi.org/10.1016/j.jphotobiol.2020.111823

Mathivanan D, Kirankumar VS, Sumathi S, Suseem SR (2018) Facile Biosynthesis of Calcium Hydroxide Nanoparticles Using Andrographis echioides Leaf Extract and Its Photocatalytic Activity Under Different Light Source. J Clust Sci 29:167–175. https://doi.org/10.1007/s10876-017-1318-6

Mehata MS (2021) Green route synthesis of silver nanoparticles using plants/ginger extracts with enhanced surface plasmon resonance and degradation of textile dye. Mater Sci Eng B Solid-State Mater Adv Technol 273:115418. https://doi.org/10.1016/j.mseb.2021.115418

Miri A, Shahraki Vahed HO, Sarani M (2018) Biosynthesis of silver nanoparticles and their role in photocatalytic degradation of methylene blue dye. Res Chem Intermed 44:6907–6915. https://doi.org/10.1007/s11164-018-3529-3

Mirsadeghi S, Zandavar H, Yousefi M, Rajabi HR, Pourmortazavi SM (2020) Green-photodegradation of model pharmaceutical contaminations over biogenic Fe3O4/Au nanocomposite and antimicrobial activity. J Environ Manage 270:110831. https://doi.org/10.1016/j.jenvman.2020.110831

Muthuvel A, Jothibas M, Mohana V, Manoharan C (2020) Green synthesis of cerium oxide nanoparticles using Calotropis procera flower extract and their photocatalytic degradation and antibacterial activity. Inorg Chem Commun 119:108086. https://doi.org/10.1016/J.INOCHE.2020.108086

Nasrollahzadeh M, Mohammad Sajadi S, Rostami-Vartooni A, Alizadeh M, Bagherzadeh M (2016) Green synthesis of the Pd nanoparticles supported on reduced graphene oxide using barberry fruit extract and its application as a recyclable and heterogeneous catalyst for the reduction of nitroarenes. J Colloid Interface Sci 466:360–368. https://doi.org/10.1016/j.jcis.2015.12.036

Norfazilah Wan Ismail W, Umairah Mokhtar S (2020) Various methods for removal, treatment, and detection of emerging water contaminants. In: Emerg Contam [Working Title], IntechOpen. https://doi.org/10.5772/intechopen.93375

Noruzi M (2015) Biosynthesis of gold nanoparticles using plant extracts. Bioprocess Biosyst Eng 38:1–14. https://doi.org/10.1007/s00449-014-1251-0

Özkara A, Akyil D, Konuk M (2016) Pesticides, environmental pollution, and health. In: Environ Heal Risk—Hazard. Factors to Living Species, InTech. https://doi.org/10.5772/63094

Pandey G (2018) Prospects of nanobioremediation in environmental cleanup. Orient J Chem 34:2838–2850. https://doi.org/10.13005/ojc/340622

Patel AB, Shaikh S, Jain KR, Desai C, Madamwar D (2020) Polycyclic aromatic hydrocarbons: sources, toxicity, and remediation approaches. Front Microbiol 11:2675. https://doi.org/10.3389/fmicb.2020.562813

Patra N, Kar D, Pal A, Behera A (2018) Antibacterial, anticancer, anti-diabetic and catalytic activity of bio-conjugated metal nanoparticles. Adv Nat Sci Nanosci Nanotechnol 9:035001. https://doi.org/10.1088/2043-6254/aad12d

Perrone MG, Carbone C, Faedo D, Ferrero L, Maggioni A, Sangiorgi G, Bolzacchini E (2014) Exhaust emissions of polycyclic aromatic hydrocarbons, n-alkanes and phenols from vehicles coming within different European classes. Atmos Environ 82:391–400. https://doi.org/10.1016/j.atmosenv.2013.10.040

Philip D (2010) Rapid green synthesis of spherical gold nanoparticles using Mangifera indica leaf, Spectrochim. Acta—Part A Mol Biomol Spectrosc 77:807–810. https://doi.org/10.1016/j.saa.2010.08.008

Radini IA, Hasan N, Malik MA, Khan Z (2018) Biosynthesis of iron nanoparticles using Trigonella foenum-graecum seed extract for photocatalytic methyl orange dye degradation and antibacterial applications. J Photochem Photobiol B Biol 183:154–163. https://doi.org/10.1016/j.jphotobiol.2018.04.014

Rajabi HR, Sajadiasl F, Karimi H, Alvand ZM (2020) Green synthesis of zinc sulfide nanophotocatalysts using aqueous extract of Ficus Johannis plant for efficient photodegradation of some pollutants. J. Mater. Res. Technol. 9:15638–15647. https://doi.org/10.1016/j.jmrt.2020.11.017

Rajput S, Singh LP, Pittman CU, Mohan D (2017) Lead (Pb2+) and copper (Cu2+) remediation from water using superparamagnetic maghemite (γ-Fe2O3) nanoparticles synthesized by Flame Spray Pyrolysis (FSP). J Colloid Interface Sci 492:176–190. https://doi.org/10.1016/j.jcis.2016.11.095

Ramadhan VB, Ni'Mah YL, Yanuar E, Suprapto S (2019) Synthesis of copper nanoparticles using Ocimum tenuiflorum leaf extract as capping agent. In: AIP Conf Proc., American Institute of Physics Inc., p 20067. https://doi.org/10.1063/1.5141680

Rana A, Yadav K, Jagadevan S (2020) A comprehensive review on green synthesis of nature-inspired metal nanoparticles: Mechanism, application and toxicity. J Clean Prod 272:122880. https://doi.org/10.1016/j.jclepro.2020.122880
Rani M, Rachna, Shanker U (2020) Metal oxide-chitosan based nanocomposites for efficient degradation of carcinogenic PAHs. J Environ Chem Eng 8:103810. https://doi.org/10.1016/j.jece.2020.103810
Rani M, Shanker U (2018) Remediation of polycyclic aromatic hydrocarbons using nanomaterials. In: Remov Adsorbents Pollut., Springer, Cham, pp 343–387. https://doi.org/10.1007/978-3-319-92111-2_10
Rasheed T, Bilal M, Nabeel F, Adeel M, Iqbal MN (2019) Environmentally-related contaminants of high concern : Potential sources and analytical modalities for detection, quanti fi cation, and treatment. Environ Int 122:52–66. https://doi.org/10.1016/j.envint.2018.11.038
Ravikumar KVG, Sudakaran SV, Ravichandran K, Pulimi M, Natarajan C, Mukherjee A (2019) Green synthesis of NiFe nano particles using Punica granatum peel extract for tetracycline removal. J Clean Prod 210:767–776. https://doi.org/10.1016/j.jclepro.2018.11.108
Rawtani D, Khatri N, Tyagi S, Pandey G (2018) Nanotechnology-based recent approaches for sensing and remediation of pesticides. J Environ Manage 206:749–762. https://doi.org/10.1016/j.jenvman.2017.11.037
Razack SA, Suresh A, Sriram S, Ramakrishnan G (2020) Green synthesis of iron oxide nanoparticles using Hibiscus rosa-sinensis for fortifying wheat biscuits. SN Appl Sci 2:1–9. https://doi.org/10.1007/s42452-020-2477-x
Rissato SR, Galhiane MS, Fernandes JR, Gerenutti M, Gomes HM, Ribeiro R, De Almeida MV (2015) Evaluation of Ricinus communis L. for the phytoremediation of polluted soil with organochlorine pesticides. Biomed Res Int 2015:1–8. https://doi.org/10.1155/2015/549863
Roberto S-CC, Andrea P-M, Andrés G-O, Norma F-P, Hermes P-H, Gabriela M-P, Fabián F-L (2020) Phytonanotechnology and environmental remediation. Phytonanotechnology:159–185. https://doi.org/10.1016/b978-0-12-822348-2.00009-7
Rong K, Wang J, Zhang Z, Zhang J (2020) Green synthesis of iron nanoparticles using Korla fragrant pear peel extracts for the removal of aqueous Cr(VI). Ecol Eng 149:105793. https://doi.org/10.1016/j.ecoleng.2020.105793
Rosbero TMS, Camacho DH (2017) Green preparation and characterization of tentacle-like silver/copper nanoparticles for catalytic degradation of toxic chlorpyrifos in water. J Environ Chem Eng 5:2524–2532. https://doi.org/10.1016/j.jece.2017.05.009
Rostami-Vartooni A, Nasrollahzadeh M, Salavati-Niasari M, Atarod M (2016) Photocatalytic degradation of azo dyes by titanium dioxide supported silver nanoparticles prepared by a green method using Carpobrotus acinaciformis extract. J Alloys Compd 689:15–20. https://doi.org/10.1016/j.jallcom.2016.07.253
Rufus A, Sreeju N, Philip D (2019) Size tunable biosynthesis and luminescence quenching of nanostructured hematite (α-Fe2O3) for catalytic degradation of organic pollutants. J Phys Chem Solids 124:221–234. https://doi.org/10.1016/j.jpcs.2018.09.026
Sai Saraswathi V, Santhakumar K (2017) Photocatalytic activity against azo dye and cytotoxicity on MCF-7 cell lines of zirconium oxide nanoparticle mediated using leaves of Lagerstroemia speciose. J Photochem Photobiol B Biol 169:47–55. https://doi.org/10.1016/j.jphotobiol.2017.02.023
Sajadi SM, Nasrollahzadeh M, Maham M (2016) Aqueous extract from seeds of Silybum marianum L. as a green material for preparation of the Cu/Fe3O4 nanoparticles: A magnetically recoverable and reusable catalyst for the reduction of nitroarenes. J Colloid Interface Sci 469:93–98. https://doi.org/10.1016/j.jcis.2016.02.009
Samrot AV, Angalene JLA, Roshini SM, Raji P, Stefi SM, Preethi R, Selvarani AJ, Madankumar A (2019) Bioactivity and heavy metal removal using plant gum mediated green synthesized silver nanoparticles. J Clust Sci 30:1599–1610. https://doi.org/10.1007/s10876-019-01602-y
Sandhya J, Kalaiselvam S (2020) Biogenic synthesis of magnetic iron oxide nanoparticles using inedible borassus flabellifer seed coat : characterization, antimicrobial, antioxidant activity and

in vitro cytotoxicity analysis Biogenic synthesis of magnetic iron oxide nanoparticles usin. J Photochem Photobiol B Biol 185:262–274

Sangami S, Manu B (2017) Synthesis of Green Iron Nanoparticles using Laterite and their application as a Fenton-like catalyst for the degradation of herbicide Ametryn in water. Environ Technol Innov 8:150–163. https://doi.org/10.1016/j.eti.2017.06.003

Saravanakumar K, Chelliah R, Shanmugam S, Varukattu NB, Oh DH, Kathiresan K, Wang MH (2018) Green synthesis and characterization of biologically active nanosilver from seed extract of Gardenia jasminoides Ellis. J Photochem Photobiol B Biol 185:126–135. https://doi.org/10.1016/j.jphotobiol.2018.05.032

Seery MK, George R, Floris P, Pillai SC (2007) Silver doped titanium dioxide nanomaterials for enhanced visible light photocatalysis. J Photochem Photobiol A Chem 189:258–263. https://doi.org/10.1016/j.jphotochem.2007.02.010

Serrà A (2020) Simple Environmentally-Friendly Reduction of 4-Nitrophenol. Catalysts 10:458

Shah Z, Gul T, Ali Khan S, Shaheen K, Anwar Y, Suo H, Ismail M, Alghamdi KM, Salman SM (2021) Synthesis of high surface area AgNPs from Dodonaea viscosa plant for the removal of pathogenic microbes and persistent organic pollutants. Mater Sci Eng B 263:114770. https://doi.org/10.1016/J.MSEB.2020.114770

Shaik AM, David Raju M, Rama Sekhara Reddy D (2020) Green synthesis of zinc oxide nanoparticles using aqueous root extract of Sphagneticola trilobata Lin and investigate its role in toxic metal removal, sowing germination and fostering of plant growth. Inorg Nano-Metal Chem 50:569–579. https://doi.org/10.1080/24701556.2020.1722694

Shanker U, Jassal V, Rani M (2017) Green synthesis of iron hexacyanoferrate nanoparticles: Potential candidate for the degradation of toxic PAHs. Biochem Pharmacol 5:4108–4120. https://doi.org/10.1016/j.jece.2017.07.042

Shikha S (2016) Phytoremediation of pharmaceutical products. Innovare J Life Sci:14–17. https://innovareacademics.in/journals/index.php/ijls/article/view/13032 (Accessed 21Jun 2021)

Singh J, Dutta T, Kim KH, Rawat M, Samddar P, Kumar P (2018) "Green" synthesis of metals and their oxide nanoparticles: Applications for environmental remediation. J Nanobiotechnology 16:1–24. https://doi.org/10.1186/s12951-018-0408-4

Sorbiun M, Shayegan Mehr E, Ramazani A, Taghavi Fardood S (2018) Biosynthesis of Ag, ZnO and bimetallic Ag/ZnO alloy nanoparticles by aqueous extract of oak fruit hull (Jaft) and investigation of photocatalytic activity of ZnO and bimetallic Ag/ZnO for degradation of basic violet 3 dye. J Mater Sci Mater Electron 29:2806–2814. https://doi.org/10.1007/s10854-017-8209-3

Sreekala G, Beevi AF, Resmi R, Beena B (2021) Removal of lead (II) ions from water using copper ferrite nanoparticles synthesized by green method. Mater Today Proc 45:3986–3990. https://doi.org/10.1016/j.matpr.2020.09.087

Srivastav A, Yadav KK, Yadav S, Gupta N, Singh JK, Katiyar R, Kumar V (2019) Nano-phytoremediation of pollutants from contaminated soil environment: Current scenario and future prospects, In: Phytoremediation Manag Environ Contam. Springer International Publishing, pp 383–401. https://doi.org/10.1007/978-3-319-99651-6_16

Stan M, Lung I, Soran ML, Leostean C, Popa A, Stefan M, Lazar MD, Opris O, Silipas TD, Porav AS (2017) Removal of antibiotics from aqueous solutions by green synthesized magnetite nanoparticles with selected agro-waste extracts. Process Saf Environ Prot 107:357–372. https://doi.org/10.1016/j.psep.2017.03.003

Stroo HF (2015) Biodegradation and bioremediation of contaminated sites: the role of soil science, role soil sci. Interdiscip Res:37–55. https://doi.org/10.2136/SSSASPECPUB45.C3

Titus D, James Jebaseelan Samuel E, Roopan SM (2019) Nanoparticle characterization techniques. Elsevier Inc. https://doi.org/10.1016/b978-0-08-102579-6.00012-5

Vanaja M, Paulkumar K, Baburaja M, Rajeshkumar S, Gnanajobitha G, Malarkodi C, Sivakavinesan M, Annadurai G (2014) Degradation of methylene blue using biologically synthesized silver nanoparticles. Bioinorg Chem Appl 2014. https://doi.org/10.1155/2014/742346

Vázquez-Núñez E, Molina-Guerrero CE, Peña-Castro JM, Fernández-Luqueño F, de la Rosa-Álvarez MG (2020) Use of nanotechnology for the bioremediation of contaminants: A review. Processes 8:1–17. https://doi.org/10.3390/pr8070826

Vinayagam R, Selvaraj R, Arivalagan P, Varadavenkatesan T (2020) Synthesis, characterization and photocatalytic dye degradation capability of Calliandra haematocephala-mediated zinc oxide nanoflowers. J Photochem Photobiol B Biol 203:111760. https://doi.org/10.1016/j.jphotobiol.2019.111760

Wang Y, O'Connor D, Shen Z, Lo IMC, Tsang DCW, Pehkonen S, Pu S, Hou D (2019) Green synthesis of nanoparticles for the remediation of contaminated waters and soils: Constituents, synthesizing methods, and influencing factors. J Clean Prod 226:540–549. https://doi.org/10.1016/j.jclepro.2019.04.128

Weng X, Jin X, Lin J, Naidu R, Chen Z (2016) Removal of mixed contaminants Cr(VI) and Cu(II) by green synthesized iron based nanoparticles. Ecol Eng 97:32–39. https://doi.org/10.1016/j.ecoleng.2016.08.003

Weng X, Ma L, Guo M, Su Y, Dharmarajan R, Chen Z (2018) Removal of doxorubicin hydrochloride using Fe3O4 nanoparticles synthesized by euphorbia cochinchinensis extract. Chem Eng J 353:482–489. https://doi.org/10.1016/j.cej.2018.07.162

Whang TJ, Huang HY, Hsieh MT, Chen JJ (2009) Laser-induced silver nanoparticles on titanium oxide for photocatalytic degradation of methylene blue. Int J Mol Sci 10:4707–4718. https://doi.org/10.3390/ijms10114707

Wilson JT, Jawson MD (2015) Science needs for implementation of bioremediation. Bioremediation Sci Appl (2015):293–303. https://doi.org/10.2136/SSSASPECPUB43.C17

Wu J, Lin Z, Weng X, Owens G, Chen Z (2020) Removal mechanism of mitoxantrone by a green synthesized hybrid reduced graphene oxide @ iron nanoparticles. Chemosphere 246:125700–125700. https://doi.org/10.1016/j.chemosphere.2019.125700

Xiao C, Li H, Zhao Y, Zhang X, Wang X (2020) Green synthesis of iron nanoparticle by tea extract (polyphenols) and its selective removal of cationic dyes. J Environ Manage 275:111262. https://doi.org/10.1016/j.jenvman.2020.111262

Yan S, Yao B, Lian L, Lu X, Snyder SA, Li R, Song W (2017) Development of fluorescence surrogates to predict the photochemical transformation of pharmaceuticals in wastewater effluents. Environ Sci Technol 51:2738–2747. https://doi.org/10.1021/acs.est.6b05251

Yi Y, Tu G, Tsang PE, Xiao S, Fang Z (2019) Green synthesis of iron-based nanoparticles from extracts of Nephrolepis auriculata and applications for Cr(VI) removal. Mater Lett 234:388–391. https://doi.org/10.1016/j.matlet.2018.09.137

F. Yousefi, Removal of Organic Pesticides by Carbon Nanoparticle Synthesized from Pomegranate Peel, Int. J. Bioorganic Chem. 2 (2017) 70–76. https://doi.org/10.11648/j.ijbc.20170202.14.

Yu P, Yu H, Sun Q, Ma B (2019) Filter paper supported nZVI for continuous treatment of simulated dyeing wastewater. Sci Rep 9:1–8. https://doi.org/10.1038/s41598-019-47863-5

Zayed SMAD, Earghaly M, Mahdy F, Taha H, Soliman SM (2001) Mineralization of 14 C-labelled aromatic pesticide molecules in Egyptian soils under aerobic and anaerobic conditions. Int. Nucl. Inf. Syst. 33:83–88

Zhu F, Ma S, Liu T, Deng X (2018) Green synthesis of nano zero-valent iron/Cu by green tea to remove hexavalent chromium from groundwater. J Clean Prod 174:184–190. https://doi.org/10.1016/j.jclepro.2017.10.302

Zuo R, Liu H, Xi Y, Gu Y, Ren D, Yuan X, Huang Y (2020) Nano-SiO2 combined with a surfactant enhanced phenanthrene phytoremediation by Erigeron annuus (L.) Pers. Environ Sci Pollut Res 27:20538–20544. https://doi.org/10.1007/s11356-020-08552-3

Chapter 6
Biogenic Synthesis of Nanomaterials: Bioactive Compounds as Reducing, and Capping Agents

Shashank Yadav, Tirth Nadar, Jaya Lakkakula, and Nilesh S. Wagh

Abstract Nanotechnology is the work done at a nano level to create or modify a biological compound for greater good. There is a wide range of bioactive compounds derived from algae bacteria and plants like phenolic compounds, carotenoids, tannins, organic acids, omega-3 fatty acids, etc. which are used for the production of nanoparticles. Secondary metabolites, phytochemicals and various compounds can act as bio-compounds and are being widely used in different fields like pharmaceutical, medicinal, and environmental products. Bioactive compounds (BACs) can modulate metabolic processes and promote better health. They exhibit antioxidant activity in which these bioactive compounds can scavenge free radicals in the body and can prevent oxidative damages and heart problems. Bioactive compounds can also inhibit receptor systems to block certain signals and stimulation or inhibition of enzymes and gene expressions. They can be also used in treatments for cancer, inflammatory diseases, microbial infections and other medical conditions. These days Nanoparticles are also used in drug delivery, they promote controlled release of BACs improving distribution of prescribed drug in the blood stream for a longer period of time and improve beneficial effects on target tissue or organ. The use of capping agents for controlling growth, aggregation, and physical and chemical properties is the key area of concern in the synthesis of nanoparticles. In our work, we have majorly

S. Yadav · T. Nadar · J. Lakkakula (✉) · N. S. Wagh
Amity Institute of Biotechnology, Amity University, Mumbai-Pune Expressway, Bhatan, Panvel, Mumbai 410206, India
e-mail: jrlakkakula@mum.amity.edu

S. Yadav
e-mail: shashank.yadav2@s.amity.edu

T. Nadar
e-mail: nadartirth@gmail.com

N. S. Wagh
e-mail: wnshirish@mum.amity.edu

S. Yadav · T. Nadar · J. Lakkakula
Centre for Computational Biology and Translational Research, Amity Institute of Biotechnology (AIB), Amity University, Mumbai, India

M. P. Shah et al. (eds.), *Biogenic Nanomaterials for Environmental Sustainability: Principles, Practices, and Opportunities*, Environmental Science and Engineering,
https://doi.org/10.1007/978-3-031-45956-6_6

focused on green synthesis of Au, Cu, Ag and Fe nanoparticles which display exceptional attributes and higher surface area and have potential application in numerous fields including food, sensors, optics and most importantly in medicine, they have attracted much attention of the researchers due to their promising bioactivates and non-toxic nature.

Graphical Abstract

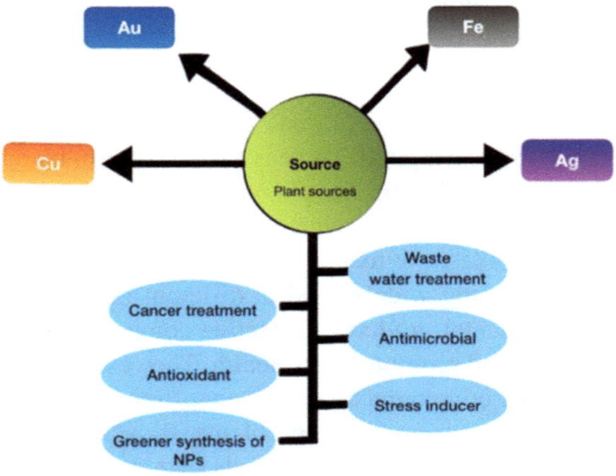

Keywords Bioactive compounds · Capping agent · Secondary metabolites · Green synthesis

6.1 Introduction

Nanoscience and technology are concerned with the scientific analysis of material properties and their applications at the nanoscale. Richard Feynman, a respected physicist and Nobel winner, pioneered the idea of nanotechnology in 1959 (Karvekar et al. 2022). The biological features of Nanoparticles (NPs) have been the topic of recent studies; nonetheless, there are risks about their long-term damage. Significant advancements in nanoparticle production and surface modification, as well as the use of unique biogenic capping agents, have enabled the biosynthesis of surface-functionalized, nontoxic, monodispersed nanoparticles for therapeutic purposes. These capping agents function as stabilizers or binding molecules, preventing aggregation and steric hindrance, altering bioactivity and surface properties, and stabilizing nanoparticle interaction within the preparatory media (Sidhu et al. 2022).After all, bioactive compounds originating from a diverse range of plants and microorganisms, especially algae and bacteria have huge promise for application as cancer, microbial infection, inflammatory illnesses, and other medical problems therapies. Encapsulated bioactive substances and pharmaceuticals in nanoparticle drug delivery systems are protected from degradation and allow for regulated release, enhancing biodistribution and allowing them to target regions effected by biological abnormalities (Abbas et al. 2022).

Nanoparticles having at least one dimension ranging from 1 to 100 nm display distinct attributes due to their extremely small size and high surface area to volume ratio, which has resulted in considerable variations in properties compared to their bulk counterparts. Plant extracts are utilized in the green biosynthesis of NPs. Biosynthesis of NPs is a popular field of research due to its various applications in pharmaceutics, biomedicine, agriculture, and industrial domains, as well as being an environmentally responsible strategy that employs nontoxic precursors to limit waste production (Shalaby et al. 2022). Many plants have been investigated as potential NP precursors like mushrooms which contains numerous medicinally important bioactive compounds such as polysaccharides, proteins, lipids, ash, glycosides, alkaloids, volatile oils, tocopherols, phenolics, flavonoids, carotenoids, folates, and ascorbic acid. So, the use of edible and medicinal mushrooms to synthesize nanoparticles has emerged as an interesting subject in the field of medical research (Pathak et al. 2022). Similarly floral nanoparticles have received significant interest as a potential environmentally friendly and sustainable pathway. *Rosa floribunda charisma* is a contemporary rose with brilliant yellow and red blossoms and a wonderful rose aroma. The flower aroma was extracted using several procedures, including hexane, microwave, and solid-phase micro-extraction. This is the most effective process for extracting phenyl ethyl alcohol, which gives roses their distinct perfume. Inas Y. et al. used *Rosa floribunda charisma* petals to produce magnesium nanoparticles (RcNPs), which offer advantages beyond chemical and physical approaches (Younis et al. 2021). Furthermore, marine resources are the most potential source for developing a new generation of biological nanoparticles. So, for the first time, biogenic CeO_2 NPs are using marine oyster extract as an effective and abundant source of bio reducing and capping compounds in a one-pot composition (Safat et al. 2021). The majority of nanoparticle characteristics are size dependent because the unique features of nanoparticles do not emerge until the size is lowered to the nanometer scale.

So, various characterization methods used in the research papers, including Transmission electron microscopy (TEM), X-Ray Diffraction (XRD), Selected Area Electron Diffraction (SAED), Ultraviolet–visible spectrophotometer (UV–Vis), Energy Dispersive X-ray Analysis (EDAX), Zeta Potential (ZP), Surface Plasmon Resonance (SPR),Fourier transform infrared (FTIR), and Scanning electron microscope (SEM), techniques to demonstrated the ability to create stable nanoparticles and to determine their size. Following are a few examples of synthesizing gold nanoparticles (AuNPs), silver nanoparticles (AgNPs), iron nanoparticles (FeNPs), and copper nanoparticles (CuNPs) using bioactive compounds as capping and reducing agents (Table 6.1).

Table 6.1 Different sources and parts used for the synthesis of Ag, Au, Cu, Fe Nanoparticles and their applications

Sr no	Source of bioactive compound	Scientific name	Part of source used	Name of alkaloids/ secondary metabolites	Type of nanometal	Characterization	Size and shape	Application	References
1	Seaweed	*Lophophora variegate*	Seaweed leaf	Anthropogenic organic pollutants, 4-nitrophenol (4NP), 3-nitrophenol (3NP), and 2-nitrophenol (2NP) to the corresponding aminophenols using NaBH4	Gold nanoparticles (AuNPs)	UV–vis, TEM, XRD, SAED, FTIR	Spherical in shape with an average size of 11.69 + 2.38 nm	Reduction of nitrophenols	Kaithavelikkakath Francis et al. (2020)
2	Plant	*Angelica, Hypericum And Hamamelis*	Leaves of *Angelica, Hypericum* And *Hamamelis*	Chloroauric acid (HAuCl4)	AuNPs	UV–Vis, FTIR, TEM, AFM	Spherical, to ovals, heart or polyhedral forms. Diameters ranging from about 4 nm to 8 nm,	Catalysis, biology, optics, electronics, medicine	Pasca et al. (2014)
3	Plant	*Aspalathus linearis*	Leaves	Aspalathin	AuNPs	XRD, SEM and TEM	1.6–6.7 nm for AgNPs and 7.5–12.5 nm for the AuNPs	Targeted drug delivery for cancer cells treatments	Akinfenwa et al. (2021a)
4	Bacteria	*Arthospira platensis exopolysaccharides* (EPS)	Cells	EPS and l-ascorbic acid	AuNPs	TEM analysis, FTIR analysis, UV–Vis	Spherical shape particles size ranges from AuNPs1–9.0 to 30.0 nm, AuNPs2 - 8.0 to 35.0 nm, and AuNPs3–6.0 to 40.0 nm	Breast cancer therapeutics	El-Deeb et al. (2022)

(continued)

Table 6.1 (continued)

Sr no	Source of bioactive compound	Scientific name	Part of source used	Name of alkaloids/ secondary metabolites	Type of nanometal	Characterization	Size and shape	Application	References
5	Plant	3,4,5-trihydroxybenzoic acid	Antioxidant, gallic acid (GA)	Natural phytochemicals such as terpenoids, flavonoids and phenols	AuNPs	UV–vis, DLS analysis, Zeta potential	Ranged between 15 nm and 80 nm,	Supplementary targeting materials	Khan et al. (2020)
6	Plant	*Cymbopogon fexuosus* (CF)	Leaf	Aromatic essential oils	AuNPs	UV–Vis, TEM, XRD, FTIR	The average crystallite size ranged from 10 to 32 nm	Antimicrobial and photocatalytic efcacies	Pathania et al. (2022)
7	Plant	*Citrus*	Aqueous sodium citrate solution	Flavonoids (quercetin (3,5,7,3'-4' pentahydroxy flavone)	AuNPs	TEM, FTIR, UV–Vis	Showed sizes less than 100 nm and a slight spherical shape	Antioxidant, antimicrobial and cytotoxic activities	Milanezi et al. (2019)
8	Plant	*Dalbergia ecastaphyllum*	Honey comb	Brazilian Red Propolis (BRP)	AuNPs	FTIR, TGA, UV–Vis spectroscopy, TEM, EDXS	Mostly spherical shapes, range of 8–15 nm	Antibacterial and antifungal activities,	Botteon et al. (2021)

(continued)

Table 6.1 (continued)

Sr no	Source of bioactive compound	Scientific name	Part of source used	Name of alkaloids/ secondary metabolites	Type of nanometal	Characterization	Size and shape	Application	References
9	Plant	*Galaxaura elongata*	Leaf	Andrographolide, Alloaromadendrene oxide, glutamic acid, hexadecanoic acid, oleic acid, 11-eicosenoic acid, stearic acid, gallic acid, Epigallocatechin Catechin and Epicatechin gallate of the algal extract	AuNPs	TEM analysis, FTIR analysis, UV–Vis	Spherical, a few rod, triangular, truncated triangular and hexagonal shaped nanoparticles. Size range of 3.85–77.13 nm	Antibacterial activity	Abdel-Raouf et al. (2017)
10	Plant	*Citrus maxima*	Peel of *Citrus maxima*	*C. maxima* extract	AuNPs	UV-Vis, TEM, XRD, FTIR	Nearly spherical in shape with the size of 8–25 nm	Antibacterial	Yuan et al. (2017)
11	Plant	*Fagopyrum esculentum* and *lichen Certaria islandica*	Lichens	Hydrolyzed aqueous extracts *C. islandica* and *F. esculentum*	Silver nanoparticles (AgNPs)	Scanning and transmission electron microscopy (SEM and TEM), Fourier transform infrared spectroscopy (FTIR), Total phenolic content (TPC),	Spherical in shape and well dispersed, with average sizes ranging from 10 to 50 nm	Antimicrobial activities against opportunistic pathogenic bacteria	Balčiūnaitienė et al. (2022)
12	Plant	*Tinospora cordifolia*	Tinospora cordifolia leaves	5 mM silver nitrate (AgNO3) solution	AgNPs	XRD, SEM and TEM	Range of 30–50 nm having spherical morphology	Anti-microbial efficacy	Mittal et al. (2017)

(continued)

Table 6.1 (continued)

Sr no	Source of bioactive compound	Scientific name	Part of source used	Name of alkaloids/ secondary metabolites	Type of nanometal	Characterization	Size and shape	Application	References
13	Plant	*Asystasia gangetica*	Leaf	*Asystasia gangetica* leaf extract	AgNPs	UV–vis, Raman and (FTIR), TEM, SEM, EDS, XRD, TGA/DTG	Spherical, size ranging from 10–15 nm	Antibacterial	Jayeoye et al. (2022)
14	Plant	*Xanthoria parietina, Flavopunctelia faventior*	Lichens	Lichen's methanolic extracts	AgNPs	UV–Vis, TEM, DLS, Zeta potential, EDS, FTIR and Gas Chromatography–Mass Spectrometry	Spherical particles of 1–40 nm size	To combat the multi-drug resistance organisms and some cancer cells	Alqahtani et al. (2020a)
15	Plant	*Ligustrum ovalifolium*	Fruit extract	*L.ovalifolium* fruit extract	AgNPs	UV–Vis, TEM, XRD and FTIR	Spherical in shape, with an average size of 7 nm	Ovarian carcinoma cells	Moldovan et al. (2018)
16	Plant	*Origanum onites*	*Origanum onites* extract	Hexane and methanol sequentially and methanol extract	AgNPs	UV–Vis, FTIR, XRD, and SEM, DPPH, ABTS and reducing power assays	Spherical-shaped nanoparticles with the average size of 54.2 nm	Pharmacy, food industry, bio-medicinal field	Genc (2021)
17	Plant	*Citrus limon*	Aqueous citrus limon zest extract,	Aqueous *Citrus limon* zest extract	AgNPs	Zeta potential, DLS, SEM, EDX, XRD, Tem	A spherical particle, relatively face-centered cubic shape, sizes range from 7–28 nm,	Used in biomedical and pharmacological fields	Khane et al. (2022)

(continued)

Table 6.1 (continued)

Sr no	Source of bioactive compound	Scientific name	Part of source used	Name of alkaloids/ secondary metabolites	Type of nanometal	Characterization	Size and shape	Application	References
18	Honey	Honey	Honey	Glucose and fructose	AgNPs	UV–Vis, FTIR, TEM, X-ray analysis	Spherical and monodispersed without major agglomeration, the particle size ranging from 5 to 25 nm,	Catalytic degradation of methylene blue	Al-Zaban et al. (2021)
19	Plant	*Viburnum opulus L*	Fruit extract	Bioactive phytocompounds, (especially phenolics)	AgNPs	UV–Vis, FTIR, TGA, TEM, XRD	Spherical shaped, with an average diameter of 16 nm, monodispersed, face centered cubic nanoparticles	Useful in environmental pollution control	David and Moldovan (2020)
20	Plant	*Caulerpa serrulata*	Green marine algae	*Caulerpa serrulata* aqueous extract	AgNPs	UV–Vis, FTIR, XRD, and HR-TEM	Spherical, average particle size of 10 ± 2 nm	Catalysis and wastewater treatment	Aboelfetoh et al. (2017)
21	Plant	*Artocarpus heterophyll*	Peel extract	Flavonoids, Fernlicht acid	FeNPs	FTIR, SEM, EDX, TEM, XRD	Spherical shape 33 nm	Pollutant removal	Jain et al. (2021)
22	Plant	*Plantain*	Peel extract	Polyphenols and carbohydrates	Fe_3O_4	TEM, EDX, XRD	Size below 50 nm	Pollutant removal	Venkateswarlu et al. (2013)
23	Plant	*Camelia Sinensis*	Leaf extract	Polyphenols	FeNPs	SEM EDX	10–100 nm	Pollutant removal	Ulker ogutveren et al. (2016)

(continued)

Table 6.1 (continued)

Sr no	Source of bioactive compound	Scientific name	Part of source used	Name of alkaloids/ secondary metabolites	Type of nanometal	Characterization	Size and shape	Application	References
24	Plant	*Mimosa pudica*	Root extract	Tannin, ash, calcium oxalate crystals and mimosine	FeNPs	(VSM). (PDA) (SEM), (XRD), (FTIR) spectroscopy UV–Visible	67 nm	Pollutant removal	Niraimathee et al. (2016)
25	Plant	Carica papaya	Leaf extract:	Phytochemicals such as polyphenols, flavonoids, terpenoids, phenolic acids,	(α-Fe_2O_3)	FTIR, XRD, FESEM, EDX and TGA studies	Their average diameter was found to be 21.59 nm	Pollutant removal/ dye treatment	Bhuiyan et al. (2020)
26	Plant	*Couroupita guianensis Aubl*	Fruit extract	10-Octadecenoic acid, methyl ester	Fe_3O_4	UV–visible spectroscopy, FT-IR, XPS, DLS and zeta potential analysis	Mean diameter of 17 ± 10 nm	Drug delivery	Sathishkumar et al. (2018)
27	Plant	*Fragaria ananassa*	Fruit extract	Flavonoids, anthocyanin, quinones	Mesopourous Fe3O4@SiO_2	HRTEM, FESEM, DLS, BET, XRD, FT-IR, EDS and VSM techniques	The mean pore size, surface area, and pore volume of mFSH-SW were 63.2 m^2 g^{-1} , 14.1 nm and 0.24 cm^3 g^{-1} , respectively	Drug delivery	Orooji et al. (2020)
28	Sea weed	*Kappaphycus alvarezii*	Seaweed extract	Alkaloids, saponin, phytosterols, tannins	Fe_3O_4NPs	(XRD), (FTIR) UV–Vis TEM	Spherical Shape average size of 14.7 nm	Anti-inflammatory	Yew et al. (2016)

(continued)

Table 6.1 (continued)

Sr no	Source of bioactive compound	Scientific name	Part of source used	Name of alkaloids/ secondary metabolites	Type of nanometal	Characterization	Size and shape	Application	References
29	Plant	*Plumeria obtusa*	Leaves extract	Quercetin, gallic acid, kaempferol, chrysophanol, naringenin	FeNPs	UV/Vis, FTIR, TEM, SEM, XRD	Spheroidal shape with average size of 50 nm	Antioxidant	S-6 Perveen et al. (2022)
30	Plant	*Psidium Guajava*	Leaf extract	Tannins, spanin, flavonoids, steroids, carbohydrates, polyphenol, glycosides	FeNPs	XRD, Cyclic Voltammetry, SEM-EDAX, FTIR spectroscopy, UV–Visible	27 nm	Antibacterial activity	Jeyasundari et al. (2017)
31	Plant	*Terminalia bellirica*	Fruit extract	Beta-sitosterol, gallic and ellagic acids, ethyl gallate, galloyl glucose, chebulagic acid and a cardiac glycoside, bellaricanin	CuNPs	XRD FTIR HR-TEM TGA	Spherical shape, Diameter 2–7 nm	Antimicrobial	Viswadevarayalu et al. (2016)
32	Plant	*Azadirachta indica*	Flower extract	Nimbin, nimbidin, nimbolide, and limonoids polyphenolic flavonoids,	CuNPs	UV–Vis, ATR-FTIR, SEM, HRTEM, and XRD analyses	Regularly spherical 5 nm	Antimicrobial	Gopalakrishnan and Muniraj (2021)
33	Plant	*E. Caucasicum Trautv*	Leaf extract	Phenols and flavonoids	CuNPs	FTIR XRD SEM	Spherical shape <40 nm	Antibacterial	Hasheminya and Dehghannya (2020a)
34	Plant	*Piper retrofractum Vahl*	Fruit extract	Flavonoid, alkaloid, tannin, and steroid/ terpenoid	CuNPs	UV–Vis, FT-IR, SEM–EDS, TEM, and XRD	Spherical shapes 2–10 nm	Antibacterial	Amaliyah et al. (2020)

(continued)

Table 6.1 (continued)

Sr no	Source of bioactive compound	Scientific name	Part of source used	Name of alkaloids/ secondary metabolites	Type of nanometal	Characterization	Size and shape	Application	References
35	Plant	*Azadirachta Indica*	Leaf extract	Flavonoids, proteins, terpenoids, polyphenols etc.,	CuNPs	SEM, TEM, XRD, FTIR, and Zetasizer analysis	Cubical shape 48 nm stable for 2 months	Cost effective and easy scale up	Nagar and Devra (2018)
36	Plant	*Ixiro coccinea*	Leaf extract	Alkaloids, glycoside, flavonoids, tannins, saponins, steroids, terpenoid	CuNPs	UV–Vis FTIR SEM TEM	Spherical shape <5 nm	Cost effective production of NPs and easy scale up	Vishveshvar et al. (2018)
37	Plant	*Bacopa monnieri (L.) Pennell*	Plant	Saponins, alkaloids, phenolics, flavonoids and DRSC	CuNPs	UV–visible spectrophotometry, FTIR, X-ray diffraction, high-resolution transmission electron microscopy and zeta potential	Diameter <20 nm,	Stress inducer, production of secondary metabolites	(Lala 2020)
38	Plant	*Allium cepa*	White leaves extract	Alkaloids, flavonoids, saponins	CuNPs	Zeta analyzer and scanning electron microscopy	15-30 nm with spherical shape	Stress inducer	(Hussain et al. 2017)
39	Plant	*Blumea balsamifera*	Leaf extracts	Phytochemicals such as flavonoids and terpenoids	CuNPs	EDX, SEM, UV-Vis	Spherical shape with size 30–55 nm	Drug delivery	(Ginting et al. 2020)
40	Plant	*Cardiospermum halicacabum*	Leaf extract	Phenols, Flavonoids	CuNPs	UV–Vis TEM XRD FTIR	30–40 nm with hexagonal shape	Drug delivery	(Punniyakotti et al. 2020)

6.2 Bioactive Compounds as Reducing and Capping Agents

The green chemistry approach has continuously been applied for the synthesis of functional nanomaterials to reduce waste, environmental hazards, and the use of toxic chemicals among other reasons. Bioactive natural compounds have been found great potential in this regard and are used to improve the stability, activity, and biodistribution of metal nanoparticles (MNPs) (Akinfenwa et al. 2021b). Plants, among other biological sources, have received a lot of interest for nanoparticle synthesis. Alkaloids, terpenes, saponins, phenols, alcohols, and proteins are bioactive compounds that function as reducing and capping agents. The extracted bioactive compounds aid in the reduction of size and shape-controlled nanoparticles. These bioactive nanoparticles have a wide range of biological and pharmacological uses. A vivid explanation of the function of isolated bioactive compounds in green nanoparticle synthesis as well as bioactivity and biocompatibility of green biosynthesized nanoparticles is given in this article (Saravanan et al. 2021).

6.2.1 Gold Nanoparticles

Gold nanoparticles may be coated with various materials such as small molecules, biomolecules, and polymers, so they have a wide variety of uses in many domains such as catalysis, sensory probes, drug transport, and therapeutic agents. So far, gold NPs have been produced utilizing secondary metabolites from several plants. There are various publications on gold NPs synthesis available; one research indicates that *Hibiscus rosasinensis* was employed at room temperature to synthesized AuNPs, and the NPs were stable. Another work employed aqueous leaf extract of *Callistemon viminalis* as a reducing agent to generate gold NPs, which were then studied using physicochemical techniques (Nasr-Esfahani & Ensafi 2022) (Vanti & Kurjogi 2021).

In this study, Princy Kaithavelikkakath Francis et al. used the aqueous brown seaweed *L. variegata* extract as a bio reductant to synthesize monodispersed and extremely stable AuNPs in a sustainable manner. Using TEM, XRD, SAED, UV–Vis, EDAX, and FTIR techniques, the characterizations of the biosynthesized AuNPs were studied. The spherical nanoparticles, which had a Surface Plasmon Resonance (SPR) band at 530 nm in the UV–Vis, had a size range of 11.69 ± 2.38 nm. SAED pattern showed bright rings, TEM images clear lattice fringe, and in the XRD pattern (111), (200), and (220) Bragg's reflections showed that the Au nanoparticles were crystalline. It can be noted that the AuNPs contained all significant peaks in the aqueous seaweed extract, revealing the contribution of the seaweed extract to the reduction and capping of the AuNPs. O–H stretching vibrations from phenols and alcohols are also what cause the strong intensity broadband to emerge between 3400 and 3450 cm^{-1}, C-N stretching vibration is thought to be accountable for the band at 1385 cm^{-1}, carboxylic acid functional group is indicated by the absorption peak at 1028 cm^{-1}. The carbonyl group of proteins –$C_{14}O$ group may be the reason of another high peak at 1624 cm^{-1}. The vibrations produced by the C-H stretching might be attributed to a small peak at 2928 cm^{-1}. The research focuses on efficiency of synthesized AuNPs as eminent catalysts in the degradation of contaminants like 4-nitrophenol (4NP), 3-nitrophenol (3NP), 2-nitrophenol (2NP) and using $NaBH_4$ to the specific amino phenols, rate constants noticed were in the reducing manner k2NP > k4NP > k3NP. Due to their outstanding SPR in the visible region, AuNPs are of particular interest among noble metal nanoparticles and have numerous applications in a variety of industries, including drug delivery, surface-enhanced Raman scattering, optoelectronics, bio-imaging, biological labelling, catalysis, and antimicrobial (Kaithavelikkakath Francis et al. 2020).

Furthermore, Roxana-Diana Pasca et al. for the green synthesis of Gold NPs in this research used plant extracts from *hypericum, hamamelis, angelica* for the reduction of Chloroauric acid ($H[AuCl_4]$). The UV–Vis absorption spectra ranged from 190–900 nm. The UV–Vis spectral data of the colored plant extracts reveal absorption peaks near UV and also in the visible region, whereas $H[AuCl_4]$ absorbs in 220–230 nm range, further in FTIR, spectra in the 4000–400 cm^{-1} region were acquired for each plant extract and the AuNPs using an FTIR spectrometer and the KBr pellet technique. *Angelica* extract exhibits the maximum absorption at 3370 cm^{-1} that correlates to the hydroxyl group in alcohols and phenols also the intensity of this band is much lessened once Au(III) is reduced in an alcoholic solution, band at 1055 cm^{-1} is due to the C-O groups of the polyols contained in *Angelica* extract, such as polysaccharides, flavones, and terpenoids, similarly reduces after the Au(III) reduction. The symmetry and asymmetry of the two bands of AuNPs higher than original extract was seen at 1410 cm^{-1} and 1655 cm^{-1}, COO stretching indicated the formation of carboxylic acids by the oxidation of components in *Angelica* extract, C-H stretching were seen at band 2925 cm^{-1} due to methylene, are not totally different either of these two samples. The FTIR spectra of *hypericum and hamamelis* extracts indicated the same properties before and after AuNPs biosynthesis. The sizes of the AuNPs, which ranged from around 4 nm to 8 nm, were obtained with a pH of about 8 and at ambient temperature and they display different forms from round, oval, heart-shaped and polyhedral shapes. With plant extracts from *Angelica, Hamamelis*, and *Hypericum*, Au(III) was quickly reduced to Au(0) at normal room temperature and for high dilutions of the plant extract, indicating the presence of a significant amount of a reducing agent. The AuNPs systems were most stable at an alkaline pH of 8–10. The spontaneous assembly of AuNPs is rarely observed, and the colloidal dispersions of AuNPs are usually very stable over time. The probability for the AuNPs to self-aggregate, however, increased with lower concentrations of the plant extract. The plant extracts contain reducing agents, substances that stabilize AuNPs, as well as elements that facilitate their self-assembly. Considering at least two properties, their size and water stability, the AuNPs synthesized by these biogenic syntheses provide many biological and medicinal applications (Pasca et al. 2014).

Similarly, Neveen Abdel-Raouf et al. used *Galaxaura elongata* powder or extract to synthesis AuNps in this study. It was found that stable AuNPs develop fast in an aqueous medium of *Galaxaura elongata* extract at standard air conditions. Most of the particles were spherical, with a some of hexagonal, triangular, rods and truncated shaped nanoparticles as well, according to TEM analysis. The size range of the Au nanoparticles according to zeta potential studies is 3.85–77.13 nm. The use of FTIR demonstrated that alga chemicals were used to cap the nanoparticles. Bands which are associated with amide bonds of proteins, are present at nearly 2158.12, 332.49 and 1635.09 cm^{-1} in aq. extract state and 2158.28, 3414.96, and 1640.60 cm^{-1} in powder state and are caused, respectively, by free –N–H-stretch and carbonyl stretch vibrations in the amide linkages of the proteins. The carbonyl group forms amino acid residues, and peptides of proteins have a better propensity to link metal, according to the FTIR spectroscopic analysis. As a result, AuNPs will be capped to avoid particle aggregation and stability in the medium. With these results, it is conceivable that biological molecules have a contribution to the stabilization and formation of AuNPs in aqueous medium. The OH stretching band sharpened with no visible displacement seen in the FTIR spectrum because it overlaps with other bands. It was anticipated that the infrared bands associated with the carboxyl groups would shift. The change in color from light yellow to red in the reaction solution of metal ion solution and algal extract or powder was seen clearly. The solution was then analyzed under UV–Vis, measured at 500–600 nm using a quartz cuvette in order to observe the bio reduction of gold ions in aqueous solution. Using the GC–MS QP5050A system, the chemical composition of the crude ethanolic extract was examined for the concentrations of fatty acids, flavonoids and amino acids were studied using an HPLC system. The chemical components of the algal extract, including andrographolide, glutamic acid, hexadecanoic acid, alloaromadendrene oxide, stearic acid, oleic acid, gandoic acid, gallic acid, epigallocatechin catechin, and epicatechin gallate, which may function as capping and stabilizing agent, were identified. The biologically produced AuNPs were tested for their antibacterial efficiency against five harmful bacteria and shown effective bactericidal activity. They tested for their antibacterial properties was against *Escherichia coli*, MRSA, and *Klebsiella pneumoniae*, AuNPs synthesized by ethanolic extract showed better antibacterial properties, with inhibition zones of 17–16 mm. *Staphylococcus aureus* and *Pseudomonas aeruginosa* came in second and third. Furthermore, it was discovered that the NPs synthesized with the powder of *Galaxaura elongata* were quite effective against *K. pneumoniae* and *E. coli.* However, *Galaxaura elongata*'s free ethanolic extract only shows strong effectiveness against MRSA. The findings of this work indicate that this technology for synthesizing Gold NPs from *Galaxaura elongata* cell extracts is an appealing green procedure that is economical, ecologically friendly, and beneficial for producing a large quantity of Gold NPs (Abdel-Raouf et al. 2017).

Moreover, Akeem O. Akinfenwa et al. used Green Roobios and *Aspalathus linearis* aq. extracts to biosynthesis gold NPs. The research also found that when compared to whole plant extracts, the pure component had a relative benefit for lower sizes of nanoparticle forms. The finding shows that ASP and GR might function as stabilizer and reducing agents in the nano particle synthesis, the dispersity of crystals were in the ranges of 7.5–12.5 nm. The crystalline nature of the AuNPs is confirmed by X-ray crystallography at room temperature. The AuNPs for both GR and ASP showed four unique Bragg peaks located at 38.185°, 44.393°, 64.578°, and 77.549° (2θ), corresponding to crystallographic reflections (111), (200), (220), and (311) from crystalline gold planes. The prominent peaks found at 1050–1076, 1698–1715, 2950–3000, and 3300–3340 cm^{-1} corresponded to their respective C–O (phenolic), C = O (α, β-unsaturated ketone), C–H (aromatics), and a wide O–H vibrational stretching bonds in the FTIR of the GR and ASP and their biosynthesized nanoparticles. The functional groups observed in GR and ASP were likewise detected in AuNPs, according to the IR spectra. In the two NPs, the C-H vibrational frequencies of ASP in the AuNPs were altered from 2995 to 3000 cm^{-1}. Furthermore, the C-O bands of GR and biosynthesized NPs were blue-shifted by around 3 cm^{-1} in the NPs. The spectra also revealed that all of the NPs had greater peak intensities prior to production. These differences in absorptions before and after synthesis indicated that they were involved in the bio-reduction process. In the UV–Vis ranges of 400–500 nm (yellowish-brown) for the AuNPs is known to exhibit distinctive SPR. According to a cytotoxic potency comparison of the IC50 values of the original and the synthesized AgNPs from both the plants, the synthesized AgNPs of Green Roobios and ASP were more efficient in stopping growth of cancer cells than the pure substance or the plant extract alone. The cellular uptake study revealed considerable AuNP uptake and indicated that rooibos AgNPs at a lower concentration (1.3–1.5 g/mL) are effective for their anticancer properties. The biosynthesized AuNPs can be utilized in medical applications such as medication delivery to cancer cells (Akinfenwa et al. 2021a).

Diksha Pathania et al. used an economically beneficial medicinal plant called *Cymbopogon fexuosus* (CF) which function as a capping, reducing, and stabilizing agent, yields essential oil that has exceptional physicochemical, anti-tumor, anti-inflammatory, antioxidant, photocatalytic characteristics, antifungal, and antibacterial properties. The application of aromatic essential oils in the bio fabrication of AuNPs is emphasized due to its ease of use, cost, low toxicity, and renewability. The UV–Visible peaks of AuS1, AuS2, and AuS3 at 558, 559, and 559 nm, respectively, confirmed the change in color from yellow to wine red (Fig. 6.1). FTIR spectra showed the presence of primary amide, amine, aldehyde, ketone, alcohols, alkene, and ethers in synthesized samples. The presence of hydroxyl and amino groups is shown by the appearance of distinctive peaks at 3311, 3251, 3636, and 3457 cm $^{-1}$. The occurrence of carboxylate groups is shown by the peaks at 1429, 1426, and 1465 cm $^{-1}$. C-O stretching is corelated by the peaks at 1080, 1090, 1081, 1047, and 952 cm $^{-1}$. C-N stretching measurements at 1185, 1187, and 1085 cm^{-1} were obtained. The NH, OH, C = O, and CH-OH functional groups are represented by significant peaks in the FTIR of essential oil-mediated AuNPs at 3457, 2924, 1664, and 1085 cm^{-1}. As a result of its antioxidant characteristics, the FTIR data reveal that the essential oil has a remarkable capacity to produce AuNPs. The crystalline planes (111), (200), (220), and (311) is assigned to the 38.22°, 44.52°, 64.54°, and 77.69° diffraction peaks lying on 2θ, respectively, since they indicate the face-centered cubic arrangement of Gold NPs. It is obvious that (111) dominates in intensity, showing that (111) is the primary direction in which particles are orientated. The results of the SEM scanning showed that the AuNPs that were mediated by the essential oils have a diffused morphology and an irregular nano-disc shape. The branched structure was generated by the self-assembly of smaller crystallites, according to the TEM images.

The presence of essential oil in synthesized AuNPs can be clearly predicted as the cause of high aggregation with diffused morphology. According to studies, crystallite sizes typically fall between 4 and 11 nm. The XRD research is inadequately in accordance with the average crystalline size, which ranged from 10 to 32 nm. The recently created AuNPs had a remarkable photodegradation efficacy of about 91.8% in three hours against the methylene blue dye. *Staphylococcus aureus, Escherichia coli*, and *Fusarium oxysporum* all were examined individually, and the synthesized AuNPs demonstrated remarkable antibacterial activity against all of them, indicating their ability to treat infectious diseases as an effective antibacterial. AuNPs are also useful for a range of green technology applications backed up by pure water, plant-extracted essential oils, and mostly harmless compounds (Pathania et al. 2022).

Since honey exhibits significant pharmacological activities such as anti-inflammatory, anti-oxidant, and antimicrobial properties it can also be utilized as a bioactive compound to reduce the gold NPs. C.E.A. Botteon et al. used Brazilian Red Propolis, which is a bee product that has anti-inflammatory, antioxidant, anti-tumor, antibacterial effects, and is regarded as an example of an affordable, simple, and ecofriendly technique for the synthesis of AuNPs utilizing plant extract. Brazilian Red Propolis extract were processed with fractions of ethyl acetate, hexane and dichloromethane for biosynthesis of AuNPs, and their structural characteristics and potential for use against cancer cells and microbes were assessed. Maximum absorption was observed between 500 and 550 nm via SPR Band, using UV–Visible for confirmation of AuNPs biosynthesis. The hydroethanolic extract and fractions of Brazilian Red Propolis exhibited a high potential for producing AuNPs with sizes

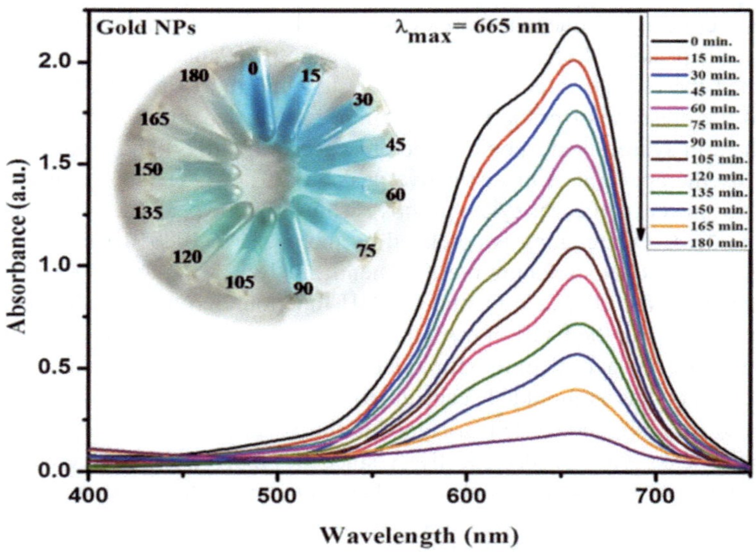

Fig. 6.1 UV–Vis absorption spectra of methylene blue dye using AuNPs. Reprinted with permission from (Pathania et al. 2022)

ranging from 8 to 15 nm, thus AuNPs with varying morphologies were seen. Ethyl acetate fraction and BRP crude extract and yielded spherical AuNPs, but dichloromethane and hexane fractions yielded AuNPs of varying forms also crystalline structure of all AuNPs was a FCC lattice. Further confirmation of crystalline nature was done via SAED pattern which shows circular rings that attributed to (220) (111), (311) and (200) Bragg's reflection planes, exhibited diffraction peak was about 1700 cm^{-1}, and the spectra of dichloromethane fractions are related to carbonyl groups. The peaks at 1600 cm^{-1} are due to COO^- stretching vibration, which is most likely caused by polyphenol oxidation during Au^{+3} reduction, all of these bands suggest that plant extract contain significant amounts of phenol and alcohol. It is noted that hydroxyl groups were being changed to carbonyl groups, because of thinner band at 3400 cm^{-1} in Au nanoparticles than in extract FTIR spectra, which also suggests that it is the OH group from the extract that cause the reduction of Au ions. The absorption peaks at 1500 cm^{-1} are most likely generated by unbound NH groups present in proteins. The fact that the strength of this band dropped as AuNPs formed shows that the extract's proteins were also used to cap AuNPs, thus enhancing their stability. The lower peak intensity at 800 cm^{-1} supports the coupling between the AuNPs and C-H group of phenolic acids. AuNPs produced using the hexane fractions and dichloromethane exhibited strong antifungal and antibacterial properties as well as cytotoxicity in the two cell types under study. AuNPs in vitro results show the biogenic Gold NPs' therapeutic potential in nanomedicine (Botteon et al. 2021).

The potential of *Arthospira platensis's* exopolysaccharides to reduce gold ions into three different types of AuNPs has been shown in this study by Nehal M. El-Deeb et al., and the synthesized nanoparticles were stable for more than three months. Algal polysaccharides can be used to biosynthesize AuNPs in an easy, economical, and environmentally friendly manner. The three different kinds of biologically generated $AuNPs_1$ which is 1:1 molar ratio of $NaAuCl_4$, $AuNPs_2$ which is 2:1 molar ratio of $NaAuCl_4$, and $AuNPs_3$ which is 1:1 molar ratio $NaAuCl_4$. The stability of AuNPs was examined using UV–Vis analysis. The peak Plasmon absorptions were used to normalize the three spectra, which were 530.0, 540.0, and 550.0 nm. The aggregation state was identified visually by observing colour change to purple from red. Here, a blue shift was seen between 550 and 530 nm. The morphology of the biogenic AuNPs is a spherical shape was confirmed utilizing TEM scanning. It also showed that the $AuNPs_1$ particles ranged in size from 9.0 to 30.0 nm, $AuNPs_2$ particles ranged between 8–35 nm, and Au $Nanoparticles_3$ ranged from 6–40 nm. The synthesized biogenic AuNPs showed absorption peaks at 1540.0, 1035.0, 1540.0, 1450.0, 1640.0, and 3445.0 cm^{-1} upon FTIR examination, OH group may be identified by the large peak in the nanoparticles' spectra at around 3445.0 cm^{-1}. This shows that the hydroxyl groups are responsible for the reduction of Au ions, absorption band around 1640 cm^{-1} can be attributed to the $C = O$ stretching vibration of secondary amide groups, detected peaks at about 1540 and 1450 cm^{-1} is due to the stretching –COO group vibrations. Amide groups of aromatic amines and $S = O$ of sulfated extracted polysaccharides are responsible for the band at around 1035.0 cm^{-1}. In the microplate experiment, $AuNPs_3$ and $AuNPs_1$ mostly affected

E. coli, with inhibition of 88.92% and 83.13%, respectively. Similarly, $AuNPs_3$ had the lowest MIC of 100.0 g/ml against *E. coli.* With regard to *C. albicans,* treatment with $AuNPs_2$ significantly slowed its development, with a reduction percentage of 82.83%. Moreover, with values of 65.51 and 58.57%, respectively, both $AuNPs_1$ and $AuNPs_2$ demonstrated moderate to significant growth inhibition percentages for *C. tropicalis.* $AuNPs_3$ was active against *S. enterica and E. faecalis*, with inhibition of 75.35% and 73.76%, respectively. The strain of *S. mutans* that responded the least favorably to AuNPs treatment had a maximum inhibition percentage of 40.0%. With a value of 90.0 g/ml, the *C. tropicalis* MIC was found to be the minimum after treatment with AuNPs. The evaluated cell lines and microbial strains were very susceptible to the biosynthesized nanoparticles powerful antibacterial and cytotoxic activities. This study paves the way for future cancer and microbial medicines that use biogenic Au nanoparticles either alone or in addition with chemotherapeutics and antibiotics (El-Deeb et al. 2022).

Similar to the *Arthospira platensis's* exopolysaccharides' antibacterial and cytotoxic activities, free quercetin also has anti-inflammatory and antihistamine effect. So, Felipe Guzansky Milanezi et al. synthesized AuNPs using quercetin as capping agent and to assess its antioxidant, cytotoxic, antimicrobial activities in contrast with free quercetin. By reducing AuCl4 utilizing sodium citrate as the reducing agent, Gold NPs were synthesized. At 520 nm, the recognizable localized surface plasmon resonance band was seen, and the sample's color changed from red to violet hues, indicating that AuNPs had successfully been synthesized. The AuNPs are geometrically Quasi spherical, exhibit diameters less than 100 nm, and exhibit highly monodisperse particle sizes under these growth and nucleation environments, according to TEM images. The characteristic peaks of quercetin at 3248, 1670, and 1500 cm^{-1}, inferring to the O–H stretching, $C = O$ stretching, and $C = C$ stretching respectively was provided by FTIR analysis which corroborated quercetin capping on the AuNPs. Additionally, it was found that the absorption bands between 650 and 1000 cm^{-1} linked to the distortion of angles of the aromatic compounds' $C = CH$ were detected, which is consistent with earlier work. The zeta potential test was used to verify the colloidal stability of the AuNPs that were synthesized with a 79% entrapment efficacy. Quercetin's molecular structure remained unaltered, and the phenolic hydroxyls that are crucial to its principal pharmacological effects as an antioxidant were retained, according to the characterization investigations. The findings of the antioxidant experiments, which demonstrated robust action, supported this theory. Additionally, the activity for AuNPs was much greater than that of free quercetin by the nitric oxide free radical scavenging technique. Both free quercetin and AuNPs at the studied concentrations had no harmful effects on the L929 fibroblast cells. The strains of *A. fumigatus* isolated from patients with aspergillosis were confirmed to have a significant antifungal effect (Milanezi et al. 2019).

Apart from the leaf extracts of a plant, its peel extract can also be utilized to cap AuNPs. In this research Chun-Gang Yuan et al. described a straightforward and environmentally friendly method for synthesizing AuNPs utilizing *C. maxima* peel extract as reducing agent. With a resolution of 1 nm, UV–vis from 300 to 800 nm were used to scan the synthesized AuNPs. The produced AuNPs were found to be evenly dispersed, mostly sub spheroidal in form, and ranging in size from 8 to 25 nm according to the TEM images. The diffraction peaks at (2θ) 38.30°, 44.28°, 64.62°, 77.57° and 81.75° were assigned to (111), (200), (220), (311) and (222) planes of the FCC lattice of Au, confirming the crystalline nature of the AuNPs. The characteristic peaks at 3411, 2925, 1619, 1051, and 622 cm^{-1} were also seen in the biosynthesized AuNPs. The stretching vibrations of C = C were a contributing factor in the peak at 1619 cm^{-1}. The stretching vibrations of O–H of phenols and carboxylic acids were responsible for the peak at around 3411 cm^{-1}. Aliphatic acid asymmetric and symmetric C-H stretching vibrations were attributed to the prominent signal at 2925 cm^{-1}. Other peaks at 1051 and 622 cm^{-1} might be attributed to bending vibrations of the C–O and C–H chains. It has been established that the main active ingredients in the C. maxima peel were naringin and hesperidin. The synthesized AuNPs shown excellent antibacterial activity against both *E. coli* gram negative bacteria and *S. aureus* gram positive bacteria as well as high catalytic activity in the breakdown of 4-NP (Yuan et al. 2017).

Nonetheless acids can also be used to cap AuNPs. Musammir Khan et al. produced AuNPs utilizing an environmentally friendly process that included the antioxidant Gallic acid (GA), which acted as a reducing agent. The SPR peak was clearly red-shifted and broadened in the UV–Vis spectra of GA-AuNPs with a higher concentration of GA. This SPR shift had a maximum shift from 518 to 531 nm and was linear in the examined GA-AuNPs concentration range (0.625–2.5 mM). With the addition of GA, AuNPs' colour changed from red to dark purple, indicating an increase in particle size. The size of the AuNPs grew linearly with the GA content, according to the Nanosizer/DLS study. This size gain for AuNPs was from 18 to 59 nm in the GA concentration range that was studied, which was between 0 mM and 2.5 mM. AuNPs first noticed a drop in zeta potential from 36 to 42 mV before rising to 36 mV. The GA-AuNPs had a spherical shape, a monodisperse size distribution, and an average particle size of 24 nm, according to the SEM images. The human fibroblast cell line (WI-38) cytotoxicity assay for 24 h and 3 days revealed that they were totally safe for biomedical use. Consequently, it is possible to predict that AuNPs will have significant uses in the field of nano biotechnology in the future based on these obtained results (Khan et al. 2020).

6.2.2 Silver Nanoparticles

Since earlier times, silver salts have been found to have positive effects to increase a material's biocompatibility, it is effective and dependable to reduce the size of its particles. There are numerous ways to create nanoparticles, including physical, chemical, and biological processes.

Enzyme-assisted extract is a valuable approach for producing sugars and bioactive compounds, both of which are reducing and capping agents required for the green synthesis of AgNPs. Viktorija Puzeryte˙ et al. used Lichen *Certaria islandica* and pseudo-cereal *Fagopyrum esculentum* extracts as eco-friendly catalysts for the synthesis while being heated in an aqueous solution. When AgNPs are present, there is a progressive change in colour from a mild yellow to a dark yellow. AgNPs made from enzyme-assisted *C. islandica* and *F. esculentum* extracts displayed several stretching vibration bands linked to organic functional groups in their FTIR spectra. Round rings of *C. islandica* are seen at 1050–1025 cm^{-1} indicating C–O phenolic stretch and 1380–1360 cm^{-1} indicating O–H stretch. Alkanes, C-H, are also present in the 2917–2833 cm^{-1} stretch. Aldehydes, carboxylic, ketone, or ester-containing compounds were found when the stretching mode was seen at 1640–1620 cm^{-1} indicating C = O stretch. It is significant to note that the green produced AgNPs' FTIR spectra failed to reveal the nitro bond at 1280 cm^{-1}, confirming the lack of observable nitrate residue. These findings support the effective production of the Silver NPs, which were synthesized using a process that reduced secondary metabolites called Phyto molecules. The produced nanoparticles had spherical shapes, were evenly scattered, and ranged in size from 10 to 50 nm on average, as per TEM analysis. This study tested the total phenolic content (TPC) and in vitro antioxidant activity in both substances using standard procedures. The results showed that there are significant differences in the amounts of TPC, ABTS·+ , FRAP, and DPPH· radical scavenging activity amongst the samples. The AgNPs synthesized from the enzymatic hydrolyzed aqueous extracts of *F. esculentum* and *C. islandica* demonstrated more antibacterial activity against the tested bacterial pathogens than their respective crude extracts. As a viable addition to technological processes, the results show that the biomolecules covering the Silver NPs in the extracts may increase their biological activity as AgNPs and aid in the synthesis of essential media components by assisting enzymes. AgNPs also showed very good antibacterial activity against all tested Gram-negative and Gram-positive bacterial strains as well as fungi, which are harmful pathogens frequently causing contagious illness, according to the antimicrobial evaluation of AgNPs. Future research is also required to thoroughly understand how enzymes affect the stability and toxicity of AgNP (Balčiūnaitienė et al. 2022).

Similarly, Mona A. Alqahtani et al. objective was to find out how well lichen can produce AgNPs, as well as how effective they are at killing MDR bacteria and other harmful microorganisms. For the transformation of Ag ions into AgNPs, lichen *(Xanthoria parietina, Flavopunctelia faventior)* methanolic extracts were used as a capping and reducing agent. Surface plasmon resonance peaks for Xa-AgNPs and Ff-AgNPs were around 412 and 405 nm, respectively according to an assessment of UV–Vis spectroscopy. TEM scans revealed spherical and well-dispersed nanoparticles with 1–40 nm size for Ff-AgNPs and Xa-AgNPs. Average DLS diameters for Ff-AgNPs and Xa-AgNPs were 69 nm and 145 nm, respectively, with polydispersity index of 0.458 and 0.291. Zeta Potential values for Ff-AgNPs and Xa-AgNPs were –20 mV and –24, respectively. Both lichen species had absorbance peaks in the FTIR spectra at 3421, 2066, 1634, and 593 cm^{-1}. Xa-AgNPs and Ff-AgNPs showed absorbance maxima at 3332, 3421, 2070, 541, and 1637 cm^{-1}, suggesting the play of different functional groups in the bio-reduction of $AgNO_3$. The existence of -OH group and N–H stretching of an amine was detected in the spectrum at 3300–3500 cm^{-1}. On the other hand, it was discovered that the C = C stretch might be responsible for the bands at 2066 and 2070 cm^{-1}. Amide may be responsible for the strong absorption bands at 1600–1650 cm^{-1} since proteins C = O stretch causes carbonyl stretch. Little difference was seen in the positions and intensities of the absorption bands when the spectra of AgNPs and lichen extracts were compared, confirming the use of lichen secondary metabolites in AgNps synthesis. Pathogenic bacteria, Vancomycin-resistant Enterococcus (VRE), and gram-negative *Pseudomonas aeruginosa* and *Escherichia coli,* including gram-positive Methicillin-resistant *Staphylococcus aureus* (MRSA), as well as the reference strains (ATCC), were evaluated for antibacterial activity of AgNPs using the agar disc diffusion method. *P. aeruginosa* demonstrated the maximum antibacterial activity of AgNPs, followed by *Staphylococcus aureus*, Enterococcus, and *E. coli.* Gram-negative bacteria were more sensitive to the effects of AgNPs than gram-positive bacteria, and they worked in combination with some antibiotics to combat the pathogens under research. Furthermore, compared to MDA-MB-231, biogenic AgNPs against the FaDu and HCT 116 cell lines exhibited increased cytotoxicity. According to the most recent findings, lichen-mediated biogenic AgNPs displayed potent antibacterial, synergistic, and cytotoxic properties. As a result, they might be viewed as a strong contender to fight off some cancer cells and multi-drug resistant organisms. AgNPs synthesized for the current study may therefore be suggested for use in pharmaceutical and biomedical applications. This research is regarded as the first to examine and validate the activity of synthesized AgNPs made from the lichens *Xanthoria parietina* and *Flavopunctelia faventior* against harmful bacteria as well as their cytotoxicity against three cell lines (Alqahtani et al. 2020b).

The focus of Yasmina Khane et al. was on the environmentally friendly biosynthesis of AgNPs using an extract of aqueous *Citrus limon* zest while optimizing various experimental parameters necessary for AgNP production and stability. The formation of nanoparticles was founded by detecting the SPR band at 535.5 nm using UV–Vis study, which was supported by the analysis that the colour of the mixture of silver nitrate changed to a reddish-brown colloidal suspension from yellow just after addition of the plant extract. The plant extract's FTIR spectra revealed the occurrence of absorption peaks at 2723 cm^{-1}, 1042.18, 1157 that were related with the stretching of the N–O and C-H methylene groups, respectively, and the carbonyl groups of polyols such as the polysaccharides and flavones. Peaks at 1630.121 cm $^{-1}$ and 1412.26 were related with primary amine carbonyl stretch N–H and C = O bonds, respectively. 1634 cm $^{-1}$ and the 3325 cm^{-1} peak were found to be caused by the amide functional groups of different aromatics and the protein carbonyl groups. The presence of aliphatic hydrocarbon chains, C-H group stretching from alcohol, carboxylic acid, and phenolic compounds is indicated by a peak at 2925 cm $^{-1}$, whereas the (OH) presence in the molecule is indicated by a band about 3431 cm $^{-1}$. The C-N stretching of amines appears around 1057 cm $^{-1}$ and the existence of amines is confirmed by the band at 3431 cm^{-1}. The FTIR spectra showed a substantial change in absorption peaks at 1634 1021, 1443 and 3428 cm^{-1} after the addition of $AgNO_3$ aqueous solution with bio extract, indicating that the functional groups interacted with the surface of the AgNPs. Using the lemon zest aqueous extract, the XRD technique was used to verify the UV–Vis study and highlight the crystal structure of the synthesized AgNPs. The crystallographic planes (111), (200), (220), and (311) correspond to the diffraction pattern at 2θ of 38.18°, 44.34°, 64.53°, and 77.46°, respectively. The avg. size of the Ag nanoparticles was found to be 82.51 nm by DLS, and the pdi index came to be 0.248. It was also used to calculate the thickness of the capping or material that encased the nanoparticles. Flavonoids and phenolic compounds, among the phytochemicals found in lemon zest aqueous solution, may be responsible for the reduction to silver ions from silver nitrate, which then aggregate to form AgNPs at the nanoscale range and have a stronger ability to stabilize the forming AgNPs. As they surround the AgNP core, these biomolecules have been exploited as a natural reducing agent. Silver NPs created in this study demonstrated an extremely intriguing capability to kill pathogenic bacteria and fungi, recognizing their therapeutic value as antioxidant and antibacterial agents against antibacterial drugs drug-resistant strains, as well as utilization in pharmacological and biological fields. *E. coli* and *S. aureus* pathogen; were both successfully eradicated by the AgNPs, which also shown excellent antifungal efficacy against *Candida albicans* (Khane et al. 2022).

The focus of Nusret Genc was to assess the antioxidant activity of AgNPs, which were synthesized in a sustainable way utilizing *Origanum onites* extract. Hexane and methanol were used to successively extract the origanum onites, and the methanol extract was then utilized to synthesize the AgNPs. The plant leaf was utilized to produce a large number of AgNPs by being boiled in water. Both secondary metabolites and fatty acid components are present in the water extract. Hexane was used in this study to extract the plant material and get rid of the fatty acids. As a result, significant bioactive substances like alkaloids and flavonoids were employed to cap the silver to create Silver NPs that might boost effectiveness. The reaction mixture's transition from bright yellow to dark brown verified the reduction of Ag^{+1} ions. AgNP formation was also confirmed by UV–Visible analysis, which was used to track the reaction's development during the Ag^{+1} reduction. The visible range between 350 and 550 nm displays a prominent band in the AgNPs' UV–Vis spectra. Additionally, the presence of spherical nanoparticles is indicated by the discovery of an absorption peak in UV–Visible spectra ranging from 410–450 nm. AgNPs were found to be crystalline by x-ray diffraction. The XRD with 2θ at 38.1°, 44.3°, 64.4°, and 77.4° is responsible for the (111), (200), (220), and (311) Bragg's reflections of the FCC crystalline structure of AgNPs, respectively. The UV–Vis scanning ranged from 300 to 800 nm. At 450 nm, the greatest absorption was recorded. AgNPs' crystalline nature was confirmed by an XRD analysis. The surface of the AgNPs included highly dense, spherical nanoparticles that were evenly scattered. The creation of AgNPs was discovered via energy dispersive x-ray analysis. The measured average particle size was 54.2 nm according to SEM analysis. The structure of possible natural compounds that may have coated the AgNPs surface and effectively stabilized it was identified using FTIR. The synthesis of AgNps was identified by comparing the FTIR spectra of extract and Silver NPs. A wide band at 3239 cm^{-1} in the FTIR spectra of Silver NPs might be ascribed to the OH stretching. Asymmetric stretching of C-H bonds was the cause of the signals at 2923 and 2851 cm^{-1}. Carbonyl groups might be attributed to the peaks at 1703 and 1594 cm^{-1}. The signal also belonged to the functional groups C = C and O–H at 1512 cm^{-1} as a stretching vibration and 1372 cm^{-1} as a bending vibration, respectively. Amine was given credit for the peak at 1255 cm^{-1} in the extract, which was not seen in the spectra of nanoparticles. This may be the result of the matching amine group being oxidized. Alcohol's C–O stretching may be the cause of the peaks at 1159 cm^{-1} found solely in the spectra of nanoparticles. It was determined that $S^{14}O$ stretching was the cause of the peak at 1033 cm^{-1}. The extract and synthesized Silver NPs showed remarkable antioxidant activity. They can therefore be utilized in the food industry as an antioxidant and may prove to be useful components in the bio-medical industry (Genc 2021).

Eman F. Aboelfetoh et al. focused on a usable, environmentally friendly method to synthesize stable AgNPs from silver nitrate solution by using an aqueous extract from the green marine algae *Caulerpa serrulata*, which is an effective reducing and capping agent. The biosynthesized AgNPs at ideal conditions [20% (v/v) extract at 95 °C], displayed a strong SPR band at 412 nm. The Ultraviolet–visible spectroscopy and color change were used to examine the production of AgNPs. The solution's hue changed from pale yellow to reddish brown, indicating the formation of Silver NPs. The XRD patterns show prominent diffraction peaks for AgNPs at $2 = 38.13°, 44.31°$, 65.16°, and 77.25°; these angles correlate to the (111), (200), (220), and (311) lattice planes of a FCC crystal arrangement. HR-TEM analysis was used to gather further details regarding the shape and size of the produced AgNPs. It was found that the conc. of the extract has a substantial effect on the size and structure of nanoparticles. The majority of the AgNPs had spherical and elliptical forms, with small numbers of silver nano cubic and nanorods, at 95 °C and at a lower concentration of extract (10%). Silver NPs remained spherical in form, but their particle sizes dropped as extract concentration increased to 20% at a steady temperature, according to TEM examination. The results indicated that AgNPs were produced over a period of 24 h and that Ag ion reduction rates were accelerated at higher temperatures (95 °C) and higher pH levels. This study demonstrates the potential of creating a catalyst that is extremely effective, stable, economical, and easily recoverable and used to remove harmful dyes and other pollutions from the environment. This may be accomplished by synthesizing AgNPs on a wide range of solid surfaces, including graphene oxide, carbon nanotubes, silica, TiO_2, and Fe_3O_4. As a conclusion, the AgNPs synthesized by the current method will find great use in catalysis and wastewater treatment (Aboelfetoh et al. 2017).

For the first time the utilization of plants in the synthesis of silver NPs is described in this paper. Jayeoye et al. utilized *Asystasia gangetica* phenolic extract (AGPE) as the reducing agent to synthesize highly stable, distributed Konjac glucomannan silver NPs. The plant's bioactive substances hold considerable potential for further research into *A. gangetica's* probable pharmaco-biological applications. The particles exhibit spherical morphologies and are uniformly mono-dispersed. The average particle size acquired from TEM showed the particles are dispersed in a range of 10–15 nm. Using DLS, it was determined that the hydrodynamic diameter was 60.2 ± 1.5 nm and the ZP value was − 28.8 ± 2.1 mV. Further indicating the stability of the particles and their –ve charges, which may originate from the COOH group of the KgM polymeric shell or spheres. Zeta potential profiling to determine how pH affects the stability of KgM-AgNPs in aqueous solution. As demonstrated, the ZP values rise as the pH of the solution rises from 2 to 4, stabilizing at 12, and then decreasing. This confirms that KgM Silver NPs is stable over a broad pH range, which may be linked to the KgM COOH group that has been deprotonated on the surfaces of the nanoparticles. The crystalline diffraction lines of AgNPs are responsible for the numerous concentric circles shown in the SAED. The morphology of the KgM film was smooth, but the morphology of the KgM-AgNPs had a great deal of interspersed AgNPs and was rough. These demonstrate how the polymeric layers produced by KgM biopolymer effectively protect the manufactured AgNPs. FTIR and Raman spectroscopy were used to reveal how the synthetic materials like AGPE, KgM, and KgM-AgNPs interacts on the level of functional groups. There are significant peaks in the FTIR spectra of AGPE at 611 cm^{-1}, 3301 cm^{-1}, 1393 cm^{-1}, 2930 cm^{-1}, 1041 cm^{-1} and 1592 cm^{-1}. They attributed to a variety of stretching vibrations, including C–OH stretching of OH groups from phenolic compounds, alkane groups, C = O or C = C groups of amides for the peak at 1592 cm^{-1}, C–C stretching of the ring structure, C–O–C and C–OH stretching of 2° alcohols, and C–Cl, respectively, for the peaks at 1041 and 611 cm $^{-1}$. According to KgM and the (111), (200), (220), and (311) FCC diffraction planes of AgNPs44, respectively, the peaks at 22.9°, 38.2°, 44.2°, 64.9°, and 77.2° in KgM-AgNPs. The XRD patterns of KgM and KgM-AgNPs indicated the major diffraction peak of KgM at 19.7°. Within 3 min, KgM-AgNPs displayed good Hg^{2+} detection capability in a Hg^{2+} concentration dependent manner. With an estimated LOD of 3.25 nM, the absorbance ratios A360/A408 were linear with Hg^{2+} concentrations from 0.010–10.0 to 10.0–60.0 M. The probe's precision in applying to a sample of lake water was excellent (Jayeoye et al. 2022).

Moreover Jitendra Mittal et al. also combined *Tinospora cordifolia* leaf aqueous extract and a 5 mM silver nitrate ($AgNO_3$) soln. to create an efficient technique for the biosynthesis of Silver NPs. This study showed that the extract's bioactive compounds serve as an agent to stabilize and cap AgNPs. The AgNPs' structure and surface morphology are verified by SEM and TEM study. AgNPs that were synthesized had a spherical shape and had a size range of 30 to 50 nm. The XRD pattern confirmed the crystalline nature of nanoparticles. FTIR was used to identify the reducing and capping agents and functional group which was recorded between 4000 and 400 cm^{-1}. Starches were clearly visible in AgNPs solution at 2924, 2854, 1685, 1462, 1377, 1076, 725, 493, and 455 cm^{-1}. The observed peaks at 2924 and 2854 cm^{-1} are indicative of alkyl C–H, whereas 1685 cm1 is indicative of the C = O group of amides, 1377 cm^{-1} is indicative of the C–F of alkyl halides, 1076 cm^{-1} is indicative of the C–O of alcohol, and 725 cm^{-1} is indicative of the C–Cl stretch of an alkyl halide. These starches give rise to bioactive substances including terpenoids and alkaloids, which serve as stabilising and capping agents. By using XRD, synthesized Silver NPs were further characterized and was found that the diffraction peaks were showing a range of 2θ (20–80°), which corresponded to the (2 6 4) (1 1 1) (2 0 0) plane. The finding demonstrated the presence of FCC structure in the AgNPs. The use of Silver NPs as an antifungal and antibacterial agent is preferable to other techniques for controlling harmful germs and can be crucial in the development of new medications to treat a variety of fatal diseases. The nanoparticles can be used as a substitute for antibiotics because they are efficient against *Pseudomonas aeruginosa, Escherichia coli, Fusarium oxysporum, Sclerotinia sclerotiorum, and Staphylococcus aureus.* But it is necessary to investigate the negative effects of AgNPs and define the ideal AgNP concentrations for utilization in medical studies (Mittal et al. 2017).

For the biosynthesis of Silver NPs during this study, honey was chosen by Mayasar I. Al-Zaban et al. as an environmental stabilizing and reducing agent. Honey is said to promote human longevity because of its high energy, enzymes, vitamins, and availability of chemical components. Honey is a great source of vitamin C, vital minerals, and compounds that can function as antioxidants, which are important in preventing cancer. This study demonstrated the AgNPs' effective catalytic activity in the reduction of the methylene blue dye. By studying double-beam UV–Vis spectra at various wavelengths between 200 and 1000 nm, the silver ion reduction was observed. Color intensity is increased as a result of reduced silver ions. The absorption band that characterized monodispersed spherical AgNPs is seen at 417 nm. The TEM picture displays how AgNP sizes decreased as pH was increased, concluding in almost spherical Silver NPs with a average size ranging from 5–25 nm. The presence of phenolic compounds was detected by the strong and broad peak at 3350.87 cm^{-1} in the FTIR bands. The O–H stretch with a hydrogen bond is also responsible for this. It has been observed that the stretching of C–H aromatic compounds caused the 2957.76 and 2882.84 cm^{-1} area bands. In addition, based on the observed strong peak at 1634.25 cm^{-1}, the functional group of $C = C$ stretching, which is an alkene, occurred in the Silver NPs. The bands were also linked to the proteins' 1418.21 cm^{-1} N–H stretching vibration. The peaks given to the C–O single bond at 1248.41 cm^{-1} and 1044.16 cm^{-1} might also represent the vibrational frequencies related to amide proteins, with the range of frequencies depending on the kind of molecule being between 1000 cm1 and 1300 cm^{-1}. AgNPs in the colloidal solution were measured using ZETA Sizer to determine their size distribution. The (111), (2 0 0), (22 0), and (311) planes' intense XRD peaks were observed at (2θ) angles of 36.45°, 44.31°, 65.41°, and 76.25°, respectively. According to the findings, 92% of the methylene blue had been broken down after 72 h. In addition, after the samples were treated with HPLC, a couple of additional peaks occurred (Al-Zaban et al. 2021).

Bianca Moldovan et al. provides the first report on the efficiency of bioactive compounds from the extract of *Ligustrum ovalifolium* fruit as reducing agent of Silver NPs, providing an eco-friendly, economic, simple, and efficient process for the manufacture of AgNPs. The AgNPs were spherical in shape, crystalline in nature, and 7 nm in size on average according to the TEM characterization. AgNPs in the solution's SPR, which caused the synthesis mixture to turn yellowish brown, marked the bio reduction of the silver ions. In accordance with the characteristic SPR band of AgNPs, which is normally in the 400–500 nm range, the Surface Plasmon Resonance peaks of Silver NPs occurred between 402 and 413 nm. When the environment was acidic (pH 6), the absorption peak turned up at 413 nm. The Surface Plasmon Resonance band of Silver NPs was reduced to lower wavelengths, 402 nm at pH 7, as the pH value of the reaction medium increased, indicating the synthesis of smaller-diameter AgNPs. The spherical AgNPs produced under these conditions ranged in size from 6 to 13 nm, with a mean diameter of 7 nm at pH 10. The X-ray diffraction pattern supported the Silver NPs' crystalline arrangement. Four distinct peaks can be seen in the X-ray diffractogram at 2θ angle of 37.59°, 44.07°, 64.17°, and 77.22°, which correspond to the (111), (200), (220), and (311) Bragg's reflections attributed to the lattice planes of the FCC structure of silver particles. When compared to the spectrum of the fruit extract, the FTIR spectrum of the AgNPs reveals multiple shifts of the absorption bands as well as certain variations in these bands' intensities. The bands that emerge at 2933, 1560, 1383, and 1078 cm^{-1} demonstrate the existence of several functional groups and were attributed to the vibrations of aromatic rings' C = C stretching, C-H stretching, and either O–H deformation or C-O stretching, respectively. The majority of the IR bands most likely originate from the fruit extract's flavonoids, anthocyanins, and phenolic chemicals, showing that these biomolecules stabilized the photosynthesized AgNPs. Thermogravimetric analysis (TGA) also supported the detection of biomolecules capping the surface of the synthesized AgNPs from privet extract. The biosynthesized AgNPs inhibits the growth of cancerous cells, providing them effective anti-cancer treatments. AgNPs have been shown in past research to be capable to stop growth of a number of cancer cell lines, including glioblastoma, breast, lung, ovarian, and hepatic. In vitro testing of the cytotoxicity of the biosynthesized AgNPs against ovarian cancer cells showed that they were effective at low concentration levels (Moldovan et al. 2018).

Similarly, the bioactive chemicals found in the extract of *Viburnum opulus* fruit were used to synthesize Silver NPs in an efficient, green, one-pot technique by Luminita David and Bianca Moldovan in this study. The ability of the produced nanoparticles to remove harmful organic dyes catalytically was also tested. The existence of bioactive phytocompounds, especially phenolics, as reducing agent of the Silver NPs was verified using FTIR and TGA tests. To begin, UV–Vis spectral analysis was utilized to confirm the biogenesis of AgNPs, which confirmed the existence of a unique absorption peak at 415 nm matching the surface plasmon vibration of colloidal Ag. The bands that were considerably displaced at 3398 cm1 in the FTIR spectra of photosynthesized AgNPs and fruit extract are the stretching vibration of the O–H groups of the polyphenols detected in the opulus fruit extract. This broad band measures around 3398 cm^{-1}. The C = O stretching at 1733 cm^{-1} disappears from the extract spectra. The bands observed at 2921, 1380, and 1030 cm^{-1} were inferring to the C-H stretching vibration, C-O stretching vibration, and C-O bending respectively. By utilizing TEM analysis to characterize the morphology and size of the synthesized AgNPs, it was found that the AgNPs were spherically shaped, evenly dispersed, and ranged in size from 7 to 26 nm. The XRD technique was used to analyze the crystalline structure of the nanoparticles, and the diffraction peaks obtained correlate to (111), (200), (220), (311) and (222) of the FCC structure of Silver NPs, and they were found at 38.33°, 44.56°, 64.62°, 77.44°, and 82.41°. Moreover, the catalytic potential of NaBH4 was tested in the deterioration of carmoisine, tartrazine, and brilliant blue FCF dyes. The outcomes showed that brilliant blue FCF was degraded with exceptional efficiency against all of the dyes under study. New instruments for reducing environmental pollution can be developed using this environmentally friendly synthetic method. As a result of these findings, it is possible to effectively use the synthetic nanoparticles that were mediated by *Viburnum opulus* in the field of ecological remediation. Their extraordinary activity may also be utilized to efficiently break down hazardous organic dyes present in a variety of industrial pollutants (David & Moldovan 2020).

6.2.3 Copper Nanoparticles

Punniyakotti et al. synthesized Cu nanoparticles with *Cardiospermum halicacabum* (balloon plant) leaf extract as a green reducing agent. Plant leaves were treated into a hot air oven, converted in powdered form then boiled with deionized water and filtered with Whatman filter paper no. 1 after centrifugation at 6000 rpm for 15 min then for the CuNPs synthesis $CuCl_2.2H_2O$ solution was added to water with addition of extract solution and heated while stirring magnetically then left at room temp. for 1 h and centrifuged at 8000 rpm for 15 min giving dark brown appearance thus Copper nanoparticles were synthesized. Characterization by UV–Vis spectrophotometry revealed SPR and highest absorption peak at 553 nm, under XRD analysis pure crystalline nature of NPs weren't found due to capping by biomolecules, dehydrated CuNPs gave sharp peaks at 43.3°(111), 50.5°(200) and 74.1°(220) thus indicating no impurities such as copper oxides and hydroxides are there, further various functional groups were found in extract and NPs by conducting FTIR analysis which shows, 1383 cm^{-1} of alkane stretching, 1120 cm^{-1} for C–O stretching, 2368 cm^{-1} assigned to C–N, 1629 cm^{-1} is due to $C = O$ and 2924 cm^{-1} indicates C–H similarly for the plant extract the data were analyzed and it was clear that plant metabolites are playing crucial role in reducing Cu NPs. FESEM graph showed homogeneity and spherical shape of nanoparticles also no agglomeration was seen, EDS data showed main peak of Cu with atomic percentage 79.6% and also some other elements such as Carbon, Phosphorus and oxygen, therefore biomolecules were also present on particle surface, TEM analysis was done for more insight which showed hexagonal and spherical shape of the NPs with avg. diameter of 30 nm which was then confirmed by DLS study confirming avg. particle size around 30–40 nm. Microdilution susceptibility assay was done to understand antibacterial activity where minimum inhibitory conc. of green synthesized CuNPs was found 50 μg/mL against *P. aeruginosa* and *S. aureus* whereas for *E. coli* it was 25 μg/mL also minimum bactericidal concentration was measured to be 100 μg/mL for *P. aeruginosa* and *S. aureus* and 50 μg/mL for *E. coli* further study for the nanoparticles to prevent biofilm formation, *P. aeruginosa* show 79%, *S. aureus* 72%, and *E. coli* 78% as maximum biofilm inhibition activity at 100 μg/mL the concentrations were a bit higher maybe due to the rapid colony formation over the substrate. Cu NPs affects cell wall and enters into the cell, increasing the intracellular enzyme activity which increases the cell permeability which affects biofilm development and cell growth. This property of Cu NPs can be further utilized in medicinal applications and drug delivery including biofilm diseases and infections (Punniyakotti et al. 2020). Another study with similar application in biomedical field was conducted by Ginting et al. CuNPs were synthesized through green route using *Blumea balsamifera* leaf extracts and further its antioxidant and cytotoxicity was studied. This plant contains flavonoids such as corylin, xanthine etc. which can show wound healing effects. For the extract, leaves were dried and grinded further macerated in ethyl acetate and filtered, $CuSO_4$ was used as a precursor of CuNPs. GC–MS was done to analyze phytochemicals in *Blumea balsamifera* total 18 were found in which 8 were having highest composition peak like 8-phenyloctanoic acid (7.18%), naphthalene (39.93%), eugenol (10.92%), 2-(1-phenyl-ethylamino)-2-thioxo-acetamide and cryptomeridiol (11.44%), under UV–Vis analysis SPR band appeared at 480–620 nm and max absorption peak observed was at 540 nm thus confirming formation of CuNPs, increase in size of NPs give blue shift in the absorption peak, XRD analysis showed diffraction peaks at 16.2°, 23.8°, 34.5°, 39.2° and 53.4° were observed to (130), (020), (002), (111) and (021) plane of $Cu(OH)_2$ as well for Cu_2O peaks found $2\theta = 29.5°$, 42.4° and 61.7° corresponding to (110), (200) and (220) crystal plane. Sharp peaks for Cu were found at 41.1° and 49.7° thus depicting Face centered cubic lattice. Broad absorption spectra of Cu nanoparticles is mainly due to presence of impurities like CuO, Cu_2O and $Cu(OH)_2$, further size of NPs found was 30–55 nm range as seen in SEM images doing TEM images also matched the data giving size of around 30 nm after that the FTIR data showed peak of 3378 cm^{-1} which is assigned to OH group of phenols, 1100 cm^{-1} for C–O and 1700 cm^{-1} corresponds to $C = O$ stretching vibrations and peak at 610 cm^{-1} corresponds to the formation of copper nanoparticles. DPPH method was used to analyze antioxidant activity, it was found that Cu nanoparticles have higher antioxidant activity compared to *B. balsamifera* leaf extracts, the IC50 values of leaf extracts was 64 mg/ml, Copper NPs 63 mg/ml and ascorbic acid 3 mg/ml further brine shrimp lethality assay for cytotoxicity study was done the measure of lethality was directly proportional to the CuNPs concentration. These Nanoparticles can be used in nano drug delivery system and in medical applications (Ginting et al. 2020).

Leaf extract were also utilized by researchers to synthesized Cu nanoparticles other than drug delivery applications like one study by Dehghannya et al. in which *E. caucasicum* Trautv leaves extract was utilized to synthesize copper nanoparticles (CuNPs) that are novel, green, and eco-friendly. In the synthesis of CuNPs, the functional groups in phenols and flavonoids reduce the Cu^{2+} to Cu nanoparticles. FTIR study was done using the KBr pellet method in 4000–400 cm^{-1} spectral range. The phenolic group stretching can be seen in 3100–3700 cm^{-1} range under FTIR spectra. Moreover, the 2800–3000 cm^{-1} peaks were attributed to alkanes stretching. Furthermore, the peaks at 1277, 1643, 1103 and 1561 cm^{-1} indicated the carbonyl group ($C^{14}O$), $C^{14}C$, vibrations of the COH stretching, CO, respectively. Using an X-ray diffractometer. Cu Ka radiation with a wavelength of 1.540 Angstroms was used to expose CuNPs to X-rays, and at room temperature the diffraction pattern was up to 70° (diffraction angle), the CuNPs were crystalline, as seen by the XRD pattern, and SEM analysis revealed that they have almost spherical forms and less than 40 nm size. Using the DPPH method, the antioxidant activity of the aq. extract, CuNPs and BHT was measured and results were 58.98, 40.02 and 90.12%, respectively. Furthermore *S. typhimurium, E. coli, S. aureus and B. cereus* were all successfully fought off by the generated nanoparticles' remarkable antibacterial characteristics. The advantages of synthesizing copper NPs were eco-friendly nature, simplicity, and safety of the synthesis process in order to the removal of poisonous and dangerous substances. The results of this research showed that CuNPs with sufficient antibacterial and antioxidant characteristics can be useful in several nanotechnology fields (Hasheminya & Dehghannya 2020a). In another research secondary metabolite synthesis was induced by preparing nanoparticles from onion whit leaves, Hussain et al. in this experiment has aimed to produce sufficient secondary metabolites in the plant by administration of synthesized nanoparticles from *Allium cepa* and in vitro germination of *Citrus reticulata* nucellus. The copper nanoparticles synthesized from white leave extract of onion which worked as reducing and keeping agent, the solution was boiled until the colour was changed to light green then for 10 min centrifugation at 14,000 RPM further for the germination of nucellus tissues fruit of *Citrus reticulata* were taken washed and nucellus were separated these were then sterilized with chlorax and sodium hypochlorite solution, the addition of Copper nanoparticles were through sonication in water, Nucellus tissues of the plant were inoculated in this solution having 30 ml MS media along with sucrose, agar and 3 mL of copper nanoparticles. Through zeta potential the size of a copper nano particle found to be in 15–30 nm range and spherical in shape by SEM image analysis further on analyzing it was found that the 30 μg/ml suspension of copper nanoparticles exhibit inhibitory effect on nucellus tissues germination, and reduces root and shoot length as well Seed Vigour Index thus concluded as Cu nanoparticles shows phytotoxicity. The phenolic and flavonoid content production, 11.2 μg/mg and 0.63 μg/mg was recorded max also the protein content 381 μg BSA Eq/mg FW was observed with media having Cu NPs after 21 days it may be due to the stress caused by nanoparticles but these results were varied for different size of NPs, furthermore studies can be done to exactly get the applications of these stress causing CuNPs (Hussain et al. 2017).

Similar study by this research of Sanchaita Lala focuses on secondary metabolite production in *Bacopa monnieri (L.) Pennell* plant by administering CuNPs. The nanoparticles made were starch stabilized where starch acted as a capping agent, ascorbic acid and cupric sulphate pentahydrate acted as reducing agent and centrifuged at 5000 rpm further NPs were stored in a vacuum desiccator after freeze drying. Tips of the plant were sand cultured for two months with hoagland's nutrient solution, acclimatized plants were shifted into medium having CuNPs at different concentrations, kept under natural light for 10 days replenishing the medium after five days finally the plants were dried at room temperature and secondary metabolites were extracted. The total saponin, flavonoid, phenolic and alkaloid content was estimated spectrophotometrically where at 5 mg/L, TSC reached 89%, TAC about 204% for 40 mg/L, TFC about 140% at 40 mg/L, TPC about 116% at 20 mg/L and in DRSC antioxidant assay absorbance was measured spectrophotometrically at 517 nm using ascorbic acid as standard, on administration of NPs DRSC increased significantly about 70% at 40 mg/L no absorbance during UV–Vis in the visible range was observed also no SPR was seen maybe due to a layer of Cu oxide around the nanoparticles and small size preventing agglomeration, HRTEM micrographs gives spherical shape and diameter of 2–20 nm, average crystallite size came 14.09 nm from Scherrer's equation continuing the characterization FTIR spectrum results showed 3412 cm^{-1} band for OH stretching vibrations possibly, 1383 cm^{-1} to Cu^{2+}–O^{2-},

668, 611, 576, 527 and 472 cm^{-1} band to Cu(II)–O, XRD data hints two crystalline phases Copper(I) oxide and metallic Cu existing at same time showing peaks of 29.63°, 36.38°, 42.32°, 61.37° and 73.57° corresponding to (110), (111), (200), (220) and (311) and 43.25°, 50.44° and 74.19°correspond to (111), (200) and (220) respectively further zeta potential of the CuNPs came as 7.57 ± 8.64 mV this low value is maybe due to capping agent (starch) and because of this no force is present to prevent the flocculation of particles. Antioxidant enzymes such as APX, CAT, glutathione reductase and SOD were activated by CuNPs. It is found that CuNPs exert their effects on *B. monnieri* by the generation of ROS like H_2O_2 and MDA which activates the antioxidant defense mechanism (non-enzymatic and enzymatic) as well secondary metabolic pathways thus increasing metabolite concentrations which can be medicinally valuable (Lala 2020).

Applications like antimicrobial, antibiotic activity etc., include fruit extract from the plants like one research with *Piper retrofractum Vahl* commonly called java chilli in Indonesia has various medicinal properties and compounds such as piperidine alkaloids, phenylpropanoids and amides acts as capping as well reducing agents. Fruits of the plant were washed and dried for 5 days in oven then grinded and water was added finally the mixture was filtered with cloth and centrifuged, storage was done in 4 °C, for the CuNPs formation the $CuSO_4$ solution was added to the extract and the color change to dark green confirmed the formation. Further characterization of the NPs by various methods was done, FT-IR spectra revealed the Cu–O band in 550–570 cm^{-1} range, Cu–O–H in 870–880 cm^{-1}, 2900 cm^{-1} band due to alkane, 3000–3500 cm^{-1} because of hydroxyl group,1078 cm^{-1} responsible for C–O stretching and 1600–1700 cm^{-1} (O–H bending), these values are used to identify the functional groups and metabolites present in the extract, EDS analysis proved the high purity of CuNPs the major elements of CuNPs were Oxygen (15.0%), Copper (70.3%) and Carbon (12.2%) and TEM results give particle size in the range of 2–10 nm where optimal pH conditions were acidic at 4pH, XRD analysis of NPs formed at 10 pH showed diffraction peaks of 37.079°, 42.693°, 61.922° and 74.133° and the size measured was 56.8 nm whereas crystallinity 26.4%, these NPs were more stable in ethanol than water. The antibacterial activity was tested using disk diffusion assay where larger inhibition zone was seen for *E. coli* than *S. aureus* i.e. 2 mm and 1.4 mm respectively against 0.2 mg/mL CuNPs, the positively charged nanoparticles react with negatively charged bacterial cell and produces ROS which inhibits bacterial growth, positive control using gentamicin (0.2 mg/mL) for *S. aureus* was 0.8 mm and 1.3 mm for *E. coli*. These promised antibacterial effects can further be used in medicinal applications (Amaliyah et al. 2020).

Viswadevarayalu et al. where they synthesised ultrasmall CuNPs using *Terminalia bellirica* fruit extract, which has phytoconstituents like bellaricanin, gallic and ellagic acids, beta-sitosterol, ethyl gallate that acted as capping and reducing agents. Fruits were dried for 10 days under shade, fine powder was created with fruit pulp, this powder then boiled at 80 °C with deionized water and filtered. $CuSO_4.5H_2O$ was used as a precaution for ultra-small copper and a particle 30 ml of plant extract mixed with 70 ml of the solution which lead to reduction of cupric ion to copper by visualization of carmine red appearance. Small size no surface plasmon resonance peak was seen, while agglomerated nanoparticles showed peak of 550–580 nm then FTIR was done for examining the presence of functional groups absorption bands at 3391 cm^{-1} corresponds to OH, 2926 cm^{-1} to CH stretching, 1615 cm^{-1} represented carbonyl stretching vibrations, 1209 cm^{-1} for C-O vibrations also band for C-F was seen at 1350 cm^{-1} indicating polyphenols, tannins, terpenoids, flavonoids and protein compounds in *T. bellirica.* XRD data showed crystalline nature of the synthesized copper NPs particles with peaks at 43.24°, 5.37°, 74.02° corresponding to (111), (200) and (220) respectively indicating FCC, and no other impurity was seen affecting the lattice. TEM images showed average diameter of 7 nm and the size obtained was 40–50 nm well dispersed and spherical in shape EDS spectra also confirmed the same, Zeta values found was –54.6 mV i.e., highly negative thus stable further antibacterial activity was studied against E. coli colonies on MH agar media with 10, 20, 30 and 40 μL conc. Of CuNPs, 5 μL of DMSO was also added for negative control for which inhibition zone came 0 mm and kanamycin for positive control for which inhibition zone came ± 21 mm and for *E. Coli* it was ±24 mm, Anti-fungal study done against *Candida tropicalis* for which inhibition diameter was ±16 mm and ±14 mm against chloramphenicol used as positive control (Viswadevarayalu et al. 2016).

Flower extract can also function as source for CuNPs, *Azadirachta indica* has tremendous therapeutic properties due to nimbin, nimbolide, limonoids etc. Flower extract was prepared by drying flower and crushing them into fine powder and adding deionized water and given hot water bath for 1 h. finally filtered with Whatman filter paper no.1 then $CuSO_4.5H_2O$ addition and magnetic stirring was done at 700 rpm, brown precipitate was formed which was then taken. Under characterization UV–Vis results gave SPR peak at 560 nm, the optimum pH found was 9 and temp. 80 °C further FTIR spectra revealed absorption peak of 3561 cm^{-1} of N–H stretching, 3390 cm^{-1} for H–O bonding, 1618 cm^{-1} for C = O and 1122 cm^{-1} refers to aldehyde stretching, 622 cm^{-1} is due to C–Cl stretching these readings confirmed the presence of terpenoids which acted as both reducer and stabilizer. The cubic phase of Cu was examined using XRD analysis where absorption peaks observed were $2\theta = 43.30°$, 50.50° and 74.30° and the size of NPs were calculated using Scherrer's equation which came out to be 44.9 nm further spherical shape was noticed in HRTEM images, the ringlike diffraction in the SAED patterns signifies crystallinity of the particles which is in common with XRD data finally the antibacterial activity was tested via inhibition zone assay against *Proteus mirabilis* with 40 μg/ml of Cu nanoparticles which showed max efficiency. This green, cost-effective approach to produce copper NPs can replace costly antibiotic drugs in the industry (Gopalakrishnan & Muniraj 2021).

Nagar et al. using green synthesis created eco-friendly CuNPs from cupric chloride salt and *Azadirachta indica*, leaf broth was added to a heating aqueous solution of $CuCl_2.2H_2O$ and intense brown coloration confirmed the formation of CuNPs, after 15 min of 6000 RPM centrifugation the supernatant was collected. The characterization of nanoparticle by UV–Vis spectroscopy showed Max absorbance peak at 560 nm, Leaf broth percentage also affected the absorption peaks at percentage of 5 a week absorption peak was observed at 506 nm and when increased to 20% the Surface plasmon resonance peak increased the shape observed by TEM images was cubical and also revealed that the nanoparticles were coated with capping agents from neem leaf extract further purity of copper atom was confirmed by EDS spectra where peak at $2\theta = 43.5°$, $49.9°$ and $74.01°$ corresponds to (111), (200) and (220) further SEM images showed the particle size to be 48 nm (Fig. 6.2) which was calculated by the Scherer's equation. It was also found that concentration of salt when increased from 7.5×10^{-3} the size of nanoparticles decreases, also the optimal reaction temperature for conversion rate of copper ions was found 85 °C. it was also found that biomolecules were inactive under acidic pH and NPs could not be formed and 6.6pH is optimal. Under FT-IR analysis the NPs display peaks at 2922 cm^{-1}, 2371 cm^{-1}, 1631 cm^{-1}, 1456 cm^{-1}, 1384 cm^{-1},1077 cm^{-1} for phenols, aromatic amine, carbonyl stretching, alkene stretching, aldehyde and CO stretching respectively which hints the presence of flavonoids, polyphenols and terpenoids which is responsible for stabilization and reduction finally the high zeta potential value (−17.5 mV) defines the highly stable Cu NPs (Nagar and Devra 2018).

Vishveshvar et al. prepared Cu Oxide nanoparticles using leaf extract of *Ixoro coccinea* which have wound healing properties and further characterization studies were done. Leaves from the plant were taken into the flask and boiled until the color turned to brown then the extract was filtered once cooled down, for the synthesis of CuO NPs 500 mL of 3 μM $CuSO_4$ solution was prepared adding different volumes of extract and left for bio reduction, centrifugation was done at 10,000 rpm and then NPs were removed and moved into hot air oven obtaining solid CuO nanoparticles. After the synthesis of both extract and nanoparticles characterization was done, UV–visible

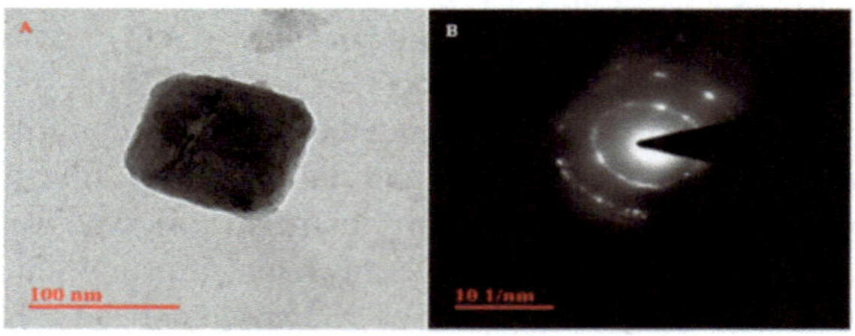

Fig. 6.2 **A** TEM image of synthesized Copper Nanoparticle, **B** SAED pattern of Copper Nanoparticles (Nagar and Devra 2018)

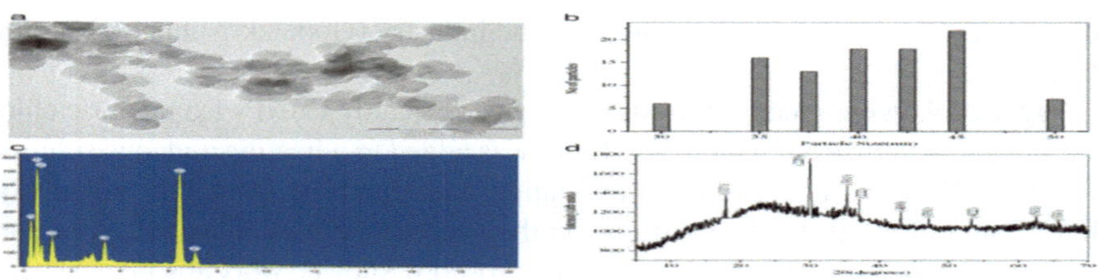

Fig. 6.3 TEM (**a** and **b**), EDS (**c**) images and XRD **d** pattern of dried powder of Fe_3O_4 MNPs respectively)

spectrum showed high absorbance in 200–300 nm range, due to SPR peak formed at 191 nm, SEM images revealed high tendency of NPs to agglomerate thus showing avg. size of 300 nm and distributed randomly, for morphological analysis TEM was done which showed size of NPs less than 5 nm with spherical morphology. It was noticed that by absorbing moisture NPs can agglomerate which can be undone by ultrasonication to make less sized nanoparticles, FTIR analysis of CuO NPs showed peak obtained at 488 cm^{-1} assigned to Cu–O and 658 cm^{-1} to Cu_2–O stretching, 1026 cm^{-1} was referred to H–OH bond stretches, peak near 700 cm^{-1} for phenolic and alcohol groups while 1300 cm^{-1} for C-N bond, thus indicating alkaloids, glycoside, flavonoids, tannins, saponins, steroids, terpenoid etc. presence which maybe acting as a reducing and capping agents. As this process was cost effective and eco-friendly, further study is required for development of more enhanced route for greener synthesis of nanoparticles (Vishveshvar et al. 2018).

6.2.4 Iron Nanoparticles

Jeyasundari et al. synthesized zero valent Fe NPs from *Psidium guajava* plant leaf extract, in which polyols are mainly responsible for reduction of Fe ions, process started through cutting of *Psidium guajava* leaves into small parts and boiling for 20 min and the extract was filtered using filter paper to get leaf extract, then $FeCl_3$ solution was added and heated for 15 min at 60–70 °C and preparation of iron nanoparticle was completed. UV–Vis analysis revealed absorption band at 254.12 nm, this peak was due to SPR. Further FTIR data was observed, band at 3466 cm^{-1} is due to OH stretching vibration of phenolic groups, band at 2075 cm^{-1} refer to C–O stretching, 1637 cm^{-1} and 670 cm^{-1} due to carbonyl groups in esters, ethers, alcohol, acids and revealed that that flavonoid, tannins, steroids, glycosides, polyphenol found in *Psidium guajava* leaf extract play a significant role in the reduction of Fe^{3+}. Under XRD analysis the diffraction peaks were 42° (100), 44° (002), 48° (101), 77° (110) and the avg. size of NPs solved by Scherrer's equation was 27 $\times 10^{-9}$ nm. By SEM images agglomerated FeNPs were seen and the EDAX profile of capped FeNPs showed strong crystalline property. The antibacterial activity of the FeNPs were also good the zone of inhibition obtained was 14 mm against *Bacillus cereus*, 17 mm against *Escherichia coli*, 10 mm in case of *Klebsiella pneumonia* and 14 mm in case of *Staphylococcus aureus*. These NPs can be used in removal of environmental contaminants and this green route is non chemical producing thus not affecting the environmental aura (Jeyasundari et al. 2017).

Similarly another studies were done using leaf extract, Perveen et al. boiled the *Plumeria obtusa* leaves, biofabrication by adding ferrous acetate into plant extract and heating up to 2 h at 70°C in the presence of a buffer maintaining pH at 9, while characterization by UV–Vis spectroscopy the maximum absorbance observed was 331 nm, also confirming the presence of aromatic compounds then FTIR analysis confirmed the presence of phenol, NH/CH/OH in amines, carbonyl group, amides,

COOH, aliphatic ether, C-N groups, S = O were observed through peaks, reduction and stabilization of the NPs were done when these biomolecules attaches to the surface of metal ion further XRD analysis showed highly crystalline nature of FeNPs and the average size of 50 nm by Scherer's approximation, TEM images gave spherical shape of NPs. The antibacterial activity was studied against B.subtilis and E.coli using the agar well diffusion method where streptomycin was used as a (+) control further the inhibition zone observed was 16 ± 0.03 mm and 18 ± 0.03 mm for *P. obtusa* extract against *E. coli* and *B. subtilis*, respectively further the inhibition zone noticed was 24 ± 0.04 mm against *B. subtilis* and 29 ± 0.03 mm against *E. coli* in case of FeNPs then antifungal activity was studied against *Schizophyllum commune* and *Aspergillus niger* by using disc-diffusion method, for plant extract inhibition zones were 17 ± 0.02 mm against *S. commune* and 14 ± 0.02 mm against *A. niger* and for the NPs these were 28 ± 0.03 mm against *S. commune* and 24 ± 0.02 mm against *A. niger* at 100 μg/mL concentration in which Fluconazole and DMSO served as positive and negative controls. For finding MICs (Minimum inhibitory conc.) microdilution method was utilized, different concentrations of NPs ranging 50–500 μg/mL were used for the study and the values observed were 250 ± 0.13,125 ± 0.11, 125 ± 0.08, 250 ± 0.09 μg/mL for *P. obtusa* and, 62.5 ± 0.11,62.5 ± 0.9, 62.5 ± 0.12, 31.2 5 ± 0.07 μg/mL for FeNPs against, *A. niger*, *E. coli, S. commune* and *B. subtilis* microbial strains, respectively and for standard drugs (streptomycin and fluconazole) the values obtained were 63.4 ± 0.1, 32.31 ± 0. 09, 61.95 ± 0.13 and 62.15 ± 0.08 (μg/mL), results of MIC study came with a probability value of < 0.005. For antioxidant property of the FeNPs, the spectrophotometry was done, in the result, DPPH radical scavenging potential values were found 70.23 ± 0.02% for FeNPs and 56.34 ± 0.03% for *Plumeria obtusa.* Standard potential was 66 ± 0.02% for ascorbic acid, thus FeNPs showed notable antioxidant activity. This approach for green synthesis of FeNPs would lead to better and eco-friendly practices also the antibacterial and antifungal properties of NPs are promising (Perveen et al. 2022).

Again, with green tea leaves extract from *Camellia sinensis* production of green iron NPs for the cause of wastewater treatment, and for that Ogutveren et al. heated them with water and then $FeCl_3.6H_2O$ was added as Fe source. Zetasizer analysis was performed to determine FeNP size distribution that came in 10–100 nm range. EDX analysis was done, and the main elements were found to be Fe, Si, O, whereas C, Na, Mg, Cl, P, K, Al were also found. SEM analysis results showed that the size of FeNPs was less than 100 nm which was done at 20 keV and 11 mm distance. Also the fraction of polyphenolic compounds in tea leaves extract came to be 10%(v/v) (Ulker ogutveren et al. 2016).

In one more research Papaya leaves were utilized, Where Habiba et al. aimed to treat reactive azo dyes widely being used in textile industries by iron oxide (α-Fe_2O_3) NPs and degrade them which pollutes water bodies and also enters human food chain. The *Carica papaya* leaves were washed then oven dried to make fine powder and finally 20 g of powder is boiled with water and filtered using filter paper. Then extract solution was added with $FeCl_3.6H_2O$ and pH 11 was maintained by NaOH, and finally iron oxide NPs were separated by centrifugation and dried after washing with ethanol further FTIR analysis was done and peaks observed at 678.94 cm^{-1},

474.49 $cm^{-1}$and 621.08 cm^{-1} represents the presence of Fe–O bond while the peaks 3357.57 cm-1 hints the presence of OH bond, 3691.57 cm^{-1} and 1433.11 cm^{-1} also refers to OH bond stretching in carboxylic group and phenolic, the ensuring the presence of alkane, secondary alcohol and alkene in the plant extract confirmed by the peaks at 1625.99 cm^{-1} 1101.35 cm^{-1} and 2927.94 cm^{-1} denoting C¼C stretching, C–O stretching and C–H stretching then characterization by x-ray diffractometer gave data indicating the crystal planes 25.16°, 35.12°, 36.63°, 40.64°, 49.97°, 57.08° and 59.42° which indicates the formation of α-Fe_2O_3 nanoparticles also the avg. size came 4.58 nm, the FESEM images showed average diameter of 21.59 nm and nonuniform nanoparticles then through EDX analysis the mass percentage for iron 32.70%, Carbon 25.05% and 31.03% for oxygen was seen. If asked about the reason, degradation of Remazol yellow dye is due to the hydroxyl ion produced by FeO nanoparticles in the presence of sunlight 0.50, 140, 245 and 340 mg of dye is removed by NPs in case of 10, 30, 50 and 70 ppm of initial dye concentration, while each gram of photocatalyst remove 120, 240, 60, 80 mg and for 0.4, 0.2, 0.8 and 0.6 g/L of catalyst. Well diffusion method was used for the antibacterial study of the nano particles, the inhibition zone for *Staphylococcus aureus* was 7 $\pm$ 1 mm for 5 mg/ml doses but when the dose was increased to 30 mg/ml inhibition zone of 12.5 $\pm$ 0.5 was noted also *Klebsiella spp.* at 5 mg/ml, 20 mg/ml doses showed resistance but at 30 mg/ml it grew and gave an inhibition zone of 9 $\pm$ 1 mm the NPs showed average effect on *Pseudomonas SPP*. and *E.coli* with inhibition zones of about 10 $\pm$ 0.5 mm and 9 $\pm$ 1 mm respectively. During the cytotoxicity study the α-Fe_2O_3 nanoparticles showed a high toxicity towards the BHK-21 and Hela cell lines, and it revealed that survival rate is only 5% after 48 h, about 10–20% of the Vero cells survived for the same doses of nanoparticles. $\propto$ Fe_2O_3 NPs showed degradation of dye in sun radiation, also high toxicity was seen so cancer can be treated under certain dosage, but further studies are required for exact procedure (Bhuiyan et al. 2020).

Not just leaves, plant peel extracts have also been used in many experiments. One of which is done by Venkateswarlu et al. utilizing peel extract of plantain that composed of lignin cellulose, pectin, carbohydrates and polyphenols that mainly function as both capping and reducing agents in synthesizing Fe_3O_4 nanoparticles, for that Plantain peel were washed and dried for 14 days, then kept in distilled water until the color becomes yellow then filtered, preparation of NPs completed after adding $FeCl_3.6H_2O$ and sodium acetate. TEM images showed the size of nanoparticle in 30–50 nm range. Through FT-IR, bands in the region of 3550–3350 cm^{-1} were of OH stretching, band 890 cm^{-1} refer to amine groups, 2930–2854 cm^{-1} for alkane and band appearing at 1725 cm^{-1} is indicative of the CO group these results indicated the presence of different biomolecules and polyphenols. Further XRD data reveals the crystalline nature of Fe_3O_4 NPs and the peaks were seen at (111), (220), (311), (222), (400), (422), (331), (422), (511) and (531) (Figure 3). Using BET method, pore size and surface area were determined which came to be 1.8 nm and 11.31 m^2g^{-1} respectively. Further the magnetic values derived from the hysteresis loop are 15.8 emu/g (saturation magnetization), 387.91 G of coercive force, remnant magnetization 0.62 emu/g and shows ferromagnetic properties. These kinds of NPs can be used in applications involving metal pollutants removal and other biomedical fields (Venkateswarlu et al. 2013).

As the previous research focused on pollutant removal the next one also utilizes peel extract but application include drug delivery, Sathishkumar et al. aimed for ecofriendly and nontoxic production of iron nanoparticles utilizing fruit extract of *Couroupita guianensis* for which outer peel of *Couroupita guianensis Aubl.* were removed, then for 10 days little pieces were sun dried and converted into powder, mixing this with water and boiling for 20 min at 60 °C then filtering with Whatman no. 1 filter paper gives the extract. $FeCl_3.6H_2O$ was then added and centrifugation was done keeping pH at 10.5 with ammonia solution. The amine and hydroxyl groups of bioactive molecules from plant extract form FeH_6O_3, which then produce Fe_3O_4 particles by reduction through other biomolecules. UV–Vis spectrum shows absorbance in a range of 25–35 nm further under FTIR spectroscopic analysis 3600–3100 cm^{-1} peak is the OH and NH stretching vibrations, 1640 cm^{-1} band is assigned to carbonyl groups and 2925 cm^{-1} for CH presence, function of polyphenols and metabolites was confirmed by FTIR as reducing sugars and stabilizers then XPS spectra of $CGFe_3O_4$ nanoparticles showed ranges for O1s-529.6 eV, C1s-283.8 eV and N1s-398.5 eV, the crystalline structure of NPs characterized by XRD shows 2θ as 35.54°, 43.07°, 53.43°, 56.95°, 62.54° for each Bragg reflection (220), (311), (400), (422), (511) and (440) and mean size of NPs calculated through Scherer's equation was 7.2 nm. The nanoparticles exhibit a negative zeta value of -26 mV which is quite large due to the surface coating of biomolecules creating an opposing force among NPs. Further the TEM analysis gives the size of nanoparticles in the 7–80 nm range, and the mean size was found to be 17 ± 10 nm and the 6.2 keV region in EDX spectrum showed the stability and purity of formed NPs. Surface plasmon resonance was also noticed at 0.5 keV peak. The synthesized $CGFe_3O_4$ nanoparticles exhibit superparamagnetic at room temperature. For cytotoxicity analysis MTT assay was done on human hepatocellular carcinoma cells to find IC50 value which came to be 120 μg/mL for CGFe and 44.51 μg/ml for $CGFe_3O_4$, results shows that the synthesized $CGFe_3O_4$ nanoparticles trigger the death of HepG2 cells by apoptosis. In hemocompatibity study results were that Fe_3O_4 NPs show lesser hemoglobin release than crude CGFe. Antibacterial activity was tested against gram negative *E. coli, S. typhi, K. pneumonia* and gram-negative *S. aureus* (Sathishkumar et al. 2018).

Another such research resulting applications in biomedical field, drug delivery specifically was done by Orooji et al. aiming for the production of Mesoporous $Fe_3O_4@SiO_2$ NPs via green route. The FeNPs are synthesized by using $FeCl_2$.4H2O and $FeCl_3$.6H2O as precursors (molar ratio 1:2) in water and stirring vigorously for half hour under nitrogen atmosphere and room temp. Fe-SiO_2 NPs were prepared by dispersing 0.1 g of FeNPs in a flask having 0.18 L ethanol and 20 ml ammonium hydroxide then adding 1.2 ml of Si $(OC_2H_5)_4$and mixing for 24 h. Strawberries were collected then dried for 2 days at room temperature then 20 g extracted by maceration process using 150 ml CH_2O. Finally, the extract solution was evaporated after filtering using a rotary machine. Strawberry extract (10 ml) was then treated with 1.67 mmol of calcium nitrates solution under ultrasonic wave, putting this into

FSNP solution and 30 ml deionized water with 1 mmol $(NH4)_2HPO_4$ with NH3 till pH is 11. After 24 h precipitate was centrifuged, washed then dried for 8 h at 70 °C and finally calcined at 500°C for 2 h. For SLN loading 1:1 proportion of SLN and mFSH-SW (150 mg) was mixed. Optical density was measured at 450 nm by Elisa reader performing XTT cell proliferation assay. X-ray diffraction patterns indicate that crystalline FeNPs and mFSH-SW nanoparticles has no notable impurities. The results of FTIR spectra confirm that APTES groups found on the mFSH-SW surface and FESEM images showed magnetic ferrous nanoparticles of spherical shape and a structure with 38 nm particle size. Small particle size of mesoporous silica results in better drug absorption and side effect reduction due to less dosage. But the size was approximately increased to 50 mm due to the SiO_2 layer coated successfully on the surface of Fe_3O_4 nanoparticles further EDX analysis showed presence of Si, P, Fe and Ca. Size distribution ranges between 70 to 110 nm as DLS data showed further TEM images display that spherical compounds with a diameter of 80 mm were synthesized. VSN analysis was done to determine magnetic properties of pure magnetite NPs, also according to BET (Brunauer Emmett Teller) and BJH (Brunauer Joyner Halenda) analysis average pore size came 14.1 mm, pore volume 0.24 cm^3/ g and surface area of mFSH-SW was 63.2 m^2/g. The percentage of Sulphur loaded in mFSH-APTES and mFSH were found by UV–Vis spectra, entrapment efficiency for pure mFSH-APTES and mFSH sample were calculated as 28.4, 49.3, 37.7 and 59.1%. No mortality was seen during the 48 h of study in histopathological study of liver, spleen and kidney after injecting SLN-loaded mFSH-SW. The Nanocarriers prepared via experimentations can be utilized as drug carriers in drug delivery and also they will reduce administration time and dosage (Orooji et al. 2020).

The large-scale pollution of water bodies by fuchsin dye released by cotton, paper and dye industries, this toxic dye enters into crops if irrigated through the polluted water that led to nausea, vomiting, diarrhea and may also cause cancer when ingested, Jain et al. in their research have come with green synthesized FeNPs due to their promising results and low cost and mainly high adsorption capabilities. They have used peel extract of *Artocarpus heterophyllus* which have biomolecules that act as capping agents. For extract, peels were dried in hot air oven and turned into fine powder, this powder was then boiled until solution changed to pale yellow further filtered and refrigerated then $FeCl_2$ was added to extract solution while maintaining pH 6 by NaOH doing that suddenly change of solution color into dark black hints the formation of Fe NPs, after that the solution was heated to obtain a fine precipitate. To exactly know about the properties of nanoparticles characterization was done by various techniques, XRD analysis shows peaks at 31.58°, 40.44°, 50.18° and 56.41° which corresponds to α-Fe_2O_3, 35.35° corresponds to Fe_3O_4, 15.05° and 28.30° peak corresponds to α-FeOOH and the 45.32° peak confirms the presence of zerovalent Fe nanoparticles and the size of NPs calculated using the 31.58° peak was 33 nm further TEM images reveals spherical shape and similar size of 33 nm, the NPs surface were found to be rough via SEM images maybe due to the presence of polyphenols from the peel extract, similar results through EDX data, presence of C(54.23%), O(29.27%), Fe(15.25%) and Cl(1.25%) were noted, through FTIR analysis different functional groups were identified, like absorption peak of 3450–3200 cm^{-1} correspond to OH bond vibrations, which reduces the Fe ions and then get adsorbed over the NPs surface protecting them from atmospheric Oxygen thus increasing shelf life, 2969 cm^{-1} refers to CH_2, 1664 cm^{-1} refers to pie bond between carbon carbon in a ring, 1042 cm^{-1} for C–O–C stretching vibrations which hints water soluble biomolecules. Catalytic activity study showed that NPs are able to oxidise the Fuchsin dye and exhibits removal efficiency of 87.5% at 318 K, this efficiency increase on increasing temperature. So, if we talk about the mechanism, when the FeNPs are added, the formation of iron oxide takes place this leads to the leaching of Fe(ll), Fe(lll) and Fe(zerovalent) ions, the Fe(ll) and Fe(lll) combine with H_2O_2 and yield Oxyhydroxide which adsorb the dye also these Oxyhydroxide reacts with Fe3 to further decompose H_2O_2 (Jain et al. 2021).

Niraimathee et al. rather than common parts like fruit and leaves, utilized roots of *M. pudica*, dried and converted them into powdered form again added in water and stirred magnetically then filtered with Whatman filter paper. In that extract ferrous sulfate solution was added and pH was set to 9, this mixture was centrifuged for 15 min at 6000 rpm and precipitate was dried giving the iron oxide NPs. This plant contains a compound called mimosine which behaves as a reducing agent. UV–Vis study showed a sharp peak at 294 nm thus indicating FeNPs presence further in FTIR analysis of mimosa root extract 3417 cm^{-1} peak refer to OH group in water, 2922 cm^{-1} peak is due to CH_2, 1415 cm^{-1} for C–C, 1074 cm^{-1} for C-N group, 2395 cm^{-1} is also due to OH stretching, peak at 1628 cm^{-1} is for C–C bond and 2354 cm^{-1} refers to C-H stretching vibrations which lead to the prediction of mimosine compound(alkaloid) in the root extract of mimosa. The peaks obtained for nanoparticles were 3422 cm^{-1} (OH), 2366 cm^{-1} (CH), 1648 cm^{-1} (CC), 1096 cm^{-1} (SO_4). On comparing both the peaks obtained from both the analysis It was found that the ability to act as a reducing and stabilizing agent for FeO nanoparticles is in *M. pudica* root extract. Further characterization by XRD revealed the NPs were crystalline in nature that the peaks obtained were at 18.2°, 23.6°, 28.7°, 33.9° and 39° marked by their indices (202), (132), (322), (016) and (151) with average particle size of 25.6 nm using Debye Scherrer formula and SEM data give particle size around 67 nm. PD analyzer shows mean size of 147 nm polydispersity index of about 0.240. Vibrating sample magnetometry revealed saturation magnetisation of nanoparticles to be 55.402 emu/g thus showing superparamagnetic behavior (Niraimathee et al. 2016).

At last other than plants and leaves, seaweeds were also manipulated for FeNPs production. *Kappaphycus alvarezii* is a seaweed known for its gelling properties, the compound carrageenan gives the jelling properties and is used in this experiment by Shameli et al. as a green stabilizer in synthesizing nanoparticles. The civil was then then bleached to make it colorless, dried under sun and this dried sample with deionized water gives extract solution. After that, Fe_3O_4 nanoparticles were synthesized by adding ferrous and ferric solution with NaOH to maintain the pH 11 and stirred for 1 h. While characterizing purity of the synthesized particles identified by XRD analysis where peaks observed were $2\theta = 30.56°$, 35.86°, 43.46°, 54.01°, 57.38°, 63.00°, and

74.46° for crystal planes (200), (311), (400), (422), (511), (440), and (533), respectively also the broad refraction peak observed was 21.50° in UV–Vis spectroscopy no strong absorption peak was seen and under XRD analysis using Scherrer equation size of the crystal NPs came out to be 16.79 nm then TEM images gave spherical shaped particles and mean size of 14.7 nm. FT-IR spectroscopy revealed functional groups as OH, CH, CO and sulphate group at 3439 cm^{-1}, 2916 cm^{-1}, 1636 cm^{-1}, 1227 cm^{-1} respectively. In synthesized Fe_3O_4 nanoparticles the peak 423 cm^{-1} and 556 cm^{-1} belong to FeO stretching vibrations. These non-toxic NPs can be further used in biomedical applications (Yew et al. 2016).

6.3 Conclusion

As we have gone through many of the research articles, we saw what are the different processes that lead to the synthesis of respective nanoparticles. This chapter consisted of Au(gold), Ag(silver), Cu(copper), Fe(iron) nanoparticles which vary widely in size and shape, may it be spherical, cubic or heart shaped, various characterization techniques were made in use to elude properties of the nanoparticles and different sizes were noticed through SEM and TEM images. No pathogenicity was seen in most of the nanoparticles, still NPs which showed pathogenicity can be utilized in biomedical fields and cancer treatment antibiotic, antioxidant, antimicrobial activities, Drug delivery was also possible due to ultra-small size of nanoparticles. Also, there were applications that were successfully applied, and results were observed like dye degradation which is useful in wastewater treatment.

References

Abbas HS, Mahmoud AM, Wahed RA, Elsantawy MAA, Hamdy NM, Ismail SES, Nabil MA (2022) Prospects of using bioactive compounds in nanomaterials surface decoration and their biomedical purposes. Int Nano Lett 12(2):125–138. https://doi.org/10.1007/s40089-021-00355-9

Abdel-Raouf N, Al-Enazi NM, Ibraheem IBM (2017) Green biosynthesis of gold nanoparticles using Galaxaura elongata and characterization of their antibacterial activity. Arab J Chem 10:S3029–S3039. https://doi.org/10.1016/j.arabjc.2013.11.044

Aboelfetoh EF, El-Shenody RA, Ghobara MM (2017) Eco-friendly synthesis of silver nanoparticles using green algae (Caulerpa serrulata): reaction optimization, catalytic and antibacterial activities. Environ Monit Assess 189(7):1–15

Akinfenwa AO, Abdul NS, Docrat FT, Marnewick JL, Luckay RC, Hussein AA (2021a) Cytotoxic Effects of Phytomediated Silver and Gold Nanoparticles Synthesised from Rooibos (Aspalathus linearis), and Aspalathin. Plants 10(11):2460. https://doi.org/10.3390/plants10112460

Akinfenwa AO, Abdul NS, Docrat FT, Marnewick JL, Luckay RC, Hussein AA (2021b) Cytotoxic Effects of Phytomediated Silver and Gold Nanoparticles Synthesised from Rooibos (Aspalathus linearis), and Aspalathin. Plants 10(11):2460. https://doi.org/10.3390/plants10112460

Alqahtani MA, Al Othman MR, Mohammed AE (2020a) Bio fabrication of silver nanoparticles with antibacterial and cytotoxic abilities using lichens. Sci Rep, 10(1):1–17

Alqahtani MA, Al Othman MR, Mohammed AE (2020b) Bio fabrication of silver nanoparticles with antibacterial and cytotoxic abilities using lichens. Sci Rep, 10(1), 1–17

Al-Zaban MI, Mahmoud MA, AlHarbi MA (2021) Catalytic degradation of methylene blue using silver nanoparticles synthesized by honey. Saudi J Biol Sci 28(3):2007–2013

Amaliyah S, Pangesti DP, Masruri M, Sabarudin A, Sumitro SB (2020) Green synthesis and characterization of copper nanoparticles using Piper retrofractum Vahl extract as bioreductor and capping agent. Heliyon 6(8):e04636

Balčiūnaitienė A, Štreimikytė P, Puzerytė V, Viškelis J, Štreimikytė-Mockeliūnė Ž, Maželienė Ž, Sakalauskienė V, Viškelis P (2022) Antimicrobial Activities against Opportunistic Pathogenic Bacteria Using Green Synthesized Silver Nanoparticles in Plant and Lichen Enzyme-Assisted Extracts. Plants 11(14):1833. https://doi.org/10.3390/plants11141833

Bhuiyan MSH, Miah MY, Paul SC, Aka T, das, Saha, O., Rahaman, M. M., Sharif, M. J. I., Habiba, O., & Ashaduzzaman, M. (2020) Green synthesis of iron oxide nanoparticle using Carica papaya leaf extract: application for photocatalytic degradation of remazol yellow RR dye and antibacterial activity. Heliyon 6(8):e04603–e04603

Botteon CEA, Silva LB, Ccana-Ccapatinta G, v, Silva, T. S., Ambrosio, S. R., Veneziani, R. C. S., Bastos, J. K., & Marcato, P. D. (2021) Biosynthesis and characterization of gold nanoparticles using Brazilian red propolis and evaluation of its antimicrobial and anticancer activities. Sci Rep 11(1):1–16

David L, Moldovan B (2020) Green synthesis of biogenic silver nanoparticles for efficient catalytic removal of harmful organic dyes. Nanomaterials 10(2):202

El-Deeb NM, Khattab SM, Abu-Youssef MA, Badr AMA (2022) Green synthesis of novel stable biogenic gold nanoparticles for breast cancer therapeutics via the induction of extrinsic and intrinsic pathways. Sci Rep 12(1):11518. https://doi.org/10.1038/s41598-022-15648-y

Genc N (2021) Biosynthesis of silver nanoparticles using Origanum onites extract and investigation of their antioxidant activity. Part Sci Technol 39(5):562–568

Ginting B, Maulana I, Karnila I (2020) Biosynthesis copper nanoparticles using Blumea balsamifera leaf extracts: characterization of its antioxidant and cytotoxicity activities. Surfaces and Interfaces 21:100799

Gopalakrishnan V, Muniraj S (2021) Neem flower extract assisted green synthesis of copper nanoparticles–optimisation, characterisation and anti-bacterial study. Materials Today: Proceedings 36:832–836

Goudarzi M, Mir N, Mousavi-Kamazani M, Bagheri S, Salavati-Niasari M (2016) Biosynthesis and characterization of silver nanoparticles prepared from two novel natural precursors by facile thermal decomposition methods. Sci Rep 6(2):32539. https://doi.org/10.1038/srep32539

Hasheminya S-M, Dehghannya J (2020a) Green synthesis and characterization of copper nanoparticles using Eryngium caucasicum Trautv aqueous extracts and its antioxidant and antimicrobial properties. Part Sci Technol 38(8):1019–1026

Hasheminya S-M, Dehghannya J (2020b) Green synthesis and characterization of copper nanoparticles using *Eryngium caucasicum* Trautv aqueous extracts and its antioxidant and antimicrobial properties. Part Sci Technol 38(8):1019–1026. https://doi.org/10.1080/02726351.2019.1658664

Hussain M, Raja NI, Mashwani Z, Iqbal M, Ejaz M, Yasmeen F (2017) In vitro germination and biochemical profiling of Citrus reticulata in response to green synthesised zinc and copper nanoparticles. IET Nanobiotechnol 11(7):790–796

Iqbal J, Abbasi BA, Yaseen T, Zahra SA, Shahbaz A, Shah SA, Uddin S, Ma X, Raouf B, Kanwal S, Amin W, Mahmood T, El-Serehy HA, Ahmad P (2021) Green synthesis of zinc oxide nanoparticles using Elaeagnus angustifolia L. leaf extracts and their multiple in vitro biological applications. Scientific Reports 11(1):20988. https://doi.org/10.1038/s41598-021-99839-z

Jain R, Mendiratta S, Kumar L, Srivastava A (2021) Green synthesis of iron nanoparticles using Artocarpus heterophyllus peel extract and their application as a heterogeneous Fenton-like catalyst for the degradation of Fuchsin Basic dye. Curr Res Green Sustain Chem 4:100086

Jayeoye TJ, Eze FN, Olatunji OJ, Tyopine AA (2022) Synthesis of biocompatible Konjac glucomannan stabilized silver nanoparticles, with Asystasia gangetica phenolic extract for colorimetric detection of mercury (II) ion. Sci Rep 12(1):1–15

Jeyasundari J, Praba PS, Jacob YBA, Vasantha VS, Shanmugaiah V (2017) Green synthesis and characterization of zero valent iron nanoparticles from the leaf extract of Psidium guajava plant and their antibacterial activity. Chem Sci Rev Lett 6(22):1244–1252

Kaithavelikkakath Francis P, Sivadasan S, Avarachan A, Gopinath A (2020) A novel green synthesis of gold nanoparticles using seaweed *Lobophora variegata* and its potential application in the reduction of nitrophenols. Part Sci Technol 38(3):365–370. https://doi.org/10.1080/02726351.2018.1547340

Karvekar OS, Vadanagekar AS, Sarvalkar PD, Suryawanshi SS, Jadhav SM, Singhan RD, Jadhav JP, Sharma KKK, Prasad NR (2022) Bos taurus (A-2) urine assisted bioactive cobalt oxide anchored ZnO: a novel nanoscale approach. Sci Rep 12(1):15584. https://doi.org/10.1038/s41598-022-19900-3

Khan M, Ahmad F, Koivisto JT, Kellomäki M (2020) Green synthesis of controlled size gold and silver nanoparticles using antioxidant as capping and reducing agent. Colloid Interface Sci Commun 39:100322. https://doi.org/10.1016/j.colcom.2020.100322

Khane Y, Benouis K, Albukhaty S, Sulaiman GM, Abomughaid MM, Al Ali A, Aouf D, Fenniche F, Khane S, Chaibi W (2022) Green synthesis of silver nanoparticles using aqueous Citrus limon zest extract: Characterization and evaluation of their antioxidant and antimicrobial properties. Nanomaterials, 12(12)

Lala S (2020). Enhancement of secondary metabolites in Bacopa monnieri (L.) Pennell plants treated with copper-based nanoparticles in vivo. IET Nanobiotechnology 14(1):78–85.

Milanezi FG, Meireles LM, de Christo Scherer MM, de Oliveira JP, da Silva AR, de Araujo ML, Endringer DC, Fronza M, Guimarães MCC, Scherer R (2019) Antioxidant, antimicrobial and cytotoxic activities of gold nanoparticles capped with quercetin. Saudi Pharm J 27(7):968–974. https://doi.org/10.1016/j.jsps.2019.07.005

Mittal J, Singh A, Batra A, Sharma MM (2017) Synthesis and characterization of silver nanoparticles and their antimicrobial efficacy. Part Sci Technol 35(3):338–345. https://doi.org/10.1080/02726351.2016.1158757

Moldovan B, Sincari V, Perde-Schrepler M, David L (2018) Biosynthesis of silver nanoparticles using Ligustrum ovalifolium fruits and their cytotoxic effects. Nanomaterials 8(8):627

Nagar N, Devra V (2018) Green synthesis and characterization of copper nanoparticles using Azadirachta indica leaves. Mater Chem Phys 213:44–51

Nasr-Esfahani P, Ensafi AA (2022) Noble metals and nonnoble metal oxides based electrochemical sensors. In: Functionalized nanomaterial-based electrochemical sensors. Elsevier, pp. 115–140. https://doi.org/10.1016/B978-0-12-823788-5.00023-5

Niraimathee VA, Subha V, Ravindran RSE, Renganathan S (2016) Green synthesis of iron oxide nanoparticles from Mimosa pudica root extract. Int J Environ Sustain Dev 15(3):227–240

Ulker Ogutveren, Fadime Karaer, Esra Kacar, Savas Koparal A (2016) In: Poceedings of 53rd the IRES international conference, Cape Town, South Africa. ISBN: 978–93–86291–52–3

Orooji Y, Mortazavi-Derazkola S, Ghoreishi SM, Amiri M, Salavati-Niasari M (2020) Mesopourous Fe3O4@ SiO2-hydroxyapatite nanocomposite: Green sonochemical synthesis using strawberry fruit extract as a capping agent, characterization and their application in sulfasalazine delivery and cytotoxicity. J Hazard Mater 400:123140

Pasca R-D, Mocanu A, Cobzac S-C, Petean I, Horovitz O, Tomoaia-Cotisel M (2014) Biogenic syntheses of gold nanoparticles using plant extracts. Part Sci Technol 32(2):131–137. https://doi.org/10.1080/02726351.2013.839589

Pathak MP, Pathak K, Saikia R, Gogoi U, Ahmad MZ, Patowary P, Das A (2022) Immunomodulatory effect of mushrooms and their bioactive compounds in cancer: A comprehensive review. Biomed Pharmacother 149:112901. https://doi.org/10.1016/j.biopha.2022.112901

Pathania D, Sharma M, Thakur P, Chaudhary V, Kaushik A, Furukawa H, Khosla A (2022) Exploring phytochemical composition, photocatalytic, antibacterial, and antifungal efficacies of Au NPs supported by Cymbopogon flexuosus essential oil. Sci Rep 12(1):1–15

Perveen S, Nadeem R, ur Rehman S, Afzal N, Anjum S, Noreen S, Saeed R, Amami M, Al-Mijalli SH, Iqbal M (2022) Green synthesis of iron (Fe) nanoparticles using Plumeria obtusa extract as a reducing and stabilizing agent: Antimicrobial, antioxidant and biocompatibility studies. Arab J Chem, 15(5):103764

Punniyakotti P, Panneerselvam P, Perumal D, Aruliah R, Angaiah S (2020) Anti-bacterial and anti-biofilm properties of green synthesized copper nanoparticles from Cardiospermum halicacabum leaf extract. Bioprocess Biosyst Eng 43(9):1649–1657

Safat S, Buazar F, Albukhaty S, Matroodi S (2021) Enhanced sunlight photocatalytic activity and biosafety of marine-driven synthesized cerium oxide nanoparticles. Sci Rep 11(1):14734. https://doi.org/10.1038/s41598-021-94327-w

Saravanan M, Barabadi H, Vahidi H (2021) Green nanotechnology: isolation of bioactive molecules and modified approach of biosynthesis. In: Biogenic nanoparticles for cancer theranostics. Elsevier, pp 101–122. https://doi.org/10.1016/B978-0-12-821467-1.00005-7

Sathishkumar G, Logeshwaran V, Sarathbabu S, Jha PK, Jeyaraj M, Rajkuberan C, Senthilkumar N, Sivaramakrishnan S (2018) Green synthesis of magnetic Fe3O4 nanoparticles using Couroupita guianensis Aubl. fruit extract for their antibacterial and cytotoxicity activities. Artif Cells, Nanomedicine, Biotechnol, 46(3):589–598

Shalaby EA, Shanab SMM, El-Raheem WMA, Hanafy EA (2022) Biological activities and antioxidant potential of different biosynthesized nanoparticles of Moringa oleifera. Sci Rep 12(1):18400. https://doi.org/10.1038/s41598-022-23164-2

Sidhu AK, Verma N, Kaushal P (2022) Role of biogenic capping agents in the synthesis of metallic nanoparticles and evaluation of their therapeutic potential. Front Nanotechnol, 3. https://doi.org/10.3389/fnano.2021.801620

Vanti G, Kurjogi M (2021) Green nanotechnology: A promising tool for agriculture disease management. In Advances in nano-fertilizers and nano-pesticides in agriculture. Elsevier, pp 505–534. https://doi.org/10.1016/B978-0-12-820092-6.00021-5

Venkateswarlu S, Rao YS, Balaji T, Prathima B, Jyothi NVV (2013) Biogenic synthesis of Fe3O4 magnetic nanoparticles using plantain peel extract. Mater Lett 100:241–244

Vishveshvar K, Krishnan A, Haribabu K, Vishnuprasad S (2018) Green synthesis of copper oxide nanoparticles using Ixiro coccinea plant leaves and its characterization. BioNanoScience 8(2):554–558

Viswadevarayalu A, Venkata Ramana P, Sreenivasa Kumar G, Sumalatha J, Adinarayana Reddy S (2016) Fine ultrasmall copper nanoparticle (UCuNPs) synthesis by using Terminalia bellirica fruit extract and its antimicrobial activity. J Cluster Sci 27(1):155–168

Yew YP, Shameli K, Miyake M, Kuwano N, Khairudin BA, N B, Bt Mohamad SE, & Lee KX (2016) Green synthesis of magnetite (Fe3O4) nanoparticles using seaweed (Kappaphycus alvarezii) extract. Nanoscale Res Lett 11(1):1–7

Younis IY, El-Hawary SS, Eldahshan OA, Abdel-Aziz MM, Ali ZY (2021) Green synthesis of magnesium nanoparticles mediated from Rosa floribunda charisma extract and its antioxidant, antiaging and antibiofilm activities. Sci Rep 11(1):16868. https://doi.org/10.1038/s41598-021-96377-6

Yuan C, Huo C, Gui B, Cao W (2017) Green synthesis of gold nanoparticles using Citrus maxima peel extract and their catalytic/antibacterial activities. IET Nanobiotechnol 11(5):523–530

Chapter 7
Mycofabrication of Silver Nanoparticles: Synthesis, Characterization and Its Biological Applications

N. Prabhu, K. Deepika, S. M. Nandhini, and S. Arsheya Begam

Abstract In the developing field of myconanotechnology, fungi can be used to create nanomaterials or nanostructures with desired shape and size. The word "myco-nanotechnology" was coined by Rai M of India. It is described as the provision of nanoparticles by fungus and their subsequent application, mainly in health-care, environmental, and medical sectors. It investigates silver nanoparticle production by important fungi, such as mushrooms, *Fusarium, Trichoderma, Endophytic* fungus, and yeast. Silver nanoparticles synthesized using fungi enable the control of pathogens, with low toxicity and good biocompatibility. The section focuses on experiments that used fungus to make silver nanoparticles synthesis, creation of nanoparticles of various sizes, surface charges, and morphologies and the synthesis can be optimized by adjusting parameters such as temperature, pH and silver precursor concentration. This chapter reviewed the potential of fungal mediated biosynthesis of silver nanoparticles and its importance in biological functions.

Keywords Myconanotechnology · Biogenic synthesis · Biocompatibility · Parameters

7.1 Introduction

Nanotechnology is raising area of technology which includes synthesis and improvement of substances at nanoscale (Rai et al. 2009). It has opened new avenues with interdisciplinary area of technology known for innumerable packages (Rai et al. 2009). These nanomaterials are utilized in diverse fields together with digital devices, sensor technology, sign enhancers, optical sensors, biomarkers, magnetic, catalysis, optical polarizability, electric conductivity, antimicrobial activity and drug transport to tumour cells (Jain et al. 2010a).

N. Prabhu (✉) · K. Deepika · S. M. Nandhini · S. A. Begam
Department of Biotechnology, Vivekanandha College of Engineering for Women (Autonomous), Tiruchengode, Namakkal 637 205, Tamil Nadu, India
e-mail: prabhu.aut.26@gmail.com

M. P. Shah et al. (eds.), *Biogenic Nanomaterials for Environmental Sustainability: Principles, Practices, and Opportunities*, Environmental Science and Engineering,
https://doi.org/10.1007/978-3-031-45956-6_7

Hence nanoparticle studies have received main use of silver nanoparticles (Ag NPs) in biomedical area with huge wide variety of sources already in market place together with ointments, dressing substances and packaging substances (Gade et al. 2010). Antimicrobial properties of silver nanoparticles are said to be effective against a variety of pathogenic bacteria (Pattabi and Uchil 2000). Mode of movement of silver nanoparticles as in line with the clinical facts shows that silver nanoparticles have unique mode of movement as an instance they're regarded to engage with the thiol groups of important enzyme, motivepit at the molecular wall and harm the DNA of the organism (Jo et al. 2009).

The famous prescient lecture "There is plenty of room at the bottom" given by American physicist Richard Feynman in 1959 is considered the conceptual birth of nanotechnology. It is now an unavoidable problem in almost every field of science, including medicine, the environment, agriculture and engineering (Qian et al. 2013). Humans have used metals such as silver and copper for centuries for their antibacterial properties (Wang et al. 2008). It is currently being investigated for application in consumer products such as textiles, shampoos, hygiene products, contraceptives and especially in waste water treatment (Misra and Mohanty 2011).

Due to their surface resonance properties, Ag NPs find applications in sensors such as colorimetric, surface-enhanced Raman spectroscopy, fluorescence, and chemiluminescence sensors (Kalimuthu 2010). These sensors detect pollutants released into the environment such as ammonia, heavy metals and pesticides. Similarly, treatment of diseases such as cancer is protracted due to late detection of the disease. Ag NPs are useful for early detection. In addition, Ag NPs can be used for remediation of contaminants as they are excellent catalysts for the decomposition of contaminants such as nitro arenes and organic dyes due to their high surface area-to-volume ratio (Kapoor and Viraraghavan 1996).

In addition, Ag NPs are less toxic to mammalian cells than other metal nanoparticles and their relatively small size allows them to easily enter cells through cell membranes and serve as potential antibacterial agents (Kim et al. 2012). Due to their considerable surface energy, Ag NPs tend to aggregate, reducing their antibacterial potency. This can be resolved by loading the Ag NPs onto a solid support system. The high electrical conductivity of Ag NPs finds applications as electronic devices that mimic inks (Ingle et al. 2009). When applied to a compound surface, it enhances the conductivity of 2D sensors like graphene oxide. Additionally, this property is useful for detecting heavy metals (Choudary et al. 2002).

Depending on the reducing agent used, AgNPs can be synthesized using physicochemical and biological approaches. Previous approaches have used chemicals, light, lasers, electricity, sound waves, microwaves (Ankamwar et al. 2005) as substrates to synthesize Ag NPs. However, for the latter, secondary metabolites or enzymes from microorganisms, fungi, and mainly polyphenol-rich natural substances are used (Rajeshkumar and Bharath 2017). Physicochemical approaches to nanoparticle synthesis are neither economical nor environmentally friendly (Kapoor and Viraraghavan 1996). On the other hand, biosynthesis can overcome these drawbacks and obtain Ag NPs with various structures and crystallinities. The material chemistry of Ag NPs is determined by their shape, size, surface charge, aggregation tendency, and dissolution rate (Kogej and Pavko 2001).

When released into the environment, the toxicity of Ag NPs increases and can affect communities such as microorganisms, nematodes, insects, plants and animals. Similarly, human inhalation of Ag NPs is also unpleasant (Lee et al. 2011). It can damage organs such as the liver, spleen, lungs, and kidneys. Moreover, microbial resistance to Ag NPs may pose problems in applying Ag NPs as broad-spectrum antifungal agents.

Microfabrication accommodates using a fixed of producing equipment primarily based totally on photolithographic strategies typically used for making IC (Inspirtory Capacity) with inside the electronics industry (Das et al. 2009). Microfabrication is likewise not un usual place key era with inside the promising new fields consisting of MEMS (micro electro-mechanical structures), μ TAS (micro overall analytical structures), and μ FLUMS (micro fluidics and molecules) and has been carried out to immune isolation, drug delivery, biochemical engineering and analytical fields inclusive of capillary electrophoresis (CE) (Aziz et al. 2015).

The capability blessings of micro-sized apparatus, relative to structures of traditional size, encompass decreased intake of samples and reagents, shorter evaluation times, extra sensitivity, portability (Jang et al. 2021) that permits in situ and real-time evaluation and disposability. A numerous fungal strain can be synthesized using silver nanoparticles (AgNPs) with the help of *Pestalotiopsis sp., Phoma sp., Humicola sp., Aspergillus Niger, Fusarium oxysporum, Penicillium fellutanum.* Optimizing the cultural and physical conditions 1 in the synthesis of silver nanoparticles from *Fusarium oxysporum* (Birla et al. 2013).

One of the most important factors affecting the mycosynthesis of nanoparticles is pH (Balakumaran et al. 2022a). It influences the size and nature of the nanoparticles synthesized. The isolation of *Penicillium fellutanum* from the coastal mangrove sediments have the ability to produce nanoparticles under controlled parameters of temperature, pH, silver ion concentration, and time of exposure of silver nanoparticles (Kathiresan et al. 2009). Using *Aspergillus flavus* (KF934407) and *Fusarium solani*, silver nanoparticles can be synthesized extracellularly in the size of 50 nm and 5–35 nm (Fatima et al. 2016; Ingle et al. 2009).

7.1.1 Why We Choose Silver Nanoparticles

Silver nanoparticles are the most well-known type of nanoparticles, with extensive production and commercialization along with significant applications (Filipenko et al. 2007). A number of examples discovered that amalgamation of silver nanoparticles will be completed with the aid of using the usage of organic organisms. Extracellular synthesis of solid silver nano crystals has been documented in fungus *Aspergillus flavus*. Similarly, endophytic fungi *Epicoccumnigrum* turned into determined to synthesize Silver nanoparticles at one-of-a-kind pH and temperature. Similar to this, well-known producers of silver nanoparticles include bacteria like

Bacillus species and *Brevibacterium casei.* Also, it was claimed that Curry leaf was used to produce rounded silver nanoparticles of length 10 to 25 nm (*Murrayakoenigii*) (Adebayo et al. 2021).

7.1.2 Synthetic Versus Mycofabrication of Silver Nanoparticles

Synthetic procedures of silver nanoparticle synthesis presently there are pretty a whole lot of techniques hired in the fabrication of silver nanoparticles (Moghaddam et al. 2015). But those procedures involve the packages of decreasing dealers like hydrazine, sodium borohydride, thiourea, thiophenol, mercaptoacetateetc (Babu and Gunasekaran 2009) which can be unsafe and unfavorable to the environment. Such decreasing dealers make the synthesis system costly. Consequently, organic synthesis of silver nanoparticle is now the maximum eco- pleasant and value powerful system etc. (Chinnamuthu and Boopathi 2009). Bio fabrication of silver nanoparticles can be accomplished via way of means of using exclusive kinds of organic organisms such as bacteria, flowers and fungi (Lu et al. 2010) (Table 7.1).

7.2 Mycofabrication of Silver Nano Particles

7.2.1 Optimization of Silver Nanoparticle Synthesis

Synthesis of silver nanoparticles using fungi is simple and effective, but the parameters used in the procedure need to be optimized for good particle monodispersity, stability and biocompatibility (Balakumaran et al. 2015). Since the most diverse fungi are potentially available for synthesis, it is important to consider individual characteristics and optimize synthesis conditions accordingly (Ottoni et al. 2017).

Parameters such as agitation, temperature, light, incubation time, and synthesis time vary depending on the fungus used and can also be adjusted to obtain the desired properties of the nanoparticles. Controlling the size and shape of nanoparticles requires tuning the parameters used in both the fungal culture and synthesis processes (Birla et al. 2013). Studies have shown that it is possible to obtain nanoparticles with different physicochemical properties using changes in temperature, metal precursor concentration, pH, medium and amount of biomass (Birla et al. 2013; Rajput et al. 2016; Saxena et al. 2016; Wu et al. 2017). Table 7.1 shows several studies in which silver nanoparticles were prepared using different types of fungi and synthesis conditions.

Table 7.1 Synthesis of nanoparticles

Fungi/Yeast	Confinement	References
Fusarium oxysporum	Extracellular	Duran et al. (2005)
Aspergillus fumigatus	Extracellular	Bhainsa and D'souza (2006)
Phanerochaete chrysosporium	Extracellular	Vigneshwaran et al. (2007)
Aspergillus flavus	Extracellular	Vigneshwaran et al. (2007)
Fusarium oxysporum	Extracellular	Mohammadian et al. (2007)
Pleurotussajor-caju	Extracellular	Vigneshwaran et al. (2007)
Aspergillus niger	Extracellular	Gade et al. (2008)
Fusariuma cuminatum	Extracellular	Ingle et al. (2008)
Cladosporium cladosporioides	Extracellular	Balaji et al. (2009)
Coriolus versicolor	Extracellular	Sanghi and Verma (2009)
Penicillium brevicompactum	Extracellular	Shaligram et al. (2009)
Fusariumsolani	Extracellular	Ingle et al. (2009)
Penicillium fellutanum	Extracellular	Kathiresan et al. (2009)
Penicillium sp.	Extracellular	Maliszewska and Sadowski (2009)
Phomaglomerata	Extracellular	Birla et al. (2009)
Penicillium sp.	Extracellular	Hemanth et al. (2010)
Aspergillus niger	Extracellular	Jaidev and Narasimha (2010)
Aspergillus clavatus	Extracellular	Kumar et al. (2018)
Fusarium oxysporum	Extracellular	Dhillon et al. (2012)
Aspergillus flavus	Extracellular	Jain et al. (2010a, 2010b)
Aspergillus fumigatus	Extracellular	Ranjbar Navazi et al. (2010)
Bipolarisnodulosa	Extracellular	Rao et al. (2010)
Fusarium solani	Extracellular	El-Rafie et al. (2012)

7.2.2 *Effect of pH*

Adjustment of synthesis pH can be used to control specific properties of nanoparticles (Lee et al. 2011) reported that the conformation of the nitrate reductase enzyme can change depending on the proton concentration in the reaction medium, leading to changes in nanoparticle morphology and size. At higher pH values, there is greater competition between protons and metal ions for binding to negatively charged regions, leading to greater synthetic success at alkaline pH (Sintubin et al. 2009).

Qian et al. (2013) observed that alkaline pH favored the synthesis of silver nanoparticles when $AgNO_3$ was added to the filtrate of the fungus *Epicoccumnigrum.* It may find that higher alkaline pH resulted in shorter synthesis times and smaller nanoparticle size distributions and polydispersity index values. These properties indicate enhanced stability due to electrostatic repulsion of anions present in the dispersion (Gurunathan et al. 2009). Synthesis using *Coleotrichum sp.* ALF2-6 at alkaline pH and elevated temperature of 50 °C was completed in about 20 min, faster than at lower pH (Azmath et al. 2016).

7.2.3 *Effect on Tempreture*

The rate of synthesis increases when the temperature is raised to 40 °C, which he considered to be the ideal temperature. In another study using *Fusarium oxysporum* filtrate, higher protein secretion by fungal biomass was observed at temperatures between 60 and 80 °C, with synthesis rate and surface uptake increasing gradually with increasing temperature. found that the reaction rate increased at higher temperatures and that the synthesis was completed within 20 min at temperatures above 50 °C (Rai et al. 2016). According to a research by Phanjom and Ahmed using *Aspergillus oryzae*, synthesis progressed more quickly at higher temperatures, finishing in six hours at 30, 50, 70, and 90 °C.

Although most studies report faster synthesis rates at higher temperatures, it is important to consider the quality of the nanoparticles. In addition to affecting synthesis rate, temperature can affect nanoparticle size and stability. Nanoparticles synthesized using the fungus *Aspergillus fumigatus* reached a size of 322.8 nm at 25 °C, increased with increasing temperature, and reached 1073.45 nm at 55°C using the fungus *Fusarium oxysporum* found that the nanoparticles size decreased with increasing temperature to 50 °C, with the smallest size (30.24 nm) at that temperature also found that using the fungus *Trichoderma viride* reduced the size of the nanoparticles with increasing synthesis temperature.

7.2.4 *Effect of $AgNO_3$ Concentration*

Most studies using fungi for the extracellular synthesis of silver nanoparticles used $AgNO_3$ at a concentration of 1 mM (Saxena et al. 2016; Xue et al. 2016). In some cases, lower metal precursor concentrations resulted in smaller nanoparticle sizes and improved distribution (Kaviya et al. 2011; Phanjom and Ahmed 2017). However, other studies found that using a moderate concentration of his $AgNO_3$ reduced the size (Abdel-Hafez et al. 2017) using the fungus *Rhizopus stolonifera.* He obtained the smallest nanoparticle size (2.86 nm) at 10 mm AgNO, while at 100 and 1 mm the sizes were 54.67 and 14.23 nm, respectively. Got it. Similar results were reported by Husseiny et al. (2015) with *Fusarium oxysporum.*

Phanjom and Ahmed studied the synthesis of nanoparticles with *Aspergillus oryzae* and his $AgNO_3$ concentration different from 1 to 10 mm. The nanoparticles were observed to have sizes from 7.22 to 17.06 nm at $AgNO_3$ concentrations up to 8 mm, while the sizes increased to 45.93 and 62.12 nm at $AgNO_3$ concentrations of 9 and 10 mm, respectively. This effect was attributed to the lack of functional groups available for reaction when the metal precursor concentration was increased.

7.2.5 Extracelluar Production of Silver Nanoparticles

Number of possible mechanisms have been proposed for the formation of metallic nanoparticles, but to date no single generalized mechanism has been identified (Kogej and Pavko 2001). Cell walls and cell wall sugars are likely to play an important role in the reduction of metal ions (Mukherjee et al. 2001). Nanoparticles are formed on cell wall surfaces, and the first step in bioreduction is the trapping of the metal ions on this surface.

This probably occurs due to the electrostatic interaction between the metal ions and positively charged groups in enzymes present at the cell wall (Gajbhiye et al. 2009) This may be followed by enzymatic reduction of the metal ions, leading to their aggregation and the formation of nanoparticles. Proteins have been implicated in nanoparticle formation in a number of different studies (Gericke and Pinches 2006) found that, during the formation of zirconia nanoparticles, the fungus secreted proteins capable of extracellularly hydrolyzingcomplexesby means ofzirconium ions and this was confirmed in subsequent studies with silica and titania (Bansal et al. 2005).

They found that their fungi were also capable of hydrolyzing metal halide precursorsunder acidic conditions. Their studies indicated that the proteins involved in the reduction of metal nanoparticles were cationic proteins with molecular weights of around 21–24 kDa (Bharde et al. 2006) also suggested that a cationic protein of around 55 kDa found in extracellular extracts of *Verticillium sp.* might be responsible for the hydrolysis of $[Fe(CN)_6]^3$ and $[Fe(CN)_6]^4$ (Bhainsa and D'Souza 2006) concluded that the absorption peaks they obtained at 220 and 280 nm corresponded to amide bonds and to tryptophan and tyrosine residues present in the proteins, respectively, and that this indicated proteinaceous compounds in fungal biomass which had a major role in the bioreduction of metal nanoparticles (Gade et al. 2008).

7.2.6 Characteristics of Silver Nanoparticles

Characterisation is an important step in the green synthesis of nanoparticles. It is a pivotal step to determine the morphology, surface chemistry, surface area and disparity in the nature of any silver nanoparticle. The most popular method for identifying metallic nanoparticles by analysing their stability and synthesis is UV–Vis spectrophotometry (Sastry et al. 1998). The next technique is X-ray diffraction analysis (XRD), which uses X-rays that are absorbed deeply by the material to observe the structure of crystalline metallic nanoparticles (Rajeshkumar et al. 2019)

The another technique is Fourier transform infrared spectroscopy (FTIR), which can be used to examine the surface chemistry of synthesised metal nanoparticles, observe the participation of biomolecules in nanoparticle synthesis, and analyse various capping agents. Energy-dispersive X-Ray spectroscopy (EDX) It has been established that EDX is a key method for determining a sample's elemental composition, and that it can be used in nanotechnology (Juibari et al. 2011).

The topography and morphology of nanoparticles can be viewed using scanning electron microscopy (SEM), which is also used to determine the sizes of different nanoparticles at the micro-(106) and nano (109) scales (Sundrarajan and Gowri 2011). Transmission Electron Microscopy (TEM) is a very helpful method for characterizing nanoparticles because it can reveal details about their size and shape (Ghosh et al. 2012). Compared to SEM,

TEM has a 1,000-fold better resolution, and its images provide more precise information about the size, shape, and crystallography of the nanoparticles (Eppler et al. 2000). A method that is widely used to determine the size and size range of molecules is dynamic light scattering (DLS). It is frequently used to describe nanoparticles and has been used to gauge the size of nanoparticles. Additionally, DLS has been widely used to size magnetic nanoparticles in liquid form (Phenrat et al. 2009).

An excellent analytical technique for nanotechnology is auger electron spectroscopy (AES), a surface-sensitive analytical technique that results from the interplay of an electron beam and atoms in residence at a sample's surface (Hadrup et al. 2020). The widely used surface analysis method known as low-energy ion scattering (LEIS) is renowned for its exceptional surface sensitivity. This method allows for the deduction of the structure and elemental makeup of a specific sample (Table 7.2).

7.3 Applications

7.3.1 Waste Water Treatment

Silver nanoparticles (Ag NPs) are highly toxic to microorganisms and thus have strong antimicrobial effects against a wide range of microorganisms, including viruses, bacteria, and fungi (Sintubin et al. 2009). As a good antimicrobial agent, silver nanoparticles have been widely used for the disinfection of water (Kitching et al. 2015).

The mechanism of the antimicrobial effects of Ag NPs is not clearly known and remains under debate. In recent years, several theories have been put forward (Moghaddam et al. 2015). Ag NPs have been reported to be able to adhere to the bacterial cell wall and subsequently penetrate it, resulting in structural changes of the cell membrane and thus increasing its permeability (Kumar et al. 2011).

Besides, when Ag NPs are in contact with bacteria, free radicals can be generated. They have the ability to damage the cell membrane and are considered to cause the death of cells (Degenkolb et al. 2018). In addition, as DNA contains abundant sulfur and phosphorus elements, Ag NPs can act with it and thus destroy it. This is another explanation for the death of cells caused by Ag NPs (Mukherjee et al. 2001) What is more, the dissolution of Ag NPs will release antimicrobial Ag^{+} ions, which can

Table 7.2 Synthesis nanoparticles using fungal species

Fungal species	Nanoparticle produced	Synthesis mode	Size (nm)	References
Aspergillus flavus (KF934407)	Silver	Extracellular	50	Fatima et al. (2016)
Fusarium solani	Silver	Extracellular	5–35	Ingle et al. (2009)
Ascomycota sp.	Silver	Intracellular and extracellular	10	Jain et al. (2012)
Duddingtonia flagans	Silver	Extracellular	30–409	Costa Silva et al. (2017)
Trichoderma longibrachaitum	Silver	Extracellular	24.43	Elamawi et al. (2018)
Penicillium oxalicum	Silver	Extracellular	10–40	Rose et al. (2019)
Aspergillus fumigatus	Silver	Extracellular	322.8	Shahzad et al. (2019)
Sclerotiniasclerotiorum	Silver	Extracellular	5–50	Tomah et al. (2020)
Aspergillus oryzae	Silver	Extracellular	7–27	Phanjom and Ahmed (2017)
Colleotrichum sp.	Silver	Extracellular	5- 13	Azmath et al. (2016)
Fussarium oxysporum	Silver	Extracellular	13–15	Husseiny et al. (2015)
Trichoderma viride	Silver	Extracellular	2–4	Fayaz et al. (2009)

interact with the thiol groups of many vital enzymes, inactivate them, and disrupt normal functions in the cell (Kapoor et al. 1999).

With the development of nanotechnology, Ag NPs have been successfully applied in water and wastewater disinfection in recent years (Juibari et al. 2011). Direct application of Ag NPs might cause some problems, such as their tendency to aggregate in aqueous media that gradually reduces their efficiency during long-term use (Merrin et al. 1998). Ag NPs attached to filter materials have been considered promising for water disinfection due to their high antibacterial activity and cost-effectiveness (Narayanan and Sakthivel 2010).

Via the in-situ reduction of silver nitrate, Ag NPs have been deposited on the cellulose fibers of an absorbent blotting paper sheet (Olaniran et al. 2013). The Ag NPs sheets showed antibacterial properties towards suspensions of *Escherichia coli* and *Enterococcus faecalis* and inactivated bacteria during filtration through the sheet (Ahmad et al. 2005). Moreover, the silver loss from the Ag NPs sheets was lower than the standards for silver in drinking water put forward by Environmental Protection Agency (EPA) and World Health Organization (WHO).

Therefore, for water contaminated by bacteria, filtration through paper deposited with AgNPs could be an effective emergency water treatment (Phenrat et al. 2009, Prasad and Jha 2010). Besides, Ag NPs synthesized by chemical reduction have been incorporated into polyethersulfone (PES) microfiltration membranes. The activity of microorganisms nearby the membranes was observed to be remarkably suppressed (Rajeshkumar and Bharath 2017). The PES-Ag NPs membranes exhibited strong antimicrobial properties and held great potential in application for water treatment (Zhang et al. 2005).

7.4 Nanoscale Silver for Bone Mending

Calcium and phosphate in the form of crystalline hydroxyapatite makes up human bones. It is a widely recognised and utilized substance for body implants (Bharti et al. 2016). Silver-hydroxyapatite nanocomposite: structural characterisation of a bone healing biomaterial. For the creation of antibacterial and bioactive bone implants, biocompatible hydroxyapatite mixed with metallic or ionic silver forms is an appropriate choice (Hirsch et al. 2003). Gram positive and Gram-negative bacteria are successfully inhibited by such hydroxyapatite coverings that include silver nanoparticles (Kora and Sashidhar 2018). In order to produce improved biomaterials, the additive should also have additional optical, mechanical, and chemical capabilities (Kora and Sashidhar 2018).

7.4.1 Biomedical Application

Some researchers argue that nanoparticles release silver cations in a controlled manner, so that the antimicrobial activity is carried out in small doses of silver ions released into the medium, not constituting a toxic threat to the patient (Nethravathi et al. 2009). However, another study has shown that the AgNP concentration is indeed a concern, showing that there is a threshold between concentrations considered toxic to microorganisms and those considered toxic to the patient's cells (Christensen et al. 2011). In addition to using the correct concentration, the use of capping with organic molecules, such as chitosan, or the use of biosynthesized nanoparticles, increases biocompatibility and decreases cytotoxicity (Bhainsa and D'souza 2006).

It was already noted that the intragastric administration of AgNPs did not result in rats' lethality or pronounced toxic effects, and did not influence the hematological and biochemical parameters (Hadrup et al. 2020). The more dangerous silver is probably ionic silver, which has been utilised extensively in dentistry. A study conducted with human volunteers (Munger et al. 2014) showed that a fourteen-day oral dosing of silver nanoparticles did not cause evident metabolic or hematologic changes, nor changes in urinalysis parameters, overall physical state, or imaging morphology. But it is noteworthy that the toxicity of 5 nm nanoparticles was described on human endothelial and bronchial epithelial cells (Jang et al. 2014) and this study used microarray analysis to show that AgNP-treated cells presented significant variations in cell death-, apoptosis-, and cell survival-related gene expression; however, 100 nm silver AgNPs did not induce cell death even at high concentrations, showing that AgNP toxicity is highly affected by the size of the nanoparticle (Arora et al. 2009).

One of the major concerns regarding the utilization of silver nanoparticles is their use in pregnant women and animals (Giner-Lamia et al. 2014). It was already demonstrated that when pregnant mice were exposed to 18–20 nm AgNP, silver-containing nanoparticles could be detected in the placenta and in the head of the fetus (Deshmukh et al. 2019). In the fetus, silver was detected in the ionic form or as nanoparticles with a size less than 13 nm (Tejada et al. 2017), and this situation points to precautions with respect to acute exposure to nanoparticles during pregnancy (Srivastava et al. 2012). This accumulation of silver in the central nervous system has already been shown to induce long-term memory impairments in a mice model (Antsiferova et al. 2021). In a similar way, phytoreduced silver nanoparticles with polyphenols from *Viburnum opulus* fruit extract presented testicular toxic effects in offspring during the embryological development of the murine gonad (Bidian et al. 2021). Therefore, more studies are necessary to evaluate if nanoparticles can be safely administered to pregnant women.

7.4.2 Health Application

Several studies have described the use of biogenic silver nanoparticles in healthcare applications to combat bacteria and fungi (Rai et al. 2009). Nanoparticles that come into contact with the cell wall and cause progressive metabolic reactions with the production of reactive oxygen species directly inhibit bacterial growth (Gudikandula et al. 2017).

Nanoparticle size is one of the factors that determine antimicrobial activity, with smaller nanoparticles being more effective (Lu et al. 2010). Small nanoparticles can penetrate bacterial cell membranes to damage the respiratory chain, alter permeability, cause DNA and RNA damage, affect cell division, and cause cell death (Morones et al. 2005). Nanoparticles also interact with thiol groups of essential enzymes by releasing $Ag+$ ions that form complexes with nucleotides that damage microbial DNA and inhibit the activity of DNAs (Tomah et al. 2020).

7.5 Conclusion

Mycosynthesis of nanoparticles has created a great attention among the researchers because of their desirable characteristics and nontoxicity. This chapter exhibits the different reports illustrated by the researchers on the intracellular and extracellular synthesis of mycogenic nano particles. Thus, we can conclude that biological organisms such as Fungi, Bacteria, and plants could be employed as suitable Nano-factories for the biological synthesis of nanoparticles of silver. There is an increasing interest in the use of fungi for these processes, and fungi may have the potential to provide relatively quick and ecologically 'clean' biofactories for metallic nanoparticles.

Simple protocols have been used for the production of metallic nanoparticles of gold, silver, platinum, silica, titania and zirconia from fungi, and these processes have been observed in both intra and extracellular preparations. Many industrial areas, including food, enzyme and pharmaceutical production and processing, currently use fungal biomass, and so downstream handling and processing procedures for fungal material are already established. There are a number of challenges that need to be undertaken before the potential of myconanotechnology can be fully evaluated, and these include determining the detailed chemical processes involved in the formation of the particles, and optimizing these to provide sufficient quantities of a consistent product that can be easily extracted or separated.

References

Abdel-Hafez SI, Nafady NA, Abdel-Rahim IR, Shaltout AM, Daròs JA, Mohamed MA (2017) Biosynthesis of silver nanoparticles using the compound curvularin isolated from the endophytic fungus *Epicoccum nigrum*: characterization and antifungal activity. J Pharm Appl Chem 3(2):135–146

Adebayo EA, Azeez MA, Alao MB, Oke AM, Aina DA (2021) Fungi as veritable tool in current advances in nanobiotechnology. Heliyon 7(11):E08480. ISSN 2405–8440

Ahmad A, Senapati S, Khan MI, Kumar R, Sastry M (2005) Extra-/intracellular biosynthesis of gold nanoparticles by an alkalotolerant fungus Trichothecium sp. J Biomed Nanotechnol 1:47–53

Ankamwar B, Damle C, Ahmad A, Sastry M (2005) Biosynthesis of gold and silver nanoparticles using *Emblica officinalis* fruit extract, their phase transfer and transmetallation in an organic solution. J Nanosci Nanotechnol 5(10):1665–1671

Antsiferova AA, Kopaeva MY, Kochkin VN, Kashkarov PK, Kovalchuk MV (2021) Disturbance in mammalian cognition caused by accumulation of silver in brain. Toxics 9:30

Arora S, Jain J, Rajwade JM, Paknikar KM (2009) Interactions of silver nanoparticles with primary mouse fibroblasts and liver cells. Toxicol Appl Pharm 236(3):310–318

Aziz N, Faraz M, Pandey R, Sakir M, Fatma T, Varma A, Barman I, Prasad R (2015) Facile algae derived route to biogenic silver nanoparticles: synthesis, antibacterial and photocatalytic Properties. Langmuir 31:11605–11612

Azmath P, Baker S, Rakshith D, Satish S (2016) Mycosynthesis of silver nanoparticles bearing antibacterial activity. Saudi Pharm J 24(2):140–146

Babu GMM, Gunasekaran P (2009) Production and structural characterization of crystalline silver nanoparticles from *Bacillus cereus* isolate. Colloids Surf B: Biointerfaces 74:191–195

Balaji DS, Basavaraja S, Deshpande R, Mahesh DB, Prabhakar BK, Venkataraman A (2009) Extracellular biosynthesis of functionalized silver nanoparticles by strains of *Cladosporium cladosporioides* fungus. Colloids Surf, B 68(1):88–92

Balakumaran MD, Ramachandran R, Kalaichelvan PT (2015) Exploitation of endophytic fungus, Guignardia mangiferae for extracellular synthesis of silver nanoparticles and their in vitro biological activities. Microbiol Res 178:9–17

BalaKumaran MD, Ramachandran R, Balashanmugam P, Jagadeeswari S, Kalaichelvan PT (2022a) Comparative analysis of antifungal, antioxidant and cytotoxic activities of mycosynthesized silver nanoparticles and gold nanoparticles. Mater Technol 37(6):411–421

BalaKumaran MD, Ramachandran R, Balashanmugam P, Jagadeeswari S, Kalaichelvan PT (2022b). Extracellular biosynthesis of functionalized silver nanoparticles by strains of *Cladosporium Cladosporioides* fungus. Biointerfaces 1:88–92

Bansal V, Rautaray D, Bhadre A, Ahire K, Sanyal A, Ahmad A, Sastry M (2005) Fungus-mediated biosynthesis of silica and titaniaparticle. J Mater Chem 15:2583–2589

Besinis A, Peralta T, Handy RD (2014) The antibacterial effects of silver, titanium dioxide and silica dioxide nanoparticles compared to the dental disinfectant chlorhexidine on *Streptococcus* mutans using a suite of biossays. Nanotoxicology 8:1–16

Bhainsa KC, D'souza SF (2006) Extracellular biosynthesis of silver nanoparticles using the fungus *Aspergillus fumigatus*. Colloids Surf B: Biointerfaces 1:160–164

Bharde A, Rautaray D, Bansal V, Ahmad A, Sarkar I, Yusuf SM, Sanyal M, Sastry M (2006) extracellular biosynthesis of magnetite using fungi. Small 2:135–141

Bharti A, Singh S, Meena VK, Goyal N (2016) Structural characterization of silver-hydroxyapatite nanocomposite: a bone repair biomaterial. Mater Today Proc 3:2113–2120

Bhattacharyya A, Duraisamy P, Govindarajan M, Buhroo AA, Prasad R (2016) Nanobiofungicides: Emerging trend in insect pest control. In: Prasad R (ed) Advances and applications through fungal nanobiotechnology. Springer International Publishing, Switzerland, pp 307–319

Bidian C, Filip GA, David L, Florea A, Moldovan B, Robu DP, Olteanu D, Radu T, Clichici S, Mitrea DR (2021) The impact of silver nanoparticles phytosynthesized with *Viburnum opulus L.* extract on the ultrastrastructure and cell death in the testis of offspring rats. Food Chem Toxicol 150:112053

Birla SS, Gaikwad SC, Gade AK, Rai MK (2013) Rapid synthesis of silver nanoparticles from by optimizing physico cultural conditions. Sci World J 2013:1–12

Birla SS, Tiwari VV, Gade AK, Ingle AP, Yadav AP, Rai MK (2009). Fabrication of silver nanoparticles by Phomaglomerata and its combined effect against *Escherichia coli, Pseudomonas Aeruginosa* and *Staphylococcus aureus*. Lett Appl Microbiol 48:173–179

Bramhanwade K, Shende S, Bonde S, Gade A, Rai M (2016) Fungicidal activity of Cu nanoparticles against *Fusarium* causing crop diseases. Environ Chem Lett 14(2):229–235

Campagnolo L, Massimiani M, Vecchione L, Piccirilli D, Toschi N, Magrini A, Bonanno E, Scimeca M, Castagnozzi L, Buonanno G et al (2017) Silver nanoparticles inhaled during pregnancy reach and affect the placenta and the foetus. Nanotoxicology 11:687–698

Chhipa H (2019) Mycosynthesis of nanoparticles for smart agricultural practice: a green and ecofriendly approach. Elsevier, pp 87–109. ISBN 9780081025796

Chinnamuthu CR, Boopathi PM (2009) Nanotechnology and agroecosystem. Madras Agric J 96:17–31

Choudary BM, Madhi S, Chowdari NS, Kantam ML, SreedharB B, Saker, (2002) Layered double hydroxide supported nanopalladium catalyst for Heck-, Suzuki, Sonogashira, and Stille-type coupling reactions of chloroarenes. J Am ChemSoc 124:14127–14136

Christensen L, Vivekanandhan S, Misra M, Mohanty AK (2011) Biosynthesis of silver nanoparticles using Murrayakoenigii (curry leaf): An investigation on the effect of broth concentration in reduction mechanism and particle size. Adv Mat Lett 2:429–434

Costa Silva LP, Oliveira JP, Keijok WJ, da Silva AR, Aguiar AR, Guimarães MCC, Ferraz CM, Araújo JV, Tobias FL, Braga FR (2017) Extracellular biosynthesis of silver nanoparticles using the cell-free filtrate of nematophagous fungus *Duddingtonia flagrans*. Int J Nanomed 6373–6381

Das SK, Das AR, Guha AK (2009) Gold nanoparticles: microbial synthesis and application in water hygiene management. Langmuir 25:8192–8199

Degenkolb L, Metreveli G, Philippe A, Brandt A, Leopold K, Zehlike L, Vogel HJ, Schaumann GE, Baumann T, Kaupenjohann M, Lang F (2018) Retention and remobilization mechanisms of environmentally aged silver nanoparticles in an artificial riverbank filtration system. Sci Total Environ 645:192–204

Deshmukh SP, Patil SM, Mullani SB, Delekar SD (2019) Silver nanoparticles as an effective disinfectant: a review. Mater Sci Eng: C 97:954–965

Dhillon GS, Brar SK, Kaur S, Verma M (2012) Green approach for nanoparticle biosynthesis by fungi: current trends and applications. Crit Rev Biotechnol 32:49–73

Durán N, Marcato PD, Alves OL, Desouza G, Esposito E (2005) Mechanistic aspects of biosynthesis of silver nanoparticles by several *Fusarium oxysporum* strains. J Nanobiotechnol 3:8–14

Elamawi RM, Al-Harbi RE, Hendi AA (2018) Biosynthesis and characterization of silver nanoparticles using *Trichoderma longibrachiatum* and their effect on phytopathogenic fungi. Egypt J Biol Pest Control 28(1):1–11

El-Rafie MH, Shaheen THI, Mohamed AA, Hebeish A (2012) Bio-synthesis and applications of silver nanoparticles onto cotton fabrics. Cotton Polym 90:915–920

Eppler AS, Rupprechter G, Anderson EA, Somorjai GA (2000) Thermal and chemical stability and adhesion strength of Pt nanoparticle arrays supported on silica studied by transmission electron microscopy and atomic force microscopy. J Phys Chem B 104(31):7286–7292

Fatima F, Verma SR, Pathak N, Bajpai P (2016) Extracellular mycosynthesis of silver nanoparticles and their microbicidal activity. Environ Chem Lett 14(2):229–235

Fayaz AM, Balaji K, Girilal M, Kalaichelvan PT, Venkatesan R, (2009) Mycobased synthesis of silver nanoparticles and their incorporation into sodium alginate films for vegetable and fruit preservation. J Agric Food Chem 57:6246–6252

Filipenko EA, Filipenko ML, Deineko EV, Shumnyi VK (2007) Analysis of integration sites of T-DNA insertions in transgenic tobacco plants. Cytol Genet 41:199–203

Gade AK, Bonde P, Ingle AP, Marcato PD, Duran N, Rai MK (2008) Exploitation of *Aspergillus niger* for synthesis of silver nanoparticles. J Biobaased Mater Bioenergy 2:243–247

Gade A, Ingle A, Whiteley C, Rai M (2010) Mycogenic metal nanoparticles: progress and applications. Biotechnol Lett 32:593–600

Gajbhiye B, Kesharwani JG, Ingle AP, Gade AK, Rai MK (2009) Fungus mediated synthesis of silver nanoparticles and its activity against pathogenic fungi in combination of fluconazole. Nanomedicine: NBM 5:282–286

Gericke M, Pinches A (2006) Biological synthesis of metal nanoparticles. Hydrometallurgy 83:132–140 Gericke M, Pinches A (2006) Microbial production of gold nanoparticles. Gold Bull 39:22–28

Ghosh S, Patil S, Ahire M, Kitture R, Kale S, Pardesi K, Cameotra SS, Bellare J, Dhavale DD, Jabgunde A, Chopade BA (2012) Synthesis of silver nanoparticles using *Dioscorea bulbifera tuber* extract and evaluation of its synergistic potential in combination with antimicrobial agents. Int J Nanomed 483–496

Giner-Lamia J, López-Maury L, Florencio FJ, Janssen PJ (2014) Global transcriptional profiles of the copper responses in the *Cyanobacterium synechocystis sp. PCC 6803*. PLoS ONE 9(9):e108912

Goebl D, Bruckner B, Roth D, Ahamer C, Bauer P (2015) Low-energy ion scattering: a quantitative method? Nucl Instrum Methods Phys Res, Sect B 354:3–8

Gudikandula K, Vadapally P, Charya MS (2017) Biogenic synthesis of silver nanoparticles from white rot fungi: Their characterization and antibacterial studies. OpenNano 2:64–78

Gurunathan S, Kalishwaralal K, Vaidyanathan R, Venkataraman D, Pandian SR et al (2009) Biosynthesis, purification and characterization of silver nanoparticles using *Escherichia coli*. Colloids Surf A: Biointerfaces 74:328–335

Hadrup N, Sharma AK, Loeschner K, Jacobsen NR (2020) Pulmonary toxicity of silver vapours, nanoparticles and fine dusts: a review. Regul Toxicol Pharmacol 115:104690

Hemanth NKS, Kumar G, Karthik L, Bhaskara Rao KV (2010) Extracellular biosynthesis of silver nanoparticles using the filamentous fungus *Penicillium sp.* Arch Appl Sci Res 2:161–167

Hendrickson OD, Klochkov SG, Novikova OV, Bravova IM, Shevtsova EF, Safenkova IV, Zherdev AV, Bachurin SO, Dzantiev BB (2016) Toxicity of nanosilver in intragastric studies: biodistribution and metabolic effects. Toxicol Lett 241:184–192

Hirsch LR, Stafford RJ, Bankson JA, Sershan SR, Rvera B, Price RE et al (2003) Nanoshellmediated near infrared thermal therapy of tumors under magnetic resonance guidance. PNAS 100:13549–13554

Husseiny SM, Salah TA, Anter HA (2015) Biosynthesis of size controlled silver nanoparticles by *Fusarium oxysporum*, their antibacterial and antitumor activities. Beni-Suef Univ J Basic Appl Sci 4(3):225–231

Hwang IS, Hwang JH, Choi H, Kim KJ, Lee DG (2012) Synergistic effects between silver nanoparticles and antibiotics and the mechanisms involved. J Med Microbiol 61(12):1719–1726

Ingle A, Gade A, Pierrat S, Sonnichsen C, Rai MK (2008) Mycosynthesis of silver nanoparticles using the fungus *Fusarium acuminatum* and its activity against some human pathogenic bacteria. Curr Nanosci 4:141–144

Ingle A, Rai M, Gade A, Bawaskar M (2009) *Fusarium solani*: a novel biological agent for the extracellular synthesis of silver nanoparticles. J Nanopart Res 11:2079–2085

Jaidev LR, Narasimha G (2010) Fungal mediated biosynthesis of silver nanoparticles, characterization and antimicrobial activity. Colloids Surf B: Biointerfaces 2:430–433

Jain D, Kachhwaha S, Jain R, Srivastava G, Kothari SL (2010a) Novel microbial route to synthesize silver nanoparticles using spore crystal mixture of *Bacillus thuringiensis*. Indian J Exp Biol 48:1152–1156

Jain N, Bhargava A, Majumdar S, Tarafdar JC, Panwar J (2010b) Extracellular biosynthesis and characterization of silver nanoparticles using *Aspergillus flavus* NJP08: a mechanism perspective. Nanoscale 3:635–641

Jain N, Bhargava A, Tarafdar JC, Singh SK, Panwar J (2012) A biomimetic approach towards synthesis of zinc oxide nanoparticles. Appl Microbiol Biotechnol 97:859–869

Jang MH, Kim WK, Lee SK, Henry TB, Park JW (2014) Uptake, tissue distribution, and depuration of total silver in common carp (*Cyprinus carpio*) after aqueous exposure to silver nanoparticles. Environ Sci Technol 48(19):11568–11574

Jang J, Park S, Choi IH (2021) Increased Interleukin-11 and stress-related gene expression in human endothelial and bronchial epithelial cells exposed to silver nanoparticles. Biomolecules 11:234

Jo YK, Kim BH, Jung G (2009) Antifungal activity of silver ions and nanoparticles on phytopathogenic fungi. Plant Dis 93:1037–1043

Juibari MM, Abbasalizadeh S, Jouzani GS, Noruzic M (2011) Intensified biosynthesis of silver nanoparticles using a native extremophilic *Ureibacillus thermosphaericus* strain. Mater Lett 65:1014–1017

Kalimuthu K (2010) Biosynthesis of silver and gold nanoparticles using *Brevibacterium casei*. Colloids Surf, B 77:257–262

Kapoor A, Viraraghavan T (1996) Biosorption of heavy metals on fungal biomass. In: Proceeding, 4th CSCE environmental engineering specialty conference (Edmonton, Alberta), pp 333–343

Kapoor A, Viraraghavan T, Cullimore DR (1999) Removal of heavy metals using the fungus *Aspergillus niger*. Bioresour Technol 70:95–104

Kathiresan K, Manivannan S, Nabeel MA, Dhivya B (2009) Studies on silver nanoparticles synthesized by a marine fungus, Penicilliumfellutanum isolated from coastal mangrove sediment. Colloids Surf B: Biointerfaces 71:133–137

Kaviya S, Santhanalakshmi J, Viswanathan B, Muthumary J, Srinivasan K (2011) Biosynthesis of silver nanoparticles using *Citrus sinensis* peel extract and its antibacterial activity. Spectrochim Acta Part A: Mol Biomol Spectrosc 79(3):594–598

Kim SW, Jung JH, Lamsal K, Kim YS, Min JS, Lee YS (2012) Antifungal effects of silver nanoparticles (AgNPs) against various plant pathogenic fungi. Mycobiology 40(1):53–58

Kitching M, Ramani M, Marsili E (2015) Fungal biosynthesis of gold nanoparticles: mechanism and scale up. Microb Biotechnol 8(6):904–917

Kogej A, Pavko A (2001) Laboratory experiments of lead biosorption by self-immobilized *Rhizopus nigricans* pellets in the batch stirred tank reactor and the packed dead column. Chem Biochem Eng 15:75–79

Kora AJ, Sashidhar RB (2018) Biogenic silver nanoparticles synthesized with rhamnogalacturonan gum: antibacterial activity, cytotoxicity and its mode of action. Arab J Chem 11(3):313–323

Kumar D, Karthik L, Kumar G, Roa KB (2011) Biosynthesis of silver nanoparticles from marine yeast and their antimicrobial activity against multidrug resistant pathogens. Pharmacol Online 3:1100–1111

Kumar PK, Krishna SV, Verma K, Pooja K, Bhagawan D, Himabindu V (2018) Phycoremediation of sewage wastewater and industrial flue gases for biomass generation from microalgae. South Afr J Chem Eng 25:133–146

Lee CH, Lu HH, Wang CA, Nayak PK, Huang JL (2011) Influence of conductive nano-TiC on microstructural evolution of Si_3N_4-based nanocomposites in spark plasma sintering. J Am Ceram Soc 94(3):959–967

Lu W, Senapati D, Wang S, Tovmachenko O, Singh AK, Yu H, Ray PC (2010) Effect of surface coating on the toxicity of silver nanomaterials on human skin keratinocytes. Chem Phys Lett 487(1–3):92–96

Maliszewska I, Sadowski Z (2009). Synthesis and antibacterial activity of silver nanoparticles. J Phys: Conf Ser 146(1):012024. IOP Publishing

Merrin JS, Sheela R, Saswathi N, Prakasham RS, Ramakrishna SV (1998) Biosorption of chromium (VI) using *Rhizopus arrhizus*. Indian J Exp Biol 36:1052–1055

Mishra A, Tripathy SK, Yun SI (2011) Biosynthesis of gold and silver nanoparticles from *Candida guilliermondii* and their antimicrobial effect against pathogenic bacteria. J Nanosci Nanotechnol 11:243–248

Moghaddam AB, Namvar F, Moniri M, Tahir PM, Azizi S, Mohamad R (2015) Nanoparticles biosynthesized by fungi and yeast: a review of their preparation, properties, and medical applications. Molecules 20:16540–16565

Mohammadian A, Habibi RM, Shoja AS (2007) *Fusarium oxysporum* mediates photo generation of silver nanoparticles

Morones JR, Elechiguerra JL, Camacho A, Holt K, Kouri JB, Ramírez JT, Yacaman MJ (2005) The bactericidal effect of silver nanoparticles. Nanotechnology, 16(10), 2346

Mukherjee P, Ahmad A, Mandal D, Senapati S, Sainkar SR, Khan MI, Parishcha R, Ajaykumar PV, Alam M, Kumar R, Sastry M (2001) Fungus-mediated synthesis of silver nanoparticles and their immobilization in the mycelial matrix: a novel biological approach to nanoparticle synthesis. Nano Lett 1(10):515–519

Munger MA, Radwanski P, Hadlock GC, Stoddard G, Shaaban A, Falconer J, Grainger DW, Deering-Rice CE (2014) In Vivo human time-exposure study of orally dosed commercial silver nanoparticles. Nanomedicine 10:1–9

Narayanan KB, Sakthivel N (2010) Biological synthesis of metal nanoparticles by microbes. Adv Colloid InterfSci 156:1–13

Nethravathi C, Nisha T, Ravishankar N, Shivakumara C, Rajamathi M (2009) Graphene–nanocrystalline metal sulphide composites produced by a one-pot reaction starting from graphite oxide. Carbon 47(8):2054–2059

Olaniran AO, Balgobind A, Pillay B (2013) Bioavailability of heavy metals in soil: impact on microbial biodegradation of organic compounds and possible improvement strategies. Int J Molecular Sci 14(5):10197–10228

Ottoni CA, Simões MF, Fernandes S, Dos Santos JG, Da Silva ES, de Souza RFB, Maiorano AE (2017) Screening of filamentous fungi for antimicrobial silver nanoparticles synthesis. Amb Express 7(1):1–10

Pattabi M, Uchil J (2000) Green synthesis of CdS nanoparticles and effect of capping agent concentration on crystallite size. Solar Energy Mater Solar Cell 63:309–314

Phanjom P, Ahmed G (2017) Effect of different physicochemical conditions on the synthesis of silver nanoparticles using fungal cell filtrate of *Aspergillus oryzae* (MTCC No. 1846) and their antibacterial effect. Adv Nat Sci: Nanosci Nanotechnol 8(4):045016

Phenrat T, Kim HJ, Fagerlund F, Illangasekare T, Tilton RD, Lowry GV (2009) Particle size distribution, concentration, and magnetic attraction affect transport of polymer-modified FeO nanoparticles in sand columns. Environ Sci Technol 43(13):5079–5085

Prasad K, Jha AK (2010) Biosynthesis of CdS nanoparticles: an improved green and rapid procedure. J Colloid Interface Sci 342(1):68–72

Qian Y, Yu H, He D, Yang H, Wang W et al (2013) Biosynthesis of silver nanoparticles by the endophytic fungus *Epicoccumnigrum* and their activity against pathogenic fungi. Bioprocess Biosyst Eng 36:1613–1619

Rai M, Yadav A, Gade A (2009) Silver nanoparticles as a new generation of antimicrobials. Biotechnol Adv 27:176–183

Rai M, Ingle AP, Birla S, Yadav A, Santos CAD (2016) Strategic role of selected noble metal nanoparticles in medicine. Crit Rev Microbiol 42(5):696–719

Rajeshkumar S, Bharath LV (2017) Mechanism of plant-mediated synthesis of silver nanoparticles–a review on biomolecules involved, characterisation and antibacterial activity. Chem Biol Interact 273:219–227

Rajeshkumar S, Bharath LV, Geetha R (2019). Broad spectrum antibacterial silver nanoparticle green synthesis: characterization, and mechanism of action. In: Green synthesis, characterization and applications of nanoparticles. Elsevier, pp 429–444

Rajput S, Werezuk R, Lange RM, McDermott MT (2016) Fungal isolate optimized for biogenesis of silver nanoparticles with enhanced colloidal stability. Langmuir 32(34):8688–8697

Ranjbar Navazi Z, Pazouki M, Halek FS (2010) Investigation of culture conditions for biosynthesis of silver nanoparticles using *Aspergillus fumigatus*. Iran J Biotechnol 8(1):56–61

Rao YN, Banerjee D, Datta A, Das SK, Guin R, Saha A (2010) Gamma irradiation route to synthesis of highly re-dispersible natural polymer capped silver nanoparticles. Radiat Phys Chem 79(12):1240–1246

Rose GK, Soni R, Rishi P, Soni SK (2019) Optimization of the biological synthesis of silver nanoparticles using *Penicillium oxalicum GRS-1* and their antimicrobial effects against common food-borne pathogens. Green Process Synth 8(1):144–156

Sanghi R, Verma P (2009) Biomimetic synthesis and characterisation of protein capped silver nanoparticles. Bioresour Technol 100(1):501–504

Sangiliyandi G, Kalishwaralal K, Vaidyanathan R, Venkataraman D, Pandian SR et al (2009) Biosynthesis, purification and characterization of silver nanoparticles using *Escherichia coli*. Colloids Surf, B 74:328–335

Sasabe E, Tomomura A, Kitamura N, Yamamoto T (2020) Metal nanoparticles-induced activation of NLRP3 inflammasome in human oral keratinocytes is a possible mechanism of oral lichenoid lesions. Toxicol Vitro 62:e104663. https://doi.org/10.1016/j.tiv.2019.104663

Sastry M, Patil V, Sainkar SR (1998) Electrostatically controlled diffusion of carboxylic acid derivatized silver colloidal particles in thermally evaporated fatty amine films. J Phys Chem B 102(8):1404–1410

Saxena J, Sharma PK, Sharma MM, Singh A (2016) Process optimization for green synthesis of silver nanoparticles by Sclerotinia sclerotiorum MTCC 8785 and evaluation of its antibacterial properties. SpringerPlus 5(1):1–10

Shahzad A, Saeed H, Iqtedar M, Hussain SZ, Kaleem A, Abdullah R, Sharif S, Naz S, Saleem F, Aihetasham A, Chaudhary A (2019) Size-controlled production of silver nanoparticles by *Aspergillus fumigatus* BTCB10: likely antibacterial and cytotoxic effects. J Nanomater 2019

Shaligram NS, Bule M, Bhambure R, Singhal RS, Singh SK, Szakacs G, Pandey A (2009) Biosynthesis of silver nanoparticles using aqueous extract from the compactin producing fungal strain. Process Biochem 44(8):939–943

Sintubin L, De Windt W, Dick J, Mast J, Van der Ha D et al (2009) Lactic acid bacteria as reducing and capping agent for the fast and efficient production of silver nanoparticles. Appl Microbiol Biotechnol 84:741–749

Srivastava M, Singh S, Self WT (2012) Exposure to silver nanoparticles inhibits selenoprotein synthesis and the activity of thioredoxin reductase. Environ Health Perspect 120(1):56–61

Sundrarajan M, Gowri S (2011) Green synthesis of titanium dioxide nanoparticles by *Nyctanthes arbortristis* leaves extract. Chalcogenide Lett 8(8):447–451

Tejada G, Piccirilli GN, Sortino M, Salomón CJ, Lamas MC, Leonardi D (2017) Formulation and in-vitro efficacy of antifungal mucoadhesive polymeric matrices for the delivery of miconazole nitrate. Mater Sci Eng: C 79:140–150

Tomah AA, Abd Alamer IS, Li B, Zhang JZ (2020) A new species of *Trichoderma* and *Gliotoxin* role: A new observation in enhancing biocontrol potential of *T. virens* against *Phytophthora capsici* on chili pepper. Biol Control 145:104261

Vigneshwaran N, Kathe AA, Varadarajan PV, Nachane RP, Balasubramanya RH (2007) Functional finishing of cotton fabrics using silver nanoparticles. J Nanosci Nanotechnol 7(6):1893–1897

Wang Y, Du MH, Xu JK, Yang P, Du YK (2008) Antifungal activity of silver nanoparticles. J Dispersion Sci Technol 29:891–894

Wu YX, Liang P, Dong QM, Bai Y, Yu Z, Huang J, Zhong Y, Dai YC, Ni D, Shu HB, Pittman Jr CU (2017) Design of a silver nanoparticle for sensitive surface enhanced Raman spectroscopy detection of carmine dye. Food Chem 237:974–980

Xue B, He D, Gao S, Wang D, Yokoyama K, Wang L (2016). Biosynthesis of silver nanoparticles by the fungus *Arthroderma fulvum* and its antifungal activity against genera of *Candida, Aspergillus* and *Fusarium.* Int J Nanomed 1899–1906

Zhang H, Li Q, Lu Y, Sun D, Lin X et al (2005) Biosorption and bioreduction of diamine silver complex by *Corynebacterium.* J Chem Technol Biotechnol 80:285–290

Chapter 8
Mycosynthesis of Nanoparticles and Their Application in Medicine

N. Prabhu, S. Nivetha, G. Pooja, B. Nivedhitha, M. Baskamary, and S. Pooja

Abstract Recently emerging field in science is myconanotechnology, where the creation and exploitation of nanoparticles is done by using fungi. Fungi used as biofactories for cost effective, ecofriendly production of nanoparticles. They have resistance to many harsh conditions and also their extracellular reductive protein makes downstream process easier. Myconanotechnology represents a new paradigm in the development of antimicrobial nanomedicine. Mostly silver nanoparticles have been discovered to have antimicrobial activity. It is used against pathogenic fungi infecting plants as an alternative to pesticides. It also has antimicrobial property. Antimicrobial property of nanoparticles becomes an alternative to antibiotics. It is more advantageous in treating bacterial infection. Implantable devices and medicinal materials are coated using nanoparticles to prevent infection, promote wound healing and used to deliver antibiotics to treat diseases. This chapter deals with antimicrobial application of myconanotechnology which includes antibacterial, antifungal and antiviral activity of myconanoparticles. Also explains about currently accepted antibacterial mechanisms of nanoparticles which are oxidative stress, metal ion release and non-oxidative mechanism and about critical factors affecting antimicrobial mechanism and their antibacterial applications.

Keywords Myconanotechnology · Antimicrobial · Eco-friendly · Silver nanoparticles

8.1 Introduction

Myconanotechnology is a branch of nanotechnology in which nanoparticles are synthesized using fungi. It is an interface between mycology and nanotechnology (Mahendra Rai et al. 2009). Nanotechnology has wide range of applications in

N. Prabhu (✉) · S. Nivetha · G. Pooja · B. Nivedhitha · M. Baskamary · S. Pooja
Department of Biotechnology, Vivekanandha College of Engineering for Women (Autonomous), Tiruchengode, Namakkal, Tamilnadu 637 205, India
e-mail: prabhu.aut.26@gmail.com

M. P. Shah et al. (eds.), *Biogenic Nanomaterials for Environmental Sustainability: Principles, Practices, and Opportunities*, Environmental Science and Engineering,
https://doi.org/10.1007/978-3-031-45956-6_8

science and technology to create novel material in nano level. Nanoparticles which are produced from fungi are known as myconanoparticles. Synthesis of nanoparticles using microbes is a green chemistry approach.

Fungi have numerous advantages to synthesize nanoparticles than other microbes and plant material. Top-down process and bottom-up process are two types of strategies for synthesis of nanoparticles (Marchiol 2012). Mycosynthesis of nanoparticles is a kind of bottom-up process (Moghaddam 2010). Myconanoparticles are synthesized by hydrolyzing metal ions using fungal proteins. They are capable of hydrolyzing metal ions and converts into metal nanoparticles. Fungi can secrete high amount of proteins than other microbes thus resulting in high productivity of nanoparticles. They are easy to isolate and culture and requires simple nutrient for growth.

Downstream processing and handling of fungal biomass are easier than any other chemical methods (Gade et al. 2008). Produced nanoparticles will be of good monodispersity and well dimensioned. Moreover, it is an eco-friendly method and it plays a significant role in remediation of toxic metals by reduction of metal ions. Nanoparticles are synthesized both in intracellular and extracellular manner (Narayanan and Sakthivel 2011).

Mechanism of intracellular synthesis involves trapping of metal ions over fungal surface by electrostatic interaction which is initiated by enzymes present in the cell wall and then metal ions are condensed by the enzymes within the cell wall and then precipitation of metal ions takes place to produce nanoparticles. Mechanism of extracellular synthesis involves arrangement of culture filtrate with desired metal-solution; metal ions from the salt solution transported in cytoplasmic membrane and cytoplasm then the ions are reduced by reductase enzyme to produce nanoparticles (Subashini and Bhuvaneswari 2018).

Some examples of produced nanoparticles are silver nanoparticles by *Aspergillus niger*, *Aspergillus terreus*, *penicillium sp.* etc. and gold nanoparticles by *Fusarium oxysporum, penicillium sp. Rhizopus oryzae* etc. and *Fusarium oxysporum* produce Pt, Si, Ti, CdS, CdSe nanoparticles. Wide range of applications of myconanoparticles involve antimicrobial agents because it provides antifungal, antibacterial and antiviral activity and used as nano fertilizers in agriculture, nano fungicide (Ingle and Rai 2011), pesticides, nanowires, nano-fibrous mat, quantum dots (Shao et al. 2011) smart delivery system (Kukowska-Latallo et al. 2005), nano scaffolds, wound treatment (Sundaramoorthi et al. 2009), bio separation, anticancer activity (Mishra et al. 2014) molecular detection, food preservation etc. This chapter explains about antimicrobial application of myconanoparticles.

8.2 Antimicrobial Application of Myconanoparticles

Myconanotechnology is the study about developing antimicrobial nanoparticles. Silver nanoparticles produced from *Fusarium oxysporum* can be integrated into materials such as cloth. These integrated cloths show antibacterial activity against pathogenic bacteria such a *Staphylococcus aureus* (Dura´n et al., 2007). Silver nanoparticles produced by *Lecanicillium lecanii* shows antibacterial activity against *S.aureus* and *E.coli* on cotton bleach fabrics (Namasivayam and Avimanyu 2011). Extracellularly produced gold nanoparticles by *Rhizopus oryzae* is used for making nano gold bio combat structure. Antifungal effect is shown by double capsulized nano silver against rose powdery mildew caused by *Sphaerotheca pannosa* (Kim et al. 2008). Silver nanoparticles with amphiphilic macromolecule are an effective antimicrobial agent. Silver nanoparticles show efficient antibacterial effect against *Vibrio cholera, E.coli, Salmonella typhi, Pseudomonas fluorescens.*

8.3 Antibacterial Activity of Myconanoparticles

Chronic infections are caused by bacterial infections. Though antibiotic is preferred for treatment of bacterial infections, widespread of antibiotics caused emergence of multidrug resistant bacterial strains. It is a major threat to modern medicine. A recently super bacterium which has resistant to almost most of the antibiotics has developed due to the abuse of antibiotics. Here comes, the antibacterial effect of nanoparticles, nanoparticles are the best alternative to antibiotics to prevent bacterial infections. Nanoparticle shows broad spectrum antibacterial properties against both gram-positive and gram-negative bacteria. For example, silver nanoparticles show concentration dependent antibacterial effect against *Pseudomonas aeruginosa* and *Escherichia coli* (Ramalingam et al. 2016), *Staphylococcus aureus* is inhibited by zinc oxide nanoparticles. Antimicrobial mechanism of nanoparticles follows three models: oxidative stress induction (Gurunathan et al. 2012), metal ions release (Nagy et al. 2011), non-oxidative mechanism (Leung et al. 2014). All the three types can occur simultaneously. Certain studies revealed that surface electric charge of bacterial membrane is neutralized by Ag nanoparticles thus change its penetrability, which leads to bacterial death (Jung et al. 2008). Antioxidant defense system is inhibited by generation of Reactive Oxygen Species (ROS) and causes mechanical damage to cell membrane. Major processes underlying the antibacterial effect of nanoparticles such as:

- Disruption of cell membrane
- Generation of ROS
- Penetration of cell membrane
- Interaction with DNA and proteins

8.4 Antibacterial Mechanisms of Nanoparticles

More and more studies are going on for exploring antibacterial mechanisms of nanoparticles after the increase in use of nanoparticles medicine (Huh and Kwon 2011). For example, metabolic activity of bacteria can be changed by metal nanoparticles (Chatzimitakos and Stalikas 2016) this advantage eliminates pathogenic bacteria to cure diseases. Biofilm formation can be inhibited by entering of nanoparticles based on the Ag inhibited expression of genes (Zhao and Ashraf 2015). To exhibit antibacterial function nanoparticles need to be in contact with bacterial cells. The contact is exhibited by electrostatic attraction (Chen et al. 2015), van der Waals forces (Armentano et al. 2014) and receptor–ligand (Gao et al. 2014) and hydrophobic interactions (Luan et al. 2016) nanoparticles crosses the membrane gather along the metabolic pathway and influences shape and function of the cell. Nanoparticles interact with the bacterial cell's basic components, such as DNA, lysosomes, ribosomes, and enzymes, leading to oxidative stress, heterogeneous alterations, changes in cell membrane permeability, electrolyte balance disorders, enzyme inhibition, protein deactivation, and changes in gene expression (Xu et al. 2016; Shrivastava et al. 2007; Yang et al.2009). The most frequently proposed mechanisms are: oxidative stress, metal ions release and non-oxidative mechanism.

8.5 Oxidative Stress

The important antibacterial mechanism of nanoparticles is ROS-induced oxidative stress. ROS—Reactive Oxygen Species, which is a generic term of reactive intermediates and molecules that possess strong positive redox potential. Different types of ROS are produced by different types of nanoparticles by reducing oxygen molecules. Following are the four types of ROS, the superoxide radical (O_2^-), hydrogen peroxide (H_2O_2), the hydroxyl radical ($OH^·$), and singlet oxygen (O_2). All these types exhibit different levels of dynamics and activity. All the types of reactive oxygen can be generated by copper oxide nanoparticles. H_2O_2 and OH can be generated by Zinc oxide nanoparticles but it cannot generate O_2^-, whereas O_2^- can be generated by calcium oxide and magnesium oxide nanoparticles. OH and O_2 cause acute microbial death whereas O_2- and H_2O_2 cause less acute stress reactions and endogenous antioxidants such as catalase and superoxide enzymes can neutralize it. Oxygen vacancies, restructuring and defect sites are the main cause for ROS production (Malka et al. 2013). In normal conditions, ROS production and clearance are balanced in bacterial cells.

When ROS production is excess, the redox balance of cell favors oxidation. This unbalance leads to oxidative stress that damages all components in the bacterial cell (Li et al. 2016; Peng et al. 2013). It has been confirmed that oxygen stress is the key reason for changing the permeability of cell membrane, which results in membrane damage (Cheloni et al. 2016).

To become intracellular, Al_2O_3 nanoparticles crossed the membrane which leads to loss of integrity of cell membrane that is ultimately due to intracellular oxidative stress, this is confirmed by Ansari analysis (Ansari et al. 2015).

Nano-silver ions are used as center of catalytic activity which activates oxygen in air or water that causes the production of reactive oxygen ions and hydroxyl radicals that prevent the proliferation of bacteria (Shrivatsava et al. 2007; Yang et al. 2009).

ROS can also increase the gene expression levels of oxidative proteins; this will be the key mechanism in bacterial cell apoptosis (Wu et al. 2011). ROS can also attack proteins and suppress the activity of periplasmic enzymes that are required to maintain the morphology and physiology of bacterial cells (Padmavathy and Vijayaraghavan 2011). ROS can be generated by nanoparticles using different mechanisms the photo-catalytic hypothesis is the current mainstream. Metal oxide nanoparticles such as zinc oxide and titanium oxide accept energy of light irradiation greater than or equal to band gap, then stimulates electrons (e^-) in the valence band and transition to the conduction band, which results in corresponding holes in valence band (H^+) and produce electron hole pairs on the surface and inside of catalytic material. After the interaction between H_2O or OH^- and ZnO, H^+ adheres to the surface of ZnO, then it is oxidized to hydroxyl radical ($OH\cdot$). In the same way, hydroxyl radical is reduced to super oxide radical after the electronic interaction with O_2 and adherence to the surface of ZnO. ROS degrades active components which are responsible for maintaining normal morphological and physiological functions of the microorganisms (Yu et al., 2014). In particular electron hole pairs are generated after absorbing light radiation by TiO_2 nanoparticles (Deepan and Misra 2014).

Water and air on the surface of nanoparticles reacts with electron hole pairs to produce highly chemically active ROS which attacks intracellular organic matter in bacteria. Example zinc activates under Ultraviolet (UV) and visible light to produce highly reactive ROS. H_2O_2 passes through the cell membrane whereas negatively charged radicals such as superoxide and hydroxyl radicals do not penetrate through cell membrane. A zinc oxide nanoparticle split H_2O into H^+ and reacts with dissolved oxygen to produce H_2O_2. These ROS penetrates into the membrane of cell to kill bacteria. Ultrasonic treatment with polymer nanoparticles and colony forming units dissociates nanoparticles and promote them to penetrate through cell membrane and antimicrobial metal ions are also released from the surface to prevent the proliferation of bacteria under ultrasonic conditions that may be caused by the increased rate of transport of bacterial oxygen, waste and nutrients caused by ultrasound (Seil and Webster 2012). Nanoparticles have also shown antibacterial activity in the dark (Xu et al. 2013).

8.6 Dissolved Metal Ions

The metal oxide slowly releases the metal ions and this is followed by the absorption of metal ions through the cell membrane. This is further followed by the direct interaction with the functional groups of nucleic acids and proteins. During the antimicrobial process of metal oxide, the impact of the metal ions on the pH inside the lipid vesicles is very small and the antimicrobial activity is very weak during metal oxide suspension. Hence, for the mechanism of metal oxide nanoparticles, dissolved metal ions are not the main mechanism (Hussein-Al-Ali et al. 2014).

8.7 Non-oxidative Mechanisms

For the study of antibacterial mechanism of MgO nanomaterial certain scholars under UV light, natural light or complete darkness used Fourier Transform Infrared Analysis (FTIR), liquid chromatography-mass spectrometry, proteomics tools, electron spin resonance, Transmission Electron Microscopy (TEM). MgO nanoparticles have good antibacterial effects. These are completely unrelated to the mechanism of oxidative stress.

8.8 Interactions of Nanoparticles with Cell Membrane

Cell wall plays a significant role in resistance of bacteria to the external environment. Different adsorption pathways for nanoparticles are provided by composition of cell wall and gram-positive and gram-negative bacteria (Lesniak et al. 2013). Gram negative bacteria has unique structure LPS of cell wall which provides negatively charged region that attracts nanoparticles. Whereas in gram positive bacteria teichoic acid is expressed hence nanoparticles are distributed along molecular chain of phosphate preventing their aggregation. Nanoparticles have greater activity against gram positive than gram negative bacteria because in gram negative cell wall it is composed of LPS, lipoproteins and phospholipids which does not allow macromolecules to penetrate whereas in gram positive bacteria cell wall contains teichoic acid as well as peptidoglycans and abundant pores which allows foreign molecules thus results in cell wall damage and cell death.

8.9 Regulation of Expression of Metabolic Gene by Nanoparticles

Bacterial metabolism takes place for growth and reproduction of bacteria. Metabolic pathways are integrated into complex activity of living cells. For the effect of nanoparticles in bacterial metabolism, various mechanisms have been proposed such as reactive oxygen mechanism and metal ion dissolution mechanism (Padmavathy and Vijayaraghavan 2011; Yu et al. 2014). Magnesium oxide nanoparticles alters the expression of many metabolic proteins which includes up -regulation of a weak thiamine ester-binding protein and riboflavin metabolic protein and the down regulation of a protein mapped to the critical path of bacterial cell metabolism, which results in reducing metabolic activity of bacteria. This nanoparticle regulates metabolic processes of bacteria by acting on specific protein. Expression of proteins related to bacterial nitrogen metabolism is regulated by CuO nanoparticles which inhibit the activity of nitrate reductase and nitrite reductase. TiO_2 nanoparticles affects formation of biofilm by affecting adhesion rate of bacteria. (Peng et al. 2013; Roguska et al. 2015).

8.10 Inhibition of Formation of Biofilms by Nanoparticles

Biofilms makes bacteria highly resistant to external environment, antibiotics and foreign chemicals. Production of extracellular polymeric substances is inhibited by silver nanoparticles, which leads to action against biofilms of drug resistant bacteria such as *E.coli* and *Klebsiella pneumoniae* (Ansari et al. 2012).

Crucial activity for biofilm formation is bacterial metabolism, D-alanine metabolism is crucial for *Streptococcus* formation and growth. Addition to that long distance electrical conduction of bacteria in a biofilm is established by potassium ions which coordinates metabolic activities inside and outside of biofilm. Magnesium nanoparticles diffuse into biofilms and leads to disruption of the membrane potential, enhanced lipid peroxidation, and DNA binding. Formation of biofilm decreases due to disturbance in the normal functioning (Lellouche et al. 2012). Dominant antimicrobial targets of nanoparticles are metabolic pathways, thus the mechanism of inhibition of NPs of the formation of bacterial biofilms is related to the regulation of bacterial metabolism.

8.10.1 Critical Factors Affecting the Antimicrobial Mechanism of Metal Ions

To regulate the actions of nanoparticles on bacterial cells size, charge, zeta potential, surface morphology and crystal structure are the most important physiochemical properties.

Size: The effect of antimicrobial activity is greatly affected by the size of the metal of nanoparticles. In the formation of bacterial biofilms, bacterial adhesion is a well-known process. It makes the organisms to be much more resistant. The nanoparticles which are smaller have a greater probability of being in touch with the cell membrane. The effect of size is not a dominant factor.

Shape: Related to the antimicrobial property of nanoparticles shape is the most important factor. Varying shapes of nanoparticles can cause bacterial cell damage in various degrees. Bacterial cell damage occurs through the interactions of periplasmic enzymes (Cha et al. 2015). On micro biota susceptibility silver nanoparticles did not have a high effect. To prevent the degradation of enzymes, pyramid–shaped n–ZnO has been shown (Prasannakumar et al. 2015).

8.11 Doping Modification

Doping modifications are employed to prevent the aggregation of nanoparticles and to disperse them in hydrophilic media. It is most effective method to regulate and control Interaction of nanoparticles and bacteria. ZnO/Au nano composites formed by the combination of ZnO nanoparticles with Au, has been used to enhance ROS generation and improve photocatalytic activity. improved light absorption due to the surface plasmon resonance wavelength of Au; an altered band gap width of ZnO, which enhances the reactivity of photoinduced charge carriers; and increased efficiency of electron transport and charge carrier separation are the factors of these effects (He et al. 2014). Doping modification can alter antibacterial activity of ZnO nanoparticles. Fluorine doped ZnO nanoparticles generate more ROS than undoped ZnO nanoparticles which results in greater damage in bacterial cells (Podporska-Carroll et al. 2017; Guo et al. 2015). Key factor for regulating antimicrobial effectiveness against both gram positive and gram-negative bacteria is oxygen content at the surface of ZnO nanoparticles (Mehmood et al. 2015). Nano–TiO_2 nanoparticles have antibacterial effect which reduce biofilm formation hence it is widely used in orthopedic and dental applications. When compared with undoped nanoTiO_2, doped nano-TiO_2 shows improved photocatalytic activity can effectively extend the active spectrum to the visible light region because the valence bandwidth is increased and the forbidden bandwidth is reduced (Sangari et al. 2015; Peng et al. 2010).

8.12 Roughness

In disparity to the large-scale research with regard to the effects of contrasting nanoparticles characteristics on bacterial cells not many studies have lobbed the effect of roughness, as the roughness of nanoparticles become greater, the size and surface area to mass ratio promote the adsorption to bacterial proteins, followed by reduction in the cohesions (Ben-Sasson et al. 2014; Rajakumar et al. 2012; Sukhorukova et al. 2015).

8.13 Zeta Potential

Current studies have evidence that the Zeta potential of NP has a strong impact on the bacterial cohesions, because of the electrostatic attraction between positively charged. Nanoparticles and the bacterial cell membrane which is negatively charged $Mg(OH)_2$_Mgcl and $Mg(OH)_2$_$MgSo_4$ nanoparticles, Which have a positive surface charge are prone to being adsorbed in the bacteria, in contrast their negatively charged counterparts (Pan et al. 2013), The potential nanoparticles to selectively gather at sites of bacterial infection increases vascular permeability (Maeda 2010). Increasing of cationic nanoparticles is favorable to inhibiting bacterial growth by limiting bacterial attachment, slight penetration of nanoparticles into the outer regions of the *S.aureus* envelope somehow provides high germicidal efficacy, possibly because the nanoparticles can reach important structure through ion exchange variation with negatively charged and neutral nanoparticles positively charged counterparts have been believed to enhance ROS production. A current study showed that negatively charged nanoparticles are not adhere to bacteria due to the negative potential on both. However, the higher concentration, negatively charged nanoparticles have a certain level of antibacterial activity due to molecular growing which leads to interaction between the nanoparticles and the bacterial surface (Arakha et al. 2015).

8.14 Environmental Conditions

Studies revealed that different environmental conditions affect antimicrobial activity. Temperature of environment significantly influence generation of ROS thus affects antimicrobial activity. Antimicrobial effectiveness of ZnO nanoparticles is enhanced when temperature stimulates nanoparticles, electrons is captured at active sites then the electrons interact with oxygen to produce ROS. In-vitro antimicrobial activity is influenced by pH. Dissolution rate of ZnO nanoparticles increases with decrease in pH thus antimicrobial activity increases (Saliani et al. 2015). At low pH nanoparticles were positively charged which interacts with negatively charged groups of cell wall which induce strong multivalent electrostatic regulation (Radovic-Moreno et al. 2012). pH and osmotic pressure which are characteristics of the medium influence surface charge, aggregation, and solubility of nanoparticles. Antibacterial tests of ZnO nanoparticles with different types of media demonstrate that mainly free Zn ions and Zn complex causes antimicrobial activity. Furthermore, the medium supply nutrients to bacteria to enhance tolerance to nanoparticles (Li et al. 2011). Studies investigated that ZnO nanoparticles prepared with different stirring conditions affect antibacterial activity against Gram-positive (*B.subtilis*) and Gram-negative (*E.coli*) bacteria and a fungus (*C.albicans*) (Khan et al. 2016).

8.15 Capping and Stabilisation of Nanoparticles

Functionalization is a process which facilitates the desired substance on the surface by employing biomolecules (Mout et al. 2012). There is a requirement of additional step in the synthesis of nanoparticles using biogenic methods. The step includes polymers & surfactants to be coated on its surfaces. Here in this process, the capping formation occurs simultaneously with the employment of biomolecules that is derived from the organism is used in biogenic method as an additional step (Chowdhury et al. 2014). The biomolecules that are derived from the reducing organism consists of high capacity to bind with metals, proteins & amino acids (Basavaraja et al. 2008; Gopinath et al. 2012).

8.16 Antibacterial Applications

Nanoparticles had been stimulated by the drug-resistant bacteria and the increased rate of hospital infection. It had excellent antimicrobial resistance properties and it had been widely used in many fields. Nanoparticle has its own advantages and localization. Parameters including the mean particle size, shape, the specific surface area, and surface curvature affect the antibacterial activity and mechanism. The application of nanoparticles in fighting bacteria had decreased bacterial infection. In the following, the antibacterial applications of nanoparticles had been discussed in detail (Table 8.1).

Table 8.1 Antibacterial effect

Source	Target organism	Effective concentration	References
Aspergillus niger	*E.coli, Staphylococcus aureus, P.aeruginosa*	1.0;5.0;10;50;and 100 μg/ml	Ottoni et al. (2017)
Macrophomina phaseolina	Multi drug resistant strains of *Escherichia coli*	0.51;0.36;0.25; 0.10;μg/ml	Chowdhury et al. (2014)
Fusarium verticilliodies	*S.aureus* and *E.coli*	5 and 10 mM	Mekkawy et al. (2017)
Aspergillus terreus	*Salmonella typhi, S.aureus* and *E.coli*	0.42–0.84 μg/ml	Rani et al. (2017)
Sclerotinia sclerotiorum	*E.coli* and *S.aureus*	100,200, and 400 ppm	Saxena et al. (2016)

8.17 Nanoparticles Coated Implantable Device

Human implantable devices with antimicrobial coatings are of two types; the first type is considered as fully implantable device which consists of heart valves or dental implants. The antimicrobial coating of cardiovascular apparatuses is used to prevent thrombosis. The coating inhibits the growth of bacteria such as *Streptococcus epidermidis*, and *E.coli* and also used to prevent the occurrence of inflammation around the implants. The second type of device is considered as partially implantable devices which consists of catheters, intravenous catheters, or neurosurgical catheters which are more prone to bacterial colonization and increased the risk of infections in clinical applications (Samuel and Guggenbichler 2004; Galiano et al. 2008). Nanoparticles coatings on neurosurgical catheters could reduce the risk of bacterial infection and complications, with sustained release of nanoparticles over six days by reducing the growth of *S.aureus*.

8.18 Wound Dressings

The natural protective barrier of the body is called skin and it protects the body from pathogens and foreign bodies. The dynamic equilibrium of water and electrolytes are also maintained by the skin. The skin had been damaged due to the trauma, burns, and several types of chronic skin ulcers. The process of wound healing to rebuild the barrier function of skin is called dressing which is also used to accelerate wound healing, and to reduce the risk of wound infection. An ideal dressing would have similar characteristics to skin. Many microorganisms can also cause wound infections, and these had been divided into two types-(1) Gram-positive bacteria, e.g. *Staphylococcus* and *Streptococcus*., (2) Gram-negative bacteria, e.g. *E.coli* multiple bacterial species and antibiotic resistance are often accompanied by the chronic infections. Nanoparticles are possessed by broad-spectrum antimicrobial properties that had inhibited the bacterial growth and reproduction. Due to the high specific surface area of nano silver, it has a good contact with bacteria, hence inhibited their growth and also increased the rate of wound healing.

8.19 Bone Cement

Bone cement is composed of methyl methacrylate (MMA) and polymethyl methacrylate (PMMA) or modified PMMA and it is self-curing plastic at room temperature. Bone cement is used for filling the gap between the bone and implant to fix joint prostheses such as hip replacement knee surgery. A previous study revealed that the infection rate after total joint replacement surgery was about 3% and it may reduce to 0% by using antibiotic loaded PMMA (Nowinski et al. 2012). However, a greater number of resistant bacterial strains are increasing day by day hence incurable infectious disease occurs if effective measures are not taken and death can also be caused by small wounds. Nano particles possess strong killing on certain types of resistant bacterial strain, this has become research hotspot now. Nanosilver at a concentration less than 0.05% can reduce infections after anthroplasty surgery including *Acinetobacter baumannii, S.aureus, S.epidermidis* infections (Kose et al. 2016). To develop promising antibacterial bone cement silver nanoparticles are used to replace antibiotics (Prokopovich et al. 2015).

8.20 Dental Materials

Initiating factor of common infectious diseases in the mouth is plaque. It is an ecological environment of microbes which allows them to settle on teeth. Dental material shows enhanced performance after nanocrystallization. As an example, for root canal treatment nanodiamond-functionalized amoxicillin in combination with gutta-percha eliminates residual bacteria after root canal filling (Lee et al. 2015). Orthodontic treatment often leads to the formation of dental plaque chalk, due to the reduction of pH and proliferation of bacteria in the course of treatment and CuO and ZnO nanoparticles coated brackets can effectively inhibit the growth of *S.mutans*, however coatings affect the appearance of brackets (Ramazanzadeh et al. 2015). Biofilms develops on maxillofacial prostheses which are kept in complex environment in which it contains variety of Flora, thus results in tissue inflammation surrounding the prostheses. Addition of nano TiO_2 results in antibacterial effect with prostheses following light exposure (Aboelzahab et al. 2012).

8.21 Antibiotic Delivery System

The major disadvantage of conventional antibiotic approach includes:

- The systemic toxicity of antibiotics
- Inability of antibiotics to reach effective concentration at local site of infection.

The efficiency of drug adsorption is inversely proportional to particle size and directly proportional surface area of adsorbent. Because of their small particle size and greater surface area. Nanoparticles can be employed as a transporter to achieve targeted drug delivery. Over the existing drug delivery system, nanoparticles have their advantage in reducing side effects of antibiotics. CS/fucoid a nanoparticles release gentamicin which is used to treat pneumonia, which renders nanoparticles into multifunctional drug delivery system with antibacterial and antioxidant activities. CS-coated alginate nanoparticles increases the permeability of daptomycin into limbal epithelium of the eye and increases its ocular accumulation and effect duration. This drug delivery system has the features of biodegradability, biocompatibility, controlled drug transport, and delivery to the target tissue (Costa et al. 2015).

8.22 Antifungal Activity of Myconanoparticles

Among various pathogens crop diseases are caused by phytopathogenic fungi in agriculture. Nanoparticles are the potent alternative to synthetic fungicide. Moreover, it is safer than synthetic fungicide. Silver nanoparticles shows strong antifungal activity against *Botrytis cinerea.* And also antifungal against *Phoma glomerata, Fusarium semitectum, Trichoderma* and *C. albicans* and *C.tropicalis* are highly sensitive to silver nanoparticles (Gajbhiyeet al. 2009). Studies revealed that when the activity of 80 μg of sliver nanoparticles compared with activity of powerful antifungal agent amphotericin B, the results are same with the activity of nanoparticles against *Candida spp.* Ag-SiO_2 nanoparticles shows strong antifungal effect against *Botrytis cinerea.* The combination of fluconazole with silver nanoparticles shows antifungal effect against *Phoma glomerata, Phoma herbarum, F.semitectum, Trichoderma sp.,* and *C.albicans* by the technique called disc diffusion. Ag_2S nanoparticles on surface of amorphous silica possess antifungal activity against *Aspergillus niger* potential fungicidal efficiency of ZnO and $ZnTiO_3$ was studied. It was observed that $ZnTiO_3$ nanopowder shown higher growth inhibition efficiency than *Bipolaris sorokiniana* and *Magnaporthe grisea,* efficiency of antifungal action of silver ions and nanoparticles were evaluated both in-vivo and in-vitro evaluations shows decreased disease development of phytopathogenic fungi. Number of germinating fragments and sprouts length deeply decreased by silver nanoparticles. Growth of *B.cinerea* was inhibited by ZnO nanoparticles which affect cellular functions and causes deformation in fungal hyphae. It also inhibits the growth of conidiophores and conidia of *Penicillium expansum.* These silver nanoparticles are also much less toxic compared to synthetic fungicide to human and animals. Moreover, toxic effects caused by nanoparticles in algae, fungi and plants can be coupled with positive effect. Phytopathogen *Colletotrichum gloeosporioides,* causes anthracnose in wide range of fruits. Their growth was decreased significantly by silver nanoparticles in dose dependent manner. Ag nanoparticles with different concentrations were tested with pathogens *Alternaria alternata, Sclerotinia sclerotiorum, Macrophomina phaseolina, Rhizoctonia solani, B.cinerea* and *Curvularia lunata.* Interestingly 15 mg of nanoparticles shows higher efficiency against all the pathogens (Table 8.2).

Table 8.2 Antifungal activity

Source	Target organism	Effective concentration	References
Pleurotus aculeatum SU 1	*E.coli, P.aeruginosa, S.aureus, B.subtilis*	50, 100, and 200 μg/ml	Saxena et al. (2016)
Trichophyton rubrum	*C.albicans*	4 μg/ml	Moazeni et al. (2012)
Arthroderma fulvum	*Candida sp.* and *Aspergillus sp.*	0.125–4.00 μg/ml	Xue et al. (2016)
Fusarium oxysporum	*Candida spp.* and *Cryptococcus spp.*	0.84–1.68 μg/ml	Ishida et al. (2013)
Pleurotus cornucopiae	*Candida spp.*	20, 40, 60 μg/ml	Owaid et al. (2015)

8.23 Antiviral Activity Ofmyconanoparticles

Myconanoparticles are used to control and prevent viral diseases. They have shown antiviral activity against H1N1 influenza virus. Silver nanoparticles with dimension of 1–10 nm binds and inhibits against HIV virus (Lara et al. 2010). It exhibits antiviral activity at an early stage of viral replication. These silver nanoparticles bind to gp120 which prevents CD4 dependent virion binding, fusion and infectivity, thus acts as an effective virucidal against cell free virus such as laboratory strains, clinical isolates, T and M tropic strains and resistant strains and cell associated virus. Besides, it inhibits H1N1 virus at its post entry stage of life cycle. Gold nanoparticles are also used to treat cancer, HIV and tuberculosis (Madhusudhan et al. 2014).

8.24 Conclusion

This chapter dealt with antimicrobial application of myconanoparticles. Nanoparticles for its small size and high surface area easily interacts with microorganisms to kill them. It excellently replace antibiotics to cure diseases and also used in smart delivery system. It is the best alternative to synthetic fungicide to destroy phytopathogenic fungi and prevent diseases in major trees and food crops. They are employed for its less toxic nature to humans and animals and are eco-friendly. Morbidity and motality increases in both humans and animals by bacterial infection. Multidrug resistant bacterial strain have been developed due to the over use of antibiotics. Thus nanoparticle based antibacterial approach is simple, alternative, eco-friendly and cost effective. Among all nanoparticles, silver nanoparticles with antibiotic or alone have a broad spectrum antibacterial activity against many pathogenic organisms. Even against virus, it acts as a virucidal agent or inhibitor of viral entry against HIV during early stage of replication. It also shows antifungal activity against invasive mycosis.

References

Aboelzahab A, Azad AM, Dolan S, Goel V (2012) Mitigation of *Staphylococcus aureus* Mediated Surgical Site Infections with IR Photoactivated TiO_2 coatings on Ti Implants. Adv Healthcare Mater 1(3):285–291. https://doi.org/10.1002/adhm.201100032

Ansari MA, Khan HM, Khan AA, Sultan A, Azam A (2012) Synthesis and characterization of the antibacterial potential of ZnO nanoparticles against extended-spectrum β-lactamases-producing *Escherichia coli* and *Klebsiella pneumoniae* isolated from a tertiary care hospital of North India. Appl Microbiol Biotechnol 94:467–477

Ansari MA, Khan HM, Alzohairy MA, Jalal M, Ali SG, Pal R, Musarrat J (2015) Green synthesis of Al_2O_3 nanoparticles and their bactericidal potential against clinical isolates of multi-drug resistant *Pseudomonas aeruginosa.* World J Microbiol Biotechnol 31:153–164. https://doi.org/10.1007/s11274-014-1757-2

Arakha M, Pal S, Samantarrai D, Panigrahi TK, Mallick BC, Pramanik K, Jha S (2015) Antimicrobial activity of iron oxide nanoparticle upon modulation of nanoparticle-bacteria interface. Sci Rep 5(1):14813

Armentano I, Arciola CR, Fortunati E, Davide F, Samantha M, Concetta FA, Livia V et al (2014) The interaction of bacteria with engineered nanostructured polymeric materials: a review. Sci World J 2014:410423

Basavaraja S, Balaji SD, Lagashetty A, Rajasab AH, Venkataraman A (2008) Extracellular biosynthesis of silver nanoparticles using the fungus *Fusarium semitectum.* Mater Res Bull 43(5):1164–1170. https://doi.org/10.1016/j.materresbull.2007.06.020

Ben-Sasson M, Zodrow KR, Genggeng Q, Kang Y, Giannelis EP, Elimelech M (2014) Surface functionalization of thin-film composite membranes with copper nanoparticles for antimicrobial surface properties. Environ Sci Technol 48(1):384–393. https://doi.org/10.1021/es404232s

Cha SH, Hong J, McGuffie M, Yeom B, VanEpps JS, Kotov NA (2015) Shape-dependent biomimetic inhibition of enzyme by nanoparticles and their antibacterial activity. ACS Nano 9(9):9097–9105. https://doi.org/10.1021/acsnano.5b03247

Chatzimitakos TG, Stalikas CD (2016) Qualitative alterations of bacterial metabolome after exposure to metal nanoparticles with bactericidal properties: a comprehensive workflow based on 1H NMR, UHPLC-HRMS, and metabolic databases. J Proteome Res 15(9):3322–3330. https://doi.org/10.1021/acs.jproteome.6b00489

Cheloni G, Marti E, Slaveykova VI (2016) Interactive effects of copper oxide nanoparticles and light to green alga *Chlamydomonas reinhardtii.* Aquat Toxicol 170:120–128. https://doi.org/10.1016/j.aquatox.2015.11.018

Chowdhury S, Basu A, Kundu S (2014) Green synthesis of protein capped silver nanoparticles from phytopathogenic fungus *Macrophomina phaseolina* (Tassi) Goid with antimicrobial properties against multidrug-resistant bacteria. Nanoscale Res Lett 9(1):1–11. https://doi.org/10.1186/1556-276X-9-365

Costa JR, Silva NC, Sarmento B, Pintado M (2015) Potential chitosan-coated alginate nanoparticles for ocular delivery of daptomycin. Eur J Clin Microbiol Infect Dis 34:1255–1262. https://doi.org/10.1007/s10096-015-2344-7

Depan D, Misra RDK (2014) On the determining role of network structure titania in silicone against bacterial colonization: mechanism and disruption of biofilm. Mater Sci Eng, C 34:221–228. https://doi.org/10.1016/j.msec.2013.09.025

Durán N, Marcato PD, De Souza GI, Alves OL, Esposito E (2007) Antibacterial effect of silver nanoparticles produced by fungal process on textile fabrics and their effluent treatment. J Biomed Nanotechnol 3(2):203–208. https://doi.org/10.1166/jbn.2007.022

Gade AK, Bonde PP, Ingle AP, Marcato PD, Duran N, Rai MK (2008) Exploitation of *Aspergillus niger* for synthesis of silver nanoparticles. J Biobased Mater Bioenergy 2(3):243–247. https://doi.org/10.1166/jbmb.2008.401

Gajbhiye M, Kesharwani J, Ingle A, Gade A, Rai M (2009) Fungus-mediated synthesis of silver nanoparticles and their activity against pathogenic fungi in combination with fluconazole. Nanomedicine: Nanotechnol, Biol Med 5(4):382–386. https://doi.org/10.1016/j.nano.2009.06.005

Galiano K, Pleifer C, Engelhardt K, Brössner G, Lackner P, Huck C, Obwegeser A (2008) Silver segregation and bacterial growth of intraventricular catheters impregnated with silver nanoparticles in cerebrospinal fluid drainages. Neurol Res 30(3):285–287. https://doi.org/10.1179/016164107X229902

Gao W, Thamphiwatana S, Angsantikul P, Zhang L (2014) Nanoparticle approaches against bacterial infections. Wiley Interdiscip Rev: Nanomedicine Nanobiotechnology 6(6):532–547. https://doi.org/10.1002/wnan.1282

Gehlot P, Singh J (Eds) (2018) Fungi and their role in sustainable development: current perspectives. Springer Singapore, Singapore. https://doi.org/10.1007/978-981-13-0393-7_39

Gopinath V, MubarakAli D, Priyadarshini S, Priyadharsshini NM, Thajuddin N, Velusamy P (2012) Biosynthesis of silver nanoparticles from *Tribulus terrestris* and its antimicrobial activity: a novel biological approach. Colloids Surf, B 96:69–74. https://doi.org/10.1016/j.colsurfb.2012.03.023

Guo BL, Han P, Guo LC, Cao YQ, Li AD, Kong JZ, Wu D (2015) The antibacterial activity of Ta-doped ZnO nanoparticles. Nanoscale Res Lett 10:1–10. https://doi.org/10.1186/s11671-015-1047-4

Gurunathan S, Han JW, Dayem AA, Eppakayala V, Kim JH (2012) Oxidative stress-mediated antibacterial activity of graphene oxide and reduced graphene oxide in *Pseudomonas aeruginosa*. Int J Nanomedicine:5901–5914. https://doi.org/10.2147/IJN.S37397

He W, Kim HK, Wamer WG, Melka D, Callahan JH, Yin JJ (2014) Photogenerated charge carriers and reactive oxygen species in ZnO/Au hybrid nanostructures with enhanced photocatalytic and antibacterial activity. J Am Chem Soc 136(2):750–757

Huh AJ, Kwon YJ (2011) "Nanoantibiotics": a new paradigm for treating infectious diseases using nanomaterials in the antibiotics resistant era. J Control Release 156(2):128–145

Hussein-Al-Ali SH, El Zowalaty ME, Hussein MZ, Geilich BM, Webster TJ (2014) Synthesis, characterization, and antimicrobial activity of an ampicillin-conjugated magnetic nanoantibiotic for medical applications. Int J Nanomed 9:3801

Ingle A, Rai M (2011) Genetic diversity among Indian phytopathogenic isolates of *Fusarium semitectum* Berkeley and Ravenel. Adv Biosci Biotechnol 2(03):142–148. https://doi.org/10.4236/abb.2011.23023

Ishida K, Cipriano TF, Rocha GM, Weissmüller G, Gomes F, Miranda K, Rozental S (2013) Silver nanoparticle production by the fungus *Fusarium oxysporum*: nanoparticle characterisation and analysis of antifungal activity against pathogenic yeasts. Mem Inst Oswaldo Cruz 109:220–228. https://doi.org/10.1590/0074-0276130269

Jung WK, Koo HC, Kim KW, Shin S, Kim SH, Park YH (2008) Antibacterial activity and mechanism of action of the silver ion in *Staphylococcus aureus* and *Escherichia coli*. Appl Environ Microbiol 74(7):2171–2178. https://doi.org/10.1128/AEM.02001-07

Kalaiselvan V, Rajasekaran A (2009) Biosynthesis of silver nanoparticles from *Aspergillus niger* and evaluation of its wound healing activity in experimental rat model. Int J Pharm Tech Res 4:1523–1529

Khan MF, Ansari AH, Hameedullah M, Ahmad E, Husain FM, Zia Q, Aliev G (2016) Sol-gel synthesis of thorn-like ZnO nanoparticles endorsing mechanical stirring effect and their antimicrobial activities: potential role as nano-antibiotics. Sci Rep 6(1):27689

Kim HS, Kang HS, Chu GJ, Byun HS (2008) Antifungal effectiveness of nanosilver colloid against rose powdery mildew in greenhouses. In: Solid state phenomena, vol 135, pp 15–18. Trans Tech Publications Ltd. https://doi.org/10.1038/srep27689

Kose N, Çaylak R, Pekşen C, Kiremitçi A, Burukoglu D, Koparal S, Doğan A. (2016) Silver ion doped ceramic nano-powder coated nails prevent infection in open fractures: in vivo study. Injury 47(2):320–324

Kukowska-Latallo JF, Candido KA, Cao Z, Nigavekar SS, Majoros IJ, Thomas TP, Baker JR Jr (2005) Nanoparticle targeting of anticancer drug improves therapeutic response in animal model of human epithelial cancer. Can Res 65(12):5317–5324. https://doi.org/10.1158/0008-5472.can-04-3921

Lara HH, Ayala-Nuñez NV, Ixtepan-Turrent L, Rodriguez-Padilla C (2010) Mode of antiviral action of silver nanoparticles against HIV-1. J Nanobiotechnology 8(1):1–10.10 https://doi.org/10.1186/1477-3155-8-1

Lee DK, Kim SV, Limansubroto AN, Yen A, Soundia A, Wang CY, Ho D (2015) Nanodiamond–gutta percha composite biomaterials for root canal therapy. ACS Nano 9(11):11490–11501. https://doi.org/10.1021/acsnano.5b05718

Lellouche J, Friedman A, Lellouche JP, Gedanken A, Banin E (2012) Improved antibacterial and antibiofilm activity of magnesium fluoride nanoparticles obtained by water-based ultrasound chemistry. Nanomedicine: Nanotechnol, Biol Med 8(5):702–711. https://doi.org/10.1016/j.nano.2011.09.002

Lesniak A, Salvati A, Santos-Martinez MJ, Radomski MW, Dawson KA, Åberg C (2013) Nanoparticle adhesion to the cell membrane and its effect on nanoparticle uptake efficiency. J Am Chem Soc 135(4):1438–1444. https://doi.org/10.1021/ja309812z

Leung YH, Ng AM, Xu X, Shen Z, Gethings LA, Wong MT, Leung FC (2014) Mechanisms of antibacterial activity of MgO: non-ROS mediated toxicity of MgO nanoparticles towards *Escherichia coli*. Small 10(6):1171–1183. https://doi.org/10.1002/smll.201302434

Li H, Chen Q, Zhao J, Urmila K (2015) Enhancing the antimicrobial activity of natural extraction using the synthetic ultrasmall metal nanoparticles. Sci Rep 5(1):11033. https://doi.org/10.1038/srep11033

Li M, Zhu L, Lin D (2011) Toxicity of ZnO nanoparticles to *Escherichia coli*: mechanism and the influence of medium components. Environ Sci & Technol 45(5):1977–1983.https://doi.org/10.1021/es102624t

Li Y, Zhang W, Niu J, Chen Y (2012) Mechanism of photogenerated reactive oxygen species and correlation with the antibacterial properties of engineered metal-oxide nanoparticles. ACS Nano 6(6):5164–5173. https://doi.org/10.1021/nn300934k

Luan B, Huynh T, Zhou R (2016) Complete wetting of graphene by biological lipids. Nanoscale 8(10):5750–5754. https://doi.org/10.1039/C6NR00202A

Madhusudhan A, Reddy GB, Venkatesham M, Veerabhadram G, Kumar DA, Natarajan S, Singh SS (2014) Efficient pH dependent drug delivery to target cancer cells by gold nanoparticles capped with carboxymethyl chitosan. Int J Mol Sci 15(5):8216–8234. https://doi.org/10.3390/ijms15058216

Maeda H (2010) Tumor-selective delivery of macromolecular drugs via the EPR effect: background and future prospects. Bioconjug Chem 21(5):797–802. https://doi.org/10.1021/bc100070g

Mahendra Rai MR, Alka Yadav AY, Bridge P, Aniket Gade AG (2009) Myconanotechnology: a new and emerging science. Appl Mycol:258–267. https://doi.org/10.1079/9781845935344.0258

Malka E, Perelshtein I, Lipovsky A, Shalom Y, Naparstek L, Perkas N, Gedanken A (2013) Eradication of multi-drug resistant bacteria by a novel Zn-doped CuO nanocomposite. Small, 9(23):4069–4076. https://doi.org/10.1002/smll.201301081

Marchiol L (2012) Synthesis of metal nanoparticles in living plants. Ital J Agron 7(3):e37–e37

Matai I, Sachdev A, Dubey P, Kumar SU, Bhushan B, Gopinath P (2014) Antibacterial activity and mechanism of Ag–ZnO nanocomposite on *S. aureus* and GFP-expressing antibiotic resistant *E. coli*. Colloids Surf, B 115:359–367. https://doi.org/10.1016/j.colsurfb.2013.12.005

Mehmood S, Rehman MA, Ismail H, Mirza B, Bhatti AS (2015) Significance of postgrowth processing of ZnO nanostructures on antibacterial activity against gram-positive and gram-negative bacteria. Int J Nanomed 10:4521. https://doi.org/10.2147/IJN.S83356

Mekkawy AI, El-Mokhtar MA, Nafady NA, Yousef N, Hamad MA, El-Shanawany SM, Elsabahy M (2017) In vitro and in vivo evaluation of biologically synthesized silver nanoparticles for topical applications: effect of surface coating and loading into hydrogels. Int J Nanomed 12:759. https://doi.org/10.2147/IJN.S124294

Mishra A, Kumari M, Pandey S, Chaudhry V, Gupta KC, Nautiyal CS (2014) Biocatalytic and antimicrobial activities of gold nanoparticles synthesized by *Trichoderma sp.* Biores Technol 166:235–242. https://doi.org/10.1016/j.biortech.2014.04.085

Moazeni M, Rashidi N, Shahverdi AR, Noorbakhsh F, Rezaie S (2012) Extracellular production of silver nanoparticles by using three common species of dermatophytes: *trichophyton rubrum, Trichophyton mentagrophytes* and *Microsporum canis.* Iran Biomed J 16(1):52. https://doi.org/10.6091/IBJ.1001.2012

Moghaddam K (2010) An introduction to microbial metal nanoparticle preparation method. J Young Investig, 19(1)

Mout R, Moyano DF, Rana S, Rotello VM (2012) Surface functionalization of nanoparticles for nanomedicine. Chem Soc Rev 41(7):2539–2544. https://doi.org/10.1039/c2cs15294k

Nagy A, Harrison A, Sabbani S, Munson Jr RS, Dutta PK, Waldman WJ (2011) Silver nanoparticles embedded in zeolite membranes: release of silver ions and mechanism of antibacterial action. Int J Nanomedicine:1833–1852. https://doi.org/10.2147/IJN.S24019

Namasivayam SKR, Avimanyu (2011) Silver nanoparticle synthesis from *Lecanicillium lecanii* and evalutionary treatment on cotton fabrics by measuring their improved antibacterial activity with antibiotics against *Staphylococcus aureus* (ATCC 29213) and *E.coli* (ATCC 25922) strains. Int J Pharm Pharm Sci 3:190–195

Narayanan KB, Sakthivel N (2011) Facile green synthesis of gold nanostructures by NADPH-dependent enzyme from the extract of *Sclerotium rolfsii.* Colloids Surf, A 380(1–3):156–161

Nowinski RJ, Gillespie RJ, Shishani Y, Cohen B, Walch G, Gobezie R (2012) Antibiotic-loaded bone cement reduces deep infection rates for primary reverse total shoulder arthroplasty: a retrospective, cohort study of 501 shoulders. J Shoulder Elbow Surg 21(3):324–328. https://doi.org/10.1016/j.jse.2011.08.072

Ottoni CA, Simões MF, Fernandes S, Dos Santos JG, Da Silva ES, de Souza RFB, Maiorano AE (2017) Screening of filamentous fungi for antimicrobial silver nanoparticles synthesis. AMB Express 7(1):1–10. https://doi.org/10.1186/s13568-017-0332-2

Owaid MN, Raman J, Lakshmanan H, Al-Saeedi SSS, Sabaratnam V, Abed IA (2015) Mycosynthesis of silver nanoparticles by *Pleurotus cornucopiae var. citrinopileatus* and its inhibitory effects against *Candida sp.* Mater Lett 153:186–190. https://doi.org/10.1016/j.matlet.2015.04.023

Padmavathy N, Vijayaraghavan R (2011) Interaction of ZnO nanoparticles with microbes—a physio and biochemical assay. J Biomed Nanotechnol 7(6):813–822. https://doi.org/10.1166/jbn.2011.1343

Pan X, Wang Y, Chen Z, Pan D, Cheng Y, Liu Z, Guan X (2013) Investigation of antibacterial activity and related mechanism of a series of nano-Mg(OH) 2. ACS Appl Mater Interfaces 5(3):1137–1142. https://doi.org/10.1021/am302910q

Peng YP, Lo SL, Ou HH, Lai SW (2010) Microwave-assisted hydrothermal synthesis of N-doped titanate nanotubes for visible-light-responsive photocatalysis. J Hazard Mater 183(1–3):754–758. https://doi.org/10.1016/j.jhazmat.2010.07.090

Peng Z, Ni J, Zheng K, Shen Y, Wang X, He G, Tang T (2013) Dual effects and mechanism of TiO_2 nanotube arrays in reducing bacterial colonization and enhancing C3H10T1/2 cell adhesion. Int J Nanomedicine:3093–3105. https://doi.org/10.2147/IJN.S48084

Podporska-Carroll J, Myles A, Quilty B, McCormack DE, Fagan R, Hinder SJ, Pillai SC (2017) Antibacterial properties of F-doped ZnO visible light photocatalyst. J Hazard Mater 324:39–47. https://doi.org/10.1016/j.jhazmat.2015.12.038

Pramanik A, Laha D, Bhattacharya D, Pramanik P, Karmakar P (2012) A novel study of antibacterial activity of copper iodide nanoparticle mediated by DNA and membrane damage. Colloids Surf, B 96:50–55. https://doi.org/10.1016/j.colsurfb.2012.03.021

Prasannakumar JB, Vidya YS, Anantharaju KS, Ramgopal G, Nagabhushana H, Sharma SC, Nagaswarupa HP (2015) Bio-mediated route for the synthesis of shape tunable Y_2O_3: Tb^{3+} nanoparticles: photoluminescence and antibacterial properties. Spectrochim Acta Part A: Mol Biomol Spectrosc 151:131–140. https://doi.org/10.1016/j.saa.2015.06.081

Prokopovich P, Köbrick M, Brousseau E, Perni S (2015) Potent antimicrobial activity of bone cement encapsulating silver nanoparticles capped with oleic acid. J Biomed Mater Res Part B: Appl Biomater 103(2):273–281. https://doi.org/10.1002/jbm.b.33196

Radovic-Moreno AF, Lu TK, Puscasu VA, Yoon CJ, Langer R, Farokhzad OC (2012) Surface charge-switching polymeric nanoparticles for bacterial cell wall-targeted delivery of antibiotics. ACS Nano 6(5):4279–4287. https://doi.org/10.1021/nn3008383

Rajakumar G, Rahuman AA, Roopan SM, Khanna VG, Elango G, Kamaraj C, Velayutham K (2012) Fungus-mediated biosynthesis and characterization of TiO_2 nanoparticles and their activity against pathogenic bacteria. Spectrochim Acta Part A Mol Biomol Spectrosc 91:23–29. https://doi.org/10.1016/j.saa.2012.01.011

Ramazanzadeh B, Jahanbin A, Yaghoubi M, Shahtahmassbi N, Ghazvini K, Shakeri M, Shafaee H (2015) Comparison of antibacterial effects of ZnO and CuO nanoparticles coated brackets against *Streptococcus mutans*. J Dent 16(3):200

Rani R, Sharma D, Chaturvedi M, Yadav JP (2017) Green synthesis, characterization and antibacterial activity of silver nanoparticles of endophytic fungi *Aspergillus terreus*. J Nanomed Nanotechnol, 8(8). https://doi.org/10.4172/2157-7439.1000457

Roguska A, Belcarz A, Pisarek M, Ginalska G, Lewandowska M (2015) TiO_2 nanotube composite layers as delivery system for ZnO and Ag nanoparticles—An unexpected overdose effect decreasing their antibacterial efficacy. Mater Sci Eng, C 51:158–166. https://doi.org/10.1016/j.msec.2015.02.046

Saliani M, Jalal R, Goharshadi EK (2015) Effects of pH and temperature on antibacterial activity of zinc oxide nanofluid against *Escherichia coli* O157: H7 and *Staphylococcus aureus*. Jundishapur J Microbiol 8(2):e17115. https://doi.org/10.5812/jjm.17115

Samuel U, Guggenbichler JP (2004) Prevention of catheter-related infections: the potential of a new nano-silver impregnated catheter. Int J Antimicrob Agents 23:75–78. https://doi.org/10.1016/j.ijantimicag.2003.12.004

Sangari M, Umadevi M, Mayandi J, Pinheiro JP (2015) Photocatalytic degradation and antimicrobial applications of F-doped MWCNTs/TiO_2 composites. Spectrochim Acta Part A Mol Biomol Spectrosc 139:290–295. https://doi.org/10.1016/j.saa.2014.12.061

Saxena J, Sharma PK, Sharma MM, Singh A (2016) Process optimization for green synthesis of silver nanoparticles by *Sclerotinia sclerotiorum* MTCC 8785 and evaluation of its antibacterial properties. Springerplus 5(1):1–10. https://doi.org/10.1186/s40064-016-2558-x

Seil JT, Webster TJ (2012) Antibacterial effect of zinc oxide nanoparticles combined with ultrasound. Nanotechnology, 23(49):495101

Shao L, Gao Y, Yan F (2011) Semiconductor quantum dots for biomedicial applications. Sensors 11(12):11736–11751

Shrivastava S, Bera T, Roy A, Singh G, Ramachandrarao P, Dash D (2007) Characterization of enhanced antibacterial effects of novel silver nanoparticles. Nanotechnology 18(22):225103. https://doi.org/10.1088/0957-4484/18/22/225103

Ramalingam B, Parandhaman T, Das SK (2016) Antibacterial effects of biosynthesized silver nanoparticles on surface ultrastructure and nanomechanical properties of gram-negative bacteria viz. *Escherichia coli* and *Pseudomonas aeruginosa*. ACS Appl Mater Interfaces 8(7):4963-4976

Subashini G, Bhuvaneswari S (2018) Nanoparticles from fungi (myconanoparticles). Fungi and their role in sustainable development: current perspectives, pp 753–779. https://doi.org/10.1007/978-981-0393-7_39

Sukhorukova IV, Sheveyko AN, Kiryukhantsev-Korneev PV, Zhitnyak IY, Gloushankova NA, Denisenko EA, Shtansky DV (2015) Toward bioactive yet antibacterial surfaces. Colloids Surf, B 135:158–165. https://doi.org/10.1016/j.colsurfb.2015.06.059

Sundaramoorthi C, Kalaivani M, Mathews DM, Palanisamy S, Kalaiselvan V, Rajasekaran A (2009) Biosynthesis of silver nanoparticles from *Aspergillus niger* and evaluation of its wound healing activity in experimental rat model. Int J PharmTech Res 1(4):1523–1529

Wu B, Zhuang WQ, Sahu M, Biswas P, Tang YJ (2011) Cu-doped TiO_2 nanoparticles enhance survival of *Shewanella oneidensis* MR-1 under Ultraviolet Light (UV) exposure. Sci Total Environ 409(21):4635–4639. https://doi.org/10.1016/j.scitotenv.2011.07.037

Xu X, Chen D, Yi Z, Jiang M, Wang L, Zhou Z, Hui D (2013) Antimicrobial mechanism based on H_2O_2 generation at oxygen vacancies in ZnO crystals. Langmuir 29(18):5573–5580. https://doi.org/10.1021/la400378t

Xu Y, Wei MT, Ou-Yang HD, Walker SG, Wang HZ, Gordon CR, Mironava T et al (2016) Exposure to TiO_2 nanoparticles increases *Staphylococcus aureus* infection of HeLa cells. J Nanobiotechnol 14(1):1–16https://doi.org/10.1186/s12951-016-0184-y

Xue B, He D, Gao S, Wang D, Yokoyama K, Wang L (2016) Biosynthesis of silver nanoparticles by the fungus *Arthroderma fulvum* and its antifungal activity against genera of *Candida, Aspergillus* and *Fusarium.* Int J Nanomed 11:1899. https://doi.org/10.2147/IJN.S98339

Yang W, Shen C, Ji Q, An H, Wang J, Liu Q, Zhang Z (2009) Food storage material silver nanoparticles interfere with DNA replication fidelity and bind with DNA. Nanotechnology 20(8):085102. https://doi.org/10.1088/0957-4484/20/8/085102

Yu J, Zhang W, Li Y, Wang G, Yang L, Jin J, Huang M (2014) Synthesis, characterization, antimicrobial activity and mechanism of a novel hydroxyapatite whisker/nano zinc oxide biomaterial. Biomed Mater 10(1):015001

Zhao L, Ashraf MA (2015) Influence of silver-hydroxyapatite nanocomposite coating on biofilm formation of joint prosthesis and its mechanism. West Indian Med J 64(5):506. https://doi.org/10.7727/wimj.2016.179

Chapter 9
"Nanomaterials Induced Cell Disruption: An Insight into Mechanism"

Anjali Raghav, Simran Kaur, Gunjit Setia, and Saroj Kumar

Abstract In research and development, cell disruption is vital to the acquisition of internal components, including metabolites and proteins. This is essential for producing many biological products, such as enzymes and antigens used to produce vaccines and nucleic acids. For cell lysis, various mechanical techniques exist, including high-pressure homogenization, sonication, bead milling, and many others. The growing importance of nanoparticles (NPs) in theranostics (therapeutic diagnostics), target modulation of cellular processes, etc., has been serving as a novel approach towards biotechnological applications such as silica-based nanoparticles (SNs) being used in drug delivery, imaging, and other therapeutic majors. The disruption of the cell caused by the biophysical interlinkage among the cell and nanoparticle allows researchers to use nanoparticles as drug carrier systems. This chapter discusses the biophysiochemical interactions between the cell and the nanomaterials, which serves as an essential factor for the disruption of cells, and also explains the methods used for the cell disruption along with current trends associated.

Keywords Nanoparticles · Cell disruption · Molecular mechanism · Intracellular components

9.1 Introduction

Nanotechnology is recognized as a new and emerging domain in research wherein researchers thrive to reduce the size of atoms and molecules at the nanometer scale. Nanomaterial refers to something that has a crystalline or amorphous particulate on

---

Anjali Raghav, Simran Kaur, Gunjit Setia—contributed equally to this work.

A. Raghav · S. Kaur · G. Setia · S. Kumar (✉)
School of Biosciences, Apeejay Stya University, Gurugram 122103, Haryana, India
e-mail: sarojbiotech@gmail.com

M. P. Shah et al. (eds.), *Biogenic Nanomaterials for Environmental Sustainability: Principles, Practices, and Opportunities*, Environmental Science and Engineering,
https://doi.org/10.1007/978-3-031-45956-6_9

a nanometer scale. Nanomaterials consist of nanoparticles, nanowires, nanomembranes, and nano blocks of material in 3D. The precision in size and shape of nanomaterials make them unique structures that are not found in many traditional materials. In today's time, nanotechnology has been widely used in biomedicine, mechanical engineering, environmental protection, and other fields and has contributed to the full development of these industries (Srikanth and Kessler 2012). Such novel and advanced properties make nanoscale materials enhance their application methods, leading to scientific and technological advances in several fields, especially environment, biology, pharmacy, health care, and medical care (Patel et al. 2022).

Nanomaterials have shown various potential applications, including bio-catalysis, protein purification, and cell disruption, followed by drug delivery wherein the nanomaterial penetrates the cell membrane and disrupts it. To accomplish this, nanomaterials can be designed with specific properties, including size, shape, charge, and surface chemistry, which enables them to interact and penetrate the cell membrane. Various mechanisms have been employed to disrupt the cell membrane, some of them being mechanical stress, chemical reactions, and electrostatic interactions.

A new field of study, nanomedicine, has emerged due to the rapid expansion of the applications of nanotechnology in biomedicine, particularly in early and accurate biosensing followed by efficient treatment of various diseases. This vision is supported by various organizations, including industrial dynamics, educational research, and government entities (Bhardwaj and Kaushik 2017).

As of now, a total of 250 nanomedicine products are currently available in the market, 50 of which are being used by physicians (Bhardwaj and Kaushik 2017; Etheridge et al. 2013). Development of nanomedicine and biomaterials for therapeutic use are primarily designed according to the factors such as size, stability, percent drug entrapment, and continuous drug release from nanoparticles during drug delivery (Kaushik and Dixit 2016; Rosario-Meléndez et al. 2012). Nanotechnology has enormous potential to improve efficiency in pollution prevention, treatment, and cleaning (Iravani 2020). During nano remediation, the nanoparticle agent is introduced to contact with the target contaminants under conditions that allow for a detoxifying reaction, and this process usually involves a pumping and treatment process (Del Prado-Audelo et al. 2021). Nanotechnology interventions in the fight against covid-19 are one of the ways to deal with the disease through the development of masks with nano-coatings, nano-diagnostic equipment, nano-sanitizers, and nanomedicine (Singh et al. 2021). The size of nanomaterial ranges to half a billionth of a meter, i.e., 10^{-9} m, and can be compared to viruses and other pathogens (Singh et al. 2021). In order to combat COVID-19, nanotechnology interventions played a significant role in the easy diagnosis, monitoring, surveillance, treatment, and development of vaccines. Utilizing these nanotechnology interventions, COVID-19 infection can be easily, detected, treated and prevented. The antimicrobial properties of nanomaterials can be used to manufacture vaccines, immune modulators, personal equipment covers, disinfectants, etc. Chen et al. (2020).

In order to prevent viral replication in the lungs, the targeted antibacterial drug is designed using a combination of nanoparticles with other drugs and thus reduces the viral load in the infected area. Biosensors are developed as point-of-care diagnostic tools by modifying nanomaterials' electrical and optical properties (Weiss et al. 2020). Integrating nanotechnology, food science, and physical chemistry has resulted in the development of sensors for real-time approaches for detecting adulterants/ metabolites using nano-sensors, specific food delivery systems, and new methods of food processing to produce nano-enabled food. Examples include the development of 3-dimensional SERS (surface-enhanced Raman spectroscopy) for detecting the presence of food adulterants even at a traceable amount, creating unique nanostructured foods and encapsulation programs using new processing methods and activating enzyme reactions in the presence of selected biomedical components. Nanotechnology has the potential to enhance food security and human health, although a better understanding of the threats associated with incorporating nanoparticles into food systems is needed (Sun-Waterhouse and Waterhouse 2016).

With the endless number of advantages and possible applications of nanomaterials, this book chapter discusses the mechanisms involved in NP-induced cell disruption. Furthermore, it outlines the benefits and limitations of utilizing nanoparticles in the medical and healthcare domain.

9.2 Introduction to Cell Types

All living organisms are made up of cells that perform a certain function in the body; hence the cells are described as the building blocks of living organisms. Cells are categorized into two categories, i.e., prokaryotic and eukaryotic cells (Khan and Farhana 2022). These cell types show some common components among each other. It includes the plasma membrane, the outer covering, the cytoplasm, a semi-fluid matrix where all the organelles are embedded, DNA, genetic material, and ribosomes. The classification is determined on basis of cells complexity (Vellai and Vida 1999).

A prokaryotic cell is identified as a single-celled organism devoid of a membrane-bound nucleus indicating the absence of a nuclear membrane. Another key feature includes the lack of membrane-bound organelles. Only the ribosome, which is a non-membrane-bound organelle, is found in prokaryotes, e.g., prokaryotes, include bacteria and archaea.

A eukaryotic cell is identified as a cell that contains the membrane-bound nucleus, i.e., the nucleus is enclosed in the nuclear envelope. Compartmentalization of cytoplasm is observed in eukaryotes due to the presence of membrane-bound organelles, which include ER, mitochondria, lysosomes, golgi complex, etc.

9.2.1 Cellular Barriers

Both eukaryotic and prokaryotic cells are differentiated on the basis of cellular barriers. The cellular barriers are of three types which are as follows.

9.2.1.1 Cell Membrane

A cell membrane, or the plasma membrane, is a 5 nm thick layer separating the external environment from the internal one (Yang and Hinner 2015). The membrane comprises of carbohydrates, lipids, and proteins. The lipids exist in the form of phospholipids. They are arranged in a bilayer where the hydrophobic fatty acids are on the inside, pointing to each other, and the hydrophilic glycerol moieties are on the outer surface, thereby protecting the fatty acids from the external environment. The proteins are found to be embedded on lipid bilayers' surface, depending on their size and the ease of extraction, they are classified as integral and peripheral proteins. This layer forms a tight barrier, i.e., it only allows certain hydrophobic molecules to pass through the barrier.

Presence of sterols in the cell membrane of eukaryotic cell elevates the stability and flexibility of eukaryotic cell. In the case of prokaryotic cells, hopanoids are present in the cell membrane, which shows similarity with sterols and works by improving the stability and rigidity of the cell membrane (Belin et al. 2018).

9.2.1.2 Cell Wall

In bacterial cells, apart from the cell membrane, cell wall is also present, referred to as peptidoglycan layer which is responsible for maintaining the cell shape, imparting rigidity, promoting cell-to-cell interactions, and acting as a barrier (Dörr et al. 2019). The peptidoglycan layer consists of 2 sugar derivatives: N-acetylmuramic acid and N-acetylglucosamine, as well as small amount of amino acids, such as D-glutamic acid, L-alanine, and D-alanine. The interactions between the sugar derivatives through glycosidic bonds give rise to a thin sheet-like structure due to the formation of a glycan chain. In addition, the amino acids are cross-linked with the glycan chain, providing rigidity to the cells (Bastos et al. 2021). The strength of structure of cell wall is determined by the cross-linking between the amino acids and the frequency within bonds.

9.2.1.3 Outer Membrane

In bacteria, the cell membrane is followed by a layer known as the outer membrane, which classifies them into gram-positive and gram-negative bacteria. The outer membrane comprises of lipopolysaccharide, which consists of lipids, polysaccharides, and proteins. This layer protects the peptidoglycan layer and is not permeable to enzymes. The bacteria is categorized into two forms, based on the membrane and how they react to the staining methods, the first one being gram-positive bacteria which contain thick peptidoglycan layer but no outer lipid layer. The second one is the gram-negative bacteria, which comprises a thin peptidoglycan layer with an outer lipid layer. The presence of multiple layers of peptidoglycan layer and the intensive cross-linking in gram-positive bacteria shield the cells from disruption (Silhavy et al.

2010). On the other hand, a single layer of peptidoglycan is present in gram-negative bacteria, which makes it easier to disrupt (Mahalanabis et al. 2009).

9.3 Cell Disruption

Various industries, including chemical, food, and pharmaceutical, heavily depend on biotechnological processes to extract materials like enzymes, lipids, pigments, and proteins from microbial cells such as bacteria, microalgae, and yeast (Gomes et al. 2020). Since the downstream steps are simple and cost-effective to execute, the process of creating biomolecules outside cells shows great industrial potential. However, there are also several processes that involve the accumulation of intracellular biomolecules in cell walls, cell membranes, and cytoplasm. Such cases require cell disruption methods depending on the location of the biomolecule within the cell and the structure of the cell wall (Peternel 2013).

9.3.1 Importance of Cell Disruption

Cell Disruption or Cell lysis is a technique that involves rupturing the cell membrane to release internal components like DNA, RNA, protein, or organelles out of the cell. Cell lysis is a widely recognized method for intracellular studies (DNA, RNA, and proteins). The disruption of the cell has numerous applications, including immunoassays for point-of-care diagnostics, molecular diagnostics of pathogens, downstream processes like purification analysis for analyzing protein structure and function, drug screening, identifying the mRNA transcriptome, cancer diagnostics and analyzing the composition of proteins, nucleic acids, and lipids. Cell lysis may be full or partial, depending on the application. It envisions the elevated yield with minimum contamination and product loss, thereby increasing the subsequent cost. The major challenge is to select an appropriate disruption method, especially for industrial-scale processes. For this purpose, various factors come into play, such as energy efficiency, high selectivity, operational costs, and ease of scale-up (Kumar et al. 2019). Techniques like patch clamping, which are employed for the testing of the drug and examining intracellular ionic currents, cause partial cell lysis (Sakmann and Neher 1984). This procedure involves inserting a glass micropipette inside the cell, partially rupturing the cell membrane. To analyze DNA, RNA, and subcellular components, the cell membrane must be completely disintegrated (complete cell lysis) (Stackebrandt and Goodfellow 1991).

9.4 Types of Cell Disruption Methods

Various procedures have been employed to disrupt the cell (Fig. 9.1). Based on the application, it is broadly classified into complete and partial cell lysis. Complete cell lysis is a type of disruption method wherein the cell is completely ruptured so as to analyze DNA, RNA, and subcellular components. For example, in electrical lysis, the cell is subjected to strong electrical fields wherein the cell undergoes electric shock, which results in complete disruption of the cell (Stackebrandt and Goodfellow 1991). Partial cell lysis is another type of disruption method invole the partially rupturing the cell to study intracellular components. For example, in patch clamping, the cell is partially ruptured by inserting a glass micropipette. It holds application in drug testing and is used for the study of intracellular ionic currents (Sakmann and Neher 1984); in this technique, the cell membrane is ruptured partially by inserting a glass micropipette.

The choice of the lysis methods depends on the simplicity of the purification procedures, the target molecules to be analyzed, and the degree of the final result (Harrison 1991). For many years, cell lysis methods hold applications in both laboratory and industrial levels. Many companies have developed commercially useful cell lysis chemicals, reagents, enzymes, and hydrogenic agents, as well as analytical tools such as sonicators and homogenizers (Van Den Berg et al. 2010). All internal parts of cells are closed by cell membranes, which act as barriers. It also controls the movement of materials from inside to outside the cell and vice-versa. Therefore, it is crucial to remove or destroy the cell membranes in order to collect DNA from the cells for molecular diagnosis applications, such as pathogen identification (Mahalanabis et al. 2009).

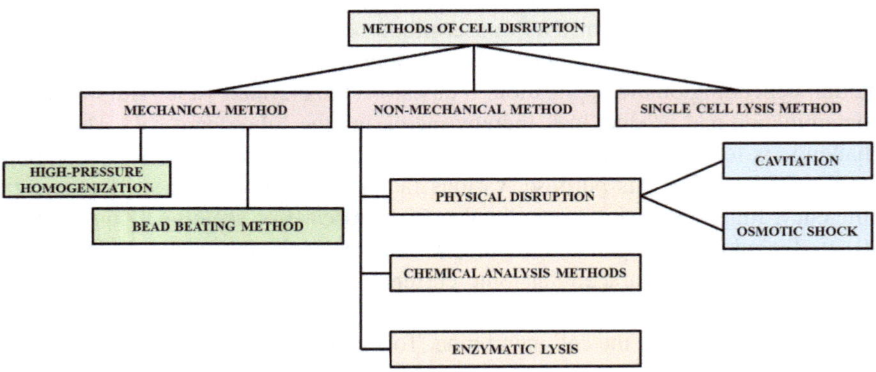

Fig. 9.1 Overview of various methods of cell disruption

9.4.1 Mechanical Methods of Cell Lysis

Mechanical lysis is the first category of lysis techniques, in which force is applied to physically rupture the cell membrane. This technique is the most popular because of its high performance and higher lysis (Islam et al., 2017). It is divided into two categories which are high-pressure homogenizers and the bread-beating method.

9.4.1.1 High Pressure Homogenizers

High pressure homogenization is identified as the prevalent technique for the disruption of microbial cells. Here, the cells present in the medium are subjected to high pressure while pushing them through the orifice. The high pressure at the orifice results in a series of compression due to which the cell membrane disrupts, followed by expansion upon discharge (Engler and Robinson 1981; Islam et al. 2017). However, the only problem with this method is the generation of heat. The use of cooling systems would reduce the amount of heat produced, thereby preventing the degradation of the enzymes (Augenstein et al. 1974). However, homogenization, combined with chemical treatment, displayed improved results (Middelberg 2000).

9.4.1.2 Bead Beating Method

The bead beating method is a common laboratory technique where the cell is disrupted by mixing the cell suspension with the minute glass, ceramic, or steel beads at a higher speed. In response to the collision between the beads and the cells, force is developed, which leads to the rupturing of the cell membrane and discharge of intracellular components. Various factors affect this process, such as cell concentration, stirrer speed, and the density and diameter of the bead. Smaller beads are more effective in 0.25–0.5 mm range and are recommended for lysis (Harrison 1991; Schütte et al. 1983). Using this technique, various cell types, including yeast and bacteria, can be lysed (Chisti and Moo-Young 1986; Taskova et al. 2006). Although this technique allows complete disintegration of the cell membrane and has high efficacy, the complete lysis results in the formation of tiny cell debris, which makes the sample separation and purification more complex. This process generates heat through the collision of beads with cells. The intracellular products like proteins and RNA can deteriorate due to the increasing heat (Goldberg 2008).

9.4.2 Non-mechanical Methods of Cell Lysis

Non-mechanical lysis is primarily classified into three categories, i.e., physical, chemical, and biological, with each category further subdivided according to the basis of the specific lysis procedure used.

9.4.2.1 Physical Methods

It is a non-contact method in which the cell membrane is ruptured by using external forces such as pressure, sound and thermal (heat) energy. Physical methods can be divided into three categories: cavitation, osmotic shock, and thermal lysis (Islam et al. 2017).

Cavitation

Cavitation refers to the formation and preservation of cavities or bubbles in a homogenous liquid medium. There are several ways to accomplish cavitation, such as local energy deposits on the fluid surface or sufficient liquid tension to form cavities (Caviglia 2015). Also, the use of ultrasonic vibrations to reduce the local pressure results in the formation of cavities, and the subsequent pressure loss results in cavities or bubbles to collapse. The pressure generated is around 1000 MPa (Harrison 1991). When the bubble collapses, a substantial amount of mechanical energy is released in the form of a shock wave, propagating through the medium and results in rupturing of the cell membrane. Cells are disrupted by the formation of cavities using hydrodynamic and ultrasonic technologies (Mahalanabis et al. 2009).

Osmotic Shock

When the salt concentration around the cell suddenly changes, the permeability of the cell membrane for water molecules increases due to osmosis. The cell swells and eventually bursts as a result of endosmosis, which is brought on by the surrounding solution's reduced salt concentration. Due to the membrane's delicate structure, this method works well with mammalian cells (Lopez and Hall 2022).

Thermal Lysis

Thermal Lysis, also known as the freezing/thawing method, wherein the ice formation on the cell membrane results in the breaking of the membrane. Moreover, it has been demonstrated that high temperatures can cause cell lysis. High temperatures induces the rupturing of membrane and results in the release of intracellular organelles by forcing the membrane proteins to denature. The DNA, however, could be damaged by prolonged heating. This approach takes a long time and cannot be utilized to extract biological components that are temperature-sensitive (Johnson and Hecht 1994). Due to its high cost (Fonseca and Cabral 2002), this approach is not frequently used in large-scale industrial applications.

9.4.2.2 Chemical Methods

Chemical methods involve a small change in the pH of lysis buffers to disrupt the membrane covering the cell. Detergent is an important component of the cell lysis solution as it helps in the solubilization of the membrane. Disruption of the cell membrane by chemical lysis involves two methods, alkaline lysis, and detergent lysis. In alkaline lysis, hydroxide ions are used to lyse the cell membrane (Tamura and Aotsuka 1988). This lysis buffer comprises of sodium dodecyl sulfate (SDS) and sodium hydroxide; OH- ions of sodium hydroxide interact with the cell membrane and facilitates the breaking of fatty acid and glycerol ester bond, thereby allowing the cell membrane to become permeable, whereas SDS is responsible for solubilizing the proteins embedded in and around the cell membrane. The pH range of 11.5–12.5 is mostly preferred for the lysis of cells (Feliciello and Chinali 1993; Harrison 1991). Although this method works with all cell types, the process is relatively slow and requires a time period of about 6–12 h, and is mainly used to isolate plasmid DNA from bacteria (Andrews and Asenjo 1987; Feliciello and Chinali 1993). Detergents are the most commonly used for the lysis of mammalian cells and can be employed to disrupt lipid and lipid or lipid and protein or protein and protein interactions. They can be classified as anionic, cationic, or nonionic detergents based on the charge capacity. In the case of bacterial cells, the disruption of the cell wall is the first step to reach the cell membrane, which is followed by cell lysis. Therefore, lysozymes are used to lyse bacteria cells (e.g., yeast).

9.4.2.3 Enzymatic Methods

Enzymatic method is defined as the use of enzymes for disrupting the cell. The enzymes commonly used are lysozyme, lysostaphin, zymolase, protease, cellulose, and glycanase. The majority of the enzymes listed above are commercially available with numerous applications. Enzymatic lysis is known for offering several benefits, specificity being one of them. For instance, plant cells, yeast cells, and bacterial cells are lysed using pectinase, chitinase, and lysozymes, respectively. Interaction between lysozyme and the peptidoglycan layer results in the breakage of the glycosidic bond. Therefore, Gram-positive bacteria can be provided with direct exposure to lysozymes, whereas Gram-negative bacteria should have their outer membranes removed before having their peptidoglycan layer exposed to enzymes. Treatment by lysozyme usually takes place at the value of pH between 6 and 7 and at a temperature of 35 °C (Harrison 1991). In gram-negative bacteria, the cell membrane and wall is disrupted by use of lysozyme and detergents together. The enzyme used to isolate genomic DNA is Proteinase K (Brown and Audet 2008; Salazar and Asenjo 2007).

9.4.3 Single Cell Lysis Method

With the increase in the advancement of technology, single cell lysis has gained a lot of importance due to its huge application. It helps researchers to decipher cellular heterogeneity in cell culture and holds importance in the fields of genomics, metabolomics, proteomics, and transcriptomics. Single cell lysis is the initial step for analyzing intracellular components of the cell, such as DNA, enzymes, proteins etc. Single cell lysis has been studied using varieties of technologies, including PCR, microfluidics, capillary electrophoresis, and high-speed imaging. Techniques used to lyse cells include laser pulse, acoustic, electrical, nanoscale barbs, and chemical such as detergents and enzymes. Single-cell RNA is extracted using single-cell lysis buffers that are readily accessible in the market. These buffers are designed to minimize the loss of sample during enzymatic procedures like reverse transcription (Svec et al. 2013).

9.5 Nanoparticle Induced Cell Disruption

Nanotechnology is an emerging field that is employed for the synthesis of materials that have dimensions of 1×10^{-9} m (Singh et al. 2021). Due to their significantly small size, the properties of nanoparticles are different from the constituent bulk matter. Upon reaching the nanoscale, the properties like reactivity and electronic structure vary greatly. Quantum and surface effects are the driving forces behind the unique properties of nanoparticles that differentiate them from bulk material. In the case of nanoparticles, a higher number of atoms are present at the surface than in the bulk of the material, allowing them to have both a large surface area and a high particle number per unit mass (Buzea et al. 2007). The enlarged surface area results in enhanced reactivity as it provides a more extensive area for chemical reactions to occur. Also, at the surface, the presence of atoms is reduced as compared to the bulk matter. Due to this, when the particle size decreases, the binding energy per atom is also reduced (Roduner 2006). These factors are responsible for the altered properties of nanomaterials, including chemical, optical, electric, mechanical, and magnetic.

Nanoparticles are referred as particles having dimension in nanometer Studies indicate that the size of the particle has a significant impact on the properties of the nanoparticles. There is not much of a difference observed between a microscale particle and its corresponding bulk material. But when the size goes down at the nanoscale, i.e., <100 nm, the properties of the nanoparticles significantly changes. The quantum size is responsible for the altered properties of the nanoparticles, thereby separating them from their bulk material. For instance, in the bulk state, the color of the gold nanoparticle is yellow, but when the particle size comes down at the nanoscale, the same gold nanoparticle exhibits purple color, which is entirely different from its bulk matter. This color change occurs when the band type of the gold nanoparticle changes from continuous to discrete, i.e., due to the confinement effect (Sun 2007).

9.5.1 Unique Properties of Nanoparticles

Each nanoparticle has its unique property, which defines its complex interactions with the target cell. There are a number of factors that influence the properties of nanoparticles, including their size, shape, surface area, charge, and the dissolution and aggregation of the nanoparticles (Gatoo et al. 2014). These factors are capable of either enhancing or limiting the cell permeability of the selected nanoparticle and its penetration ability while entering the biological system of living organisms.

9.5.1.1 Shape, Size, and Structure of Nanoparticles

The morphology of nanoparticles, including their shapes like spheres, truncated triangles, rods, cubes, particles, wires, etc., affects their movement, transport, and subsequent cell absorption. To support this claim, it has been demonstrated that triangular nanoplates suppress Escherichia coli more effectively than spherical or rod-shaped Ag nanoparticles, which may be a result of the triangle nanoparticles' high atom density (Baalousha 2009).

NPs features such as size, shape and charge differ from the bulk material in terms of physical and chemical characteristics due to small size and high surface-to-volume ratio. As shown below, the antibacterial activity of nanoparticles can be greatly enhanced by changing their physical and chemical properties.

The last three decades have seen remarkable development in the nanotechnology domain as the researchers were able to understand the characteristics of nanoparticles suitable for the diagnosis of disease and delivery of the drug at the targeted site. NPs are characterized as a set of tools to perform one or more functions simultaneously in the emerging field called nanomedicine. They offer exceptional opportunities to penetrate cell barriers and improve the delivery of a variety of pharmaceuticals, including variety of promising biomacromolecules such as anti-sense oligonucleotides, nucleic acids, plasmid DNA, and siRNAs (small interference RNAs) that can only perform their function once it has entered the cell and cannot perform otherwise. As far as polar molecules are concerned, they aren't capable of penetrating the biological membranes, including plasma membrane made of lipid bilayers, blood–cerebrospinal fluid barrier, gastrointestinal barrier, and air-blood barrier. NPs are considered as ideal drug carriers due to their ability to carry a high concentration of the intended therapeutic at the targeted site. Pharmacokinetics and biodistribution of formulated drugs are enhanced by large payloads on nanocarriers (Pal et al. 2007). Treatment and imaging can be performed simultaneously on the NP surface or core by loading multiple substances (Chen et al. 2011; Ng et al. 2011). NPs can also be used to monitor real-time cellular parameters, one such being the intracellular oxygen concentration within the cell.

NPs are made from various biological and non-biological materials, such as lipid-based (liposomes, micelles, and solid lipids or lipoprotein-based NPs) (Attama 2011; Breunig et al. 2008; Fercher et al. 2011; Gobbi et al. 2010), polymer-based (polylactic acid, poly[lactic co-glycolic acid], and poly[alkyl cyanoacrylate]), quantum dots, chitosan-based, silicon-based, magnetic NPs and gold NPs (Ng et al. 2011).

Since their size, shape, and surface are directly related to their inherent properties and applications, much work has been done to regulate NP synthesis. NPs are also created in various ways to address the stability problem in physiological media (Rippel and Seifalian 2011).

9.5.1.2 Surface Charge and Coatings

Surface charge is a critical factor in the dispersion properties of nanoparticles, which is crucial for binding to cell membranes, ion absorption, and subsequent cellular uptake. The ability for aggregation and dissolution is indirectly impacted by the surface coatings, which in turn raises the surface charge (Panariti et al. 2012).

9.6 Interaction of Nanoparticles with Cells

The increase in the porosity of the cell membrane is brought on by exposing the subjected cell to various types of nanoparticles and this holds applications in therapeutics and gene delivery (Fischer et al. 2003; Leroueil et al. 2007, 2008; Majumder 2015; Mecke et al. 2005; Verma et al. 2008; Zhang and Smith 2000). Also, it was observed that cells treated with non-cytotoxic polycation nanoparticles for an hour cause a release of lactate dehydrogenase (LDH) in the supernatant, and pigment dyes such as fluorescein diacetate (FDA) could pass through cell membranes that were previously impermeable (Majumder 2015). These studies at the cellular level were supported by the experiments conducted using atomic force microscopy (AFM) that investigated the development of nanoparticle-induced nanoholes in supported lipid-bilayer model membranes (Leroueil et al. 2008; Majumder 2015; Mecke et al. 2005). Theoretical studies validate the nanoholes formation to be thermodynamically feasible (Ginzburg and Balijepalli 2007). This gave rise to a hypothesis stating that the development of nanoholes in the plasma membrane of the cells promotes cell membrane leakage at non-cytotoxic concentrations (Chen et al. 2009).

The internalization of nanocarriers within the cell is approached by various mechanisms, which depends on the physiochemical properties of NPs, including shape, size, and its chemical properties (Huq et al. 2022). NPs work efficiently when soluble in physiological solutions. The route of administration decides the interaction of the

NPs with the cell plasma membrane, which paves the way for the nanoparticle to access the cell and reach the biological target site. Understanding how the physical and chemical properties of NPs affect their interactions and working with biological systems and their infiltration into cells is essential so as to maximize the potential value of nano-distribution.

Since cells may exhibit different levels of target receptors or show differences related to membrane environments in polarized cells, most types of cells can adopt multiple internalization methods at the same time.

9.7 Mechanism of Nanomaterials Induced Cell Disruption

The cell disruption techniques opened the doors for advancement in the healthcare domain but also had a series of setbacks. Researchers are now diverting their attention toward nanoparticle-based cell disruption due to the unique properties associated with them. Since the bio-efficacy is directly associated with the particle shape and size, the small size of the nanoparticles enhances easy cell permeability, thereby promoting cell disruption.

For instance, Due to their wide range of uses, AgNPs are one of the most versatile nanomaterials in large-scale industrial production around the world (Hillaireau and Couvreur 2009). Sometimes, AgNPs involve factors like monitoring their toxicity and biological effects to ensure their safer applications in various areas like human exposure through environmental or occupational processes (Hillaireau and Couvreur 2009).

Additionally, AgNPs are known for their antibacterial activity, and it has been proven in research on *V. parahaemolyticus* and *S. Typhimurium* (Champion and Mitragotri 2006). Some studies determine that AgNPs damage the cell membrane and result in the integration of membrane and reduction in the metabolic activity of cells which causes bacterial cells' death (Champion and Mitragotri 2006). It was confirmed by the study of field emission scanning electron microscopy that when Gram-positive and Gram-negative pathogenic bacteria were treated with AgNPs, The treated bacteria's altered shape and weakened cell wall were brought on by AgNPs, which also prompted cell death (Xu et al. 2008). AgNPs cause the inhibition of both protein and cell wall synthesis, which is compelling evidence that the outer cellular membrane is disrupted by proteins, and the increased ATP leakage leads to cell death (Beningo and Wang 2002). In addition to these, the AgNPs' size and shape affect their potential efficacy against bacterial illness because of their increased surface area, which increases the release of Ag + ions. The activity of cell disruption is further hampered by the dissolving rate of AgNPs; if the rate is high, the potential activity may increase (Banquy et al. 2009), or if the rate is low, the potential activity may decrease. AgNPs with a size range of less than 10 nm have the potential to enter the bacterial cells directly, crosses the cell membranes, and start cell lysis (Buyukhatipoglu and Clyne 2011).

Several investigations have already indicated that biosynthesized AgNPs have demonstrated notable antibacterial activity against many human diseases. The small size and large surface area of AgNPs make them easily interact and penetrate into the bacterial cell walls. They can also harm cell membranes, produce reactive oxygen species, block DNA replication and protein synthesis, and induce cell death (Heng et al. 2011).

The strong bonding of nanoparticles to bacterial membrane surfaces interferes with the normal functions of the cells and initiates the process of cell disruption by AgNPs (Buyukhatipoglu and Clyne 2011). Interaction of nanoparticle to cell wall allows the penetration of nanoparticle inside the cell, which disintegrates the organelles and alters its permeability (Fig. 9.2). AgNPs have an affinity to interact with sulfur or phosphorous groups present in an intracellular component like DNA and protein, thereby altering the structure and functions of the cell. They can also interact with the thiol groups of the ROS and free radicals produced by enzymes, damage the internal machinery, and trigger apoptotic cascades. Nanoparticles penetrate the cell walls, break them down, and change cell permeability (Heng et al. 2011).

The release of silver ions constantly from AgNPs can be considered as a mechanism of microbe destruction. Sulphur proteins have a high affinity for silver ions and are, therefore, they bound to the cell wall and the cytoplasmic membrane. The disruption of bacterial envelop is due to the increased permeability of cytoplasmic membrane which occurs due to interaction among sulphur proteins and silver ions. When silver ions are absorbed into the cells, respiratory enzymes are deactivated, which results in the production of reacting oxygen species and interrupts the production of adenosine triphosphate. Reactive oxygen species are the main causes of cell membrane disruption and DNA modification. Given the importance of phosphorus and sulphur in DNA, the interaction between silver ions and these elements can

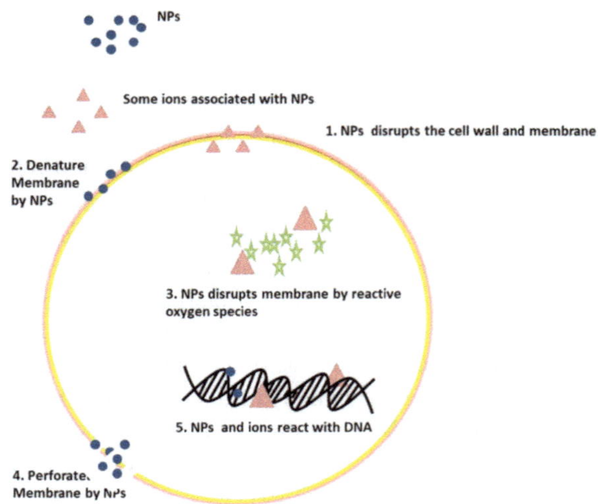

Fig. 9.2 Mechanism of nanoparticles induced cell disruption

prevent the replication of DNA, inhibit cell growth and division, and even cause the death of microbes. Additionally, silver ions denature the cytoplasmic ribosomes, thus preventing protein synthesis (Bapat et al. 2018; Khorrami et al. 2018).

9.7.1 Factors Affecting Nanomaterials Induced Cell Disruption

The various factors associated with the nanomaterials that affect the cell disruption are as follows.

9.7.1.1 Nanoparticle Size

The biodistribution, efficiency (i.e. the amount of NPs present in cells over a certain period of time), and cell absorption pathways are all influenced by particle size and their interaction and binding with target cells (Pietroiusti et al. 2018). In several applications of NP, the main objective is to prevent clearance through the reticular endothelium system, prolonging blood circulation and boost the bioavailability at the target site. NP size increases the clearance rate: it has been shown that NPs of 250 nm to 3 μm have optimal phagocytosis in vitro, whereas the size limit of about 200 nm is most likely to include other absorption pathways such as caveolin or clathrin-mediated endocytosis (Foroozandeh and Aziz 2018).

9.7.1.2 Nanoparticle Shape

Along with size, the shape is another important parameter that affects the functionality of the NPs. Although a large number of NPs industrialized for drug delivery are spherical in shape, other forms, such as cubic (cubosomes), ellipsoidal, cylindrical, and disc-shaped, have been suggested as new drug nanocarriers (Pietroiusti et al. 2018). However, it has been observed that the actin in macrophages requires a large surface area due to its complex structure, and this complexity affects the internalization of NPs. The study conducted by Hillaireau and Couvreur demonstrated spherical NPs have a faster endocytosis rate than cylinders, discs, or gold NPs, but other studies found that cylinder-shaped particles are primarily absorbed (Hillaireau and Couvreur 2009). The mobility of nanomaterials in cells is also influenced by the form of NP: rod-shaped forms move to the nucleus of microtubules, while hexagonal forms are retained in the cytoplasm (Guo et al. 2013; Laux et al. 2018).

9.7.1.3 Nanoparticle Rigidity

It has been shown that nanoparticle rigidity has a significant effect on the entry pathway. NPs rigidity refers to the nanoparticle's ability to change its shape under different physiological conditions. The NPs can be either hard or soft, meaning that they can keep the same shape and can change their shape, respectively. Banquy et al. showed that soft hydrogel NPs are internalized by macro pinocytosis, rigid NPs are internalized by a mechanism dependent on clathrin, and medium-elastic NPs have several absorption mechanisms (Banquy et al. 2009). It is generally known that when NPs are described in terms of surface charge, and as the cell plasma membrane is negatively charged, cationic NPs are absorbed more effectively than anionic and neutral NPs.

9.7.1.4 High Oxidative Stress

Oxidative stress occurs due to an imbalance between the generation of reactive oxygen species (ROS) and the detoxification capacity of cells. There are various factors involved, such as the interaction of the cell with nanoparticles which leads to oxidative stress. Various studies have discovered that nanoparticles interact with cells resulting in oxidative stress and triggering a variety of biological reactions. Moreover, in mammalian cells, the interaction between the NPs and cellular organelles results in the internalization of the NPs, followed by biological reactions (Augustine et al. 2020). Cameron et al. suggested that reactive oxygen species (ROS) are generated within cells due to interactions between nanoparticles and cells which results in oxidative stress. When nanoparticles enter the cell and interact with components of the cell, it results in the production of ROS by a variety of mechanisms, including lysosomes' destruction of the particles, interactions with cellular proteins, and the activation of inflammatory pathways (Cameron et al. 2022). High level of oxidative stress can harm cellular components like proteins, lipids, and DNA and leads to cell death and cell instability. Moreover, oxidative stress has been known to cause a number of diseases, including cancer, neurological and cardiovascular diseases (Moschini et al. 2023). Researchers are looking into ways to alter nanoparticle surfaces to make them less reactive or to provide antioxidant therapies that can block the effects of ROS to lessen the harmful effects of nanoparticles on cells (Kodali and Thrall 2015; Manke et al. 2013).

9.8 Current Trends for Nanomaterials-Induced Cell Disruption

Currently, various types of NPs are used as a drug carriers’ system, such as polymer-based NPs, solid lipid (SL) NPs, liposomal NPs, micelle-based NPs, and inorganic NPs, including silicon NPs, magnetic NPs, gold NPs, and silver NPs. Nanoparticle’s size is the main benefit as compared to traditional delivery systems, which makes it suitable for carrying antimicrobials and targeting bacteria. Due to their small size and poor membrane transport, antibiotics have little impact on intracellular bacteria. This limitation can be prevailed by using NPs for drug delivery. Since nanoparticles have unique structural characteristics and are small in size, they are ideal for drug delivery as they are easily phagocytosed by the host (Qi et al. 2013). NPs can easily penetrate into the enter host cells, leading to endocytosis and intracellular drug release which could improve serum antibiotic levels, promote drug resistance lead through bacteria, and provide security from unwanted chemical reactions (Huh and Kwon 2011). Some works suggest that different NPs could overcome the increased efflux and reduce the uptake of the drug in some bacteria, including *P. aeruginosa* and *E. coli* (Mühling et al. 2009). Antibiotics can be administered using NPs at the targeted site, thereby reducing the systemic side effects associated with conventional antibiotics. The delivery of drugs in nanocarriers can lead to the controlled release of antibiotics. Moreover, it is possible to maintain stable blood concentrations of the drug in the infected area for a prolonged period of time, thereby reducing pain. This delivery method also leads to a long release of drugs and a significant repressive effect on cell proliferation (Lee et al. 2019). In addition, to increase antibacterial effects and prevent resistance, a combination of two or more types of NPs can be performed (Liu et al. 2016). Therefore, nanomaterials are beneficial in various fields, such as antimicrobial therapy, cancer treatment, and drug delivery.

9.9 Conclusion

Within this chapter, we have summarized the types and mechanisms of cell disruption with the help of nanoparticles. Below the threshold of cytotoxicity, diverse interactions between NP and cells have an impact on various functional levels of cell physiology (mitochondria, cytoskeletal, ROS production, membrane currents, and intracellular calcium) and produce a range of tissue reactions. It is clearly shown that nanostructures not only passively but also actively alter the molecular processes that are typically necessary to regulate cellular functions.

To acquire the desired NP-mediated effect, one approach could be to mask the NP in a way that utilizes biomimicry and allow the physiological pathways to guide the pharmacological load. However, we must remember that material intake and escort between cells are mainly active processes, which could lead to increased consumption of energy. In addition, NP absorption can occur passively through cytoskeleton-induced energy gradients within cells and allowing intracellular transport to occur.

NP may stimulate certain cell functions rather than disrupt them, which can, in turn, increase the effectiveness of cellular machinery. However, the term "cellular activity" as a whole should be understood as activating or inhibiting specific pathways which are mediated by NPs. Nanoparticle-based treatment depends on how well the favorable and harmful consequences are balanced. In this case, siRNA or anti-microRNA NPs specifically suppress or disrupt cellular pathways associated with tumor development, progression, and chemotherapy resistance. The possible advantages of loaded cargo must be weighed against the potential drawbacks of NPs. Finally, the last consideration should be for the potential applications of nanoparticle-based cell disruption methods. Hereby, the fact that NPs should be viewed as playing a potential role in conciliating biological effects as target-based therapies by decreasing the shortcomings and increasing the efficiency of individual NPs should be emphasized.

References

Attama A (2011) SLN, NLC, LDC: state of the art in drug and active delivery. Recent PatS Drug Deliv Formul 5(3):178–187. https://doi.org/10.2174/187221111797200524

Andrews BA, Asenjo JA (1987) Enzymatic lysis and disruption of microbial cells. Trends Biotechnol 5(10):273–277. https://doi.org/10.1016/0167-7799(87)90058-8

Augenstein DC, Thrasher K, Sinskey AJ, Wang DIC (1974) Optimization in the recovery of a labile intracellular enzyme. Biotechnol Bioeng 16(11):1433–1447. https://doi.org/10.1002/BIT.260161102

Augustine R, Hasan A, Primavera R, Wilson RJ, Thakor AS, Kevadiya BD (2020) Cellular uptake and retention of nanoparticles: insights on particle properties and interaction with cellular components. Mater Today Commun 25:101692. https://doi.org/10.1016/J.MTCOMM.2020.101692

Baalousha M (2009) Aggregation and disaggregation of iron oxide nanoparticles: Influence of particle concentration, pH and natural organic matter. Sci Total Environ 407(6):2093–2101. https://doi.org/10.1016/J.SCITOTENV.2008.11.022

Banquy X, Suarez F, Argaw A, Rabanel JM, Grutter P, Bouchard JF, Hildgen P, Giasson S (2009) Effect of mechanical properties of hydrogel nanoparticles on macrophage cell uptake. Soft Matter 5(20):3984–3991. https://doi.org/10.1039/B821583A

Bapat RA, Chaubal TV, Joshi CP, Bapat PR, Choudhury H, Pandey M, Gorain B, Kesharwani P (2018) An overview of application of silver nanoparticles for biomaterials in dentistry. Mater Sci Engineering C, Mater Biol Appl 91:881–898. https://doi.org/10.1016/J.MSEC.2018.05.069

Bastos PAD, Wheeler R, Boneca IG (2021) Uptake, recognition and responses to peptidoglycan in the mammalian host. FEMS Microbiol Rev 45(1). https://doi.org/10.1093/FEMSRE/FUAA044

Belin BJ, Busset N, Giraud E, Molinaro A, Silipo A, Newman DK (2018) Hopanoid lipids: from membranes to plant-bacteria interactions. Nat Rev Microbiol 16(5):304–315. https://doi.org/10.1038/NRMICRO.2017.173

Beningo KA, Wang YL (2002) Fc-receptor-mediated phagocytosis is regulated by mechanical properties of the target. J Cell Sci 115(Pt 4):849–856. https://doi.org/10.1242/JCS.115.4.849

Bhardwaj V, Kaushik A (2017) Biomedical applications of nanotechnology and nanomaterials. Micromachines 8(10):298. https://doi.org/10.3390/MI8100298

Breunig M, Bauer S, Goepferich A (2008) Polymers and nanoparticles: intelligent tools for intracellular targeting? Eur J Pharm Biopharm 68(1):112–128. https://doi.org/10.1016/J.EJPB.2007.06.010

Brown RB, Audet J (2008) Current techniques for single-cell lysis. J R Soc Interface 5(Suppl 2). https://doi.org/10.1098/RSIF.2008.0009.FOCUS

Buyukhatipoglu K, Clyne AM (2011) Superparamagnetic iron oxide nanoparticles change endothelial cell morphology and mechanics via reactive oxygen species formation. J Biomed Mater Res, Part A 96(1):186–195. https://doi.org/10.1002/JBM.A.32972

Buzea C, Pacheco II, Robbie K (2007) Nanomaterials and nanoparticles: sources and toxicity. Biointerphases 2(4):MR17–MR71. https://doi.org/10.1116/1.2815690

Cameron SJ, Sheng J, Hosseinian F, Willmore WG (2022) Nanoparticle effects on stress response pathways and nanoparticle-protein interactions. Int J Mol Sci 23(14). https://doi.org/10.3390/IJMS23147962

Caviglia C (2015) Hydrodynamic cavitation as a method for cell disruption. https://digitalcommons.utep.edu/open_etd

Champion JA, Mitragotri S (2006) Role of target geometry in phagocytosis. Proc Natl Acad Sci USA 103(13):4930–4934. https://doi.org/10.1073/PNAS.0600997103

Chen J, Hessler JA, Putchakayala K, Panama BK, Khan DP, Hong S, Mullen DG, DiMaggio SC, Som A, Tew GN, Lopatin AN, Baker JR, Holl MMB, Orr BG (2009) Cationic nanoparticles induce nanoscale disruption in living cell plasma membranes. J Phys Chem B 113(32):11179–11185. https://doi.org/10.1021/JP9033936

Chen N, Zhou M, Dong X, Qu J, Gong F, Han Y, Qiu Y, Wang J, Liu Y, Wei Y, Xia J, Yu T, Zhang X, Zhang L (2020) Epidemiological and clinical characteristics of 99 cases of 2019 novel coronavirus pneumonia in Wuhan, China: a descriptive study. Lancet (london, England) 395(10223):507–513. https://doi.org/10.1016/S0140-6736(20)30211-7

Chen T, Shukoor MI, Wang R, Zhao Z, Yuan Q, Bamrungsap S, Xiong X, Tan W (2011) Smart multifunctional nanostructure for targeted cancer chemotherapy and magnetic resonance imaging. ACS Nano 5(10):7866. https://doi.org/10.1021/NN202073M

Chisti Y, Moo-Young M (1986) Disruption of microbial cells for intracellular products. Enzyme Microb Technol 8(4):194–204. https://doi.org/10.1016/0141-0229(86)90087-6

Del Prado-Audelo ML, García Kerdan I, Escutia-Guadarrama L, Reyna-González JM, Magaña JJ, Leyva-Gómez G (2021) Nanoremediation: nanomaterials and nanotechnologies for environmental cleanup. Front Environ Sci 9:645. https://doi.org/10.3389/FENVS.2021.793765/BIBTEX

Dörr T, Moynihan PJ, Mayer C (2019) Editorial: bacterial cell wall structure and dynamics. Front Microbiol 10:2051. https://doi.org/10.3389/FMICB.2019.02051/BIBTEX

Engler CR, Robinson CW (1981) Disruption of Candida utilis cells in high pressure flow devices*. Biotechnol Bioeng 23(4):765–780. https://doi.org/10.1002/BIT.260230408

Etheridge ML, Campbell SA, Erdman AG, Haynes CL, Wolf SM, McCullough J (2013) The big picture on nanomedicine: the state of investigational and approved nanomedicine products. Nanomed Nanotechnol Biol Med 9(1):1–14. https://doi.org/10.1016/J.NANO.2012.05.013

Feliciello I, Chinali G (1993) A modified alkaline lysis method for the preparation of highly purified plasmid DNA from escherichia coli. Anal Biochem 212(2):394–401. https://doi.org/10.1006/ABIO.1993.1346

Fercher A, Borisov SM, Zhdanov AV, Klimant I, Papkovsky DB (2011) Intracellular O_2 sensing probe based on cell-penetrating phosphorescent nanoparticles. ACS Nano 5(7):5499–5508. https://doi.org/10.1021/NN200807G/SUPPL_FILE/NN200807G_SI_001.PDF

Fischer D, Li Y, Ahlemeyer B, Krieglstein J, Kissel T (2003) In vitro cytotoxicity testing of polycations: Influence of polymer structure on cell viability and hemolysis. Biomaterials 24(7):1121–1131. https://doi.org/10.1016/S0142-9612(02)00445-3

Fonseca LP, Cabral JMS (2002) Penicillin acylase release from Escherichia coli cells by mechanical cell disruption and permeabilization. J Chem Technol Biotechnol 77(2):159–167. https://doi.org/10.1002/JCTB.541

Foroozandeh P, Aziz AA (2018) Insight into cellular uptake and intracellular trafficking of nanoparticles. Nanoscale Res Lett 13(1):1–12. https://doi.org/10.1186/S11671-018-2728-6

Gatoo MA, Naseem S, Arfat MY, Mahmood Dar A, Qasim K, Zubair S (2014) Physicochemical properties of nanomaterials: implication in associated toxic manifestations. BioMed Res Int. https://doi.org/10.1155/2014/498420

Ginzburg VV, Balijepalli S (2007) Modeling the thermodynamics of the interaction of nanoparticles with cell membranes. Nano Lett 7(12):3716–3722. https://doi.org/10.1021/NL072053L

Gobbi M, Re F, Canovi M, Beeg M, Gregori M, Sesana S, Sonnino S, Brogioli D, Musicanti C, Gasco P, Salmona M, Masserini ME (2010) Lipid-based nanoparticles with high binding affinity for amyloid-beta1-42 peptide. Biomaterials 31(25):6519–6529. https://doi.org/10.1016/J.BIOMATERIALS.2010.04.044

Goldberg S (2008) Mechanical/physical methods of cell disruption and tissue homogenization. Methods Mol Biol (Clifton, N.J.) 424:3–22. https://doi.org/10.1007/978-1-60327-064-9_1

Gomes TA, Zanette CM, Spier MR (2020) An overview of cell disruption methods for intracellular biomolecules recovery. Prep Biochem Biotechnol 50(7):635–654. https://doi.org/10.1080/10826068.2020.1728696

Guo M, Xu X, Yan X, Wang S, Gao S, Zhu S (2013) In vivo biodistribution and synergistic toxicity of silica nanoparticles and cadmium chloride in mice. J Hazard Mater 260:780–788. https://doi.org/10.1016/J.JHAZMAT.2013.06.040

Harrison STL (1991) Bacterial cell disruption: a key unit operation in the recovery of intracellular products. Biotechnol Adv 9(2):217–240. https://doi.org/10.1016/0734-9750(91)90005-G

Heng BC, Zhao X, Tan EC, Khamis N, Assodani A, Xiong S, Ruedl C, Ng KW, Loo JSC (2011) Evaluation of the cytotoxic and inflammatory potential of differentially shaped zinc oxide nanoparticles. Arch Toxicol 85(12):1517–1528. https://doi.org/10.1007/S00204-011-0722-1

Hillaireau H, Couvreur P (2009) Nanocarriers' entry into the cell: relevance to drug delivery. Cell Mol Life Sci: CMLS 66(17):2873–2896. https://doi.org/10.1007/S00018-009-0053-Z

Huh AJ, Kwon YJ (2011) "Nanoantibiotics": a new paradigm for treating infectious diseases using nanomaterials in the antibiotics resistant era. J Control Release: off J Control Release Soc 156(2):128–145. https://doi.org/10.1016/J.JCONREL.2011.07.002

Huq MA, Ashrafudoulla M, Rahman MM, Balusamy SR, Akter S (2022) green synthesis and potential antibacterial applications of bioactive silver nanoparticles: a review. Polymers 14(4):742. https://doi.org/10.3390/POLYM14040742

Iravani S (2020) Nanomaterials and nanotechnology for water treatment: recent advances. 51(12):1615–1645. https://doi.org/10.1080/24701556.2020.1852253

Islam MS, Aryasomayajula A, Selvaganapathy PR (2017) A review on macroscale and microscale cell lysis methods. Micromachines 8(3). https://doi.org/10.3390/MI8030083

Johnson BH, Hecht MH (1994) Recombinant proteins can be isolated from E. coli cells by repeated cycles of freezing and thawing. Bio/Technology (Nature Publishing Company) 12(13):1357–1360. https://doi.org/10.1038/NBT1294-1357

Kaushik AK, Dixit CK (2016) Nanobiotechnology for sensing applications : from lab to field

Khan YS, Farhana A (2022) Histology, cell. StatPearls. https://www.ncbi.nlm.nih.gov/books/NBK554382/

Khorrami S, Zarrabi A, Khaleghi M, Danaei M, Mozafari MR (2018) Selective cytotoxicity of green synthesized silver nanoparticles against the MCF-7 tumor cell line and their enhanced antioxidant and antimicrobial properties. Int J Nanomed 13:8013–8024. https://doi.org/10.2147/IJN.S189295

Kodali V, Thrall BD (2015) Oxidative stress and nanomaterial-cellular interactions. Stud Exp Toxicol Pharmacol 347–367. https://doi.org/10.1007/978-3-319-19096-9_18/COVER

Kumar V, Arora N, Nanda M, Pruthi V (2019) Different cell disruption and lipid extraction methods from microalgae for biodiesel production. Microalgae Biotechnol Dev Biofuel Wastewater Treat 265–292. https://doi.org/10.1007/978-981-13-2264-8_12

Laux P, Tentschert J, Riebeling C, Braeuning A, Creutzenberg O, Epp A, Fessard V, Haas KH, Haase A, Hund-Rinke K, Jakubowski N, Kearns P, Lampen A, Rauscher H, Schoonjans R, Störmer

A, Thielmann A, Mühle U, Luch A (2018) Nanomaterials: certain aspects of application, risk assessment and risk communication. Arch Toxicol 92(1):121–141. https://doi.org/10.1007/S00204-017-2144-1

Lee NY, Ko WC, Hsueh PR (2019) Nanoparticles in the treatment of infections caused by multidrug-resistant organisms. Front Pharmacol 10:1153. https://doi.org/10.3389/FPHAR.2019.01153/BIBTEX

Leroueil PR, Berry SA, Duthie K, Han G, Rotello VM, McNerny DQ, Baker JR, Orr BG, Holl MMB (2008) Wide varieties of cationic nanoparticles induce defects in supported lipid bilayers. Nano Lett 8(2):420–424. https://doi.org/10.1021/NL0722929

Leroueil PR, Hong S, Mecke A, Baker JR, Orr BG, Holl MMB (2007) Nanoparticle interaction with biological membranes: does nanotechnology present a janus face? Acc Chem Res 40(5):335–342. https://doi.org/10.1021/AR600012Y

Liu JL, Zhang WJ, Li XD, Yang N, Pan WS, Kong J, Zhang JS (2016) Sustained-release genistein from nanostructured lipid carrier suppresses human lens epithelial cell growth. Int J Ophthalmol 9(5):643. https://doi.org/10.18240/IJO.2016.05.01

Lopez MJ, Hall CA (2022) Physiology, osmosis. StatPearls. https://www.ncbi.nlm.nih.gov/books/NBK557609/

Mahalanabis M, Al-Muayad H, Kulinski MD, Altman D, Klapperich CM (2009) Cell lysis and DNA extraction of gram-positive and gram-negative bacteria from whole blood in a disposable microfluidic chip. Lab Chip 9(19):2811–2817. https://doi.org/10.1039/B905065P

Majumder N (2015) Click chemistry in nano drug delivery system and its applications in biology. Int J Res Pharm Chem 5(1):95–105. www.ijrpc.com

Manke A, Wang L, Rojanasakul Y (2013) Mechanisms of nanoparticle-induced oxidative stress and toxicity. BioMed Res Int. https://doi.org/10.1155/2013/942916

Mecke A, Majoros IJ, Patri AK, Baker JR, Banaszak Holl MM, Orr BG (2005) Lipid bilayer disruption by polycationic polymers: the roles of size and chemical functional group. Langmuir: ACS J Surf Colloids 21(23):10348–10354. https://doi.org/10.1021/LA050629L

Middelberg APJ (2000) 2 microbial cell disruption by high-pressure homogenization, pp 11–21. https://doi.org/10.1007/978-1-59259-027-8_2

Moschini E, Colombo G, Chirico G, Capitani G, Dalle-Donne I, Mantecca P (2023) Biological mechanism of cell oxidative stress and death during short-term exposure to nano CuO. Sci Rep 13(1):1–18. https://doi.org/10.1038/s41598-023-28958-6

Mühling M, Bradford A, Readman JW, Somerfield PJ, Handy RD (2009) An investigation into the effects of silver nanoparticles on antibiotic resistance of naturally occurring bacteria in an estuarine sediment. Mar Environ Res 68(5):278–283. https://doi.org/10.1016/J.MARENVRES.2009.07.001

Ng KK, Lovell JF, Zheng G (2011) Lipoprotein-inspired nanoparticles for cancer theranostics. Acc Chem Res 44(10):1105. https://doi.org/10.1021/AR200017E

Pal S, Tak YK, Song JM (2007) Does the antibacterial activity of silver nanoparticles depend on the shape of the nanoparticle? a study of the Gram-negative bacterium Escherichia coli. Appl Environ Microbiol 73(6):1712–1720. https://doi.org/10.1128/AEM.02218-06

Panariti A, Miserocchi G, Rivolta I (2012) The effect of nanoparticle uptake on cellular behavior: disrupting or enabling functions? Nanotechnol Sci Appl 5(1):87. https://doi.org/10.2147/NSA.S25515

Patel J, Patel A, Patel M, Vyas G (2022) Introduction to nanoparticulate drug delivery systems. Pharmacokinet Pharmacodyn Nanoparticulate Drug Deliv Syst 3–23. https://doi.org/10.1007/978-3-030-83395-4_1

Peternel Š (2013) Bacterial cell disruption: a crucial step in protein production. New Biotechnol 30(2):250–254. https://doi.org/10.1016/J.NBT.2011.09.005

Pietroiusti A, Stockmann-Juvala H, Lucaroni F, Savolainen K (2018) Nanomaterial exposure, toxicity, and impact on human health. Wiley Interdiscip Rev Nanomedicine Nanobiotechnology 10(5). https://doi.org/10.1002/WNAN.1513

Qi G, Li L, Yu F, Wang H (2013) Vancomycin-modified mesoporous silica nanoparticles for selective recognition and killing of pathogenic gram-positive bacteria over macrophage-like cells. ACS Appl Mater Interfaces 5(21):10874–10881. https://doi.org/10.1021/AM403940D

Rippel RA, Seifalian AM (2011) Gold revolution–gold nanoparticles for modern medicine and surgery. J Nanosci Nanotechnol 11(5):3740–3748. https://doi.org/10.1166/JNN.2011.4170

Roduner E (2006) Size matters: why nanomaterials are different. Chem Soc Rev 35(7):583–592. https://doi.org/10.1039/B502142C

Rosario-Meléndez R, Harris CL, Delgado-Rivera R, Yu L, Uhrich KE (2012) PolyMorphine: an innovative biodegradable polymer drug for extended pain relief. J Control Release: off J Control Release Soc 162(3):538–544. https://doi.org/10.1016/J.JCONREL.2012.07.033

Sakmann B, Neher E (1984) Patch clamp techniques for studying ionic channels in excitable membranes. Annu Rev Physiol 46:455–472. https://doi.org/10.1146/ANNUREV.PH.46.030184.002323

Salazar O, Asenjo JA (2007) Enzymatic lysis of microbial cells. Biotech Lett 29(7):985–994. https://doi.org/10.1007/S10529-007-9345-2

Schütte H, Kroner KH, Hustedt H, Kula MR (1983) Experiences with a 20 litre industrial bead mill for the disruption of microorganisms. Enzyme Microb Technol 5(2):143–148. https://doi.org/10.1016/0141-0229(83)90050-9

Silhavy TJ, Kahne D, Walker S (2010) The bacterial cell envelope. Cold Spring Harb Perspect Biol 2(5). https://doi.org/10.1101/CSHPERSPECT.A000414

Singh YD, Ningthoujam R, Panda MK, Jena B, Babu PJ, Mishra AK (2021) Insight from nanomaterials and nanotechnology towards COVID-19. Sens Int 2:100099. https://doi.org/10.1016/J.SINTL.2021.100099

Srikanth M, Kessler JA (2012) Nanotechnology-novel therapeutics for CNS disorders. Nat Rev Neurol 8(6):307–318. https://doi.org/10.1038/NRNEUROL.2012.76

Stackebrandt E, Goodfellow M (1991) Nucleic acid techniques in bacterial systematics. Nucl Acid Tech Bact Syst 115–175. https://doi.org/10.3/JQUERY-UI.JS

Sun-Waterhouse D, Waterhouse GIN (2016) Recent advances in the application of nanomaterials and nanotechnology in food research. Nov Approaches Nanotechnol Food 21–66. https://doi.org/10.1016/B978-0-12-804308-0.00002-9

Sun CQ (2007) Size dependence of nanostructures: impact of bond order deficiency. Prog Solid State Chem 35(1):1–159. https://doi.org/10.1016/J.PROGSOLIDSTCHEM.2006.03.001

Svec D, Andersson D, Pekny M, Sjöback R, Kubista M, Ståhlberg A (2013). Direct cell lysis for single-cell gene expression profiling. Front Oncol 274. https://doi.org/10.3389/FONC.2013.00274/ABSTRACT

Tamura K, Aotsuka T (1988) Rapid isolation method of animal mitochondrial DNA by the alkaline lysis procedure. Biochem Genet 26(11–12):815–819. https://doi.org/10.1007/BF02395525

Taskova RM, Zorn H, Krings U, Bouws H, Berger RG (2006) A comparison of cell wall disruption techniques for the isolation of intracellular metabolites from Pleurotus and Lepista sp. Zeitschrift Fur Naturforschung. C, J Biosci 61(5–6):347–350. https://doi.org/10.1515/ZNC-2006-5-608

Van Den Berg A, Craighead H, Yang P, Mark D, Haeberle S, Gu¨nter A, Roth G, Felix Von Stettenz A, Zengerlez R (2010) Microfluidic lab-on-a-chip platforms: requirements, characteristics and applications. Chem Soc Rev 39(3):1153–1182. https://doi.org/10.1039/B820557B

Vellai T, Vida G (1999) The origin of eukaryotes: the difference between prokaryotic and eukaryotic cells. Proc R Soc B: Biol Sci 266(1428):1571. https://doi.org/10.1098/RSPB.1999.0817

Verma A, Uzun O, Hu Y, Hu Y, Han HS, Watson N, Chen S, Irvine DJ, Stellacci F (2008) Surface-structure-regulated cell-membrane penetration by monolayer-protected nanoparticles. Nat Mater 7(7):588–595. https://doi.org/10.1038/NMAT2202

Weiss C, Carriere M, Fusco L, Fusco L, Capua I, Regla-Nava JA, Pasquali M, Pasquali M, Pasquali M, Scott JA, Vitale F, Vitale F, Unal MA, Mattevi C, Bedognetti D, Merkoçi A, Merkoçi A, Tasciotti E, Tasciotti E et al (2020) Toward nanotechnology-enabled approaches against the COVID-19 pandemic. ACS Nano 14(6):6383–6406. https://doi.org/10.1021/ACSNANO.0C03697

Xu ZP, Niebert M, Porazik K, Walker TL, Cooper HM, Middelberg APJ, Gray PP, Bartlett PF, Lu GQ (Max) (2008) Subcellular compartment targeting of layered double hydroxide nanoparticles. J Control Release: Off J Control Release Soc 130(1):86–94. https://doi.org/10.1016/J.JCONREL.2008.05.021

Yang NJ, Hinner MJ (2015) Getting across the cell membrane: an overview for small molecules, peptides, and proteins. Methods Mol Biol (Clifton, N.J.) 1266:29–53. https://doi.org/10.1007/978-1-4939-2272-7_3

Zhang ZY, Smith BD (2000) High-generation polycationic dendrimers are unusually effective at disrupting anionic vesicles: membrane bending model. Bioconjug Chem 11(6):805–814. https://doi.org/10.1021/BC000018Z

Chapter 10
Biogenic Zinc Oxide Nanoparticles: Mechanism and Environmental Applications

Khalida Bloch, Sirikanjana Thongmee, and Sougata Ghosh

Abstract Biogenic synthesis of nanoparticles is more rapid, eco-friendly and cost-effective technique that do not involve any harmful chemical during synthesis and stabilization. Various viruses, bacteria, fungi, algae and medicinal plants are used for developing green route for biological synthesis of nanoparticles. Although biogenic metal nanoparticles include gold, silver, copper, platinum, palladium and others, zinc oxide nanoparticles (ZnONPs) have gained more attention for their biocompatibility and photocatalytic potential. This chapter discusses about biogenic synthesis of ZnONPs along with its environmental applications. Several bacteria such as *Bacillus licheniformis, Bacillus subtilis, Pseudochrobacterum* sp., *Rhodococcus pyridinivorans*, and *Serratia nematodiphila* can synthesize ZnONPs which are flower-like, hair-like, spherical or hexagonal in shape. Similarly, fungi such as, *Acremonium potronii, Alternaria tenuissima*, and *Aspergillus niger*, can also synthesize ZnONPs which can photocatalytically degrade hazardous dyes like methylene blue and bismarck brown under ultraviolet and/or visible light. Further, algal extracts of *Chlorella*, *Pterocladia capillacea, Sargassum muticum*, and *Ulva lactuca* are employed to fabricate anisotropic ZnONPs with exotic shapes that includes rod, flower, triangle, hexagon and spheres which are mostly in size ranging from 10 to 50 nm with few exceptions. Similarly, plants such as, *Abelmoschus esculentus, Allium cepa, Artocarpus heterophyllus, Azadirachta indica, Pheonix dactylifera, Prosopis juliflora, Ruellia tubersoa*, and *Salvia officinalis* with rich phytochemical diversity can efficiently synthesize ZnONPs with high photocatalytic dye degrading potential metanil yellow, methyl orange, rhodamine B, reactive blue 21, turquoise blue, and others. The detailed mechanism behind bacteriogenic, mycogenic, phycogenic and phytogenic ZnONPs synthesis is also discussed along with the future perspectives.

K. Bloch · S. Ghosh (✉)
Department of Microbiology, School of Science, RK. University, Rajkot, Gujarat, India
e-mail: ghoshsibb@gmail.com

S. Thongmee · S. Ghosh
Department of Physics, Faculty of Science, Kasetsart University, Bangkok, Thailand

M. P. Shah et al. (eds.), *Biogenic Nanomaterials for Environmental Sustainability: Principles, Practices, and Opportunities*, Environmental Science and Engineering,
https://doi.org/10.1007/978-3-031-45956-6_10

Keywords Zinc oxide nanoparticles · Biogenic synthesis · Bacteria · Fungi · Algae · Plants · Mechanism · Applications

10.1 Introduction

With the increase in urbanization and industrialization, various organic, inorganic, and biological contaminants directly or indirectly enter the water bodies which lead to environmental pollution. This is a major concern as it has become the main cause of various diseases in humans, live stocks and plants across the globe. Various physical, chemical, and biological methods are employed to treat these polluted bodies. Due to several drawbacks in these methods, a new alternative way is required to overcome this problem (Chahal et al. 2019; Hu et al. 2020). Organic dyes are one of the major wastewater pollutants. The dumping of azo dyes leads to toxicity. These dyes accumulate in water and soil and enter the food chain. The breakdown of dyes results in the formation of carcinogenic agents. Hence traditional methods used to treat them are ineffective. The use of nanotechnology is an advanced method for the degradation of dyes. Various types of metal oxides such as titanium oxide (TiO_2), indium oxide (In_2O_3), silver (Ag), gold (Au), tin (IV) oxide, and many more are synthesized using chemical, physical and biological methods (Piccinno et al. 2012). But ZnO with wide range of radiation adsorption capability exhibits unique piezoelectric, and catalytic properties suitable for developing semiconductors (Sharma et al. 2011). Due to its nontoxic nature, it is safe to use in formulations for developing novel therapeutics for animals and humans. Hence, ZnONPs have been of greater interest. The green process of synthesizing NPs over conventional methods is accepted widely. Several biological entities such as bacteria, yeast, fungi, algae, and plants are used for the synthesis of ZnONPs as these processes are eco-friendly, less toxic, and cost-effective. Microbes act as mini-factories in the reduction of metal ions as they possess enzymes and bioactive compounds which act as reducing, capping, and stabilizing agents during formation of NPs (Kundu et al. 2014; Moghaddam et al. 2017). Different parts of plants like leaf, roots, stem, fruit, and seed are also used for the synthesis of ZnONPs as they are rich in phytochemicals which aid in oxidation and reduction of the metal ions. The water-soluble polyphenols, reducing sugars, flavonoids, amines, and proteins facilitate the stabilization of NPs (Rahman et al. 2021). ZnONPs exhibit potent photocatalytic degradation activity against several dyes such as methylene blue, methyl orange, malachite green, and eosin yellow as listed in Table 10.1. The following section emphasizes the synthesis, mechanism, and removal of hazardous pollutants from the environment using ZnONPs synthesized from bacteria, fungi, algae, and plants.

Table 10.1 Biogenic ZnONPs and their applications in photocatalytic dye degradation

Biogenic source	Shape	Size	Pollutant	Irradiation	References
Bacteria					
Bacillus licheniformis MTCC 9555	Nanoflowers	620 nm	Methylene blue	Ultraviolet light	Tripathi et al. (2014)
Bacillus subtilis	Hair like structure	Diameter 1.2–1.3 μm, width 10–15 nm	Methylene blue	Sunlight, ultraviolet light	Dhandapani et al. (2020)
Pseudochrobacterum sp. C5	–	90–110 nm	Methylene blue, brilliant blue R, brilliant yellow, reactive black 15, and reactive red 120	Sunlight	Siddique et al. (2021)
Rhodococcus pyridinivorans NT2	Quasi-spherical to hexagonal	100–120 nm	Malachite green	Ultraviolet light	Kundu et al. (2014)
Serratia nematodiphila ZTB15	Spherical	15–30 nm	Methyl orange	Ultraviolet light	Jain et al. (2020)
Fungi					
Acremonium potronii	Spherical	13–15 nm	Methylene blue	Ultraviolet light	Ameen et al. (2021)
Alternaria tenuissima AUMC10624	Spherical	15.62 ± 4.51 nm	Methylene blue	Sunlight	Abdelhakim et al. (2020)
Aspergillus niger	Spherical	61 ± 0.65 nm	Bismarck brown	–	Kalpana et al. (2018)
Algae					
Chlorella	Hexagonal	20–50 nm	Dibenzothiophene	Ultraviolet light	Khalafi et al. (2019)
Pterocladia capillacea	Rod and flower shape	1 μm	Ismate violet 2R	–	Mansour et al. (2022)
Sargassum muticum	Spherical	15–50 nm	Methylene blue	Ultraviolet light and sunlight	Subramanian et al. (2021)

(continued)

Table 10.1 (continued)

Biogenic source	Shape	Size	Pollutant	Irradiation	References
Ulva lactuca	Triangles, hexagons, rods, and rectangles	10–50 nm	Methylene blue	Sunlight	Ishwarya et al. (2018)
Plant					
Abelmoschus esculentus	Spherical	20–45 nm	Methylene blue and methyl orange	Ultraviolet light	Mirgane et al. (2020)
Allium cepa	Hexagonal	~ 50 nm	Crystal violet and methylene blue	Sunlight and ultraviolet light	Rajkumar et al. (2019)
Artocarpus heterophyllus	Hexagonal	~ 15–25 nm	Rose bengal	Ultraviolet light	Vidya et al. (2016)
Azadirachta indica	Spherical	9.6–25.5 nm	Methylene blue	Ultraviolet light	Bhuyan et al. (2015)
Citrus limon	semispherical	> 50 nm	Methylene blue, methyl orange, methyl red, turquoise blue, and reactive blue 21	Ultraviolet light	Davar et al. (2015)
Coconut husk	Hexagonal bipyramid	–	Methylene blue and metanil yellow	Ultraviolet light and visible light	Priyadharshini et al. (2022)
Kalopanax septemlobus	Flower-like shape	500 nm	Methylene blue	Ultraviolet light	Lu et al. (2018)
Pheonix dactylifera	Spherical	–	Methylene blue and eosin yellow	Ultraviolet light	Rambabu et al. (2020)
Prosopis juliflora	Spherical	< 75 nm	Paper mill effluent	Sunlight	Abbas et al. (2021)
Ruellia tubersoa	Rod shape	40–50 nm	Malachite green and methylene blue	Sunlight	Vasantharaj et al. (2021)
Salvia officinalis	–	26.14 nm	Methyl orange	Ultraviolet light	Abomuti et al. (2021)

10.2 Bacteriogenic ZnONPs

Several bacteria with nanobiotechnological potential are used for synthesis of ZnONPs (Ghosh and Webster 2021). The bacterial biomass of *Bacillus licheniformis* MTCC 9555 was used for the synthesis of zinc oxide (ZnO) nanoflowers (Tripathi et al. 2014). Wet biomass (5 g) was reacted with zinc acetate dihydrate (Zn $(CH_3COO)_2$. $2H_2O$) (0.2 M) prepared in 50 mL deionized water which was heated at 60° C for 15 min along with sodium carbonate (0.6 M). The reaction mixture was incubated for 48 h at 37 ± 1 °C under continuous stirring at 200 rpm. The formation of white precipitate indicated the formation of ZnO nanoflowers. The nanoflowers were characterized using X-ray diffraction (XRD). The diffraction peaks corresponding to (100), (002), (101), (102), (110), and (103) sets of lattice planes indicated hexagonal wurtzite structure with crystalline nature of pure ZnO. Transmission electron microscope (TEM) images showed a flower-like structure of nanoparticles with sizes ranging from 200 nm to 1 μm. The nano petals of nanoflowers were 40 nm in width and 400 nm in length. The average size was 620 nm. The nanostructures were further sonicated for 40 min and the micrograph showed a rod-like structure indicating that the formation of nanoflowers was the result of the agglomeration of rods. Energy dispersive X-ray spectroscopy (EDS) revealed the presence of zinc (Zn) and oxygen (O). Fourier transform infrared spectroscopy (FTIR) analysis showed two peaks occurring at 3371.93 and 3400.97 cm^{-1} corresponding to N–H bond stretching. The change in peaks of –N–H and COO^- as compared to peaks obtained in biomass gave evidence several functional residues and proteins present in the biomass plays important role in the stability of nanoflowers. The photocatalytic degradation activity of ZnO nanoflowers was checked against MB ($C_{16}H_{18}N_3SCI$) under UV irradiation. The MB dye reacted with ZnO nanoflowers for 60 min. The absorption spectra of MB were taken at regular time intervals in presence of a photocatalyst. The decrease in intensity was observed with an increase in irradiation time. The degradation of MB was obtained up to 83% with a reaction rate of 4.4×10^{-3} s^{-1}. The generation of electron–hole pairs by ZnO nanoflowers was responsible for the degradation of MB dye. The degradation efficiency (74%) was achieved up to 3 recycles which indicated good photostability of ZnO nanoflowers.

In another study, the synthesis of ZnONPs was carried out using the ureolytic bacteria *Bacillus subtilis* HM475276 (Dhandapani et al. 2020). The ammonium carbonate was generated from synthetic urine (calcium carbonate 0.651 g/L, magnesium chloride 0.651 g/L, potassium chloride 1.6 g/L, sodium chloride 4.6 g/L, sodium sulfate 2.3 g/L, ammonium chloride 1.0 g/L, creatinine 1.1 g/L, tryptone 10 g/L, potassium phosphate 2.8 g/L and urea 40%) using *B. subtilis* as biocatalyst. Zinc acetate (0.2 M) was prepared in 300 mL of deionized water. Further 20 mL of an activated ammonium carbonate solution was added to it until pH reached 8.5 followed by stirring at 1000 rpm for 30 min. The reaction mixture was vigorously stirred for 3 h at 80° C. Formation of white precipitate indicated the synthesis of ZnONPs. The precipitates were heated for 3 h from 37 to 250 °C for annealing. The synthesized nanoparticles showed a maximum absorption peak at 385 nm which was close to the solar spectrum hence it can show efficient photocatalysis of dyes. The increase in high-intensity diffraction peaks was observed in ZnONPs as annealing temperature was increased which was characterized using XRD analysis. It showed the hexagonal wurtzite structure of ZnO. FTIR analysis showed peaks at 1570 cm^{-1}, 1507 cm^{-1}, 1391 cm^{-1}, and 1059 cm^{-1} indicating the presence of zinc ammonium carbonate hydroxide. ZnO crystal growth was confirmed as a Zn–O stretching vibration was observed in FTIR analysis. TEM analysis revealed that the nanoparticles were hair-like with a diameter of 1.2–1.3 μm and a width of 10–15 nm. Brunauer–Emmett–Teller (BET) analysis showed that the surface area was 30.22 m^2/g. Electron paramagnetic resonance (EPR) spectroscopy was conducted to analyze the reactive oxygen species generated from ZnO particles when exposed to sunlight, UV light C (365 nm), visible light, and dark conditions for 15 min. Under sunlight, four different peaks were observed indicating the generation of hydroxyl radical from ZnO photocatalyst. There was no formation of hydroxyl radical under dark conditions. No significant changes in peaks were observed under UV light. Hence ZnO was found to be effective in sunlight which aids in the decolorization of dye. Photocurrent measurements of hairy-shaped ZnO nanoparticles were estimated at 7.76 $\mu A\ cm^{-2}$. Photocatalytic activity of ZnONPs was checked against MB dye under sunlight. Sunlight exposed reaction medium showed a decrease in MB dye absorption band. The decolorization of MB dye was obtained at 50 min. The decolorization efficiency (30%) was obtained after 120 min when exposed to visible light. The efficiency was higher in solar light as compared to UV light C.

The bacterial strain *Pseudochrobacterum* sp. C5 was grown in a nutrient broth medium and was used for the synthesis of ZnONPs (Siddique et al. 2021). Bacterial culture (50 mL) was mixed with 0.003 M zinc acetate salt. The reaction mixture was allowed to incubate at 28 °C, 180 rpm. The cell-free supernatant was dried at 85 °C and the powder was calcinated at 700 °C for 7 h in a muffle furnace. Off-white color precipitate indicated the formation of ZnONPs. UV–Vis spectra of ZnONPs showed an absorption peak at 372 nm. The synthesized NPs showed a zeta potential value of −27.41 mV indicating the monodispersed nanoparticles. The ZnONPs were agglomerated with sizes ranging from 90 to 110 nm which was confirmed using a field emission scanning electron microscope (FESEM). The high surface area in the biologically synthesized NPs may result in better efficiency in the degradation of dye. FTIR analysis showed peaks at 3740 cm^{-1}, 1644 cm^{-1}, 1429 cm^{-1}, and 1013 cm^{-1} corresponding to OH stretching, C = C stretching and C–C stretching of alkanes. ZnONPs exhibited characteristic peaks at $2\theta = 32.3, 35.2, 37, 48.3$ corresponding to (100), (002), (101), (102) planes. Catalytic degradation of MB and 4-nitrophenol (4-NP) was carried out using ZnONPs. A decrease in absorption spectra was observed when an aqueous solution of MB and 4-NP was exposed to ZnONPs. Biologically synthesized ZnONPs showed more efficient decolorization of MB and 4-nitrophenol. Due to smaller surface area and size, ZnONPs showed degradation within 480 min and 120 min in the case of MB and 4-NP, respectively. Decolorization of MB (86.9%) and 4-NP (44.9%) was obtained after 480 min and 280 min, respectively. Photocatalytic degradation of azo dyes such as brilliant blue R, brilliant yellow, reactive black 5, and reactive red 120 was carried out using

ZnONPs as biocatalysts. After 10 h, 83.2%, 83.1%, 88.8%, and 95.2% decolorization was obtained in the case of brilliant blue R, brilliant yellow, reactive red 120, and reactive black 5, respectively. Textile wastewater was also treated with ZnONPs and various parameters such as dye decolorization, chemical oxygen demand (COD) removal, total dissolved solids (TDS), pH, and electrical conductivity (EC) were tested. Decolorization of reactive black 5 and reactive red 120 using ZnONPs as catalysts were up to 78.3 ± 3.4 and 72.9 ± 2.6%, respectively. A total of 62.7 ± 3.1% of COD removal was observed when wastewater spiked with reactive black 5 was treated with ZnONPs. A decrease in pH, EC, and TDS of wastewater was observed when ZnONPs were added to it. Hence bacteriogenic ZnONPs serves as good biocatalyst for the treatment of dye-loaded textile wastewater.

Rapidly synthesized ZnONPs using actinobacteria *Rhodococcus pyridinivorans* NT2 isolated from effluent collected from the pesticide industry was reported in another study (Kundu et al. 2014). The organism was grown in mineral salt basal (MSB) broth supplemented with zinc sulfate ($ZnSO_4.H_2O$). Pure bacterial strain inoculated in MSB broth was added to 20 mL of 0.1 M $ZnSO_4.H_2O$. The reaction mixture was incubated under shaking at 180 rpm for 72 h at 30 °C. Formation of white precipitates from the colorless reaction mixture indicated the formation of NPs. It was further centrifuged, filtered, and washed with water followed by drying in a hot air oven for 4 h at 70 °C. Cell-free supernatant containing extracellular proteins were also used for the synthesis of ZnONPs. The cell-free supernatant was added to the same concentration of $ZnSO_4.H_2O$ and was incubated in the same conditions. UV–Vis spectroscopy exhibited an absorption peak at 373 nm. Photoluminescence (PL) spectroscopy was also recorded. An emission peak at 355 nm was obtained indicating the presence of tyrosine residues. Another distinct peak at 520 nm was observed indicating a string green emission band of ZnONPs. XRD analysis supported that the synthesized ZnONPs were hexagonal with wurtzite structures. FTIR analysis proved that intra and extracellular proteins were present on the surface of ZnONPs. Bioactive compounds such as amino, sulfate, carboxyl, and hydroxyl groups may play role in the formation, stabilization, and capping of ZnONPs. The presence of protein prevents agglomeration. The ZnONPs were quasi-spherical to hexagonal in shape with sizes ranging from 100 to 120 nm. EDS confirmed the purity of nanoparticles with no trace amount of impurities. The ZnONPs showed -15.5 mV of zeta potential value indicating moderate stability. The thermal and thermal gravimetric analysis confirmed the stability. At ~340 °C and 550 °C, 21% of the residual mass was removed due to the decomposition of organic groups. Further coating of ZnONPs on cotton fibers, its UV-blocking property, and photocatalytic studies were carried out. XRD analysis showed the hexagonal wurtzite phase of ZnO. FESEM was used to observe the interaction between the cotton fabric and NPs. The cotton fabric treated with ZnONPs exhibited a uniform and dense distribution of ZnONPs. EDS confirmed the elemental composition of the ZnONPs coated cotton fabric. UV protective properties of ZnONPs coated cotton fabric was checked by measuring transmission covering UV-A, UV-B, and UV- C. ZnONPs coated fabric showed better UV blocking property (<370 nm). With the increase in loading of ZnONPs, UV blocking property was also increased. Ultraviolet protection factor (UPF) value of coated cotton fiber was 50 + for 1–5 g/m^2 of ZnONPs. Photocatalytic degradation of malachite green (MG) adsorbed on ZnONPs coated textile fabrics was noted. Photocatalysis (95%) was obtained within 2 h and absorption maxima (625 nm) were decreased as time progressed. Hence ZnONPs coated cotton fabric exhibited photocatalytic and long long-term protective ability.

More recently, a zinc tolerant bacterial strain *Serratia nematodiphila* ZTB15 was used for the synthesis of ZnONPs (Jain et al. 2020). The organisms were grown in Luria Bertani (LB) broth. Overnight grown culture was mixed with 0.1 M zinc sulfate and the reaction mixture was heated at 80 °C for 10 min followed by incubation for 24 h. White precipitates were purified, washed and dried at 120 °C. UV–Vis spectroscopy showed surface plasmon resonance (SPR) at 379 nm. The structural composition was studied using XRD analysis. The intense peaks confirmed the high purity and crystalline nature of ZnONPs. The nanoparticles were polydispersed and roughly spherical in shape with sizes ranging from 15 to 30 nm with a diameter of 23.09 ± 4.23 nm as seen in Fig. 10.1. EDX data confirmed that ZnONPs contain organic elements which act as reducing and stabilizing agents in the synthesis of ZnONPs. The average particle size ranging from 10 to 30 nm was revealed by DLS analysis which showed a polydispersed index. The zeta potential value was −33.4 mV indicating high stability. Raman spectroscopic analysis was done to evaluate purity, crystallinity, and defects in ZnONPs synthesized using bacterial strain. Photon modes of hexagonal ZnO structure at 438.5 cm^{-1} were attributed to the oxygen vibrations and the band of wurtzite ZnO structure. Photocatalytic degradation of methyl orange (MO) was examined using ZnONPs. MO dye treated with ZnONPs was incubated in dark for 30 min to check the adsorption–desorption equilibrium. The catalytic degradation was carried out under UV light. The decrease in peak intensity of MO (463 nm) was observed when exposed to UV light. After 80 min, ~90% of degradation of MO was obtained.

10.3 Fungi

Mycogenic synthesis of various nanoparticles is a popular route because of its rapid and environmentally benign nature (Bloch et al. 2021a). *Acremonium potronii* was used for the synthesis of ZnONPs (Ameen et al. 2021). The fungal filtrate (100 mL) was mixed with zinc acetate hexahydrate followed by heating at 35 °C. The visible colour change from pale yellow to white indicated the formation of ZnONPs. The white pellet was heated at 150 °C for 30 min. The blue shift of ZnONPs was observed between 320 and 350 nm using UV–Vis spectroscopic technique. FTIR analysis exhibited peaks at 445 cm^{-1} corresponding to Zn–O bonds. Other peaks C–C and C-H bonds were found to be attached on the surface of ZnONPs which act as capping and stabilizing agents. The NPs were spherical in shape with an average particle size of 13–15 nm which was characterized using SEM and TEM analysis. The photocatalytic activity of ZnONPs was examined against MB dye under UV–Vis irradiation. The absorbance (420 and 620 nm) completely disappeared within 30 min showing

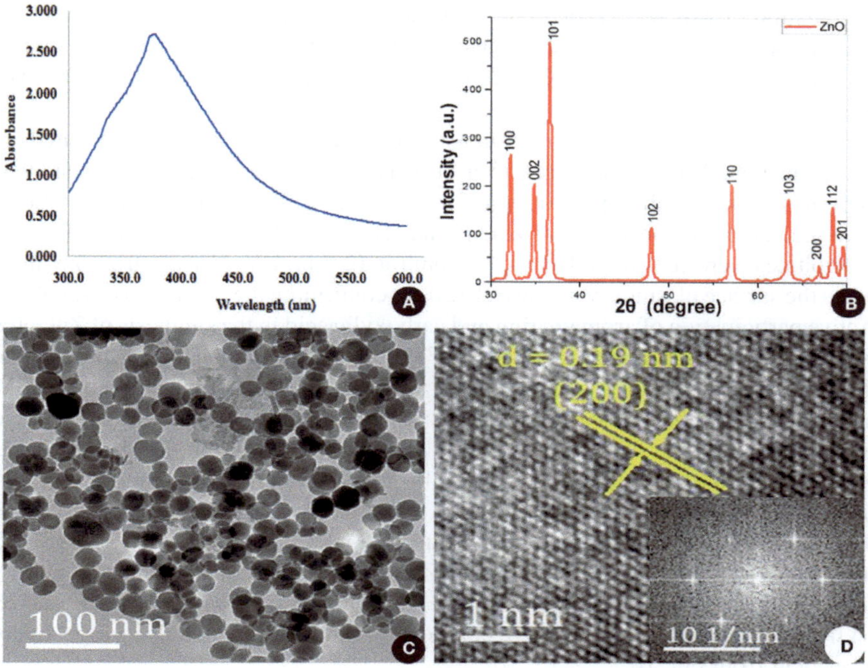

Fig. 10.1 **a** UV–VIS absorption spectra from 300 to 600 nm; **b** XRD analysis; **c** TEM micrograph; **d** SAED pattern of ZnONPs synthesized by zinc-tolerant bacteria. Reprinted from Jain et al. (2020) Microbial fabrication of zinc oxide nanoparticles and evaluation of their antimicrobial and photocatalytic properties. Front Chem 8:778. https://doi.org/10.3389/fchem.2020.00778

degradation of MB. The degradation of MB was found to be 93% with a rate constant of 0.042 min^{-1}. This suggests that the mycogenic synthesis of ZnONPs shows a potent application in the photocatalysis of dyes.

Mycogenic synthesis of ZnONPs was carried out using *Alternaria tenuissima* AUMC10624 (Abdelhakim et al. 2020). Fungal spores were harvested and 100 mL of *A. tenuissima* cell-free culture filtrate (ATCF) was mixed with 100 mL of zinc sulfate (2 mM). The reaction mixture was kept under shaking conditions at room temperature for 20 min. formation of white precipitate indicated the formation of nanoparticles. The NPs were separated using centrifugation and the pellets were washed and were kept for drying in a hot air oven at 50 °C. The precipitates were dissolved in ethanol and were further characterized. The UV–Vis spectroscopy showed surface plasmon resonance at 369 nm. XRD analysis showed that the nanoparticles were crystalline in nature with no impurities. DLS analysis showed that the synthesized NPs were monodispersed with particle sizes ranging from 10 to 30 nm. The zeta potential was −23.92 mV showing high stability. The NPs were spherical in shape with an average size of 15.62 ± 4.51 nm. FTIR analysis indicated that the phenol, primary amine, COO^-, and OH groups aided the synthesis of ZnONPs. The photocatalytic degradation of MB was carried out using ZnONPs. The reaction mixture was kept in dark for 1 h. A decrease in absorbance (664 nm) was observed after 20 min when exposed to sunlight. The ZnONPs (200 mg) showed efficient degradation (100%) of MB.

Fungal biomass of *Aspergillus niger* was used for the synthesis of ZnONPs (Kalpana et al. 2018). Zinc nitrate (5 mM) was added to the fungal culture and was kept on shaking condition at 200 rpm for 2 d at 32 °C. Production of white precipitate indicated the conversion of zinc nitrate by fungal biomass into ZnONPs. The precipitates were centrifuged at 10,000 rom for 10 min. UV–Vis spectroscopy indicated the surface plasmon resonance at 320 nm confirming the formation of ZnONPs. Active participation of aromatic ring and carboxylic acid in the synthesis of ZnONPs was observed which was revealed using FTIR analysis. XRD analysis showed that the myogenic synthesized ZnONPs were wurtzite and crystalline in nature. SEM analysis showed that the synthesized NPs were spherical in shape with an average particle size of 61 ± 0.65 nm and with a diameter of 53–69 nm. Further ZnONPs were used for the catalytic degradation of bismarck brown dye. Maximum degradation was obtained at 72 h. Decolorization of bismarck brown dye was observed up to 89% (100 μL ZnONPs), 72% (65 μL ZnONPs), 52% (50 μL ZnONPs), 38% (25 μL ZnONPs). Hence it was observed that ZnONPs showed good degradation ability. It is important to note that 100% and 40% germination of *P. sativum* was observed on treatment with the degraded dye compounds and the untreated dye, respectively as evident from Fig. 10.2.

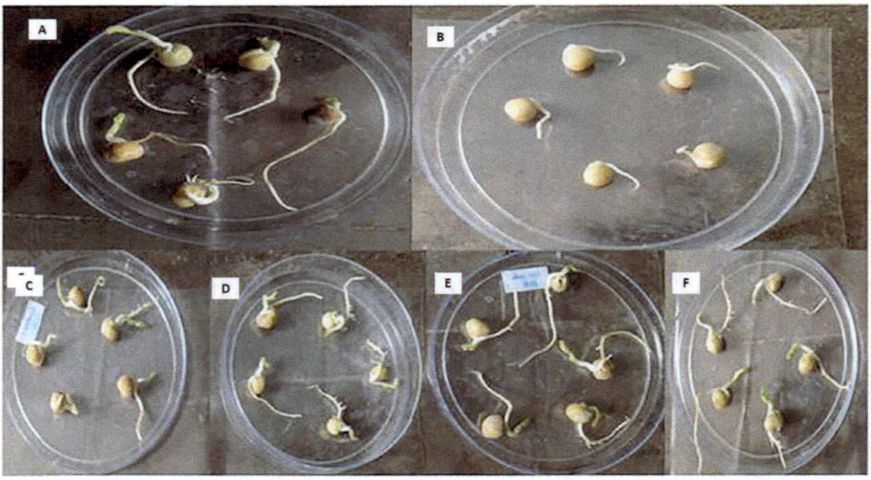

Fig. 10.2 Catalytic phytotoxicity study of synthesized ZnONPs. **a** Control; **b** Untreated dye; **c** Treated dye (25 μL of ZnONPs); **d** Treated dye (50 μL of ZnONPs); **e** Treated dye (75 μL of ZnONPs); **f** Treated dye (100 μL of ZnONPs). Reprinted with permission from Kalpana et al. (2018) Biosynthesis of zinc oxide nanoparticles using culture filtrates of *Aspergillus niger*: Antimicrobial textiles and dye degradation studies. OpenNano 3:48–55. https://doi.org/10.1016/j.onano.2018.06.001 Copyright © 2018 The Authors. Published by Elsevier Inc

10.4 Algae

Algae mediated nanoparticle synthesis is popularly known as phycogenic approach which is efficient and faster (Nitnavare et al. 2022). Synthesis of ZnONPs using microalgae *Chlorella* extract (20 mL) was carried out by mixing the same with 80 mL zinc acetate solution under stirring at 150 rpm for 60 min at 58 °C and pH 8 (Khalafi et al. 2019). The electron-rich organic biomolecule present in microalgae accelerated the reduction process at high temperature. The reaction mixture was further incubated at 85 °C for 25 min after the formation of the white milky solution. The mixture was centrifuged, washed, and dried at 50 °C. The visible colour change from light green to white showed the formation of ZnONPs which exhibited absorbance maxima at 362 nm in UV–Vis spectroscopy. XRD analysis pattern gave diffraction peaks specific to the hexagonal wurtzite crystalline nature of ZnONPs. The algal extract contained 45% proteins, 20% peptides, 20% fat, 20% carbohydrates, 10% vitamins, and 5% fibers. All biomolecules were involved in the synthesis, reduction, and stabilization of ZnONPs. SEM analysis showed that the ZnONPs were spherical in shape with sizes ranging from 20 to 50 nm. TEM results exhibited that the NPs were hexagonal in shape with an average particle size of 20 ± 2.2 nm. XRD analysis revealed the hexagonal wurtzite structure of ZnONPs. Photocatalytic degradation of an organosulfur pollutant dibenzothiophene (DBT) by ZnONPs was also checked. The experiment was carried out under UV light. ZnONPs (0.01 g) were mixed with 20 mL of wastewater and DBT (10 ppm) was added. The reaction mixture was stirred at 480 rom for half an hour. The mixture was exposed to UV light for various time intervals and the degradation process of DBT was checked through gas chromatography (GC). It was observed that 97% of DBT was decomposed in 3 h. The efficiency was also checked in real wastewater samples collected from oil refineries containing DBT. Diluted DBT solution (200 ppm) was treated with ZnONPs (10 mg). As the exposure to UV was increased the absorbance peak of DBT at 270 nm was decreased and it completely disappeared after 190 min. Hence ZnONPs showed high photocatalytic efficiency. The ZnONPs were reused for up to 5 runs and they showed 97% to 93% degradation of DBT, indicating high reusability and durability.

In another study, green synthesis of ZnONPs was reported using red seaweed *Pterocladia capillacea* (Mansour et al. 2022). Initially 2 g of the dry powder of the red seaweed was added to 100 mL of double distilled water followed by heating at 100 °C that was further filtered. The resulting aqueous extract of the red seaweed (20 mL) was added to a 2 mM solution of zinc acetate followed by drop wise addition of 2 M NaOH till pH 12 is obtained. The reaction mixture was kept at 65 °C for 2–3 h after which white precipitates appeared that was centrifuged, washed, and dried at 80 °C. The synthesized ZnONPs showed a peak at 3465 cm^{-1} assigned to N–H stretching in FTIR. Intense peaks at 1553 and 1517 cm^{-1} were attributed to $C = O$ and $C = C$ aromatic stretching. The particles were around 1 μm in size with rod and small flower-like shapes with an uneven and porous texture. BET analysis showed a surface area of 113.751 m^2g^{-1} with a pore volume and pore size of 0.143 c^2g^{-1} and 2.527 nm, respectively. The intense peak in XRD analysis indicated crystalline structure of ZnONPs with hexagonal and cubic phases. UV–Vis absorbance peaks were obtained in the range of 300–400 nm. The effectiveness of adsorption was checked at various pH levels. The pH showed an impact on the surface charge of the adsorbent and ionization of pollutants. At pH 2, 98.78% dye elimination was observed. When pH was increased from 2 to 6, adsorption was initially increased but later it was decreased. At low pH, the surface charge of the adsorbent was positive due to H^+ ions. Change in surface charge due to change in pH from 2 to 6 resulted a high percentage of Ismate violet 2R (IV2R) removal. As the pH was further increased the charge on the surface of the adsorbent became negative that helped in the electrostatic attraction aiding adsorption of positively charged cationic dye. The elimination of IV2R was increased from 96.47 to 99% when the dosage of ZnONPs was increased from 0.005 to 0.8 g. ZnONPs at pH 6, 55 °C with a contact time equivalent to 120 min. The dye adsorption capacity of ZnONPs was 72.24 mg/g. Thermodynamic studies revealed that the adsorption process was spontaneous and endothermic.

Similarly, Subramanian et al. (2021) reported on synthesis of ZnONPs using aqueous extract of brown seaweed *Sargassum muticum* that reacted with 100 mL of 5 mM zinc nitrate ($Zn\ (NO_3)_2$) for 30 min under stirring conditions (150 rpm). Further, it was incubated in a boiling water bath at 70 °C for 20 min followed by a second round of stirring (150 rpm) for 2 h followed to which 5 mL of freshly prepared 0.5 M potassium hydroxide (KOH) was added dropwise. The formation of a white precipitate indicated the synthesis of ZnONPs which was then collected by centrifugation that was washed, and kept in a hot air oven at 60 °C for drying. The color change from dark brown to pale white indicated the reduction of zinc nitrate to Zn^0 ions by functional groups present in the extract of seaweed. The UV–Vis spectrum showed an absorption peak obtained at 350 nm revealing the characteristic of ZnONPs. XRD diffraction peaks confirmed the crystalline nature of ZnONPs. Photoluminescence spectroscopy was carried out at room temperature using an excitation wavelength of 325 nm. The peak obtained at 390 nm exhibited the wurtzite hexagonal ZnONPs. SEM analysis showed well dispersed spherical ZnONPs with sizes ranging from 15 to 50 nm. FESEM revealed the polydispersed morphology of ZnONPs while the EDX analysis exhibited 92.77% of Zn^+ and 4.23% of other carbon substances. X-ray fluorescence (XRF) spectrum showed 90% ZnONPs and 10% other elements such as Zn, Al, and Fe while DLS analysis showed monodispersed NPs with particle sizes less than 25–50 nm. Zeta potential was −21.4 mV. FTIR spectra of ZnONPs synthesized from brown seaweed exhibited strong peaks at 3432 and 1609 cm^{-1} indicating the presence of O–H stretching while others at 1402 and 1609 cm^{-1} were corresponding to the C-H hydroxyl and $C = O$ carbonyl group. The peak below 500 cm^{-1} is assigned to Zn–O stretching. Photocatalytic degradation of MB using ZnONPs under UV light, visible light, and sunlight was accompanied by the change in absorbance spectra of MB dyes at different time intervals. The low photocatalytic activity (40%) was exhibited under visible light within 240 min while the decolorization efficiency was 84% (at 105 min) and 96% (at 60 min) under UV light and sunlight, respectively.

In another study, *Ulva lactuca* extract (5 mL) reacted with 95 mL of 1 mM zinc acetate for the synthesis of ZnONPs within 3–4 h at 70 °C (Ishwarya et al. 2018).

Further, it was centrifuged at 4000 rpm for 10 min and pellets were collected that were heated at 450 °C for 4 h in a hot air oven. UV–Vis spectra showed an absorbance at 325 nm while XRD pattern showed strong diffraction peaks at 31.6, 34.35, 36.2, 47.5, 56.4, 62.8, 68.9, and 76.9° that were assigned to (100), (002), (101), (102), (110), (103), and (200) planes, respectively which indicated the crystalline nature of the ZnONPs with no impurities. FTIR analysis indicated that O–H stretching played an important role in the formation of ZnONPs while the absorption band at 420 cm^{-1} corresponds to ZnO and 500 and 900 cm^{-1} correspond to oxygen. The ZnONPs were anisotropic with triangles, hexagons, rods, and rectangles in shape with a size of 10–50 nm as seen in Fig. 10.3. EDX showed 78% zinc and 48% oxygen in the ZnONPs. Catalytic degradation of MB was carried out in presence of sunlight. Absorption spectra of dye treated with ZnONPs were recorded at different time intervals and a reduction in the peak was observed. After 30 min, a rapid reduction of dye was observed. The decolorization of MB up to 90.4% was achieved when exposed to sunlight for 120 min. Different dye concentration (25, 50, 75, and 100 ppm/100 mL) was treated with 50 mg of ZnONPs at pH 7 under sunlight. The degradation of MB (90%) was achieved at an optimal concentration of 25 ppm. Further, different concentrations of biocatalysts (25, 50, 75, and 100 mg/100 mL) were used to check the degradation of MB dye (25 ppm, pH 7). In is important to note that the increase in the concentration of biocatalyst also increased the dye degradation. However, use of 100 mg of biocatalyst resulted in the reduction in degradation that was attributed to the sedimentation and agglomeration of the biocatalyst making it unavailable for the reaction.

10.5 Plant

Plants are natural treasure houses of various phytochemicals that can reduce the metal ions to nanoparticles followed by stabilization. Phytogenic synthesis of ZnONPs by reacting 10 mL leaf extract of *Abelmoschus esculentus* Linn (okra) with identical volume of 0.01 mol zinc acetate dihydrate and 0.01 mol NaOH was reported by Mirgane et al. (2020). The pH was adjusted between 9 and 11. The reaction mixture was centrifuged at 7000 rpm for 10 min followed by the pellets were collected and dried for 5 h at 100 °C. XRD analysis showed diffraction peaks at 32.77, 35.9, 37.23, 47.64, 58.35, 66.76, 66.27, 68.79 and 70.12° indicating crystalline nature of ZnONPs. FTIR analysis showed C–C, C–N, –OH, C–H, carbonyl and C–O–C stretching at 1620 cm^{-1}, 1387 cm^{-1}, 3551–3257 cm^{-1}, 2968 cm^{-1}, 1774 cm^{-1}, and 1127 cm^{-1}, respectively indicating the presence of various phytochemicals participating as capping agent for ZnONPs. SEM images showed the polycrystalline nature of the ZnONPs synthesized from okra leaves that were majorly spherical in shape with a size of 20 to 45 nm as confirmed from TEM images as well. Degradation of MB and MO dye by the phytogenic ZnONPs was confirmed by the decrease in absorbance peak specific to the dyes. The degradation of MB and MO was achieved up to 96% within 540 min.

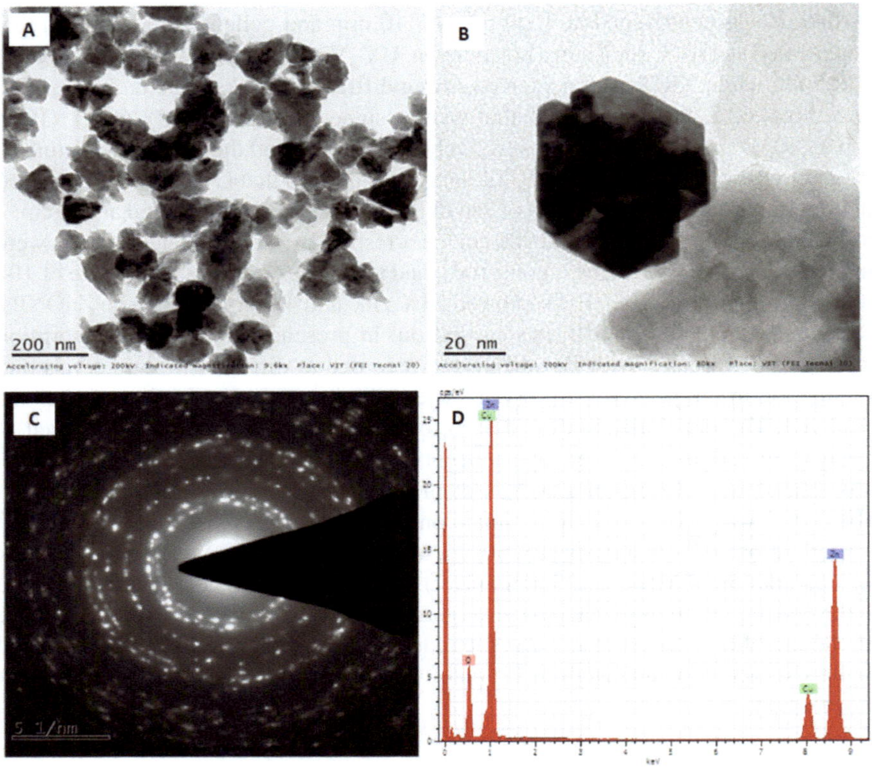

Fig. 10.3 **a** and **b** TEM images of *Ulva lactuca*-fabricated ZnONPs; **c** SAED pattern of *U. lactuca*-fabricated ZnONPs; **d** EDAX spectrum of *U. lactuca* fabricated ZnONPs showing a zinc signal. Reprinted with permission from Ishwarya et al. (2018) Facile green synthesis of zinc oxide nanoparticles using *Ulva lactuca* seaweed extract and its evaluation of photocatalytic, antibiofilm and larvicidal activity: impact on mosquito morphology and biofilm architecture. J Photochem Photobiol B 178:249–258. https://doi.org/10.1016/j.jphotobiol.2017.11.006 Copyright © 2017 Elsevier B.V

In another study, Rajkumar et al. (2019) reported a green synthesis of ZnONPs using an aqueous extract of dry onion peels (*Allium cepa* L.). Zinc nitrate hexahydrate (0.1 M, 95 mL) was mixed with 5 mL of plant extract, and pH was adjusted to 12 using 2 M NaOH. The reaction mixture was kept for 3 h under stirring conditions followed to which the resulting white precipitate was centrifuged at 8000 rpm for 5 min. UV–Vis spectroscopy was used to elucidate the optical properties of ZnONPs that indicated an absorbance peak at 370 nm. XRD analysis exhibited 2θ value at 31.87, 34.52, 36.35, 47.62, 56.68, 62.93, 66.45, 68.03 and 69.18° corresponding to (100), (002), (101), (110), (103), (200), (112), and (201) lattice planes, respectively. These planes offered face-centered cubic hexagonal wurtzite structure of ZnONPs. FTIR analysis revealed involvement of phenols, alcohols, primary amines, and carboxylic acid in the reduction and stabilization of ZnONPs. FESEM showed hexagonal ZnONPs with a diameter of about ~50 nm. EDS confirmed the presence of Zn (51.29%) and O (48.71%) in ZnONPs. A high percentage of oxygen indicated the formation of NPs in oxide form. Photocatalytic dye degradation of crystal violet (CV) and MB was done in presence of sunlight and UV light using ZnONPs as biocatalysts. The degradation of CV and MB was up to 74.82 and 94.04% under sunlight, while it was 54.10 and 36.64% under UV light, respectively.

In another study, the leaf extract of jack fruit *(Artocarpus heterophyllus)* was used for the synthesis ZnONPs (Vidya et al. 2016). Zinc nitrate hexahydrate was mixed with the required volume of leaf extract and heated for 1 h. A dark brown color liquid obtained was calcinated at 600 °C for 1 h. The formation of white precipitates indicated the synthesis of ZnONPs. Several phytochemicals present in the extract of jack fruit such as terpenoids, phenols, steroids, glycosides, flavonoids, saponins, and carbohydrates might act as strong stabilizing and reducing agents in the synthesis of ZnONPs. The polycrystalline nature of ZnONPs was determined using XRD analysis. HRTEM analysis showed the hexagonal shape of ZnONPs with a size of ~15–25 nm. The elemental composition of ZnONPs determined using EDS showed 85.32% zinc and 14.68% oxygen. The degradation of Rose Bengal dye by ZnONPs was dependent on the pH, concentration of nanoparticles and dye concentration. The degradation rate increased as pH was increased from 6 to 8 beyond which it decreased. The rate of dye degradation was enhanced when 60 mg of biocatalyst was used while it decreased at 80 mg. The degradation was up to 85% when 10–30 ppm concentration of dye was used while at 40 ppm dye concentration, the rate of degradation was reduced. As concentration increases the density of particles increases which leads to intercollision between particles and hence hinders the light incident on metal oxide which in turn reduces the release of hydroxyl radical affecting the rate of degradation. The photodegradation efficiency of phytogenic ZnONPs (0.24 g/L) against Rose Bengal dye was >80% within 1 h.

Phytogenic synthesis of ZnONPs was carried out using leaves of *Azadirachta indica* (Bhuyan et al. 2015). Leaf extract (25% w/v, 1 mL) was added to 50 mL of 2 M zinc acetate containing 2 M NaOH. The mixture was placed on a magnetic stirrer for 2 h. The resulting white precipitates were washed and filtered and kept for drying at 60 °C overnight. The aqueous solution of ZnONPs exhibited a surface plasmon resonance associated intense peak at 377 nm when analyzed by UV–Vis spectroscopy. TEM micrographs showed that the ZnONPs were spherical in shape with a diameter ranging from 9.6 to 25.5 nm. EDX confirmed the purity of NPs as it showed a single strong emission peak for zinc and oxygen. XRD patterns indicated the ZnONPs were crystalline in nature. Photocatalytic degradation of MB by ZnONPs as a biocatalyst was checked under UV light. The reaction mixture was exposed to UV light for 30–180 min. The degradation of MB obtained at 0, 30, 60, 90, 120, 150, and 180 min was 49.6, 56.6, 62.1, 68.7, 76.2, and 82.1%, respectively.

Synthesis of ZnONPs was from lemon (*Citrus limon*) juice was reported by Davar et al. (2015). Zinc acetate (0.004 mol) was dissolved in ethylene glycol (10 mL) and different volumes of lemon juice (5, 10, 30, 50, and 70 mL) were added to it. The mixture was kept under stirring for 1 h. Further, it was heated at 90 °C for 3 h under shaking conditions. The brown-colored gel was calcinated at 600 °C for 2 h.

Another reaction mixture was prepared by adding 0.012 mol sucrose in lemon juice and all reaction conditions were kept as mentioned before. FESEM micrographs revealed the semispherical shape of NPs. At 5 and 10 mL volumes of lemon juice, the particle size obtained was >50 nm. As the amount of lemon juice (50 mL) was increased the particle size was found to be in micrometer range. The ZnONPs were formed with 30 mL lemon juice and 0.002 mol sucrose were 21.5 nm in size. EDX data showed a characteristic peak around 8.63 keV for ZnONPs. XRD pattern of ZnONPs confirmed the hexagonal and highly crystalline nature. TG–DTA analysis of ZnONPs was carried out and little weight loss was observed at 100–200 °C, followed by 58% loss at 230–450 °C indicating combustion of organic compounds. After 600 °C, weight loss of 6.3% was obtained due to carbon residue present that might be preventing the agglomeration of ZnONPs. FTIR analysis showed a broad absorption band of O–H stretching at 3200–3500 cm^{-1}. The ZnONPs showed a stretching band at 666 cm^{-1}. Under UV light, high photodegradation of MB, MO, and MR was obtained in presence of ZnONPs. Decolorization of textile reactive turquoise blue, reactive blue 21 (RB 21) dye was achieved using phytogenic ZnONPs. The dye (20 ppm) was degraded (>75%) within 270 min under UV light by ZnONPs.

In another study, ZnO nanocrystals were synthesized reacting coconut husk extract (10 mL) with 0.595 g of zinc nitrate ($Zn(NO_3)_2.6H_2O$) followed by stirring for 20 min (Priyadharshini et al. 2022). The resultant gel was calcinated at 400 °C for 3 h. Within 10 min, the pale yellow powder was formed. ZnONPs exhibited surface plasmon resonance at 355 nm which was recorded using UV–Vis spectroscopy. The X-ray diffraction peaks generated at different planes attributed to stable hexagonal wurtzite structure. FTIR analysis showed several peaks of functional groups and oxides of metals. The infrared spectrum obtained at 464 cm^{-1} corresponds to spherical ZnO nanocrystals. Raman spectroscopy revealed wurtzite structure. A hexagonal bipyramid with six facet fold morphology was observed by SEM images as observed in Fig. 10.4. EDX spectra revealed elemental Zn and O up to 74.88% and 25.12% in the particles. Photocatalytic decomposition of MB and metanil yellow (MY) dye by ZnONPs was influenced by various parameters like light source, amount of biocatalyst, the initial concentration of dye, and pH. It is interesting to note that within 100 min, ~99% and 97% MB and MY degraded under UV light, while only ~22% degradation was noted under visible light. Increase in the concentration of biocatalyst from 5 to 15 mg resulted in enhancement of the degradation of MB and MY, from ~63 to 99% and ~61 to 97%, respectively. However, treatment with 20 mg of ZnONPs resulted in ~88 and ~99% degradation of MB and MY, respectively. The reduction in dye degradation efficiency at very high concentration of the ZnONPs might be attributed to the agglomeration of the particles finally leading reduction in surface area and less exposure of the photocatalytic active sites. Under UV light, maximum degradation was observed at minimum concertation of dye (5 ppm). Increasing in the dye concentration from 5 to 20 ppm resulted in decrease in degradation efficiency from ~99 to 56% and 97 to 56% for MB and MY, respectively. MB and MY were degraded up to ~38% and ~33%, respectively at pH 4. Almost ~99% of MB degradation was noted at pH 10. MB is a cationic dye with a positive charge that can get absorbed on the surface having a high amount of $OH^{•}$ radical ions. Removal of MB

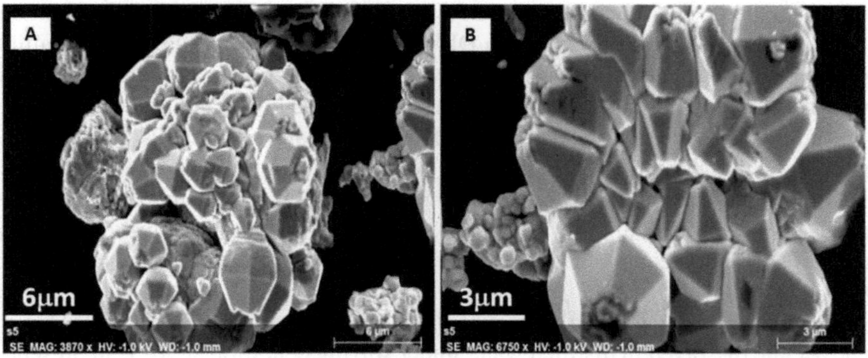

Fig. 10.4 FE-SEM images of the as-synthesized ZnO nanocrystals in **a** low and **b** high magnification. Reprinted from Priyadharshini et al. (2021) Photocatalytic degradation of methylene blue and metanil yellow dyes using green synthesized zinc oxide (ZnO) nanocrystals. Crystals 12(1): 22. https://doi.org/10.3390/cryst12010022

was due to effective adsorption and photodegradation by the phytogenic ZnONPs. However, at high pH the photocatalyst generates high $OH^{\bullet}$ radical ions which increase the efficiency of the photodegradation process of MY which is an anionic dye. The degradation rate constants (k) of MB at pH 4, 7, and 10 were 0.0049, 0.0348 and 0.0441 min^{-1}, respectively. The rate constants for MY at pH 4, 7, and 10 were 0.047, 0.0143, and 0.0217 min^{-1}. This degradation of MB and MY by ZnONPs followed a pseudo-first-order reaction kinetics.

The bark of *Kalopanax septemlobus* was also used for the synthesis of ZnONPs (Lu et al. 2018). Initially the bark of the plant was pulverized into a fine powder for preparing the aqueous extract. The plant extract (20 mL) was heated at 50 °C and zinc acetate (1 mM) was added dropwise. Precipitate was recovered after 30 min by centrifuging at 12,000 rpm for 10 min at 4 °C followed by washing the pellets and drying at 60 °C in a vacuum oven. UV–Vis spectra showed a strong peak at 372 nm. FTIR analysis showed stretching of hydroxyl group, alkane, and alcohol at 3400, 1627, and 1405.83 cm^{-1}, respectively. The stretch band of ZnONPs was obtained at 518.73 cm^{-1}. FE-TEM micrograph showed that the NPs were flower shaped with a size of around 500 nm. EDX analysis revealed 85.35% of zinc and 14.65% oxygen in the phytogenic nanoparticles. Photocatalytic degradation of MB under UV light was time-dependent when reacted with 0.5 mg/mL and 1 mg/mL concentration of ZnONPs. Photocatalytic degradation of MB was ~97% in 30 min at 0.5 mg/mL concentration of ZnONPs that followed a first order reaction.

Recently, Rambabu et al. (2020) used date pulp waste (DPW) for the synthesis of ZnONPs. The aqueous extract of DPW (50 mL) was mixed with 5 g of zinc nitrate hexahydrate. The mixture was ultra-sonicated for 30 min. The resultant slurry was calcinated at 400 °C for 1 h that gave around 89.3% yield of ZnONPs. The UV–Vis spectroscopy showed a characteristic peak at 381 nm while photoluminescence spectroscopy showed major peaks in regions of 400–500 nm. FTIR studies showed polyols, tannins, flavonols, anthocyanidins, and amino acids to be associated on the surface of ZnONPs synthesized from DPW. The ZnONPs were pure and crystalline in nature which was confirmed using XRD. Raman spectroscopy showed a hexagonal wurtzite phase exhibiting structural integrity and phase homogeneity. XPS analysis showed the presence of Zn and O atoms. The morphology of ZnONPs was spherical in shape with agglomeration which was examined using SEM and TEM analysis. EDS indicated 63.3 and 36.7% of elemental Zn and O in the nanoparticles. TGA analysis showed a minimum weight loss of 9.6%wt for the phytogenic ZnONPs. The ZnONPs photocatalytically degraded MB and eosin yellow (EY) up to 90.5 and 90.6%, respectively.

In a similar study, the leaf extract of *Prosopis juliflora* was reacted with 2.5 g of zinc acetate dissolved in distilled water under stirring condition at 50 °C in presence of sunlight (Abbas et al. 2021). Further, the reaction mixture was centrifuged at 6000 rpm for 10 min and washed with water and ethanol followed by heating at 600 °C for 2 h. The XRD diffraction peaks were obtained at 31.69, 34.34, 36.18, 47.51, 56.53, 62.78, and 68.99° exhibited the crystalline nature of ZnONPs. The TEM micrographs showed the spherical shape of NPs with sizes less than 75 nm. FTIR revealed the intra and intermolecular interactions in the synthesis of ZnONPs from *P. juliflora*. The paper mill effluent was treated with ZnONPs that showed notable reduction of % COD (chemical oxygen demand). Reusability of ZnONPs was confirmed as the % COD reduction was up to 74.30, 63.23, and 54.96% for the 1st, 2nd, and 3rd cycles.

Vasantharaj et al. (2021) synthesized ZnONPs by reacting 20 mL of *Ruellia tuberosa* leaf extract with an equivalent amount of $ZnSO_4$ (1 M) under stirring at 70 °C for 6 h. The pH was adjusted to 12 using 2 M NaOH. Development of white coloured from pale yellow indicated the synthesis of ZnONPs. The precipitate was dried at 80° C. FTIR analysis showed the presence of esters, ethers, aromatics, alkenes, alcohols, and aromatic amines in the extract that might be responsible for synthesis and capping of the ZnONPs. EDS analysis showed the presence of 75.65% Zn and 24.35% O in the particles. The TEM images revealed rod-shaped ZnONPs with size ranging from 40 to 50 nm. The ZnONPs photocatalytically degraded malachite green (MG) and MB up to 92 and 94%, respectively.

The biofabrication of ZnONPs was also reported by reacting 50 mL of *S. officinalis* leaf extract with an identical volume of 0.2 M zinc nitrate (Abomuti et al. 2021). After stirring for 2 h, the pH was adjusted to 12 using NaOH under continuous stirring at 50 °C. The mixture was centrifuged at 10,000 rpm for 20 min and was washed with ethanol and water. The washed particles were heated for 24 h at 80 °C followed by annealing at 400 °C for 2 h. The UV–Vis spectroscopy showed surface plasmon resonance at 368 nm. The UV emission band was obtained at 382 nm and the visible band at 497 nm attributing to the recombination of photogenerated holes in ZnONPs which was shown by photoluminescence spectroscopy. The FTIR analysis showed that the phenolic and flavonoid moieties present in leaf extract act as stabilizing agents in the formation of ZnONPs. The average particle size of NPs was 26.14 nm. EDS data revealed the presence of zinc metal and oxygen apart from additional peaks that indicated the presence of amino acids, flavonoids, vitamins, polyphenols, and saponins. TGA analysis showed weight loss of up to 20% at a temperature between 180 and 430 °C. Temperature above 430–780 °C showed a 10% further decrease in weight. XRD diffraction pattern showed the polycrystalline nature of ZnONPs. The surface area, pore volume, and pore diameter of ZnONPs were 53.001 m^2/g, 0.240 cc/g, and 3.052 nm, respectively. The MO was degradation by ZnONPs up to 92.47% within 120 min that followed a pseudo-first-order kinetics with a rate constant of 0.02134 min^{-1}.

10.6 Mechanism of Microorganism Mediated Synthesis of ZnONPs

Microorganisms such as bacteria, fungi, and algae play a vital role in the biological synthesis of ZnONPs. However, microbial synthesis of ZnONPs is still comparatively less explored. Microbial synthesis is more reproducible compared to plants but it is time-consuming and expensive due to use of growth medium (Kitching et al. 2015). Several enzymes, protein, and biomolecules play role in the reduction of NPs (Bloch et al. 2021b). Reaction parameters such as temperature, pH, the concentration of salt, and reaction time determine the rate of production, yield, and morphology of NPs (Ghosh et al. 2022). The ability of microbes to tolerate metals is also a very crucial part of the synthesis of metal and metal oxide NPs. The presence of metal creates stress and this result in the reduction of ions. The NPs are synthesized by microbes using extracellular or intracellular pathways. Extracellular synthesis is more advantageous as it does not need any downstream processing which is applicable in the intracellular pathway (Krol et al. 2017).

10.6.1 Intracellular Microbial Synthesis

The microbial cell walls and the charge present on them play an important role in the intracellular synthesis of NPs. Several enzymes, coenzymes, and others tend to involve in ion transportation into the microbial cell (Ghosh, 2018). Polysaccharides and proteins act as active sites for the binding of metal ions (Salvin et al. 2017). The stress created by heavy metals on microbes promotes the trapping of the ions by the negatively charged cell wall. The ions trapped are reduced into the zero valent atoms by electron transfer carried out by electron carriers such as NADH by NADH-dependent reductase embedded in the plasma membrane. After this nucleation process, the nuclei grow to form NPs which get accumulated in the cytoplasm or periplasmic space. The proteins and amino acids act as stabilizing agents in the synthesis of NPs (Iravani et al. 2014; Ghosh et al. 2022).

10.6.2 Extracellular Microbial Synthesis

Nitrate reductase actively participates in the extracellular synthesis of NPs by reducing metal ions into metal NPs (Chauhan et al. 2015). The enzymes present on the cell membrane are released into the growth medium as extracellular enzymes. The reduction of Zn^{2+} is initiated by the electron transfer carried out by NADH-dependent reductase from NADH acting as an electron carrier (Hulkoti et al. 2014). During formation of ZnONPs, various extracellular proteins act as capping agents rendering stability to the NPs. Nonenzymatic synthesis of nanoparticles is carried out by certain organic functional groups present on the cell wall which reduces the metal ions.

10.6.3 Effect of Optimization on the Synthesis of NPs

The increase in the yield of NPs synthesized using microbes is a major concern. The desired size, shape, and yield of NPs can be achieved by varying the physical parameters. The pH plays a very crucial role in the synthesis of NPs because the change in pH alters the shape of biomolecules which participates in the capping and stabilization of NPs. At acidic pH, NPs are of smaller size while they are bigger in size at basic pH. The catalytic activity of enzymes involved in the synthesis of NPs gets deactivated under alkaline pH which results in a bigger size of NPs (Hosseini and Sarvi, 2015). At high temperatures, the rate of reaction increases which decreases the size of NPs as metal ions are consumed more in the formation of nuclei (Gurunathan et al. 2009). The salt concentration also affects the size of NPs as at higher concentrations the growing particles form a coat that prevents them from aggregation. Several enzymes are produced at various growth phases of the microbes which also affects the yield of NPs. The microbial culture at the exponential phase produces enzymes and proteins which results in maximum reduction of metal NPs (Gericke and Pinches, 2006). The exposure time of metal ions to microbes also plays role in controlling the size of NPs. Prolong exposure leads to irregular shape while a small size was achieved when the reaction time was kept for 12–24 h (Moghaddam et al. 2017). Hence optimization is required to produce desired shape, size, yield, and monodispersed of NPs.

10.7 Mechanism of Plant-Mediated Synthesis of ZnONPs

Plants are most commonly used for the synthesis of NPs as phytogenic synthesis is more cost-effective, convenient, and biocompatible. The possible mechanism behind the formation of ZnONPs from plants is under research. Several phytochemicals such as phenols, flavonoids, terpenoids, saponins, tannins, and carbohydrates are responsible for the synthesis of ZnONPs (Ghosh et al. 2016a, 2016b). These active compounds are known as antioxidants which neutralize the reactive oxygen species (ROS), and free radicals. They are also involved in the reduction of metal ions leading to the formation of NPs (Shende et al. 2017, 2018). The narrowing of bandgap and shifting of the absorption toward visible light is due to the capping of NPs by phytochemicals. Several polyphenols, esters, and vitamin-based compounds are reported to act as ligation agents. The hydroxyl groups ligate with Zn^{2+} ions and form a stable complex. After calcination, this system undergoes direct decomposition which results in the formation of ZnONPs (Bandeira et al. 2020).

10.8 Mechanism Behind the Degradation of Pollutants by ZnONPs

Since past few decades, extensive research is being carried out on the photocatalytic degradation of dye and other pollutants. Several textile, furniture, and plastic industries use dyes, and 20% of the dye contaminated effluent enters the environment causing serious health issues. Using nanomaterials in phytoremediation is one of the recent methods for dye degradation. Photocatalytic degradation uses light which is an eco-friendly and less expensive technique. The nanoparticles act as a biocatalyst and increase the rate of degradation. The degradation using NPs is carried out in presence of visible and ultraviolet light. The phytogenic NPs are either directly exposed to high-energy light sources or assisted with the photosensitization pathway. In photocatalytic degradation involves direct exposure to light which results the electrons to shift from the valence band (VB) to the conduction band (CB) leading to photoexcitation. This photoexcited electron is utilized by dissolved oxygen present in the reaction mixture, which leads to free oxygen radical formation. At the same time, photoexcited holes are generated in VB which oxidizes the adsorbed water molecules and results in the formation of $\cdot$ OH radicals. This highly reactive free radical generated degrades the dye molecules adsorbed on photocatalysts (Mittal and Roy 2021).

10.9 Conclusion and Future Perspectives

The major concern regarding pollution created in the environment is increasing day by day. The use of harsh chemicals and other materials for the removal of pollutants from the environment must be avoided. So bioremediation of toxic pollutants using nanomaterials synthesized biogenically is one of the advanced methods. The process of synthesis of NPs using microorganisms and plants is simple and non-toxic. Although the lesser yield of NPs synthesized via microbes remains one of the big challenges. But the optimization process may aid in an increase in yield however more investigation is required to understand the mechanism behind the formation of NPs using different microbial species. Phytogenic synthesis of ZnONPs showed promising potential applications in the removal of environmental pollutants. The chemical, physical, structural, and optical properties of the ZnONPs can be enhanced using a variety of dopants like silver, gold, platinum and other metals. The biogenic ZnONPs can be surface functionalized with drugs for targeted delivery and gradual release.

In conclusion, biogenic ZnONPs synthesized from microbes and plants can be potential agents for treating industrial effluents. Rational optimization of the various operational parameters can help to develop community based large scale wastewater treatment process to ensure clean and safe water. The molecular and biochemical mechanism in the synthesis of ZnONPs can be elucidated by integrating multidisciplinary approaches involving genomics, proteomics and metabolomics.

Acknowledgements Dr. Sougata Ghosh acknowledges the Program Management Unit for Human Resources & Institutional Development and Innovation (PMU-B) for funding the proposal entitled "Developing High Performance workforce in Postdoc and Postmaster under Research Topic "Study and Development of Porous Nanofibers, Graphene Oxide Quantum Dot Nanofiber Composite, and Metal Doped Graphene Oxide Quantum Dot Nanofiber Composite for Carbon Dioxide Reduction and Capture" under the Program of National Postdoctoral and Postgraduate System approved by PMU-B Board Committees (Contract No. B13F660065). Dr. Sougata Ghosh is also thankful to Kasetsart University, Bangkok, Thailand for Postdoctoral Fellowship and funding under Reinventing University Program (Ref. No. 6501.0207/9219 dated 14th September, 2022).

References

Abbas AM, Hammad SA, Sallam H, Mahfouz L, Ahmed MK, Abboudy SM, Ahmed AE, Alhag SK, Taher MA, Alrumman SA, Alshehri MA, Soliman WS, Tabassum-Abbasi MM (2021) Biosynthesis of zinc oxide nanoparticles using leaf extract of *Prosopis juliflora* as potential photocatalyst for the treatment of paper mill effluent. Appl Sci 11:11394. https://doi.org/10.3390/app112311394

Abdelhakim HK, El-Sayed ER, Rashidi FB (2020) Biosynthesis of zinc oxide nanoparticles with antimicrobial, anticancer, antioxidant, and photocatalytic activities by the endophytic *Alternaria tenuissima*. J Appl Microbiol 128(6):1634–1646. https://doi.org/10.1111/jam.14581

Abomuti MA, Danish EY, Firoz A, Hasan N, Malik AM (2021) Green synthesis of zinc oxide nanoparticles using *Salvia officinalis* leaf extract and their photocatalytic and antifungal activities. Biology 10:1075. https://doi.org/10.3390/biology10111075

Ameen F, Dawoud T, AlNadhari S (2021) Ecofriendly and low-cost synthesis of ZnO nanoparticles from *Acremonium potronii* for the photocatalytic degradation of azo dyes. Environ Res 202:111700. https://doi.org/10.1016/j.envres.2021.111700

Bandeira M, Giovanela M, Roesch-Ely M, Devine DM, Crespo JDS (2020) Green synthesis of zinc oxide nanoparticles: a review of the synthesis methodology and mechanism of formation. Sustain Chem Pharm 15:100223. https://doi.org/10.1016/j.scp.2020.100223

Bhuyan T, Mishra K, Khanuja M, Prasad R, Varma A (2015) Biosynthesis of zinc oxide nanoparticles from *Azadirachta indica* for antibacterial and photocatalytic applications. Mater Sci Semicond Process 32:55–61. https://doi.org/10.1016/j.mssp.2014.12.053

Bloch K, Pardesi K, Satriano C, Ghosh S (2021b) Bacteriogenic platinum nanoparticles for application in nanomedicine. Front Chem 9:624344. https://doi.org/10.3389/fchem.2021.624344

Bloch K, Webster TJ, Ghosh S (2021a) Mycogenic synthesis of metallic nanostructures and their use in dye degradation. In: Dave S, Das J, Shah M (eds) Photocatalytic degradation of dyes: current trends and future. Elsevier, pp 509–525. eBook ISBN: 9780128242025 Paperback ISBN: 9780128238769. https://doi.org/10.1016/B978-0-12-823876-9.00014-7

Chahal S, Rani N, Kumar A, Kumar P (2019) UV-irradiated photocatalytic performance of yttrium doped ceria for hazardous Rose Bengal dye. Appl Surf Sci 493:87–93. https://doi.org/10.1016/j.apsusc.2019.06.284

Chauhan R, Reddy A, Abraham J (2015) Biosynthesis of silver and zinc oxide nanoparticles using *Pichia fermentans* JA2 and their antimicrobial property. Appl Nanosci 5:63–71. https://doi.org/10.1007/s13204-014-0292-7

Davar F, Majedi A, Mirzaei A (2015) Green synthesis of ZnO nanoparticles and its application in the degradation of some dyes. J Am Ceram Soc 98(6):1739–1746. https://doi.org/10.1111/jace.13467

Dhandapani P, Prakash AA, AlSalhi MS, Maruthamuthu S, Devanesan S, Rajasekar A (2020) Ureolytic bacteria mediated synthesis of hairy ZnO nanostructure as photocatalyst for decolorization of dyes. Mater Chem Phys 243:122619. https://doi.org/10.1016/j.matchemphys.2020.122619

Gericke M, Pinches A (2006) Microbial production of gold nanoparticles. Gold Bull 39:22–28. https://doi.org/10.1007/BF03215529

Ghosh S (2018) Copper and palladium nanostructures: A bacteriogenic approach. Appl Microbiol Biotechnol 102(18):7693–7701. https://doi.org/10.1007/s00253-018-9180-5

Ghosh S, Chacko MJ, Harke AN, Gurav SP, Joshi KA, Dhepe A, Kulkarni AS, Shinde VS, Parihar VS, Asok A, Banerjee K, Kamble N, Bellare J, Chopade BA (2016a) *Barleria prionitis* leaf mediated synthesis of silver and gold nanocatalysts. J Nanomed Nanotechnol 7(4):394. https://doi.org/10.4172/2157-7439.1000394

Ghosh S, Gurav SP, Harke AN, Chacko MJ, Joshi KA, Dhepe A, Charolkar C, Shinde VS, Kitture R, Parihar VS, Banerjee K, Kamble N, Bellare J, Chopade BA (2016b) *Dioscorea oppositifolia* mediated synthesis of gold and silver nanoparticles with catalytic activity. J Nanomed Nanotechnol 7(5):398. https://doi.org/10.4172/2157-7439.1000398

Ghosh S, Sarkar B, Kaushik A, Mostafavi E (2022) Nanobiotechnological prospects of probiotic microflora: Synthesis, mechanism, and applications. Sci Total Environt 838:156212. https://doi.org/10.1016/j.scitotenv.2022.156212

Ghosh S, Webster TJ (2021) Biologically synthesized nanoparticles for dye removal. In: Shah M, Couto SR, Biswas JK (eds) Removal of emerging contaminants from wastewater through bionanotechnology. Elsevier, pp 573–603. Paperback ISBN: 9780323855839. https://doi.org/10.1016/B978-0-323-85583-9.00008-9

Gurunathan S, Kalishwaralal K, Vaidyanathan R, Venkataraman D, Pandian SRK, Muniyandi J, Hariharan N, Eom SH (2009) Biosynthesis, purification and characterization of silver nanoparticles using *Escherichia coli*. Colloids Surf B Biointerfaces 74(1):328–335. https://doi.org/10.1016/j.colsurfb.2009.07.048

Hosseini MR, Sarvi MN (2015) Recent achievements in the microbial synthesis of semiconductor metal sulfide nanoparticles. Mater Sci Semicond Process 40:293–301. https://doi.org/10.1016/j.mssp.2015.06.003

Hu H, Li X, Wu S, Yang C (2020) Sustainable livestock wastewater treatment via phytoremediation: current status and future perspectives. Bioresour Technol 315:123809. https://doi.org/10.1016/j.biortech.2020.123809

Hulkoti NI, Taranath TC (2014) Biosynthesis of nanoparticles using microbes-a review. Colloids Surf B Biointerfaces 121:474–483. https://doi.org/10.1016/j.colsurfb.2014.05.027

Iravani S (2014) Bacteria in nanoparticle synthesis: current status and future prospects. Int Sch Res Not 2014:1–18. https://doi.org/10.1155/2014/359316

Ishwarya R, Vaseeharan B, Kalyani S, Banumathi B, Govindarajan M, Alharbi NS, Kadaikunnan S, Al-anbr MN, Khaled JM, Benelli G (2018) Facile green synthesis of zinc oxide nanoparticles using *Ulva lactuca* seaweed extract and its evaluation of photocatalytic, antibiofilm and larvicidal

activity: impact on mosquito morphology and biofilm architecture. J Photochem Photobiol B 178:249–258. https://doi.org/10.1016/j.jphotobiol.2017.11.006
Jain D, Shivani BAA, Singh H, Daima HK, Singh M, Mohanty SR, Stephen BJ, Singh A (2020) Microbial fabrication of zinc oxide nanoparticles and evaluation of their antimicrobial and photocatalytic properties. Front Chem 8:778. https://doi.org/10.3389/fchem.2020.00778
Kalpana VN, Kataru BAS, Sravani N, Vigneshwari T, Panneerselvam A, Rajeswari VD (2018) Biosynthesis of zinc oxide nanoparticles using culture filtrates of *Aspergillus niger*: Antimicrobial textiles and dye degradation studies. OpenNano 3:48–55. https://doi.org/10.1016/j.onano.2018.06.001
Khalafi T, Buazar F, Ghanemi K (2019) Phycosynthesis and enhanced photocatalytic activity of zinc oxide nanoparticles toward organosulfur pollutants. Sci Rep 9:6866. https://doi.org/10.1038/s41598-019-43368-3
Kitching M, Ramani M, Marsili E (2015) Fungal biosynthesis of gold nanoparticles: mechanism and scale up. Microb Biotechnol 8:904–917. https://doi.org/10.1111/1751-7915.12151
Krol A, Pomastowski P, Rafinska K, Railean-Plugaru V, Buszewski B (2017) Zinc oxide nanoparticles: synthesis, antiseptic activity, and toxicity mechanism. Adv Colloid Interf Sci 249:37–52. https://doi.org/10.1016/j.cis.2017.07.033
Kundu D, Hazra C, Chatterjee A, Chaudhari A, Mishra S (2014) Extracellular biosynthesis of zinc oxide nanoparticles using *Rhodococcus pyridinivorans* NT2: multifunctional textile finishing, biosafety evaluation and in vitro drug delivery in colon carcinoma. J Photochem Photobiol B Biol 140:194–204. https://doi.org/10.1016/j.jphotobiol.2014.08.001
Lu J, Batjikh I, Hurh J, Han Y, Ali H, Mathiyalagan R, Ling C, Chan Ahn J, Yang DC (2018) Photocatalytic degradation of methylene blue using biosynthesized zinc oxide nanoparticles from bark extract of *Kalopanax septemlobus*. Optik 182:980–985. https://doi.org/10.1016/j.ijleo.2018.12.016
Mansour AT, Alprol AE, Khedawy M, Abualnaja KM, Shalaby TA, Rayan G, Ramadan KMA, Ashour M (2022) Green synthesis of zinc oxide nanoparticles using red seaweed for the elimination of organic toxic dye from an aqueous solution. Materials 15:5169. https://doi.org/10.3390/ma15155169
Mirgane NA, Shivankar VS, Kotwal SB, Wadhawa GC, Sonawale MC (2020) Degradation of dyes using biologically synthesized zinc oxide nanoparticles. Mater Today Proc 37:849–853. https://doi.org/10.1016/j.matpr.2020.06.037
Mittal S, Roy A (2021) Fungus and plant-mediated synthesis of metallic nanoparticles and their application in degradation of dyes. In: Shah M, Dave S, Das J (eds) Photocatalytic degradation of dyes. Elsevier, Netherlands. pp 287–308. https://doi.org/10.1016/B978-0-12-823876-9.00009-3
Moghaddam AB, Moniri M, Azizi S, Rahim RA, Ariff AB, Saad WZ, Namvar F, Navaderi M (2017) Biosynthesis of ZnO nanoparticles by a new *Pichia kudriavzevii* yeast strain and evaluation of their antimicrobial and antioxidant activities. Molecules 22(6):872. https://doi.org/10.3390/molecules22060872
Nitnavare R, Bhattacharya J, Thongmee S, Ghosh S (2022) Photosynthetic microbes in nanobiotechnology: Applications and perspectives. Sci Total Environ 841:156457. https://doi.org/10.1016/j.scitotenv.2022.156457
Piccinno F, Gottschalk F, Seeger S, Nowack B (2012) Industrial production quantities and uses of ten engineered nanomaterials in Europe and the world. J Nanopart Res 14:1109. https://doi.org/10.1007/s11051-012-1109-9
Priyadharshini SS, Shubha JP, Shivalingappa J, Adil SF, Kuniyil M, Hatshan MR, Shaik B, Kavalli K (2021) Photocatalytic degradation of methylene blue and metanil yellow dyes using green synthesized zinc oxide (ZnO) nanocrystals. Crystals 12(1):22. https://doi.org/10.3390/cryst12010022
Rahman A, Harunsani MH, Tan AL, Khan MM (2021) Zinc oxide and zinc oxide-based nanostructures: biogenic and phytogenic synthesis, properties and applications. Bioprocess Biosyst Eng 44:1333–1372. https://doi.org/10.1007/s00449-021-02530-w

Rajkumar KS, Arun S, Babu MD, Balaji P, Sivasubramanian S, Vignesh V, Thirumurugan R (2019) Facile biofabrication, characterization, evaluation of photocatalytic, antipathogenic activity and in vitro cytotoxicity of zinc oxide nanoparticles. Biocatal Agric Biotechnol 22:101436. https://doi.org/10.1016/j.bcab.2019.101436

Rambabu K, Bharath G, Banat F, Show PL (2020) Green synthesis of zinc oxide nanoparticles using *Phoenix dactylifera* waste as bioreductant for effective dye degradation and antibacterial performance in wastewater treatment. J Hazard Mater 402:123560. https://doi.org/10.1016/j.jhazmat.2020.123560

Sharma D, Sharma S, Kaith BS, Rajput J, Kaur M (2011) Synthesis of ZnO nanoparticles using surfactant free in-air and microwave method. Appl Surf Sci 257(22):9661–9672. https://doi.org/10.1016/j.apsusc.2011.06.094

Shende S, Joshi KA, Kulkarni AS, Shinde VS, Parihar VS, Kitture R, Banerjee K, Kamble N, Bellare J, Ghosh S (2017) *Litchi chinensis* peel: A novel source for synthesis of gold and silver nanocatalysts. Glob J Nanomedicine 3(1):555603

Shende S, Joshi KA, Kulkarni AS, Charolkar C, Shinde VS, Parihar VS, Kitture R, Banerjee K, Kamble N, Bellare J, Ghosh S (2018) *Platanus orientalis* leaf mediated rapid synthesis of catalytic gold and silver nanoparticles. J Nanomed Nanotechnol 9(2):494. https://doi.org/10.4172/2157-7439.1000494

Siddique K, Shahid M, Shahzad T, Mahmood F, Nadeem H, Saifur Rehman M, Hussain S, Sadak O, Gunasekaran S, Kamal T, Ahmad I (2021) Comparative efficacy of biogenic zinc oxide nanoparticles synthesized by *Pseudochrobactrum* sp. C5 and chemically synthesized zinc oxide nanoparticles for catalytic degradation of dyes and wastewater treatment. Environ Sci Pollut Res 28:28307–28318. https://doi.org/10.1007/s11356-021-12575-9

Slavin YN, Asnis J, Häfeli UO, Bach H (2017) Metal nanoparticles: understanding the mechanisms behind antibacterial activity. J Nanobiotechnol 15:65. https://doi.org/10.1186/s12951-017-0308-z

Subramanian H, Krishnan M, Mahalingam M (2021) Photocatalytic dye degradation and photoexcited anti-microbial activities of green zinc oxide nanoparticles synthesized via *Sargassum muticum* extracts. RSC Adv 12:985. https://doi.org/10.1039/D1RA08196A

Tripathi RM, Bhadwal AS, Gupta RK, Singh P, Shrivastav A, Shrivastav BR (2014) ZnO nanoflowers: novel biogenic synthesis and enhanced photocatalytic activity. J Photochem Photobiol B 141:288–295. https://doi.org/10.1016/j.jphotobiol.2014.10.001

Vasantharaj S, Sathiyavimal S, Senthilkumar P, Kalpana VN, Rajalakshmi G, Alsehli M, Elfasakhany A, Pugazhendhi A (2021) Enhanced photocatalytic degradation of water pollutants using bio-green synthesis of zinc oxide nanoparticles (ZnONPs). J Environ Chem Eng 9:105772. https://doi.org/10.1016/j.jece.2021.105772

Vidya C, Chandra Prabha MN, Antony Raj MAL (2016) Green mediated synthesis of zinc oxide nanoparticles for the photocatalytic degradation of Rose Bengal dye. Environ Nanotechnol Monit Manag 6:134–138. https://doi.org/10.1016/j.enmm.2016.09.004

Chapter 11
Nanomaterials Prone Cell Leakiness: A Mechanistic Approach

Debangi Chatterjee and Subhasish Dutta

Abstract The growing exposure in the nanotoxicology field revealed the considerably notable toxic nature of nanomaterials. Studies have displayed that carbon nanotubes, quantum dots, metal oxide nanoparticles, metal nanoparticles (gold, silver, platinum, palladium etc.), and silica nanoparticles can cause cellular disruption. It is highly essential to understand the properties of nanomaterials and how they affect the human body. For toxicity evaluation, a thorough investigation of physicochemical properties, contamination of toxic materials, and cellular morphology of nanomaterials on intrinsic and extrinsic levels are needed. Several factors, i.e., synthesis, pH, size, shape, temperature, crystallinity index and solubility, influences nanomaterial characteristics. Inflammation and oxidative stress are two primary mechanisms illustrating the toxic nature of nanomaterials. The chapter focuses on practical methods and mechanism of nanomaterials toxicology and provides a critical hypothesis of the leading obstacles this field faces.

Keywords Cell disruption · Genotoxicity · Nanomaterials · ROS · Toxicology

11.1 Introduction

Numerous scientific studies have been conducted due to the widespread usage of nanomaterials (NMs), which promoted public concerns regarding their potential health hazards (Setyawati et al. 2013). Nanotoxicology has manifested in the last ten years to study the toxicological characteristics of nanomaterials and nanoparticles (Yan et al. 2013). While the usage of NMs for biological applications like drug delivery, biomedical imaging, and biosensing maintains to grow, the electronics,

---

D. Chatterjee · S. Dutta (✉)
Department of Biotechnology, Haldia Institute of Technology, ICARE Complex, Hatiberia, Haldia 721657, West Bengal, India
e-mail: subhasish.rahul@gmail.com

S. Dutta
Regional Centre for Biotechnology, 3rd Milestone Road, Faridabad 121001, Haryana, India

M. P. Shah et al. (eds.), *Biogenic Nanomaterials for Environmental Sustainability: Principles, Practices, and Opportunities*, Environmental Science and Engineering,
https://doi.org/10.1007/978-3-031-45956-6_11

cosmetics, coating, and even food industries are already efficiently integrating NMs into their products. More than 1600 nanoproducts are on the market, of which more than 700 falls under the "health and fitness" category (Setyawati et al. 2015). The first cellular component to come into touch with foreign particles is the plasma membrane, a lipid bilayer that surrounds all cells. One of the proposed mechanisms by which NMs produce cytotoxicity is by causing membrane damage (Farnoud and Nazemidashtarjandi 2019). A few NM may reach the bloodstream and accumulate in vital organs because of their nanoscale dimensions (Setyawati et al. 2013).

Compounds with one or more exterior dimensions at the nanoscale are referred to as nanomaterials (1–100 nm), and nanotechnology is "the design, characterization, manufacture, and use of structures, devices, and systems by regulating shape and size of materials at the nanometer scale". Nanomaterials comprise nanoparticles (NPs), nanofibers, nanotubes, composite materials, and nanostructured surfaces. Examples of NPs are gold (Au) NPs, carbon NPs, europium oxide (Eu_2O_3) NPs, titanium NPs, magnetic NPs (MNPs), biodegradable NPs poly (lactic-co-glycolic acid) (PLGA), carbon nanotubes (CNTs; singled-walled and multiwalled (SWCNT and MWCNT, respectively)), nanowires, fullerene derivatives, quantum dots (QDs), and so on. Their advantageous characteristics are due to these nanomaterials, which differ from their traditional counterparts in physicochemical and biological aspects. Due to their tiny size, nanoparticles (NPs) can engage in prolonged interactions with intracellular structures and macromolecules for extended periods. They can enter the human body through ingestion, inhalation, skin penetration, or injection. Less than 1% of publications have addressed the biological effects of NPs; the bulk has concentrated on the production and development of new nanomaterials. There is a substantial gap between data on the manufacturing of nanomaterials and assessments of their toxicity. The safe design of NPs may not be possible due to a lack of toxicology data (Sharifi et al. 2012). Although some conflicting reports exist, in vitro and in vivo investigations over the past ten years have shown that NMs have the ability to produce cytotoxicity and genotoxicity (Setyawati et al. 2015). The relation between the surface area and volume of the NMs grows as the particle size decreases, which is why their physico-chemical properties differ from those of their bulk material. Because of their vast surface area to volume ratio, NMs have more atoms on their surface, which increases their toxicity and bioactivity (Setyawati et al. 2015). The most important physicochemical properties of NM are size, surface chemistry, crystallinity, shape, solubility, aggregation propensity, homogeneity of dispersions, and turbidity. It is necessary to evaluate these characteristics to ascertain how much they contribute to toxicity (Hussain et al. 2009).

The toxicity of nanomaterials has been associated with various chemical pathways, including catalytic nano surfaces, the leaching of hazardous ions from nanoparticles, and the production of reactive oxygen species (ROS). ROS-induced oxidative stress is among the most commonly acknowledged nanomaterial-induced cell and tissue injury models. The mechanisms for intracellular ROS synthesis by nanomaterials can be categorized into (1) the direct synthesis of ROS through nanomaterial-catalyzed free-radical reactions in cells and (2) the indirect generation of ROS through disrupting the natural cell-based biochemical equilibrium. Continuous ROS production within cells can disrupt the antioxidant-antioxidant equilibrium and result in oxidative stress (Farnoud and Nazemidashtarjandi 2019). ROS toxicity may be more evident in the central nervous system (CNS) because it contains a high amount of unsaturated fatty acids (Sharifi et al. 2012).

Recently, a novel, systematic classification of cell death was developed by the Nomenclature Committee on Cell Death (NCCD) based on morphological traits, measurable biochemical properties, and functional considerations. These descriptions of cell death will be used to organise and summarise the molecular mechanisms underlying the toxicity of nanomaterials (Stefano et al. 2012). The scientific community has to make a significant effort to examine the physiological effects of acute and chronic exposure to NMs to close the enormous knowledge gap between the development and in vivo toxicity of NMs. Future safe nanotechnologies designs will depend on a fundamental understanding of how NMs interact biologically with cells, proteins, and tissues. NM-products must be proven to have a high level of biocompatibility with negligible adverse outcomes on blood elements, genetic material, and cell viability before being widely used in everyday items and in clinical settings (Sharifi et al. 2012).

11.2 Mechanisms of Nanotoxicity

Numerous body internal and external physiological mechanisms might lead to nanomaterial toxicity. The fundamental molecular mechanism of in vivo nanotoxicity is the creation of free radicals, which cause oxidative stress. Due to their ability to oxidize lipids, proteins, and DNA, free radicals harm biological components. Free radicals can come from various factors, such as the phagocytic cell reaction to foreign objects, a deficiency in antioxidants, the presence of transition metals, environmental influences etc. Aillon et al. (2009). Because several of these NMs intrinsic features can catalyze ROS generation, the physicochemical characterization of NM, including particle size, surface charge, and chemical composition, is an essential predictor of the subsequent ROS response and NM-induced damage. However, as nanoparticles are more prone to generate excessive quantities of reactive oxygen species (ROS), concerns about their potential toxicological effects still exist. The excessive ROS produced by nanoparticles can damage biomolecules and organelle structures because of their high oxidation potential. This damage includes oxidative protein carbonylation, lipid peroxidation, DNA/RNA breakage, and membrane structure destruction, which can result in necrosis, apoptosis, or even mutagenesis (Yu et al. 2020). In addition to cellular oxidative stress, several other biological processes can produce ROS in vivo. Moreover, the assumption that ROS formation is a requirement for NM-induced toxicity, however, is incorrect, given that a few investigations have documented the direct toxicity of NM without producing ROS.

The transformation of molecular oxygen to water in the mitochondria of cells produces ATP through a series of linked proton and electron transport events. Superoxide anion radicals and later other oxygen-containing radicals are created when a small portion of the oxygen is not entirely reduced during this process (Fu et al. 2014). As a result, ROS are the waste product of cellular oxidative metabolism. The production of ROS occurs mainly in organelles like the endoplasmic reticulum (ER), peroxisomes and primarily in the mitochondria. ROS are oxygen-containing chemically reactive particles, such as reactive superoxide anion radicals (O_2^-), hydroxyl radicals (•OH), and hydrogen peroxide (H_2O_2), singlet oxygen ($1O_2$), and hypochlorous acid (HOCl) (Yu et al. 2020). Within the cell, ROS are produced either intrinsically or extrinsically. ROS harms cellular membranes because of their high reactivity. A lipid bilayer makes up the cell membrane, which is crucial for both preserving the structure of the cell and carrying out cell signaling activities (the epidermal growth factor (EGF), the mitogen-activated protein kinase (MAPK) cascades, the transcription factor activator protein-1 (AP-1), and the nuclear factor-KB (NF–κB), and additionally took part in the process of mammalian growth, proliferation, and differentiation). ROS's primary target is lipid, a crucial element of the cell membrane (Yu et al. 2020). Engineered nanomaterials are smaller, have a greater specific surface area, and have a higher surface reactivity than their bulk-size equivalents. It causes more ROS to be produced, which reasons cytotoxicity and genotoxicity (Fu et al. 2014).

Examples of early inflammatory reactions from oxidative stress related to TiO_2 NPs include increased polymorph nuclear cells, decreased macrophage phagocytosis, and/or fibroproliferative alterations in rodents. Human endothelial cells can also be affected by the proinflammatory effects of TiO_2 NPs (Setyawati et al. 2013). ROS generation mainly occurs via two methods, i.e.

11.2.1 Fenton-Like Reaction

A Fenton or Fenton-like reaction is a procedure that causes the formation of hydroxyl radicals, as depicted in the reaction (Setyawati et al. 2013). It has been demonstrated that the metal ions produced by NPs can combine with redox cycling and chemo catalysis. It is demonstrated by reactions between H_2O_2 and Fenton-type chemical reagents containing metal ions like Fe^{2+} and Cu^+.

$$H_2O_2 + Mn^+ \rightarrow M^{n+1} + \cdot OH + OH^- \tag{11.1}$$

$$Ag^+H_2O_2 + H^+ \rightarrow Ag^+ + \cdot OH + H_2O \tag{11.2}$$

In addition to the deactivation of cellular enzymes, disruption of membrane structure, disturbance of the electron-shuttling process, depletion of potential redox levels, reduction of mitochondrial membrane potentials (MMP), and further enhancement of the accumulation of intracellular ROS are all additional effects of the dissociated metal ion (i.e., Ag^{+}) (Yu et al. 2020; He et al. 2014a; Unfried et al. 2007).

11.2.2 Surface Plasmon Resonance Enhancement

Localized surface plasmon resonance occurs when the oscillation frequency of electrons and the frequency of incident light photons are equivalent (Yu et al. 2020).

The following are the main factors of NM-induced ROS:

I. Prooxidant functional groups are present on the reactive surface of NM.
II. The NM's base in transition metals brings active redox cycling on the NP surface.
III. Interactions between particles and cells (Manke et al. 2013).

11.2.3 Effects of NMs on Cells via Increased ROS Production

Excessive ROS generation can cause oxidative stress, which makes it difficult for cells to sustain typical physiological, redox-regulated functions. NMs typically first adsorb on the cell's surface before being translocated through the membrane and into the cell, where they stimulate the production of ROS. Because of its oxidative potential, ROS causes a great deal of stress on cells and damages almost all sorts of biomolecules within, comprising carbohydrates, nucleic acids, unsaturated fatty acids, proteins and amino acids, and vitamins. Disruption to cellular functions includes oxidative modification of proteins to produce protein radicals, modification of nucleic acids, manipulation of gene expression through the activation of redox-sensitive transcription factors, all of which result in cell death and genotoxic effects (Fu et al. 2014). DNA and RNA, the constituent of nucleotides necessary for cell growth and development, are weak ROS targets. Because of their low redox potential, ROS can directly interact with and change nucleobases. Gene miscoding, aneuploidy, polyploidy, and the induction of mutagenesis in cells exposed to NMs are triggered by the rate of ROS production.

Furthermore, ROS has also been discovered to promote the growth of healthy and cancerous cells, triggering mutations that lead to carcinogenesis (Yu et al. 2020; Fu et al. 2014). ROS in excess can damage mitochondrial DNA (mtDNA) as well. There is evidence that mtDNA damage is linked to several clinical disorders, including ataxia, retinitis pigmentosa, neurogenic muscular weakness, mitochondrial encephalomyopathy, stroke-like episodes, retinitis pigmentosa, cardiac conduction dysfunction, and increased CSF fluid protein (Sharifi et al. 2012).

11.3 Oxidative Stress

The term "oxidative stress" refers to a redox imbalance within cells typically brought on by an increase in intracellular reactive oxygen species (ROS) and a reduction in antioxidants. It is one of the primary mechanisms hypothesized to be responsible for nanomaterials' genotoxic effects (Singh et al. 2009). In addition to being a response to cell damage, oxidative stress can also result from cellular respiration, metabolism, ischemia/reperfusion, inflammation, and the metabolism of foreign substances (Buzea et al. 2007). The increased surface area seen in nanomaterials can encourage ROS production. As a result, the amount of oxidative stress that nanoparticles cause increases with their size (Singh et al. 2009).

When exposed to NP, cells and tissues respond to increasing rates of oxidative stress via antioxidant enzyme systems. Cells can engage both enzymatic and nonenzymatic antioxidant mechanisms to combat the excessive ROS response (Manke et al. 2013). Scientists developed a hierarchical oxidative stress model to understand better the relationship between oxidative stress levels and harmful effects. Some proteins were suggested for use as biomarkers for assessing cellular oxidative stress brought on by nanomaterials (Yan et al. 2013). Nevertheless, excessive amounts of oxidative stress cause cell death by damaging the mitochondrial membrane and disrupting the electron chain. Reducing antioxidants or enhanced ROS production are two effective mechanisms favour modified NM's prooxidant effects. Due to its chemical reactivity, oxidative stress can potentially cause DNA damage, lipid peroxidation, and the activation of signalling networks linked to fibrosis, carcinogenesis, and loss of cell growth (Manke et al. 2013). Furthermore, oxidative stress can cause mutation, modification in DNA and chromosome by activating particular signalling pathways, such as mitogen-activated protein kinase (MAPK) and NF–kB (30), as well as causing inflammation and the release of pro-inflammatory cytokines. It has been demonstrated that antioxidant therapy or coating NMs with antioxidant medicines can reverse the cellular oxidative stress caused by NMs. For instance, treating cells or modifying the QDs with antioxidant medications can minimize oxidative stress in cells (Setyawati et al. 2015). Various nanomaterials, including fullerenes, CNTs, and metal oxides, have been proven to cause oxidative stress. In particular, airway inflammation and interstitial fibrosis are caused by the oxidative stressors brought on by occupational NM exposures and practical challenges with different NM (Manke et al. 2013). Finally, Improved knowledge of the nature and relevance of these oxidative (signalling) responses may open up new possibilities for regulating nanoparticle dispersion, toxicity, and clearance.

11.4 Overview of Nanotoxicity

Identifying the possible toxicity of nanoparticles to cells and organisms has received much attention. Numerous studies have found that nanoparticles of various compositions raise toxicological issues. Time, dose, and cell type all seem to play a role in how NMs affect biological processes. The interaction, localization, and toxicity of NMs are significantly influenced by their distinctive Physico-chemical characteristics-size, aggregation, shape, purity, surface coating, composition, and solubility. NMs with less than 100 nm can enter cells; NMs with less than 40 nm can enter the nucleus; NMs with less than 35 nm can penetrate the blood–brain barrier (Setyawati et al. 2015; Sohaebuddin et al. 2010). Numerous in vivo and in vitro experiments have shown that NM can cause toxicity in various organ systems. NMs are taken up by cells through a variety of processes, including passive diffusion, macropinocytosis, phagocytosis, and endocytosis (Setyawati et al. 2015).

Environment-contact points on the human body include the skin, lungs, and digestive system. The lungs and digestive system are particularly susceptible, even though the skin typically serves as an adequate barrier to external chemicals. Other exposure routes include injections and implants, primarily used with engineered materials (Migliore et al. 2015). Since they are so small, nanoparticles can move from these entry points into the lymphatic and circulatory systems to human tissues and organs. Relying on their composition and size, some nanoparticles can result in oxidative stress and organelle injury in cells, which can result in causing damage. Some investigations have demonstrated that NM is localized in either the cytoplasm or nucleus; this suggests that the mechanisms governing NM insertion into cells vary depending on the kind of cell (Hussain et al. 2009; Handy and Shaw 2007). Inhaled nanoparticles are linked to the following diseases: lung cancer, emphysema, asthma, bronchitis, and neurological conditions like Parkinson's and Alzheimer's are further examples of chronic illnesses. Colon cancer and Crohn's disease have both been associated with gastrointestinal nanoparticles. Atherosclerosis, blood clots, arrhythmia, heart disorders, and eventually cardiac mortality are linked to nanoparticles that enter the circulatory system. Also, some autoimmune diseases caused by NM exposure are rheumatoid arthritis, scleroderma, systemic lupus erythematosus etc.

11.4.1 Cytotoxicity and Genotoxicity

The interaction of NMs with biological molecules such as lipids, proteins, and nucleic acids leads to harmful effects. Numerous in vitro investigations have shown increased ROS production as the first sign of damaged cells exposed to nanoparticles. Reactive oxygen species (ROS) generation (Setyawati et al. 2013), lipid peroxidation, and altered electron transport are chemical interactions, whereas membrane rupture, changed protein folding, or protein aggregation are physical interactions. Smaller NMs cause an increased inflammatory response compared to bigger ones. Based on prior observation, it has proven challenging to define a comprehensive mechanism of nanoparticle cytotoxicity (Sohaebuddin et al. 2010). NMs can accidentally enter the nucleus during mitosis, when the nuclear membrane ruptures, by diffusion, nuclear pore complexes, or other mechanisms.

Further evidence that oxidative stress is an essential mechanism by which nanoparticles cause cytotoxicity was provided by the inverse relationship between cell viability loss and an increase in ROS levels (Yang et al. 2009). NM-induced genotoxicity can be either directly caused by intercalation or physical contact, or indirectly caused by oxidative stress or changing proteins that are involved in the cell division (Setyawati et al. 2015). Different NMs genotoxicity may be predominantly caused by the form of the particles rather than their chemical composition. The dispersion of NMs in various media is a factor in essays that assess their toxicity. This factor should be considered because various surroundings could result in different toxicities (Peralta-Videa et al. 2011). For first assessing the anticipated biocompatibility of novel NMs, straightforward in vitro approaches that generate precise and quantitative toxicity assessments are incredibly beneficial. Examples frequently referenced are the MTT (3-(4,5-dimethylthiazolyl-2)-2,5-diphenyltetrazolium bromide) (Monteiro-Riviere and Inman 2006) assay for mitochondrial activity, the LDH (lactate dehydrogenase) assay for cell membrane integrity, and immunochemistry markers for apoptosis and necrosis. Using these techniques, it is possible to assess the immunological, neurological, reproductive, cardiovascular, and developmental toxicity of nanomaterials and identify their long-term systemic toxicity (Sharifi et al. 2012).

11.5 Physiochemical Characteristics of Nanoparticles and Their Impact on Toxicity

The physicochemical characterization of NM substances is vital in understanding and forecasting the effects of molecules and particles for applications in human health and the environment. Even while NMs have been extensively studied, a complete understanding is still lacking. The prior knowledge of an NMs toxicological characteristics, dissolution, chemical composition, size, shape, agglomeration state, crystal structure, specific surface area, surface energy, surface charge, surface morphology, and surface coating influence the biological interaction of NMs and other factors have generally been assumed to be the cause of toxicity. These variables can alter the potential of inducing tissue damage, protein binding, translocation from the portal entrance to the target region, and cellular uptake. The biological effects of NM and their capacity to spread throughout the body are influenced by interactions with cells, bodily fluids, and proteins. Protein-NM binding may result in more mobile complexes that can access usually inaccessible tissue locations (Nel et al. 2006). Therefore, it is crucial to assess these characteristics when estimating the hazardous potential of nanomaterials.

11.5.1 Effect of Size

The interaction of materials with biological systems is significantly influenced by particle size and surface area. From the perspective of toxicity, encounters with living organisms and nanoparticles often happen at the NM's surface. As materials get smaller, their surface area grows exponentially faster than their volume (the number of particles per unit mass increases), making the surface of the nanomaterial more reactive both to itself and its surroundings. It is well known that several biological processes, such as endocytosis, cellular uptake, and the effectiveness of particle processing in the endocytic pathway, depend on the size of the substance (Aillon et al. 2009; Nel et al. 2006). Thus, size plays a crucial role in the physiologic response, material dispersion, and excretion.

It is possible to test the cytotoxicity of NMs of various sizes in vitro using a variety of cell types, growth conditions, and exposure times, but doing so in vivo is challenging due to the complexity of biological systems and the need for a deeper understanding of the particles. It indicates that while in vitro uses of NMs require less rigorous toxicological evaluation, in vivo applications of NMs require a thorough knowledge of the kinetics and toxicity of the particles. The development of oxidative reactions via the production of free radicals is the primary mechanism for the in vivo toxicity of NMs. In this process, size plays a crucial role, as many authors have noted: the smaller the size, the more capable it is of producing ROS. These free radicals pose risks to biological systems, most notably damaging DNA, oxidizing lipids, and inducing inflammatory reactions (Gatoo et al. 2014). The most significant way humans are exposed to NMs is by inhalation; hence the early studies on the in vivo toxicity of NMs were done in respiratory systems. In general, it has been established that lung toxicity tends to rise as particle size decreases; the lung serves as a solid barrier to the uptake and transmission of NMs. According to clinical and experimental research, small and, consequently, large surface areas are found to promote ROS formation. The superoxide anions or hydrogen peroxide are produced when the electron donor or acceptor sites on the NMs react with molecule oxygen, which then causes other molecules to undergo oxidation. This behaviour contributes to NMs' capacity to generate more tissue damage (Sharifi et al. 2012). It suggests that, despite the importance of size and surface area in determining nanoparticle toxicity, other factors, such as the chemical composition of the contents, may also impact the intrinsic toxicity of the nanoparticles.

11.5.2 Effect of Particle Shape

NM shape and aspect ratio have also significantly impacted its interactions with biological units. The uptake rate of a nanomaterial can be substantially influenced by its form. Numerous structures, including fibres, spheres, tubes, rings, and planes, can be achieved with nanomaterials. Shape-dependent nanotoxicity generally affects how membranes are wrapped in living cells during phagocytosis or endocytosis (Gatoo et al. 2014; Tay et al. 2014). It has been found that these cylindrical materials' dimensions significantly affect internalization, with high-aspect-ratio particles internalizing notably more quickly than low-aspect-ratio particles because they have more symmetry. Spherical nanoparticles exhibit higher uptake than nanorods as well as are comparatively less hazardous regardless of whether they are homogeneous or heterogeneous (Gatoo et al. 2014). The endocytosis of these particles is easier and faster than that of rod-shaped or fibre-like nanoparticles. The contact area between rod-shaped or needle-like NMs and the cell membrane receptors may be greater than that of spherical NPs when the rods' longitudinal axis contacts the receptors (Sharifi et al. 2012). Since surface atoms frequently have unfulfilled high energy bonds, the increased surface area of the material improves its catalytic activity and, as a result, has been well observed to boost its reactivity. The more surface area a material has, the easier it will be for these surface bonds to interact with other molecules and stabilize the material. Due to their larger surface area than equivalent micron-sized particles, these nanomaterials have a higher chance of coming into touch with biomolecules, causing direct cellular damage and promoting oxidative stress (Sharifi et al. 2012; Singh et al. 2009).

11.5.3 Effect of Surface Charge

Surface charge significantly influences nanomaterial toxicity, which determines how well they interact with biological systems. The surface charge of nanoparticles has a significant role in controlling the intake of specific nanoparticles, colloidal behaviour, cellular uptake, plasma protein binding, blood–brain barrier integrity, and transmembrane permeability in nanomaterials. Unfortunately, there is no established experimental method for determining the surface charge of NMs. The NM surface engages in interactions with the ions in the media. Specific ions can adsorb in a tightly bonded monolayer just at the NM surface (Tay et al. 2014). Most charged functional groups are responsible for the active engagement of nanoparticles with cells, while neutral functional groups are excellent at preventing unfavourable nanomaterial-biological interactions. Comparatively, positively charged nanoparticles have a lot more cellular uptake than neutral and negatively charged ones (Gatoo et al. 2014). Positively charged particles are the most effective in penetrating cell membranes and internalizing cells; they serve as the leading synthetic platform for drug and gene delivery (Verma and Stellacci 2010). Despite the particles' adverse interaction with the negatively charged cell membrane, there is evidence of negatively charged particle absorption. Nanoparticles with cationic surface charges have been found to have more cytotoxic effects than those with anionic charges.

Furthermore, cationic surfaces are more likely to result in hemolysis and platelet aggregation. However, it is uncertain if the enhanced absorption is frequently connected with cationic nanoparticles or if the surface charge is directly responsible for cell death. Cationic nanomaterials may potentially interact with DNA more readily since it is negatively charged (Singh et al. 2009; Verma and Stellacci 2010). Surface charge, a critical factor in colloidal behaviour, directly impacts how an organism reacts to nanoparticle exposure by altering the shape and size of the particles through the development of aggregates or agglomerates. It has been noted that blood–brain barrier integrity and permeability are affected by nanoparticle surface charge. Nanoparticle surface charge has impacted the blood–brain barrier's permeability and integrity (Sharifi et al. 2012). The scientific community has used a variety of modifications to protect or control NMs' surface properties since surface charge significantly impacts how well NMs interact with biological systems. It helped to diminish the manifestations of NMs' toxicity.

11.5.4 Effect of Composition and Aggregation

It is assumed that the toxicological effects of various nanomaterials may have been primarily influenced by their chemical composition. Despite the importance of particle size in determining nanoparticle toxicity, particle composition is more important in connection to molecular cell chemistry and oxidative stress (Sharifi et al. 2012; Sohaebuddin et al. 2010; Gatoo et al. 2014). The composition of NMs was the main contributing factor to toxicity, even though they were identical in size but had different surface charges. In all the examined organisms, nano silver, nano copper, and their soluble versions were toxic. In comparison, TiO_2 of the same size did not result in any toxicity problems. In addition to chemical composition, surface charge, size, and form, toxicity is influenced by various specific properties (Peralta-Videa et al. 2011; Zhu et al. 2013). Compared to individual nanoparticles, agglomerates are often larger and more rigid (Singh et al. 2009). The size, surface charge, and composition of NPs were among other factors that affect their aggregation states. It makes sense that the NPs will aggregate under the influence of van der Waals forces as they are simple particles with a homogeneous inorganic core. The NPs need to be stabilized to prevent accumulation. It can be achieved by employing the charged surface states of the NMs, attaching or adhering charged molecules to their surface via steric hindrances, or combining the two processes (Tay et al. 2014).

11.5.5 Effect of Medium and Purity

The biological distribution and subsequent action of nanoparticles depend significantly on their proper and stable dispersion in the delivery medium, which impacts their particle size, consequently affecting nanoparticle toxicity. The nature of the medium in which nanoparticles are suspended affects how toxic they are; similarly, the same nanoparticles exhibit distinct toxic characteristics when dissolved in other media (Sharifi et al. 2012; Gatoo et al. 2014). The dispersion of particles can be impacted by medium conditions like pH and ionic strength. Purity of nanomaterials is another crucial factor, in contrast to the actual nanomaterial, metal contaminants may be responsible for toxicological effects. Researchers have made promising attempts to purify the nanoparticles. Nevertheless, it has been demonstrated that these techniques destroy the nanomaterials, even with the attachment of functional groups to their surfaces, which may change their overall biological influence, which might acquire imperfections and alterations (Singh et al. 2009).

11.6 Limitations

In nanotoxicity, it is preferable to take a systematic strategy to investigate the relationship between NMs' physicochemical properties and biological end-point response. Because nanoecotoxicology is still a young discipline, there is a long way to go before a clear picture can be formed concerning dose estimations, exposure assessment, hazard identification, and risk characterization. Fundamentally, free radicals are highly reactive and often have a very short half-life, making it challenging to understand their dynamics at the molecular level. Most biologically significant free radicals have short half-lives, making them challenging to detect. Therefore, to catch the production of ROS from chemical processes, Electron spin resonance (ESR) spin trapping techniques are frequently employed (He et al. 2014a; Zhu et al. 2013). Only ESR spin trapping offers concrete proof of the presence of free radicals and a way to recognize them. ESR is frequently employed in investigations of free radical generation in the presence of nanomaterials, and it is increasingly used in mechanistic studies (Fu et al. 2014). Biochemical and cell-based techniques are frequently applied to understand the acute and long-term toxicity caused by nanomaterials. Determining the mechanisms of free radical formation in vivo is more difficult than determining the mechanisms of chemical reactions (Gatoo et al. 2014). The origin of the test models employed for such a pursuit is one of the drawbacks of this comparison approach. Additionally, there is still room for improvement in applying model test results to people.

11.7 Conclusion and Future Perspective

The pathways of NMs-induced cell death that have been discussed are merely the beginning. Numerous forms of NMs have been developed due to recent breakthroughs in engineering and technology, and they are now used in a wide range of industrial industries. Exposure of the general public will undoubtedly rise as more and more consumer products with NMs enter the market. Concerns over the possible human toxicity and environmental impact of these particles have grown due to the increasing exposure of workers and the general population to NMs. As a result, in recent years, a new field known as nanotoxicology has evolved. Its objective is to make precise how the physical and chemical characteristics of NMs relate to one another and the activation of toxic biological reactions. We emphasize the toxicity of NMs, concerning the oxidative stress paradigm. The two leading causes of NM-induced oxidative stress are (a) the NM's oxidative characteristics and (b) oxidant production during NM interaction with cellular components. The physicochemical characteristics of NM, such as surface reactivity, particle size, surface charge, chemical composition, and the presence of transition metals, are responsible for their direct prooxidant effects. Considering little is known about the potential threats that the growing production and usage of NMs may pose to human health and environmental safety, there appears to be a compelling need to learn more. The first line of defence for hazard mitigation is understanding the mechanism of nanomaterial-induced toxicity. In this chapter, we outlined the processes by which NMs produce ROS at the cellular level and the most current developments in the biomedical sector on NM toxicity due to ROS. The hydroxyl radical has the highest one-electron reduction potential of all physiologically relevant ROS and is reactive with almost all biomolecules, including lipids, proteins, and nucleic acids. Antioxidants are crucial in preventing or reducing the harm that ROS can do. The hydroxyl radical is the most potent ROS-inducing biological harm. However, hydroxyl radical cannot be eliminated by antioxidant enzymes. There is currently a lack of information on the genotoxic potential of nanomaterials and their possible long-term impacts on human health. Most investigations have concentrated on cytotoxicity, and a wealth of relevant data has been produced regarding the unique physicochemical characteristics of nanomaterials. However, this must also be evaluated with DNA damage, and elements like cellular accumulation, retention, and subsequent effects must be considered. Different nanoparticles' genotoxicity may be predominantly caused by the form of the particles rather than their chemical composition. Future development and safety of nanoparticles will benefit from the capacity to create nanoparticles that reduce cytotoxicity to a variety of cells that could be exposed.

Future research will have difficulty determining whether these seemingly contradictory effects depend on particular NM characteristics or the target cell type. A more thorough understanding of how NMs cause cell death requires more profound insights into hypothesis-driven molecular research. We must not ignore the fact that a single type of NM may cause a variety of cell death processes, making the study of nanotoxicology challenging yet exciting. Regarding methods and activity, it is still unclear whether and how ROS production is related to physicochemical features. There is a need to investigate further the processes by which NMs produce ROS because doing so would give researchers more knowledge on how to change the physicochemical properties of NPs to regulate ROS production. Advancement of nanotechnology and the study of nanotoxicology has raised our understanding of the environmental particulate pollution caused by both natural and man-made causes, and we expect that this new understanding will result in a considerable decrease in human exposure to these potentially toxic elements. With additional information and continued research, we will better understand the origins and mechanisms of diseases linked to nanoparticle exposure.

References

Aillon KL, Xie Y, El-Gendy N, Berkland CJ, Forrest ML (2009) Effects of nanomaterial physicochemical properties on in vivo toxicity. Adv Drug Deliv Rev 61(6):457–466

Buzea C, Pacheco II, Robbie K (2007) Nanomaterials and nanoparticles: sources and toxici-ty. Biointerphases 2(4):MR17–MR71

De Stefano D, Carnuccio R, Maiuri MC (2012) Nanomaterials toxicity and cell death modalities. J Drug Deliv. https://doi.org/10.1155/2012/167896

Farnoud AM, Nazemidashtarjandi S (2019) Emerging investigator series: interactions of engineered nanomateri-als with the cell plasma membrane; what have we learned from membrane models? Environ Sci Nano 6(1):13–40

Fu PP, Xia Q, Hwang HM, Ray PC, Yu H (2014) Mechanisms of nanotoxicity: generation of reactive oxy-gen species. J Food Drug Anal 22(1):64–75

Gatoo MA, Naseem S, Arfat MY, Mahmood Dar A, Qasim K, Zubair S (2014) Physicochemical properties of nanomaterials: implication in associated toxic manifestations. BioMed Res Int. https://doi.org/10.1155/2014/498420

Handy RD, Shaw BJ (2007) Toxic effects of nanoparticles and nanomaterials: implications for public health, risk assessment and the public perception of nanotechnology. Health Risk Soc 9(2):125–144

He W, Liu Y, Wamer WG, Yin JJ (2014a) Electron spin resonance spectroscopy for the study of nanomateri-al-mediated generation of reactive oxygen species. J Food Drug Anal 22(1):49–63

Hussain SM, Braydich-Stolle LK, Schrand AM, Murdock RC, Yu KO, Mattie DM et al (2009) Toxicity evaluation for safe use of nanomaterials: recent achievements and technical challenges. Adv Mater 21(16):1549–1559

Manke A, Wang L, Rojanasakul Y (2013) Mechanisms of nanoparticle-induced oxidative stress and toxicity. BioMed Res Int. https://doi.org/10.1155/2013/942916

Migliore L, Uboldi C, Di Bucchianico S, Coppedè F (2015) Nanomaterials and neurodegenera-tion. Environ Mol Mutagen 56(2):149–170

Monteiro-Riviere NA, Inman AO (2006) Challenges for assessing carbon nanomaterial toxicity to the skin. Carbon 44(6):1070–1078

Nel A, Xia T, Madler L, Li N (2006) Toxic potential of materials at the nanolevel. Science 311(5761):622–627

Peralta-Videa JR, Zhao L, Lopez-Moreno ML, de la Rosa G, Hong J, Gardea-Torresdey JL (2011) Nano-materials and the environment: a review for the biennium 2008–2010. J Hazard Mater 186(1):1–15

Setyawati MI, Tay CY, Leong DT (2013) Effect of zinc oxide nanomaterials-induced oxidative stress on the p53 pathway. Biomaterials 34(38):10133–10142

Setyawati MI, Tay CY, Chia SL, Goh SL, Fang W, Neo MJ, Chong HC, Tan SM, Loo SC, Ng KW, Xie JP (2013) Titanium dioxide nanomaterials cause endothelial cell leakiness by disrupting the homophilic interaction of VE–cadherin. Nat Commun 4(1):1–12

Setyawati MI, Tay CY, Chia SL, Goh SL, Fang W, Neo MJ et al (2015) Epigenetic mechanisms in nanomaterial-induced toxicity. Epigenomics 7(3):395–411

Sharifi S, Behzadi S, Laurent S, Forrest ML, Stroeve P, Mahmoudi M (2012) Toxicity of nanomaterials. Chem Soc Rev 41(6):2323–2343

Singh N, Manshian B, Jenkins GJ, Griffiths SM, Williams PM, Maffeis TG et al (2009) NanoGenotoxicology: the DNA damaging potential of engineered nanomaterials. Biomaterials 30(23–24):3891–3914.

Sohaebuddin SK, Thevenot PT, Baker D, Eaton JW, Tang L (2010) Nanomaterial cytotoxicity is composi-tion, size, and cell type dependent. Part Fibre Toxicol 7(1):1–17

Tay CY, Setyawati MI, Xie J, Parak WJ, Leong DT (2014) Back to basics: exploiting the innate physi-co-chemical characteristics of nanomaterials for biomedical applications. Adv Func Mater 24(38):5936–5955

Unfried K, Albrecht C, Klotz LO, Von Mikecz A, Grether-Beck S, Schins RP (2007) Cellular responses to nanoparticles: target structures and mechanisms. Nanotoxicology 1(1):52–71

Verma A, Stellacci F (2010) Effect of surface properties on nanoparticle–cell interactions. Small 6(1):12–21

Yan L, Gu Z, Zhao Y (2013) Chemical mechanisms of the toxicological properties of nanomaterials: generation of intracellular reactive oxygen species. Chemistry–An Asian J 8(10):2342–2353

Yang H, Liu C, Yang D, Zhang H, Xi Z (2009) Comparative study of cytotoxicity, oxidative stress and geno-toxicity induced by four typical nanomaterials: the role of particle size, shape and composition. J Appl Toxicol 29(1):69–78

Yu Z, Li Q, Wang J, Yu Y, Wang Y, Zhou Q, Li P (2020) Reactive oxygen species-related nanoparticle tox-icity in the biomedical field. Nanoscale Res Lett 15(1):1–14

Zhu M, Nie G, Meng H, Xia T, Nel A, Zhao Y (2013) Physicochemical properties determine nanomaterial cellular uptake, transport, and fate. Acc Chem Res 46(3):622–631

Chapter 12
Nanomaterials in Drug Delivery: Application of Polysaccharides and Protein-Based Nanomaterials in Modern Drug Delivery

Annesha Basu, Manasi Halder, Aishwarya Das, Ranjana Das, and Chiranjib Bhattacharjee

Abstract Drug Delivery refers to techniques that help in accomplishing greater analeptic effects of any pharmaceutical drugs by transporting it to its desired location in the body. In recent research, the benefits of using nanomaterials in drug delivery are gaining immense attention. Nanotechnology provides numerous benefits in treatment of human diseases by target specific delivery of precise medicines. Over the synthetic nanomaterials, biopolymers are getting more recognition because of their biocompatibility, biodegradability, low immunogenicity, and antibacterial activity. Also, biopolymers have the capacity of delivering drugs or other bioactive compounds to definite cells or sections of cells despite the fact that they have lower drug binding ability than synthetic polymers. This book chapter represents the wide applications of bio polymeric nanoparticles derived from proteins and polysaccharides. This chapter also provides an insight to the miscellaneous advantages of proteins and polysaccharide-based bio-polymeric nanoparticle.

Keywords Drug delivery · Nanotechnology · Biopolymeric nanoparticles · Polysaccharide · Proteins

12.1 Introduction

The process of creating a new medicinal molecule is time consuming and not economical. The safety efficacy ratio of "old" medications has been improved utilizing a variety of techniques, including dose titration, individualizing drug therapy, and therapeutic drug monitoring. Other extremely appealing strategies that have been actively researched include controlled drug delivery, delayed drug delivery, and targeted drug

A. Basu · M. Halder · A. Das · R. Das (✉) · C. Bhattacharjee
Chemical Engineering Department, Jadavpur University, Kolkata, India
e-mail: ranjanads78@gmail.com

M. P. Shah et al. (eds.), *Biogenic Nanomaterials for Environmental Sustainability: Principles, Practices, and Opportunities*, Environmental Science and Engineering,
https://doi.org/10.1007/978-3-031-45956-6_12

delivery. Pharmaceutical molecules like antibodies, peptides, vaccines, medicines, enzymes, and many more can be treated more effectively and efficiently with the use of drug delivery systems (DDS). DDS have the potential to increase millions of patients' longevity and quality of life. The economics of drug development have shifted as a result of the capacity to formulate current medications in controlled release forms (Almedia and Souto 2007).

12.1.1 Liposome Mediated Drug Delivery

Phospholipids make up liposomes, which self-enclose to form spheres of lipid bilayers and contain aqueous centres. Alec D. Bangham created the first liposomes in England in 1961. Each molecule has an end that is water soluble and an end that is water insoluble. Fat-soluble drugs were absorbed into the phospholipid layer, while water-soluble pharmaceuticals introduced to the water were trapped inside the aggregation of the hydrophobic ends. Phospholipids, which are amphiphilic in nature, produce polar shells in aqueous solutions. The formation of the polar shells is due to the hydrophobic effect. Liposomes have the potential to resemble cell membranes in both structure and chemistry, making them an appealing drug delivery mechanism. Furthermore, amphiphilic compounds that are natural, non-immunogenic, nontoxic, and biodegradable are easily converted into liposomes (Haley and Frenkel 2008). In late 1980 and 1990s, there came many advances in drug delivery to overcome the challenges. These advancements led to a better understanding in morphology of lipids, physiological mechanisms of lipids, liposome disposition in cells, lipid protein interaction etc. This development made it possible to create liposomes with superior designs for more stability both in vitro and in vivo, for improved biodistribution, and for liposomes with optimum stability in circulation. Liposomes were recognised as drug carriers for pharmaceutical purposes by the middle of the 1990s. Liposomes were recognised as drug carriers for pharmaceutical purposes by the middle of the 1990s. The creation of large unilamellar liposomes (LUV) was made possible by innovative techniques that improved the homogeneity and efficiency of the trapping process. Liposomal anthracyclines, which include forms with significantly extended circulation such liposomal daunorubicin and pegylated liposomal doxorubicin, have achieved exceptionally effective drug encapsulation, leading to considerable anti-cancer activity with reduced cardiotoxicity. Doxorubicin pegylated liposomal has demonstrated significant success in the treatment of breast cancer, both when used alone and in conjunction with another chemotherapeutics. For the delivery of various medications, further liposome constructions are being created. Immuno liposomes and other ligand-directed constructions offer a combination of biological elements capable of tumour recognition with delivery technology (Krishna and Pandit 1996). Liposomal delivery systems are still primarily experimental; their exact mechanisms of action in the body and methods of directing them to particular damaged tissues are being researched. Soon after injection, the RES quickly removed the first-generation liposomes. The first method used to create liposomes that circulate for a long time altered the content and size of liposomes. Additionally, it was found that small liposomes had a lower clearance rate by the reticuloendothelial system (RES) than bigger liposomes. Likewise, Allen and Everest noted that smaller unilamellar vesicles had circulation times that were higher than those of larger multilamellar vesicles. Several modifications were made in formulation of liposomes to avoid RES uptake. Likewise previous researches have constructed liposomes using saturated phospholipids having higher transmembrane (TM). This has led to better retention capacity and increased circulation time for large vesicles. Modifying the surface of the liposome molecules by gangliosides or sialic derivatives has increased the stability. Later, PEG, a hydrophilic polymer has also been used to form liposomes for enhanced stability. Additionally, the addition of polyethylene glycol-lipid conjugates, such as methoxy polyethylene glycol (mPEG) distearoylphosphatidylethanolamine (DSPE), to the bilayer membrane lengthened the half-life of the liposomes in blood circulation and encouraged liposome accumulation where leaky vasculature was present by the EPR effect.

Conventional liposomes, sterically stabilized liposomes, ligand-targeted liposomes, and a combination of the aforementioned are the four primary types of liposomal delivery systems. Sterically-stabilized liposomes were developed to increase the stability of liposomes and speed up blood circulation. Polyethylene glycol (PEG), a hydrophilic polymer, has been demonstrated to be the best for creating sterically-stabilized liposomes. The steric barrier aided in up gradation of the enveloped agents, by lowering in-vivo opsonization with serum components, thus preventing the encapsulated compounds from being quickly recognised and cleared by the RES. This strategy provides a number of different benefits, such as minimizing drug elimination by extending blood circulation, providing accumulation at diseased areas, and reducing adverse effects. Longer circulation periods are a benefit of PEG-coated liposomes; however, this may result in a loss in their ability to interact with the targeted molecules (Sercombe et al. 2015).

12.1.2 Transdermal Drug Delivery

One of the most often employed approaches to drug administration is parenteral delivery. Transdermal applications, which serve as a substitute for controlled (therapeutic) release of medications, show some advantages over the parenteral and oral routes and can help overcome some of their drawbacks. There are two categories of factors that can affect transdermal delivery: biological and physicochemical factors. Biological aspects to take into account skin health, skin aging, blood flow, skin region, skin metabolism, and species variations. Transdermal medication distribution is influenced by a number of physicochemical parameters, including skin moisture, temperature, pH, diffusion coefficient, drug concentration, partition coefficient, drug molecular size, and shape. Prausnitz and Langer have described TDDS in three generations. Transdermal patches are used in the first generation to achieve continuous delivery of low molecular weight compounds. The stratum corneum's structure is temporarily and irreversibly altered or damaged using procedures like ultrasound and iontophoresis in the second generation, which increases the skin's permeability and enables the transfer of greater molecular weight molecules (Benson 2005). The stratum corneum is more aggressively disrupted in the third generation as a result of the implementation of a variety of enhancers.

12.1.3 Microemulsion Drug Delivery System

Microemulsions are defined as clear, transparent, thermodynamically stable dispersions of oil and water that are frequently combined with a co-surfactant and stabilized by an interfacial surfactant coating. Microemulsions deliver drugs in many routes which includes parenteral, transdermal, nasal and oral.

12.2 Challenges in Drug Delivery

Cytotoxicity is one of the major limitations in commercializing liposomes as a drug delivery system. Certain investigations have shown charged liposomes to be hazardous is another cause for concern. Additionally, liposomes face manufacturing-related difficulties such as batch-to-batch variances, little drug entrapment, efficient sterilizing techniques, stability issues, and, most significantly, scaling issues. Transdermal drug delivery also has certain challenges. Inability of TDDS's to deliver ionic medicines. TDDS is not appropriate for medications with large molecular sizes since it cannot achieve high drug levels in blood or plasma. The inability of TDDS's to pulse-deliver medications. If a medicine or formulation irritates skin, TDDS cannot be created. Microemulsion techniques become a costly method as it requires several instruments for size reduction. Homogenizers, for example, are a crucial piece of equipment for creating micro-emulsions, and techniques like micro fluidization and ultrasonication are expensive to implement. Stability is also a major issue for long term storage of microemulsion.

Major difficulties arise when using large-scale materials for drug administration, including in vivo instability, low bioavailability and solubility, poor body absorption, problems with target-specific distribution, problems with tonic effectiveness, and possibly harmful pharmacological effects. Advancements in drug delivery systems hence have brought new dimensions to deal with the restrictions of existing systems. Nanotechnology has offered several changes in existing forms and also has formulated new systems. Advanced drug formulations, targeted areas, and their well-controlled drug release and administration all benefit greatly from the use of nanotechnology (Jahangirian et al. 2017).

12.2.1 Nano-Based Drug Delivery

Nanostructures allow the release of combined medications at the prescribed dose since they persist in the blood circulation system for a long time. Consequently, they create fewer plasma oscillations and have fewer negative consequences. Nanostructures provide advantage in penetrating the tissue and ease the uptake of drugs and also render controlled delivery to targeted regions. When creating target-specific drug delivery systems, metallic, organic, inorganic, and polymeric nanostructures, such as dendrimers, micelles, and liposomes, are commonly taken into account. These nanoparticles are specifically applied to medications with poor solubility and poor absorption capacity (Fig. 12.1).

12.3 Different Types of Nano Carrier

Nanoparticles' size range varies between 10 and1000 nm, and is novel compounds used as colloidal drug delivery systems. In modern oral drug delivery drugs are dissolved and encapsulated inside the matrix of nanoparticles to promote more proper and site-specific drug delivery. Nowadays, protein and polysaccharides along with synthetic polymers are used to build nanoparticles. Most popular or common Nano systems employed for drug delivery systems are Nano micelles, Nano liposomes, Nano spheres, Nano capsules etc.

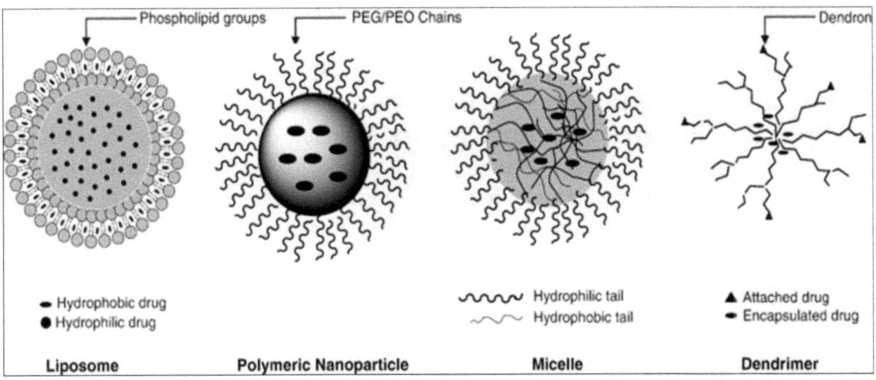

Fig. 12.1 Some nanocarriers for improved drug delivery (Ganta et al. 2008)

12.3.1 Nano Micelle

Nano micelles are amphoteric molecules meaning they have well defined hydrophobic and hydrophilic segments and they vary within a size range of 20–100 nm. The hydrophobic or lipophilic cores of micelles entrap the fat soluble drugs and hold it stable for a long time in any aqueous medium. In case of polymeric Nano micelles, they are assembled by themselves in micelle form when introduced to aqueous solution.

Usually biodegradable polymers are used in manufacturing these micelles for drug delivery systems. Polymers which are used most frequently to build Nano micelles for oral drug delivery are poly lactide, poly propylene oxide (PPO), poly glycolide, poly lactide-co-glycolide, poly ethylene glycol (PEG). Among all of these molecules PEG is most widely used for its amazing water solubility and higher biocompatibility. In case of oral drug delivery utilization of these polymeric Nano micelles increases oral bioavailability more than three times than before of lipid soluble drugs (Ganta et al. 2008).

12.3.2 Solid Lipid Nanoparticles

Solid lipid nanoparticles are great colloidal transporters with some excellent properties such as very small compact size, greater biocompatibility, stable physical framework, excellent emulsion formation techniques etc. They help in site specific delivery of water soluble drugs. Controlled release of the drug can also be achieved by using these nanocarriers.

Solid lipid nanoparticles are basically lipids that remain solid at room temperature. These solid lipid nanoparticles can be made of using fatty acids (e.g. stearic acid) or their esters (e.g. glyceryl monostearate, glyceryl behenate etc.), steroids (for example, cholesterol), waxes (e.g. cetyl palmitate), triglycerides (like tristearin, trilaurin etc.). Use of solid lipid nanoparticles enhances the effectiveness of oral drug delivery; also these are easily degradable by enzymes like pancreatic lipase or bile salts in the human body (Mehnert and Mäder 2012).

12.3.3 Dendrimers

Dendrimers are a very prominent three-dimensional branched molecule used as nanocarriers in oral drug delivery. All the branches or bonds emerge from a central core in case of these dendrimers. This well-defined core makes it an outstanding carrier of drug cause drug physical encapsulation. But in case of some drug molecules physical encapsulation does not work, molecules can be hooked inside this core.

Primarily the core entraps any molecule by the process of attaching themselves by forming non covalent bonds (e.g., hydrogen bonding) with the dendritic construct.

The large surface area of these dendrimers also helps in drug delivery by forming covalent bonding with the drug molecules. Antibodies and sugar molecules can form these covalent bonds with the surface of these dendrimers. Undoubtedly dendrimers have great potential as drug carriers but their biocompatibility and cytotoxic effects are still under study (Ganta et al. 2008).

12.3.4 Nanoliposomes

Nanoliposomes are nanosized particles made up of two main sections, namely the aqueous core and phospholipid bilayer. These nanocarriers are capable of easy entrapping of the drug, target site delivery and sustained release of the drug in the target section of the body. They can easily deliver low molecular weight drugs and other particles such as imaging agents, nucleic acids, peptides etc.

Mode of drug encapsulation in case of nanoliposomes, depends on the polarity of the drug. If the drug is polar in nature, it easily gets entrapped in the hydrophilic core of the liposome. On the other hand, in case of non-polar drugs, the hydrophobic phospholipid bilayer becomes the place where the drug can easily reside (Demirci et al. 2017).

12.3.5 Nano Capsule

Nano capsules are structures composed of two main parts. Active core, which can be solid or liquid and a protective membrane made up of synthetic or natural polymers. The core can be made as aqueous or lipid based depending on the polarity of the drug needed to be carried. A wide range of natural, semi synthetic, and synthetic polymers are used in formation of these nano capsules. Some of the polymers used in making these nano capsules are gum arabica, hydroxypropyl methyl cellulose, phthalate, poly (D L lactide), poly E-caprolactone, poly alkyl cyanoacrylate etc. Wide range of mineral and natural oils is also used in building nano capsules along with polymers. Lecithin is the most used surfactant in manufacturing nano capsules. These nano capsules are mostly used in delivery of peptides and proteins through oral route, on reaching the target location it bursts open & delivers the drug (Couvreur et al. 2002).

12.3.6 Nanospheres

Polymeric nanospheres within the size range of 2–100 nm, made up of biodegradable polymers, are used for targeted drug delivery. These nanospheres have a stable spherical structure with a hollow core. The empty cavity or the core present inside the sphere entraps the drug molecules easily. Else, the drug can attach itself to the matrix of the solid nanospheres. To manufacture these nanospheres, polymers are needed to be dissolved in organic solvents and then they are cross-linked to form the sphere shape (Ganta et al. 2008).

12.3.7 Nanogel

Nanogels or hydrogels are nanosized particles, formed by physically or chemically crosslinked polymer networks so that nanogels are swell in a good solvent. Due to the presence of a hydrophilic group of nanogel, it can hold a massive amount of water. That is why nanogels are highly biocompatible and biodegradable also. Traditional carriers for skin delivery confront various difficulties such as obstacles of drug leakage, minimal permeation and low embroilment efficiency. Nanogels have recently been recognized as renowned delivery systems to overcome such difficulty and impart peaceful application. Nanogels have outstanding advantages such as easily permeable through membranes, non-toxic, high drug loading capacity (Ghaywat et al. 2021).

12.3.8 Drug Phospholipid Complex

Many herbal and manmade drugs create difficulty of poor oral bioavailability, because these are very low water solubility or poor permeation through the biological membrane. Low bioavailability and low water soluble capacity of synthetic drugs results in inefficacy in therapeutic delivery. Nowadays, many researchers draw attention to drug phospholipid complex preparation because it can be overcome these problems. Drug phospholipid complex is also called Pharmacosome. Pharmacosomes are nano size micelles, vesicles or may be in the form of hexagonal assembly of colloidal drug dispersions attached covalently to the phospholipid. They are excellent carriers for drug delivery because of their beneficial properties such as high entrapment efficiency, high drug loading capacity, low size, stability and amphiphilic nature. They also assist in controlled release of drug at the site of action, reduced cost and increase bioavailability of drug (Pandita and Sharma 2013).

12.3.9 Nano Emulsion

Emulsions are used in drug delivery for their unique amphiphilic nature. Nanoemulsions are oil-in-water emulsions, size ranging from 50 to 1000 nm. Average particle size is 100–500 nm, and can remain as oil-in-water and water-in-oil forms. High shear stress and mechanical extrusion process are used in nanoemulsions preparation by mixing a water immiscible oil phase with an aqueous phase (Jaiswal et al. 2014). Nanoemulsions are solid spheres; their surface is amorphous and lipophilic with negative charge. Site specificity of these nanocarriers can be increased by magnetic nanoparticles. As a nanocarrier of drug delivery systems they increased therapeutic efficacy and reduced adverse effects and toxic reactions. Crucial uses of these nanocarriers include treatment of infection of the Reticuloenthelium system, enzyme replacement therapy of the liver, treatment of cancer and vaccination. Nanoemulsions can give protection of drugs, which are easily susceptible to hydrolysis and oxidation. Greater surface area of nanoemulsions can give better absorption. As nanoemulsions can be considered as safe, effective, and increased bioavailability these are used in different types of drug delivery systems such as ocular, parenteral, topical, and oral drug delivery.

12.4 Polymer Based Nano Carriers

Monomers are joined together to form macromolecules also known as polymers which are broadly found in nature. Polymers can be classified as natural polymers, synthetic polymers and semi synthetic polymers. Natural polymers are naturally found in animal and plant sources such as polysaccharides, protein, nucleic acid and lipid. Man- made polymers are called synthetic polymers such as phosphorus based polymer, acrylic based polymer etc. semi-synthetic polymers are derived from natural polymers. Use of nanoparticles in the field of drug delivery is a modernist concept. Nanoparticles are small colloidal particles which act as carriers for targeted drug delivery. Different types of nanoparticles are used in drug delivery systems but polymeric nanoparticles are the most valuable route of drug delivery systems. Different polymeric nanoparticles were discovered by researchers but biodegradable natural polymer-based nanoparticles have various advantages such as non-immunogenic, non- allergen, less toxicity etc. The soporific preparation technique of nanoparticles overcome by the newly arise natural polymer-based nanoparticles. Polymer-based nanoparticles used in the form of Nano gel, nanomicelles, nanoliposomes, nanocapsules, nanospheres etc. Among them nanocapsules and nanospheres are widely used. As polymeric nanoparticles have the unique properties such as target specificity, reduced site effects, it is used in drug delivery mostly nowadays. Treatment of some diseases such as cancer is very difficult because it requires target organ specific action. The newly developed polymeric nanoparticles have made the cancer treatment lot more facile (Sur et al. 2019).

12.5 Methods of Polymeric-Nanoparticles Synthesis

Synthesis of polymeric nanoparticles is done in two main steps. Firstly, nanoparticles are formed using basic methods of nanoprecipitation or solvent evaporation or salting out or supercritical fluid technology or dialysis. Then formed nanoparticles are polymerized to produce these polymeric nanoparticles. For polymerization, processes like emulsion polymerization or interfacial polymerization or living radical polymerization can be utilized.

12.5.1 Procedures of Polymeric Nanoparticles Formation

Several methods of polymeric nanoparticle synthesis and associated issues have been illustrated in this section. Figure 12.2 highlights various procedures of the same. Solvent displacement, emulsion-solvent evaporation, emulsions- diffusion, solvent displacement, double emulsion and evaporation, salting out and dialysis are the most necessary methods for preparation of polymeric nanoparticles from preformed polymers.

12.5.1.1 Nanoprecipitation

In the Nanoprecipitation process, the production of the nanoparticles depends on the polymer precipitation and solidification caused by interfacial deposition of the polymer.

This process is also known as interfacial deposition method or solvent displacement method or desolvation method. Fessi was the first person to introduce this technique of nanoparticles formation in 1989. The nanoparticles formation is done in two major steps, Precipitation and solidification of the polymers. As said earlier this solidification of the polymers occurs due to interfacial deposition.

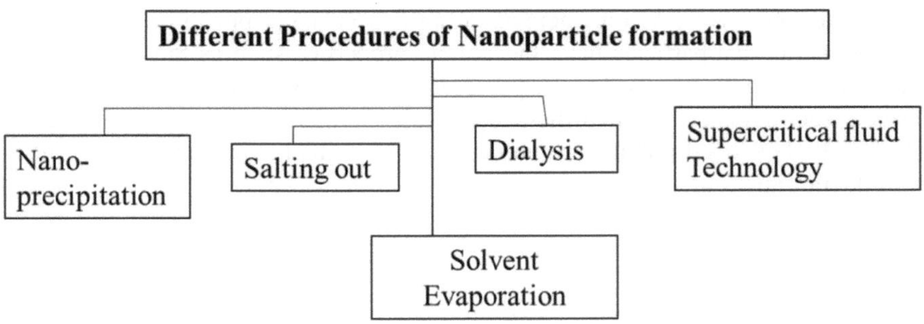

Fig. 12.2 Various procedures of nanoparticles formation

In this technique a desolvating agent or organic solvent (preferably miscible with water) is needed to be added. As organic solvents acetone or ethanol is used generally. The drug and the polymer need to be dissolved in that desolvating agent. But unstable nanoparticles are formed this way. So, a stabilizer is added and the solution is stirred continuously to form stable nanoparticles (Quérette et al. 2019).

12.5.1.2 Solvent Evaporation

Solvent evaporation is one of the easiest and widely used procedures for preparation of nanoparticles. This preparation method is carried out in four major stages.

(a) **Dispersion stage**: An organic solvent is needed in this stage. The drug and the polymer both are added one by one in the organic solvent. Then the solution is stirred continuously till both the drug and the polymer get dissolved in it.
(b) **Surfactant addition**: Then the solution containing dissolved drug and polymer is added into an aqueous solution of a surfactant. This step is necessary to form a stable emulsion. Surfactants like Poloxamer 188 (P188), Polysorbate 80, and Polyvinyl Alcohol (PVA) etc. are used commonly. This procedure is carried out inside a homogenizer.
(c) **Evaporation of the solvent**: In this stage, the emulsion is converted into a suspension of nanoparticles. This suspension is then addressed to high heat under low pressure to evaporate the solvent.
d) **Solidification of the nanoparticles**: To form solidified nanoparticles ultracentrifugation can be done (Guo et al. 2018).

12.5.1.3 Salting Out

In general, salting out is a technique in which high concentration of salts are used in a solution to initiate precipitation of the dissolved chemicals by reducing their solubility.

First, the drug and the polymer are dissolved in an organic solvent. Then, an aqueous phase containing a surfactant is added to form an emulsion. This solution is added in an aqueous solution containing a salting out agent. Sulfates and Chlorides of Magnesium, Sodium and Calcium are generally used as salting out agents. This step leads to precipitation of the polymer. The drug is generally encapsulated inside the nano-polymer matrix (Hu et al. 2019).

12.5.1.4 Dialysis

This technique is very similar to nano-precipitation or desolvation procedure. Here also the drug and the polymer are dissolved in an organic solvent. The main difference is that this water miscible organic solvent-drug-polymer mixture solution is placed inside a dialysis membrane. In general, a dialysis membrane is a semipermeable membrane that helps in separation of toxic molecules using the techniques of differential diffusion.

As a result of using this dialysis membrane, the easy diffusion of the solvent occurs in the aqueous phase. So, inside the membrane, a cluster of nanopolymers is formed (Chen and Jafvert 2018).

12.5.1.5 Supercritical Fluid Technology

This is the only technique that does not involve an organic solvent. Instead, an easily available natural solvent is used, which is environment friendly too. Usually, Carbon dioxide is used as a solvent in its supercritical state. Supercritical CO_2 is a liquid condition of CO_2 where it is kept on or over a critical temperature and pressure.

As usual the drug and the polymer both are dissolved in this supercritical solvent. This solution then expands quickly, which causes precipitation of the nanoparticles through the process of homogeneous nucleation. Nanoparticles of very fine size can be derived in this technique (Adschiri and Yoko 2018).

12.5.1.6 Some Other Processes of Nanoparticles Formation

Some other processes of nanoparticle's formation may be categorised as follows-

i. **Emulsification process**: In this procedure, first a lipid like cotton seed oil or an organic solvent is added and emulsified with the aqueous solution of the protein. Then to stabilize the emulsion high surfactant concentrations are used or through energetic stirring the emulsion is made permanent. Then these albumin droplets are needed to be hardened. For thermal hardening a high temperature of 110 ° C is introduced for 10–30 min. Else a high-pressure homogenizer or chemicals like glutaraldehyde can be used to solidify the biopolymer droplets. Lastly, these manufactured nanoparticles are washed with alcohol or ether (Demirkurt et al. 2019).
ii. **Thermal Gelation**: In this process, the solidification of nanoparticles is done by introducing heat to the aqueous or alkaline solution of albumin for a set amount of time. Generally, a temperature of 80 °C is applied for a period of time. During this procedure, the pH is maintained at pH 7 and then the solution is cooled down. The solution needs to be stirred continuously till the nanoparticles are formed (Martínez-López et al. 2020).
iii. **Self-assemblyBreak** method can also be used to form polymeric nanoparticles. High intensity ultrasound techniques are used to make nanoparticles through this process (Martínez-López et al. 2020).

12.5.2 Polymerization of Formed Nanoparticles

Polymerization is the process through which formed monomeric or single unit nanoparticles are combined with each other and form a polymeric structure. If the polymerization occurs when the monomers are present in different interfaces the procedure is called interfacial polymerization. If the polymerization is done when the encapsulation occurs using a single or double emulsion process, then it is called single/double emulsion technique polymerization. One needs to be very careful throughout performing any of these polymerization processes, because of a high chance of formation of toxic substances or toxic by-products (Sur et al. 2019).

Some advance methods of polymerization currently in use:

i. Advanced methodologies of synthesis of polymeric nanoparticles
ii. Ring-opening polymerization
iii. Electrohydrodynamic atomization
iv. Mussel-inspired chemistry for polymerization
v. Self-polymerization etc.

12.6 Polysaccharides Based Nanoparticles for Drug Delivery

In modern times, polysaccharides have come to play a major role in the modern drug delivery system. So that, there is an increased attentiveness in their isolation, synthesis, modification, characterization, and application in this relevant new drug delivery system. Natural polymer such as polysaccharides, protein, lipid and nucleic acid used in drug delivery systems. Among them polysaccharides are used extensively because they are easily available in nature and also low cost too. Polysaccharides, also known as glycan, are complex bio macromolecules that are made up chains of monosaccharides, and are joined by glycosidic linkages. The general chemical formula for polysaccharides is $C_X(H_2O)_Y$. Carbohydrates have available from various natural resources such as animal origin (chitosan, hyaluronic acid), algal origin (alginate, carrageenan, galactans), plant origin (cellulose, starch, pectin), microbial origin (fullulan, dextran). Polysaccharides are very useful in drug delivery and targeting because they are highly stable, hydrophilic in nature, biodegradable, easy to modify chemically and biochemically, safe and nontoxic (Ali and Ahmed 2018). Polysaccharides have universal medicinal applications in oral drug delivery for the following reason: -

i. Polysaccharides are opulent in nature and isolation from natural sources is very easy.
ii. As a natural polymer, polysaccharides are biodegradable, biocompatible compared to other synthetic polymers.
iii. Presence of various groups such as hydroxyl group, amino groups, carboxyl groups make these polymers easy to conjugate with other biomolecules.

12.6.1 *Classification of Polysaccharides Based on Their Varying Sources*

This section involves classification of polysaccharides based on characteristic features of resource materials. The resource materials with salient properties have been presented in subsequent segments.

12.6.1.1 Polysaccharides from Animal Origin

Chitosan: The second most useful natural carbohydrate polymer after cellulose, is chitosan in the drug delivery system. It is a linear polysaccharide, a large number of sugar units present in this polymer and obtained from plant origin. The chemical formula of chitosan is $C_{56}H_{103}N_{9039}$.Chitosan is composed of two sugar units mainly that are d-glucosamine and N-acetyl-d-glucosamine units, which are distributed randomly. Chitosan is a structurally modified form of chitin by alkaline acetylation. Chitin, is a natural polymer, an amide derivative of glucose present in crustacean's shells such as crabs, shrimps, lobster, prawns etc. Sood et al. (2021).

Chitosan is highly used in drug delivery systems, tissue engineering, and wound-healing dressing, due to the presence of the specific advantages of these such as biocompatibility, biodegradability, nontoxicity, high drug loading ability etc. Various excellent properties such as cationic nature, controlled drug release, muco-adhesion, antimicrobial, permeation enhancement are present in chitosan due to presence of the primary amino group. Chitosan plays an important role in fat metabolism in the body and enhances the aqueous solubility of poorly soluble drugs. Due to the presence of hydroxyl group and amino group of chitosan can be easily modified and prepared in various forms such as threads, nanoparticles, Nano gel beads, matrix etc. Most usable chitosan in a targeted drug delivery system are chitosan microparticles, chitosan microbeads, chitosan beads, chitosan sponges, etc. Chitosan is freely soluble in acidic solution and insoluble in water and organic compounds. Chitosan, prepared from modified deacetylated chitin nanofiber, can be used in preparation of nanogel that is used in drug delivery systems. Various investigations showed that chitosan-based nanoparticles have been used in drug delivery systems by orally or other route in several diseases such as gastrointestinal disease, pulmonary disease, dermatological disease, and brain and ocular infection (Ali and Ahmed 2018).

Hyaluronic acid: Hyaluronic acid is a naturally occurring glycosaminoglycan and contains two units N-actyl-D-glucosamine and D-glucuronic acid. Hyaluronic acid found from the body's connective tissue.

Due to the presence of unique properties such as biodegradability, biocompatibility, high viscoelasticity and non-immunogenicity, it can be used widely in controlled release and targeted drug delivery systems. The functional groups such as amino groups, carboxyl and acetyl amino groups present in the main chain of the hyaluronic acid that can be modified and used in nanoparticles drug delivery system, gel drug delivery system, nanoemulsions drug delivery systems, polyelectrolyte microcapsule drug delivery system, microsphere drug delivery system, cationic polymer gene carrier system etc. Additionally, hyaluronic acid helps in killing tumour cells because it has the ability to recognize the specific receptors that are overexpressed on the surface of tumour cells. Hence hyaluronic acid getting much attention to the researcher for drug delivery system (Huang and Huang 2018).

12.6.1.2 Algal Polysaccharides

Alginates: Alginate is a naturally occurring algal linear polysaccharide polymer, water soluble and is highly biocompatible and biodegradable in nature. Alginate is composed of two components namely β—D-mannuronic acid and α—L-guluronic acid.

Alginate used in oral drug delivery in the form of hydrogel, nanoparticles, microparticles and porous scaffolds. Ionic gelation, emulsion, covalent cross linking, complexation method and self-assembly methods are used in preparation of alginate. Among them ionic gelation and complexation methods are extensively used in alginate nanoparticles preparation. In an ionic crosslinking approach, the divalent cations bind to guluronate blocks of alginate chains from which aqueous solution forms alginate based nanogel. Alginate is well known as the one of most widely preferred natural polysaccharides for synthesizing hydrogels, due to its outstanding gel formation ability. The conventional roles of drug delivery of anticancer therapeutics can be manipulated by the novel alginate-based hydrogels (D'Ayala et al. 2008).

Carrageenan: Carrageenan are the natural linear polysaccharides obtained by the extraction of certain red and purple seaweeds of *Rhodophyceae* class. Carrageenan consists of two carbohydrates such as D-galactose and 6-anhydro-D-galactose.

Carrageenan has been used as a gelling agent and viscosity enhancing agent for controlled targeted drug delivery systems for its outstanding characteristics such as the strong negative charge. Besides a good gel formation characteristic of carrageenan, it also has shown some useful properties such as anticoagulant, anticancer, antihyperlipidemic, and immunomodulatory activities etc. In other hand carrageenan has been used for tissue regeneration with therapeutic biomacromolecules and for cell delivery. Carrageenan is used in drug delivery as nanoparticles, nanospheres, microcapsules etc. (Li et al. 2014).

12.6.1.3 Polysaccharides from Higher Plants

Cellulose: Cellulose is another natural polymer of polysaccharides. The linear homopolysaccharides cellulose contains anhydro—D–glucopyranose units, joined together with β (1–4) glycosidic linkage. Cellulose has hydroxyl functional groups. So that cellulose has a tendency to self-assemble and form a broader network via intermolecular hydrogen bonds.

Nowadays, a number of researchers used to pay attention for find out various way of synthesis and processing cellulose and its derivatives for prospective use in drug delivery system because of its excellent properties such as biodegradability, biocompatibility, low toxicity, low cost, abundant availability.6 Nanoprecipitation methods and hydrolysis in acids and alkali method used in cellulose nanomaterial preparation. These structurally modified cellulose nanoparticles have many effects and help in colloidal stability. Further, nanofibre can be used to enhance the aqueous solubility of a drug and consequently increase the sustainability of drug delivery ability (Löbmann and Svagan 2017).

Starch: Starch is a most popular, low cost, natural storage polysaccharides easily available in nature, is used as an essential food in human and animal diets. Starch has a wide range of applications in drug delivery system and biomedical field because of its distinctive physicochemical and functional properties and it can be found from natural resources such as rice, wheat, potato and green plants (Kaur et al. 2007). From the viewpoint of the structure, starch is a homopolysaccharides and consists of two units these are amylose and amylopectin made up of D glucose units joined by glycosidic linkages (Beneke et al. 2009).

Starches play an important role in bone repair and replacement and general uses of starches as a nanocarrier in controlled drug releases. Starches are used for these applications because of some special characteristics such as solubility, enzymatic digestibility, rheological and morphological characteristics, swelling and mechanical properties. The interaction of drug molecules with reactive sites of starch can be increased by chemically structure modifications and modified starch can also easily metabolized in human body and increase its biocompatibility (Chomto and Nunthanid 2017).

Pectin: Complex polysaccharides, pectin is present in cell wall materials of the higher plants such as apple, oranges, grapes etc. The average sizes of pectin are 84–107 μm in diameter. They were spark like particles, amorphous, solids, and moderately hygroscopic in nature. Molecular weight of these polysaccharides is very high and consists α (1–4) linked esterified D-Galacturonic acid. The degree of methyl esterification of D-Galacturonic acid decides the numerous properties of pectin such as solubility, viscosity, gelling properties etc.

Pectin is easily soluble in water. The presence of various functional groups such as hydroxyl, carboxyl, carbomethoxy, and acylamino groups helps pectin to formation various derivatives. The hydrophilic nature of pectin helps to modified these easily and can be used in oral drug delivery system and for the preparation of modified release formulation. Chemical modification of pectin helps in changing some properties such as hydration, swelling, erosion; these properties create problems in drug delivery. Pectin may be considered as acceptable polysaccharides based nanocarrier in drug delivery because their adaptable modifications approaches and anticancer bioactivity. In other hand pectin has showed some advantages such as a particular gelling mechanism, biocompatibility, and microbial degradation in GI tract (Singh et al. 2015).

12.6.1.4 Polysaccharides from Microbial Sources

(a) Pullulan: Pullulan, is a water-soluble polysaccharide polymer consisting repeated units of maltotriose produced by yeast like fungus *Aureobasidium pullulans*. The unique characteristic such as $\alpha - (1 \rightarrow 4)$ and $\alpha - (1 \rightarrow 6)$ linkages are present in pullulan. Due to the presence of the unique linkages, pullulan is hydrophobic and hydrophilic in nature.

In the field of drug delivery, the utility of pullulan can be increased by the chemical modification of its structure. In drug delivery pullulan plays an important role because drug conjugated pullulan shows high bioactivity with the release of cytotoxic molecules. In the treatment of various diseases such as liver, lungs, brain, and spleen, the conjugated pullulan can be used as targeted drug delivery systems (Hu et al. 2021).

(b) Dextran: Naturally available polysaccharides, dextran and its derivative have gained attention in targeted drug delivery systems because of their unique properties including biodegradability, biocompatibility, and nontoxicity. A large number of reactive hydroxyl groups present on the dextran backbone so that different derivatives have been developed by chemical modification. Nowadays, various dextran based delivery systems have been developed such as self-assembled micelles and nanoparticles, hydrogels. Nanoemulsion, magnetic nanoparticles, microparticles etc. Currently dextran based nanoparticles used in treatment in various diseases, among them play an important role in most popular diseases, cancer treatment. In spite of the fact that chemotherapy is popular in cancer treatment but most anticancer drugs have serious toxicity and poor water solubility. Dextran based Nano carrier is most popular in drug delivery system because it can increase some other processes of nanoparticles formation, the solubility and availability of drug. The physicochemical properties, release mechanism, and therapeutic effects of dextran based nanoparticles help in cancer treatment. Intravenous injection route is most appropriate route for delivery of dextran based derivatives loaded drug (Sinha and Kumria 2001).

12.7 Protein Based Nanoparticles for Drug Delivery

Proteins are biopolymers made up of a long chain of amino acids. These amino acids join themselves in a proper sequence by forming covalent peptide bonds. Depending on their configuration, proteins can be classified into three major classes—A. simple proteins, B. conjugated proteins, C. Derived proteins. When broken down, simple proteins only provide amino acids. Means simple proteins are only made up of amino acids. Albumins, Globulins, glutelins, albuminoids, Histones, protamines are some examples of simple proteins. On the other hand, conjugated proteins not only contain amino acids but they also have a prosthetic group or non-protein portion. Conjugated proteins are named after their prosthetic groups, for example: (i). Lipoproteins (when the prosthetic group is lipid by nature), (ii). Phosphoproteins (when the prosthetic group is phosphorus), (iii). Nucleoproteins (proteins combined with nucleic acids), (iv). Metallo-proteins (when the prosthetic group combined with protein is a metal ion). The last category, derived proteins are nothing else but derivatives of simple and conjugated proteins by the action of chemicals, partial hydrolysis, enzymes, heat etc. Gelatins, Caseinates and some artificially produced polypeptides also come under this class of derived proteins.

12.7.1 Albumins

Albumin is water soluble, heat labile, globular protein. It is also semi soluble in salt water. There are various sources of albumins but for pharmacological use, serum albumin and bovine α- lactalbumin are used. In human and other vertebrates, serum albumin is the major constituent of blood plasma. Liver is the main manufacturer of plasma albumin. A healthy liver can synthesize almost 10.5 g of albumin in 24 h. The turnover time for albumins is 25 days. Albumins have great physiological importance, like they can maintain pH level, osmotic pressure. Also, it has some antioxidant properties as it can entrap free radicals. Though albumin is negatively charged, it can bind with both negative and positively charged particles. In the current scenario, albumins are used as an expanding agent of blood volume to treat burn injuries, sepsis etc. On the other hand, bovine α-lactalbumin is a great source of tryptophan, leucine, lysine, isoleucine, valine, cysteine etc. amino acids. Albumin nanoparticles can be formed by performing emulsification, thermal gelation, desolvation, self-assembly etc. techniques. Benefits of using albumin nanoparticles as drug carriers can be summarised as: (i) enhancement of stability and drug concentration, (ii) reduced drug wastage in gastric juice, (iii) improvement of cellular uptake, (iv) improvement in biocompatibility and reduced cytotoxicity, (v) enhancement in anti-angiogenic effects of the entrapped drug in vivo (Merlot et al. 2014).

12.7.2 Globulins

These globular proteins are insoluble in pure water but soluble in salt water. α, β and γ are the three main fractions of globulins present in all mammals. Liver is the main manufacturer of α and β globulins, for example fibrinogens, whereas, plasma cells and lymphocytes produce γ-globulins. Immunoglobulins A, M and G are examples of γ-globulins. Not only animals, some plants are also a great source of globulins. In some plant seeds (Peas, Soybean, French bean and other legumes), Globulin is present as storage proteins. Two major types of plant-based globulins are: Vicilin and Legumin. Globulins from Pea and soybeans are mostly used for nanoparticle formation. On the other hand, β-Lactoglobulin are, the chief globulins present in whey, also use in nanoparticles formation. Easily Nanoprecipitation or Thermal gelation or

Self-assembly techniques can be performed in order to form nanoparticles of Globulins. Advantages of using Globulin Nanoparticles can be summerised as: (i) 6–7 times drug stability enhancement in aqueous medium, (ii) reduced leakage and wastage of entrapped drug in gastric juice, (iii) more controlled release of the drug and prolonged effect with reduced cytotoxicity, (iv) Target specific drug delivery (Gunasekaran et al. 2007).

12.7.3 *Prolamins*

In the seeds of cereal grains, prolamins are present as storage protein. Prolamins are insoluble in water but soluble in alcohol-water mixture. For example, they show great solubility in 60–80% ethanol solution. The main cereal sources of prolamins are wheat, corn, barley, oats, rye, sorghum etc. Prolamins from varied sources are named differently, for example, Wheat prolamin is called Gliadin, Rye prolamin is called Secalin, Hordein is Barley prolamin, Zein is prolamin from corn and Prolamin from Sorghum is known as Kafirin. Patients of Celiac disease are intolerant to Gliadin, Secalin and Hordein. But, Kafirin and zein are known as safe to be consumed by celiac disease patients. Presently, researchers are focusing on prolamin extraction from the by-products of these cereal grains. Agro based wastes are the main targeted by-products. Prolamins are a great source of proline and glutamine amino acids. But they are deficient in lysine, arginine and histidine. Various advantages of prolamin nanoparticles use are summarised as: (i) Controlled release of Vitamin E can be achieved by using Gliadin nanoparticles. (ii) Hordein nanoparticles can enhance the stability and antioxidant effect of Resveratrol drug. (iii) Zein nanoparticles can increase Oral bioavailability of the drug Resveratrol. (iv) Nanoparticles of Kafirin can enhance the photostability of the drug and also can protect it from lipid oxidation (Tapia-Hernández et al. 2019).

12.7.4 *Elastin*

Elastin is a protein present in the extracellular membrane of mammals. It has properties like rubber, meaning it stretches but comes back to its original form easily. These proteins help in tissue stretching. This hydrophobic protein is a great source of valine, proline and glycine. It helps in chemotaxis, cellular growth and homeostasis of tissues. Tropoelastin is the precursor of elastin, which is rich in lysine valine and proline. Elastin nanoparticles can be formed using the self-assembly method mainly. Also, nanoparticles of Elastin can be formed within the range of 300–400 nm by electro spraying method. Elastin nanoparticles help in (i) decreasing the pH and enzyme sensitivity of the drug, (ii) It protects the drug from high temperatures (MacEwan and Chilkoti 2014).

12.7.5 Protamines

Protamines are peptides rich in arginine amino acids. For pharmacological use mainly protamine rich fish sperms are used. Friedrich Miescher, a biologist and physician, first discovered and extracted protamines in 1874. The molecular weight of protamine is 4000–5000 Da. Protamines are basic in nature and made of 50–110 amino acids. Depending on the presence of basic amino acids, protamines can be divided into 3 groups. Group 1 is called Monoprotamines, these protamines are composed of only arginine amino acids. Group 2 or Diprotamines have one another basic amino acid (lysine or histidine) along with Arginine. And the final group or the third group will have these all three basic amino acids (arginine, lysine and histidine), so they are called Triprotamines. Nanocapsules or nanoliposomes are common drug carriers made using protamines. By using reverse phase evaporation or solvent evaporation process protamines nanoparticles are formed.

Advantages of using Protamine nanoparticles are being (i). insulin entrapped in Protamine nanoparticles help in diabetes treatment, (ii) in cardiac surgery nanoparticles of protamine sulfate prevent the anticoagulatory effect of heparin, (iii) solid lipid nanoparticles or liposomes of protamines help in better target mediated delivery of the drug, (iv) nanocarriers carry antigenic RNA molecules in vaccination (Ruseska et al. 2021).

12.7.6 Casein

Casein is a phospho-protein derived from milk of mammals. Casein is the main constituent protein present in bovine milk. α-s1 casein, α-s2 casein, β-casein, and κ-casein—are the four types of casein present in bovine milk. Casein molecules are amphoteric in nature. Caseins are relatively heat stable as they can survive pasteurization temperatures of ~62–70 °C. Caseins are widely used in nanoparticles formation because casein micelles of 50–500 nm size can be easily produced. Because of their amphiphilic property, they self-assemble into micelle form when introduced to any aqueous medium. Along with many other advantages, because casein is present in a food item it is considered as GRAS (Generally Recognised as Safe). The only problem with casein nanoparticles is, many people are allergic to casein, in fact, and casein is the potent allergen present in milk. In case of casein nanoparticles formation through nanoprecipitation process, Dimethyl Sulfoxide (DMSO) is used as a desolvating agent. DMSO is a solvent miscible in water as well as many organic solvents. Benefits of using Casein nanoparticles are being (i) Improvement in the bioavailability of the drug orally almost 10 times in case of Resveratrol, (ii) Extended drug release time (iii) in the case of Flutamide drug it prolongs the circulation in the blood plasma (Głąb and Boratyński 2017).

12.7.7 Gelatin

Gelatin is a protein derived from animal collagen (mainly from cows and pigs) or fish skins. It is a translucent, colourless, water soluble protein. Gelatin is obtained through hydrolytic reaction of collagen. Depending on the type of hydrolysis, obtained gelatin is classified in 2 categories. Type A gelatin is obtained through acidic hydrolytic process and Type B gelation is acquired by alkaline treated collagen hydrolysis. Because Gelatin is easily biodegradable and non-toxic (identified as GRAS), it is used in colloidal drug delivery systems. Most common methods of gelatin nanoparticles formation are salting out and solvent evaporation. But other techniques such as the Desolvation process or Monomer polymerization or Emulsification can also be used in preparation of gelatin nanoparticles. Benefits of Gelatin nanoparticles in drug delivery are (i) It improves cellular uptake of the drug Amphotericin B, (ii) Timolol drug when entrapped inside gelatin nanoparticles shows better bioavailability, (iii) Enhancement of controlled release of Paclitaxel drug (Sahoo et al. 2015).

12.7.8 Silk-Fibroin

Naturally silk fibroin is produced by some spiders (*Nephila clavipes* and *Araneus diadematus*) and silkworms (*Bombyx mori*). It is a very large glycoprotein, composed of 5507 amino acids. This is a prominent problem in manufacturing silk fibroin nanoparticles. It is a rich source of Glycine, Alanine and Serine. Though it has many other disadvantages along with the high molecular weight, i.e., it is highly sensitive towards high temperature, varied pH, & different enzymes; it possesses many advantages. For example, fibroin has robust mechanical properties & high cellular compatibility, surface mediated biodegradation and so on. When introduced to heat and salt, silk fibroin self-assembles into gels or fibers. In desolvation or nanoprecipitation process DMSO (Dimethyl Sulfoxide) is used as a desolvating agent in case of fibroin nanoparticles formation. Seven steps are performed to form Fibroin nanoparticles through this process: centrifugation, purification, sonication, filtration and lyo-philisation. Also micro emulsion or self-assembly processes can be performed to obtain silk fibroin nanoparticles. Silk fibroin nanoparticles are beneficial for (i) increased biocompatibility of the drug, (ii) reduced hematotoxicity, (iii) enhanced anti-inflammatory properties of the intestine (iv) increased cellular uptake of the drug (Quérette et al. 2019) (Fig. 12.3).

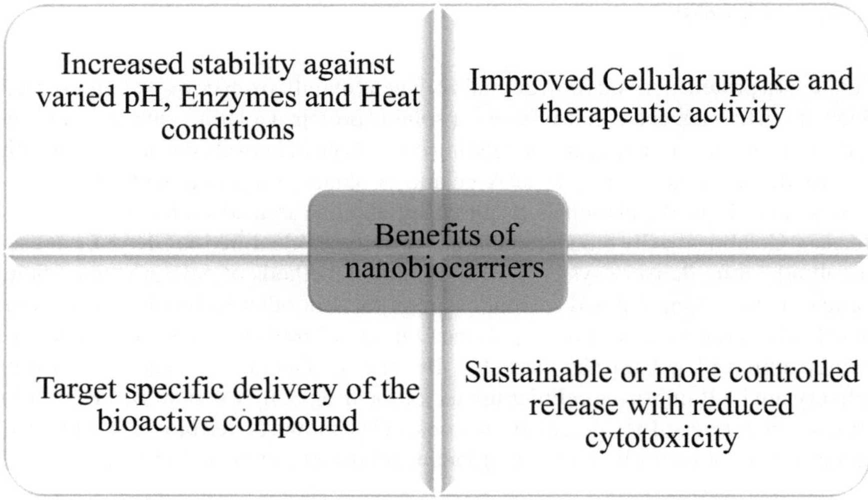

Fig. 12.3 Advantages of using nanobiocarriers in drug delivery

12.8 Conclusion

Polysaccharide and Protein based nano-biopolymers are excellent carrier of various bioactive components, nutraceuticals and other therapeutics. More importantly these polymers are nontoxic and most of them are Generally Recognized as Safe (GRAS) to consume. Also, the production procedures for both the types of polymers are quite convincing. Using these biocarriers effectively reduce wastage of the drug, provide more sustainable release and reduced cytotoxicity. Currently, large scale industrial production of these polymer-drug molecules is on. Which is in fact a great growth. Though more clinical trials of these nano-carrier-drug compounds are needed for different types of diseases. In present days, Abraxane (Albumin bound drug compound use in chemotherapy) is discontinued from cancer treatment. So, more clinical trials and experiments can only unfold the various possibilities or advantages of these nanocarriers.

References

Adschiri T, Yoko A (2018) Supercritical fluids for nanotechnology. J Supercrit Fluids 134:167–175. Elsevier BV. https://doi.org/10.1016/j.supflu.2017.12.033

Ali A, Ahmed S (2018) A review on chitosan and its nanocomposites in drug delivery. Int J Biol Macromol 109:273–286. Elsevier BV. https://doi.org/10.1016/j.ijbiomac.2017.12.078

Almedia A, Souto E (2007) Solid lipid nanoparticles as a drug delivery system for peptides and proteins. Adv Drug Deliv Rev 59(6):478–490. Elsevier BV. https://doi.org/10.1016/j.addr.2007.04.007

Beneke C, Viljoen A, Hamman J (2009) Polymeric plant-derived excipients in drug delivery. Molecules 14(7):2602–2620. MDPI AG. https://doi.org/10.3390/molecules14072602

Benson H (2005) Transdermal Drug delivery: penetration enhancement techniques. Curr Drug Deliv 2(1):23–33. Bentham Science Publishers Ltd. https://doi.org/10.2174/1567201052772915

Chen M, Jafvert CT (2018) Application of cross-linked stearic acid nanoparticles with dialysis membranes for methylene blue recovery. Sep Purif Technol 204:21–29. Elsevier BV. https://doi.org/10.1016/j.seppur.2018.04.053

Chomto P, Nunthanid J (2017) Physicochemical and powder characteristics of various citrus pectins and their application for oral pharmaceutical tablets. Carbohydr Polym 174:25–31. Elsevier BV. https://doi.org/10.1016/j.carbpol.2017.06.049

Couvreur P, Barratt G, Fattal E, Vauthier C (2002) Nanocapsule technology: a review. Crit Rev Ther Drug Carr Systs 19(2):99–134. Begell House. https://doi.org/10.1615/critrevtherdrugcarriersyst.v19.i2.10

D'Ayala G, Malinconico M, Laurienzo P (2008) Marine derived polysaccharides for biomedical applications: chemical modification approaches. Molecules 13(9):2069–2106. MDPI AG. https://doi.org/10.3390/molecules13092069

Demirci M, Caglar MY, Cakir B, Gülseren İ (2017) Encapsulation by nanoliposomes. In: Nanoencapsulation technologies for the food and nutraceutical industries. Elsevier, pp 74–113. https://doi.org/10.1016/b978-0-12-809436-5.00003-3

Demirkurt B, Cakan-Akdogan G, Akdogan Y (2019) Preparation of albumin nanoparticles in water-in-ionic liquid microemulsions. J Mol Liq 295:111713. Elsevier BV. https://doi.org/10.1016/j.molliq.2019.111713

Ganta S, Devalapally H, Shahiwala A, Amiji M (2008) A review of stimuli-responsive nanocarriers for drug and gene delivery. J Control Release 126(3):187–204. Elsevier BV. https://doi.org/10.1016/j.jconrel.2007.12.017

Ghaywat SD, Mate PS, Parsutkar YM, Chandimeshram AD, Umekar MJ (2021) Overview of nanogel and its applications. GSC Biol Pharm Sci 16(1):040–061. GSC Online Press. https://doi.org/10.30574/gscbps.2021.16.1.0196

Głąb TK, Boratyński J (2017) Potential of casein as a carrier for biologically active agents. Top Curr Chem 375(4). Springer Science and Business Media LLC. https://doi.org/10.1007/s41061-017-0158-z

Gunasekaran S, Ko S, Xiao L (2007) Use of whey proteins for encapsulation and controlled delivery applications. J Food Eng 83(1):31–40. Elsevier BV. https://doi.org/10.1016/j.jfoodeng.2006.11.001

Guo Y, Liu Q, Peng C, Wang E, Joy A, Cakmak M (2018) Colloid silica nanoparticles trapped morphology of polymer blends during solvent evaporation. Eur Polym J 107:164–172. Elsevier BV. https://doi.org/10.1016/j.eurpolymj.2018.08.013

Haley B, Frenkel E (2008) Nanoparticles for drug delivery in cancer treatment. Urol Oncol: Semin Orig Investig 26(1):57–64. Elsevier BV. https://doi.org/10.1016/j.urolonc.2007.03.015

Hu T-M, Chou H-C, Lin C-Y (2019) Facile green synthesis of organosilica nanoparticles by a generic "salt route." J Colloid Interface Sci 539:634–645. Elsevier BV. https://doi.org/10.1016/j.jcis.2018.12.080

Hu Q, Lu Y, Luo Y (2021) Recent advances in dextran-based drug delivery systems: From fabrication strategies to applications. Carbohydr Polym 264:117999. Elsevier BV. https://doi.org/10.1016/j.carbpol.2021.117999

Huang G, Huang H (2018) Application of hyaluronic acid as carriers in drug delivery. Drug Deliv 25(1):766–772. Informa UK Limited. https://doi.org/10.1080/10717544.2018.1450910

Jahangirian H, Ghasemian lemraski E, Webster TJ, Rafiee-Moghaddam R, Abdollahi Y (2017) A review of drug delivery systems based on nanotechnology and green chemistry: green nanomedicine. Int J Nanomed 12:2957–2978. Informa UK Limited. https://doi.org/10.2147/ijn.s127683

Jaiswal M, Dudhe R, Sharma PK (2014) Nanoemulsion: an advanced mode of drug delivery system. 3 Biotech 5(2):123–127. Springer Science and Business Media LLC. https://doi.org/10.1007/s13205-014-0214-0

Kaur L, Singh J, Liu Q (2007) Starch–a potential biomaterial for biomedical applications. In: Nanomaterials and nanosystems for biomedical applications. Springer Netherlands, pp 83–98. https://doi.org/10.1007/978-1-4020-6289-6_5

Krishna R, Pandit JK (1996) Carboxymethylcellulose-sodium based transdermal drug delivery system for propranolol. J Pharm Pharm 48(4):367–370. Oxford University Press (OUP). https://doi.org/10.1111/j.2042-7158.1996.tb05934.x

Li L, Ni R, Shao Y, Mao S (2014) Carrageenan and its applications in drug delivery. Carbohydrate Polym 103:1–11. Elsevier BV. https://doi.org/10.1016/j.carbpol.2013.12.008

Löbmann K, Svagan AJ (2017) Cellulose nanofibers as excipient for the delivery of poorly soluble drugs. Int J Pharm 533(1):285–297. Elsevier BV. https://doi.org/10.1016/j.ijpharm.2017.09.064

MacEwan SR, Chilkoti A (2014).Applications of elastin-like polypeptides in drug delivery. J Control Release 190:314–330. Elsevier BV. https://doi.org/10.1016/j.jconrel.2014.06.028

Martínez-López AL, Pangua C, Reboredo C, Campión R, Morales-Gracia J, Irache JM (2020) Protein-based nanoparticles for drug delivery purposes. Int J Pharm 581:119289. Elsevier BV. https://doi.org/10.1016/j.ijpharm.2020.119289

Mehnert W, Mäder K (2012) Solid lipid nanoparticles. Adv Drug Deliv Rev 64:83–101. Elsevier BV. https://doi.org/10.1016/j.addr.2012.09.021

Merlot AM, Kalinowski DS, Richardson DR (2014) Unraveling the mysteries of serum albuminâ more than just a serum protein. Front Physiol 5. Frontiers Media SA. https://doi.org/10.3389/fphys.2014.00299

Pandita A, Sharma P (2013) Pharmacosomes: an emerging novel vesicular drug delivery system for poorly soluble synthetic and herbal drugs. ISRN Pharm 1–10. Hindawi Limited. https://doi.org/10.1155/2013/348186

Quérette T, Fleury E, Sintes-Zydowicz N (2019) Non-isocyanate polyurethane nanoparticles prepared by nanoprecipitation. Eur Polym J 114:434–445. Elsevier BV. https://doi.org/10.1016/j.eurpolymj.2019.03.006

Ruseska I, Fresacher K, Petschacher C, Zimmer A (2021) Use of protamine in nanopharmaceuticals—a review. Nanomaterials 11(6):1508. MDPI AG. https://doi.org/10.3390/nano11061508

Sahoo N, Sahoo RK, Biswas N, Guha A, Kuotsu K (2015) Recent advancement of gelatin nanoparticles in drug and vaccine delivery. Int J Biol Macromol 81:317–331. Elsevier BV. https://doi.org/10.1016/j.ijbiomac.2015.08.006

Sercombe L, Veerati T, Moheimani F, Wu SY, Sood AK, Hua S (2015) Advances and challenges of liposome assisted drug delivery. Front Pharm 6. Frontiers Media SA. https://doi.org/10.3389/fphar.2015.00286

Singh RS, Kaur N, Kennedy JF (2015) Pullulan and pullulan derivatives as promising biomolecules for drug and gene targeting. Carbohydr Polym 123:190–207. Elsevier BV. https://doi.org/10.1016/j.carbpol.2015.01.032

Sinha VR, Kumria R (2001) Polysaccharides in colon-specific drug delivery. Int J Pharm 224(1–2):19–38. Elsevier BV. https://doi.org/10.1016/s0378-5173(01)00720-7

Sood A, Gupta A, Agrawal G (2021) Recent advances in polysaccharides based biomaterials for drug delivery and tissue engineering applications. Carbohydrate Polym Technol Appl 2:100067. Elsevier BV. https://doi.org/10.1016/j.carpta.2021.100067

Sur S, Rathore A, Dave V, Reddy KR, Chouhan RS, Sadhu V (2019) Recent developments in functionalized polymer nanoparticles for efficient drug delivery system. Nano-Struct Nano-Objects 20:100397. Elsevier BV. https://doi.org/10.1016/j.nanoso.2019.100397

Tapia-Hernández JA, Del-Toro-Sánchez CL, Cinco-Moroyoqui FJ, Juárez-Onofre JE, Ruiz-Cruz S, Carvajal-Millan E, López-Ahumada GA, Castro-Enriquez DD, Barreras-Urbina CG, Rodríguez-Felix F (2019) Prolamins from cereal by-products: Classification, extraction, characterization and its applications in micro- and nanofabrication. Trends Food Sci Technol 90:111–132. Elsevier BV. https://doi.org/10.1016/j.tifs.2019.06.005

Chapter 13
Recent Advancements in the Application of Nanomaterial in Modern Drug Delivery and Future Perspective

Pitambar Patel and Sachin Rameshrao Geed

Abstract Recently, nanomaterials are widely employed in the medical sector for sensitively detecting biological molecules, imaging of diseased tissues, therapeutics, and drug delivery. Nanomaterials with various biological characteristics and compositions have widely explored for drug delivery uses. It's critical to comprehend the interactions while using drugs for delivery of the nanomaterials with the bio-system, therapeutic agent's stability, surface of target receptors, and release of the drug. Different organic and inorganic based nanomaterials have designed, and their application as nanocarriers was widely investigated. The present chapter focuses on the recent updates of nanomaterial-based drug delivery methods and nanomedicine, along with complete scrutiny of nanomaterial's design, synthesis, and application for improving drug delivery and efficacy. The advantages and challenges of nanomaterials in drug delivery in their clinical applications are also addressed. Additionally, the future opportunities and perspectives of these nanomaterials are discussed.

Keywords Nanomaterials · Nanocarriers · Drug delivery · Nanomedicine · Drug targeting

13.1 Introduction

Human beings have traditionally utilized natural remedies derived from plants/herbsfor the treatment of many different illnesses/disease. Most contemporary medicines are derived from plants/herbs using conventional knowledge. About 25% of the most relevant pharmaceutical active compounds and their metabolites derived

P. Patel (✉)
Faculty of Science, Department of Chemistry, University of Allahabad, Allahabad, India
e-mail: patel.pitambar@gmail.com

S. R. Geed (✉)
CSIR-North East Institute of Science and Technology, Jorhat 785006, India
e-mail: sachingeed23@gmail.com

M. P. Shah et al. (eds.), *Biogenic Nanomaterials for Environmental Sustainability: Principles, Practices, and Opportunities*, Environmental Science and Engineering,
https://doi.org/10.1007/978-3-031-45956-6_13

from nature (Baker 2009). A recent approach in the development of drugs from natural products has synthetically feasible active ingredients which resemble the same structure and properties (Patra et al. 2018). Natural products show remarkable chemical, biological characteristics with macromolecular specificity, and lower toxicity, among other noteworthy traits. Nanomedicine and nano-based delivery systems are newlyemergingtechnologies because of the nanoscale range of size are used as a diagnostic tool or therapeutic agents to specified targeted sites (Fais et al. 2016; Sahu et al. 2021). Nevertheless, Eric Drexler's conception of new molecular altering nanoparticles has transformed the traditional science and broadened the scope of its applications. Due to advancements in modern science, research on nanoparticles is gaining speed and progressing more rapidly among specialists globally. Nanotechnology provides a platform for manipulating and enhancing the essential properties of metals-based nanomedicine in the form of nanoparticles, which have promising applications in diagnostics and treatments (Sowjanya 2015; Mitchell et al. 2021). Nanoparticles' achievement can be attributed to their distinctive biotechnology, which has developed a variety of potent drugs yet many of them have challenges during delivery of biological systems. Due to its contraindications and unique chemical constitution, their therapeutic potential is significantly reduced (Patra et al. 2018; Sahu et al. 2021). Awide range multidisciplinary approach has recently given special focus on nanotechnology to produce nanoparticles. The development of nanoparticles has affected the formulation and medicines delivery (Sahu et al. 2021). Nanomaterials' discovery and application help improve the selectivity of diagnostic tests utilizing disease-related markers to use both new and conventional plant-based drugs (Patra et al. 2018; Sahu et al. 2021).

The application of nanotechnology for drug delivery systems improves performance, effectiveness, safety and reduces health care costs. Additionally, it might improve the performance of medicine that failed to go beyond the clinical trial phases with the use of different approaches, such as targeting trans membranes, transferring chemicals to specific organelles, and reducing delivery, nanotechnology can make these technologies more practical and accessible. As a substitute approach, nanotechnology-based therapeutic agents are being developed to improve therapeutic efficacy and alter the physicochemical characteristics of antiviral treatment (Sahu et al. 2021; Mitchell et al. 2021). Nanotechnology delivers nanomedicine safely and effectively for treating diseases as diabetes, cancer, HIV, asthma, and hypertension etc. Before application, there are many crucial elements, namely immunogenicity, target selectivity, and bio-compatibility (Mitchell et al. 2021). Overview of drug delivery system is given in the Fig. 13.1. In the initial phase small interfering RNA (siRNA) delivery via a target-basednanoparticle approach was conducted in conducted in 2010 on patients with solid tumors (Sahu et al. 2021; Davis et al. 2010). A different clinical investigation revealed a better tumor, effectiveness of a dynamically targeted polymeric nanoparticle as a therapy Contains docetaxel (DTXL), a chemotherapy agent to a DTXL formulation based on solvents (Hrkach et al. 2012). Nanobiotechnology is a part of biomedicine; nanoparticles have been utilized in drug delivery, tissue engineering, and biosensors.

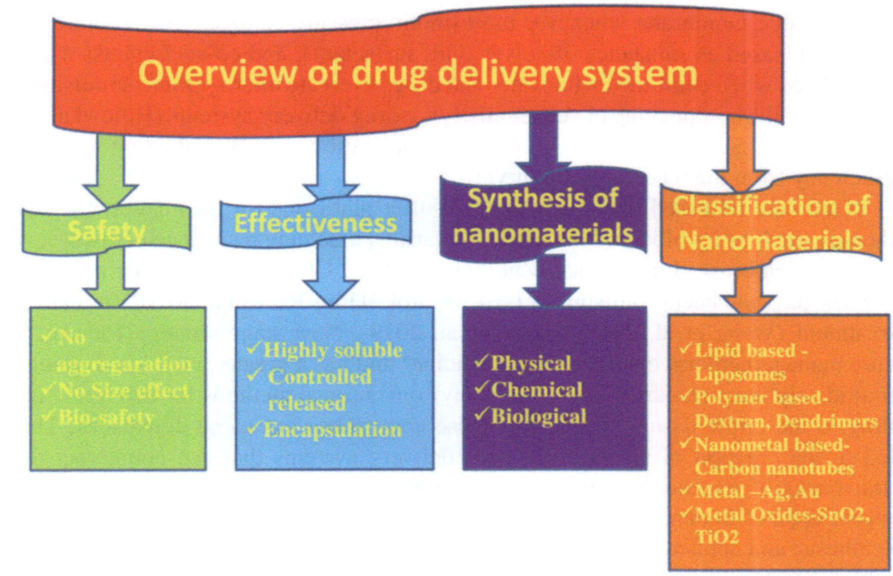

Fig. 13.1 Overview of drug delivery system the major components are safety, effectiveness, classification and synthesis of nanomaterials

Nanoparticles are typically small nanospheres as they are composed of materials manufactured employing atomic or molecular building blocks. As a result, they can move within the human body with greater freedom than more giant molecules. Due to the ability of nanoparticles, the structural, chemical, mechanical, magnetic, electrical, and biological characteristics of nanoscale-sized particles are different to be used in drug delivery by encapsulating therapeutic medicines and transporting allowing them to more accurately target tissues with a controlled release (Patra et al. 2018; Mitchell et al. 2021).

Nanomedicine integrates nanoscience techniques and knowledge to medical biology, disease treatment, and disease prevention. It suggests using nanomaterial-based nanosensors and nanorobots for drug delivery, diagnosis, sensory functions, and actuating materials in living cells (Patra et al. 2018; Mottaghitalab et al. 2019). When it came to material design, execution strategies, and clinical application, experts understood engineering concepts. The primary considerations in drug designing are diffusion, erosion, deterioration, shearing, swelling, binding, surface and the rates of both active and passive cell absorption. Self-emulsifying system and regulated drug delivery are part of the material design. The different controlled drug release systems are osmotic pump, hydrogel, matrix, reservoir, and erodible material. Drugs are released in the matrix system through interconnected pores. Drug permeates through the selectively permeable membrane of the reservoir system. When a substance is degradable, it develops a porous structure that allows a drug to be extruded. The drug is released in an osmotic pump system through hole(s) in an impermeable membrane whenever the osmotic pressure changes. An erodible the drug is released as substance dissolves. Oil, surfactants, and co-surfactants, which are utilized when combined with water and drugs to create micro- or nanoemulsions, are indeed the components of self-emulsifying drug delivery systems (Holowka and Bhatia 2016; Sultana et al. 2022).Clinical application entails evaluating a drug's efficiency from the medical point of view.

Nanomaterial based nanomedicine also offer platforms for medication cocktail therapy, such as by blocking the action of various medicine resistance pathways, to the cell membrane, for instance (Zhou et al. 2020). Utilizing nanomedicine in the present day has beenpromising to beat Multiple Drug Resistant (MDR) in cancer treatment (Wang et al. 2018; Zhang et al. 2019). New stage strategy for cancer cure management has emerged nonomedicines and has a greater attention thorough multidisciplinary research. There are many conventional methods for delivering the drug to the target tissue, but each of these approaches has some drawbacks. This led to the creation of smart medication delivery systems that can counteract the drawbacks of traditional methods. The fundamental ideas of this chapter are the recent updates in nanomedicine and drug delivery systems based on nanomaterials, synthesis, and application for improving drug delivery and efficacy. Advantages and difficulties of using nanomaterials for medication administration in clinical settings are also discussed. Additionally, the future opportunities and perspectives of these nanomaterials are discussed.

13.2 Nanomaterial Based Drug Delivery System

Numerous treatment methods have invented for nanomedicine development and drug delivery systems. Traditional clinical analytic techniques have been examined to increase medicine selectivity and diagnostic precision. For instance, novel drug delivery systems are being investigated to ensure that they act precisely where they are needed, increasing their availability while reducing their toxicity when ingested (Wang et al. 2018). Polymer based, dendrimers, dextrans, poly (lactic-coglycolic acid) (PLGA) and liposomes and are a few examples of synthetic and natural polymers that have employed to construct a number of nanocarriers for delivering therapeutic and imaging substances. Similar to carbon nanotubes, metal-based nanoparticles (Ag/Au), nanoparticles based on metal oxide (superparamagnetic), and semiconductor-based nanoparticles (quantum dots) are also employed in medical applications (Naresh and Lee 2021; Rabbani et al. 2020; Hirsch et al. 2003). To discuss physicochemical factors contributing to toxicity and approaches for evaluating toxicity related to nanoparticles, initiatives have been attempted to examine the effectiveness and safety of drugs based on nanotechnology for diagnosing and treating cancers. Nanoparticles' physicochemical characteristics are caused by their highly reactive surfaces, chemical composition, solubility, size, shape, etc. Zhou et al. (2020), Wang et al. (2018), Zhang et al. (2019), Naresh and Lee (2021), Rabbani et al. (2020), Nagpal et al. (2010).

Drug delivery systems with the use of in nanomedicine are mostly lipid-based, polymer-based, non-metallic-based, and metal oxide-based nanomaterials are utilized. When it comes to non-biological lipid-based carriers, liposomes are a popular alternative for delivering drugs that enhance the prognosis of kidney, skin, liver, lung, and prostate tumors. Liposomes are preferable to many other nanodelivery systems in a number of ways because of their lower toxicity and significant therapeutic index (Rabbani et al. 2020). Liposomes are classified into neutral, anionic and cationic, nanoparticles based on their net charge. Cationic liposomes generally have included amphipathic cationic lipid, dioleyolphosphatidylethanolamine (DOPE). Once liposome is absorbed into cells, the DOPE concentration ranges (0–50%), and its elastic response DNA by disrupting the endosome. Low transfection efficiencies were explained by liposomes lacking DOPE being discovered entrapped in endosomes and lysosomes. Furthermore, DOPE is a crucial factor that influences a liposomal formulation's capability to transfect cells. Low transmission efficiencies were explained by the finding of Liposomes deficient of DOPE entrapped in lysosomes and endosomes (Wang et al. 2018; Zhang et al. 2019; Naresh and Lee 2021; Rabbani et al. 2020).

Glucose polymers classified as dextrans are utilized in a number of pharmacological activities. High percentage of a-1,6-glucopyranosidic linkages (95%) opposed to 1,3-linkages (5%), which is a characteristic of dextrans. Pharmaceutical agents have been supplied utilizing dextrans of 40–70 kDa molecular weight. Dextrans are synthesized by partial depolymerization utilising acid hydrolysis and separation from native, high-molecular-weight dextrans(10^7–10^8).Hormones, growth factors, enzymes, and imaging compounds can all be conjugated to dextrans by reversible or irreversible processes. Drug and enzyme conjugation to dextrans improves physicochemical qualities including the solubility and stability and assists in drug discovery at the site of action (Mottaghitalab et al. 2019; Holowka and Bhatia 2016; Sultana et al. 2022; Zhou et al. 2020).

To prepare systems for drug loading, poly(amino acids), polyamides, poly orthoesters, polyurethanes, poly(alkyl-a-cyano acrylates), polyacrylamides and polyesters have been used extensively. Since the 1970s, the copolymerpoly (lactic-coglycolic acid) PLGA has been used and is the most well-known biodegradable polymer regarded as a biomaterial that is compatible with humans (Wang et al. 2018; Rabbani et al. 2020).The properties that are biological and biomechanical of indigeneous vascular tissue have been mimicked using PLGA. To extend the period that paclitaxel is delivered, Paclitaxel has been enclosed in PLGA nanomaterial to deliver controlled release drug to the both the luminal surface and interior of extended polytetrafluoroethylene tissue engineering. Dendrimers have multiple branches that resemble large, spherical formations. A dendrimer typically has a spherical three-dimensional structure and is symmetric around a core. Dendrimers have an external surface, an interior shell, and a seat. Each region has a different functionality that can be synthesized to regulate compounds' dispersion, thermal stability, and attachment for a range of diverse applications (Wang et al. 2018; Zhang et al. 2019).

A carbon nanotube is most constituents of graphite. The technique wherein graphite is formed into a tube influences the nanotubes structure. Though carbon nanotubes are insoluble in all liquids, functionalization and no covalent supramolecular sorption can improve its solubility. Nanotubes are characterize having an aspect ratio >100, sizes between 0.7 and for 1.5 nm single-walled carbon nanotubes; multi-wall range in size from 2 to 50 nm and lengths of several millimeters. The technique wherein graphite is formed into a tube influences the nanotubes structure. Though carbon nanotubes are insoluble in all liquids, functionalization and noncovalent supramolecular sorption can improve its solubility. Various biomedical applications are facilitated with use of carbon nanotubes. Applications for both cancer treatment and diagnostic tools can be developed with carbon nanotubes that can be supplied with small-molecule inhibitor and/or imaging agents owing to their enormous surface area conjugated with a variety of compounds (Zhou et al. 2020; Felgner et al. 1987).Metallic nanoparticles typically contain a metal core, which is often surrounded by a shell made of metal oxide or another inorganic substance. The core itself is composed of an inorganic metal or metal oxide. Biocompatible and stable metal oxide nanoparticles can serve as carriers for drugs and diagnostic agents. Additionally, the fluorescence and photo-stability of fluorophores and metals in metal-based nanoparticles may be amplified through interactions with one another (Hrkach et al. 2012; Farokhzad et al. 2006; Nagpal et al. 2010; Felgner et al. 1987; Hirsch et al. 2003; Hainfeld et al. 2004).

The drug design aspects for the purification and characterization of peptides, protein, and biological compounds as well as technological advancements in computer science are necessary for the growth and development of pharmaceutical industry. Finding novel leading medicines based on an understanding of a target tissue has been a promising feature of such a scenario. There are numerous studies emphasize the design of various types of molecules and significanceof looking into diverse medicines release mechanisms. Additionally, natural products development can give practical and fascinating solutions to issues concerning drug design by encouraging the discovery of novel drugs with desired physicochemical characteristics. The main targets for development in nano biotechnologies in drug delivery: reducing toxicity even while utilizing therapeutic drugs, improving both biocompatibility and safety, delivering drugs with more precision, faster development of new safe medicine (Mottaghitalab et al. 2019; Holowka and Bhatia 2016; Sultana et al. 2022; Farokhzad et al. 2006; Nagpal et al. 2010; Felgner et al. 1987; Hirsch et al. 2003).

Systems for drugs delivering have become increasingly important. These systems are easy to build and could promote the body's customized delivery of the active ingredients. It's interesting to note that different drug polarities can be delivered by a variety of methods using chemical or physical contacts, including such electrostatic, Van der Waals forces of attraction, covalent bonds, and hydrogen bond. Superparamagnetic iron oxide nanoparticles (SPION) are organic substances like fatty acids, phospholipids, peptides, polysaccharides, polymers and surfactants or that is coated with silica or gold on top of iron oxide cores that are inorganic particles. Nanoparticlesmay be heated when an applied alternating magnetic field from the exterior because they have magnetic characteristics that allow for collecting in a specific region. When coated with antibodies or peptides, nanoparticles can attach to specific cells to cure or image diseases. These nanoparticles are appealing due to their property for range of application as magnetic hyperthermia, drug delivery, MRI, and magnetically assisted cell transfection. The nanoparticles are between 50 and 160 nm in size. Clinical trials now use organic-coated particles as MRI contrast agents to diagnose liver cancers and differentiate between inflammatory and metastatic lymph nodes (Fais et al. 2016; Sahu et al. 2021; Sowjanya 2015; Mitchell et al. 2021; Davis et al. 2010; Hirsch et al. 2003).

Nanomaterials are also used as nanocarriers that interact with biological systems and how quickly the active ingredient is released into the body. Although there are many different nanocarriers with unique drug release mechanisms, profiles are being created to strengthen the distinctiveness of the nanostructures focus on selected places of the organism, to reduce their chemical functionalization or coating with various compounds, like polymers, to increase their immunogenicity natural polysaccharides, a cell's membrane, antibodies, as well as tunable surfactants, peptides, etc. In rare circumstances, medicines do not show binding and alignment with a particular target or avoid crossing. A number of barriers, including the blood–brain barrier and ligand-modified proteins, crossed the cerebrospinal fluid. Table 13.1 gives insight into Nanomaterial based drug delivery systems (Farhood et al. 1994; Farokhzad et al. 2006).

Nanocarriers were used to enter the cell and enable the distribution of a specific drug in a membrane-specific environment. Numerous nanocarriers have included hyaluronic acid as a ligand with encouraging findings for intra-vitreal delivery. An extracellular matrix polymer called hyaluronic acid boosts the anticancer activity of cancer stem-like cells, breast cancer cells, and lung adenocarcinoma cells. Although time-consuming, the creation of ligand-appended drug delivery systems necessitates a variety of targeted designs that consider physiological aspects, involving tissue structure, blood flow, and disease severity. The lesson in such situations is that we need to examine what the body actually requires; possibly working in conjunction with the body's natural processes is preferable to working against them (Hirsch et al. 2003; Hainfeld et al. 2004; Meng et al. 2009).

The important factors to be considered for the nanomaterial-based drug delivery system.

- Increased blood flow and accumulation with passive or active targeting
- Increased depth of penetration in target tissue
- Delivery of the payload into the cytoplasm under control
- Conquering drug resistance in target tissue

Table 13.1 Details about the nanomaterial-based drug delivery system, active molecules, drug administration and disease target

Nanomaterial based drug delivery system	Active molecules	Name	Drug administration	Disease target	Routinely used components	References
Liposomes	Daunorubicin	DaunoXome	Intravenous	Acquired immunodeficiency syndrome AIDS-related	Small interfering RNA /DNA	Farhood et al. (1994), Puri et al. (2009), Tran et al. (2009), Ragelle et al. (2017)
	Morphine	DepoDur	Epidural	Pain relaxation		
Protein-based F Protein nanoparticle	RSV	RSV-F vaccine	Intravenous	Respiratory syncytial	Small interfering RNA /DNA	Ragelle et al. (2017), Sharma et al. (2012)
Polymer-based-PLGA nanoparticle	Rapamycin	SEL-212	Intravenous	Gout (common/ complex arthritis)	Delivery of proteins, peptides, vaccines,/ imaging	Ragelle et al. (2017), Farokhzad et al. (2006), Nagpal et al. (2010)
PEG polyamino acid nanoparticle	Cisplatin	NC-6004	Intravenous	Solid tumors		
Lipid-based Lipid nanoparticle	TTR siRNA	Patisiran (ALNTTR02)	Intravenous	TTR-mediated amyloidosis	Small interfering RNA /DNA	Ragelle et al. (2017), Felgner et al. (1987)
	MYC siRNA	DCR-MYC	Intravenous	Hepatocellular carcinoma		
Metal-based Au metal shell silica core nanoparticle	Laser irradiation	AuroShell	Intravenous	Prostate cancer	Thermal ablation/ imaging/antimicrobial drug delivery	Hirsch et al. (2003), Hainfeld et al. (2004)
Hafnium oxide nanoparticle Silver nanoparticle	Radiotherapy	NBTXR3	Intrahepatic/ intra-arterial	Liver cancers		
Magnetic iron NP	Magnetic field	–	Intratumoral	Prostate cancer	Magnetic targeting/ thermal ablation/MRI contrast agent	Meng et al. (2009)

13.3 Basic Principle and Mechanism of Nanotechnology in Drug Delivery

Drugs with low solubility have a number of biopharmaceutical issues, including accessibility and bioavailability after intake through mouth, require high dose intake and creates unwanted side effects. However, by incorporating nanotechnology into the drug delivery process, all of these issues might be resolved. Numerous research organizations worldwide have researched the use of nanoparticles as medication carriers extensively since it is the most cutting-edge technology. Solubility, diffusivity, bioavailability, drug release patterns, and immunogenicity are some of these properties of the nano carriers are easily tunable. This makes them a convenient route of drug administration with a prolonged pharmacological life cycle, lower toxicity, and less adverse effects with improved biodistribution (Mirza and Siddiqui 2014).The formation of nanoparticle basically involves the spontaneous self-assembly of building blocks in to well-defined patterns or structure (Lu et al. 2016). The therapeutic agents may be attached to the nanoparticle surface or stored inside the spherical shape of the nanoparticle. The therapeutic chemicals may release from the nanoparticle through swelling, erosion, diffusion, or disintegration once they reach the target site.

13.3.1 Drug Delivery Strategies Using Nanostructures

The designed drugs delivery systems are either target to deliver the drug to a targeted sites are planned for controlled drug molecule release. Usually, nano carrier delivers the drugs via two ways: (i) passive and (ii) self-delivery (Patra et al. 2018; Lu et al. 2016).

13.3.1.1 Passive Delivery

In passive delivery drugs are integrated into the inner chamber of self-assembled nano structures through covalent interactions (mainly the hydrophobic effect) between the carrier and drug molecule, medicines. Many nanocarriers, including porous nanoparticles, nanomicelles, and nanocapsules, have a hydrophobicity to stabilize the medicinal molecules they are encapsulating (Chung et al. 2014). Drugs will be released at the target site after the nano carrier is disassembled. However, very less drug loading inside the hydrophobic compartments of the nano carrier is observed in physical encapsulation, mostly weight range from 2 to 5% (Lin et al. 2013).

13.3.1.2 Self-delivery

The drugs are effectively linked to the nanocarrier during self-delivery. The conjugation between the carrier and drug molecule should be easily cleavable at target sites. In this process, the releasing time of drug molecule from the carrier is very crucial. If the medicine is released before it reaches the intended place, its activity and effectiveness will be reduced. This procedure, called "burst release," will cause the body to immediately remove the drugs (Keith and Cui 2014). Therefore, it's essential to design the drug's and its nanocarrier's conjugation in a balanced way.

Notably, each drug delivery systems have their unique physical, chemical and morphological features. Depending upon their chemical interactions or physical interactions with drug molecules they may have their own affinity for different drugs. Addition to this, various other parameters, like the nanocarriers' chemical composition and the mode of association of drug with them are also important parameters for understanding the drug delivery system (Siepmann et al. 2008; Mattos et al. 2017). Further, numerous investigations were done to determine how these nanocarriers release drugs. Solvent, diffusion, and chemical reactions are a few examples of the mechanisms that might indicate the release of drugs from nanocarriers (Fig. 13.2) (Lee et al. 2010; Ding and Li 2017). For instance, neem bark extract loaded biogenic silica has a greater drug release profile than neem bark extract encapsulated silica nanoparticle (Mattos et al. 2017).

Another illustration is the cancer treatment's pH-sensitive medication release from polymeric nanomicelles. In this instance, the polymeric nanocarriers distribute the medication selectively into the tumour cell's acidic pH environment. When polymers with ionizable functional groups are exposed to pH changes, their structure changes, releasing the medicines through structural breakdown. In some cases, acid-labile linker is used between drug and copolymer to accomplishregulated drug release in tumours with an acidic pH. In controlled drug release, acid-labile bonds such oxime, hydrazone, orthoester, imine, and vinyl ether bonds were examined (Wang et al. 2018; Salim et al. 2014).

13.4 Nanomaterials Used in Drug Delivery

Owing to the importance of nanoparticle in drug delivery system, wide libraries of nanoparticles, composed of various size, shape and materials having unique chemical and surface properties have been synthesized already (Yan and Chen 2014; Mudshinge et al. 2011; Lombardo et al. 2019). The nanoparticle can be classified in two broad categories as mentioned below.

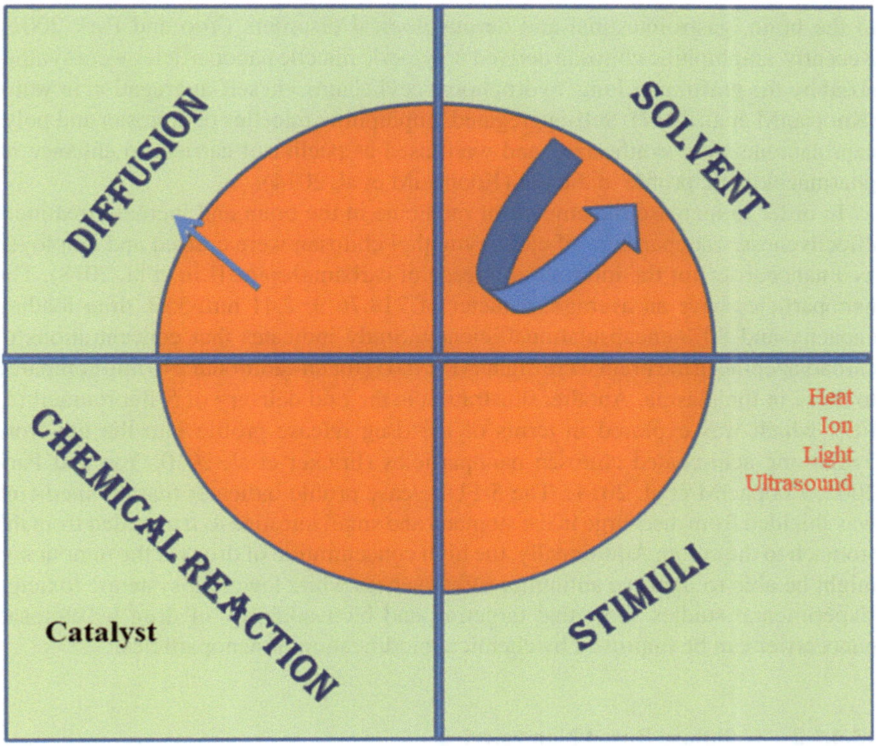

Fig. 13.2 Drug release mechanisms operated in nanocarriers

13.4.1 Organic and Polymer Based Nanocarrier

Recently, interest on organic and polymer-based nanoparticles in drug delivery systems has increased. They include chitosan, cellulose, liposomes, dendrimers, polymeric micelles, protein aggregates etc. The important advantage of organic nanoparticles over inorganic is they are biodegradable, which canovercome the chronic toxicity to cells or tissues. In addition, use of phospholipids in nanoparticles synthesis will improve the biocompatibility and ability to penetrate cell membrane (Fricker et al. 2010).

13.4.1.1 Chitosan Based Nanocarrier

A biocompatible and biodegradable polymer with surface functional groups, chitosan may readily be changed to fulfil certain tasks. Nanocarrier based on chitosan have received extensive research for use in medication delivery applications with different mode of administrations for the treatment of pulmonary diseases, drug delivery to the brain, gastrointestinal and dermatological disorders (Yoo and Park 2001). Recently, amphiphilic chitosan derived polymeric micelle nanoparticles were synthesized by the grafting of long, hydrophobic acyl chains via self-aggregation in water (KoopaeiM et al. 2014). Self-aggregated amphiphilic micelles of chitosan and polycaprolactone were synthesized and were used as paclitaxel carriers to enhance its pharmacokinetic profile in the gut (KoopaeiM et al. 2014).

In order to increase the amount of medicine in the brain and increase treatment effectiveness, nanoparticles of carboxymethyl chitosan were created and employed as a nanocarrier for the intra-nasal release of carbamazepine (Liu et al. 2018). The nanoparticles have an average diameter of 218.76 ± 2.41 nm, 35% drug loading capacity and 80% encapsulationefficiency. Study indicates that concentrations of carbamazepine in the brain were higher ($P < 0.05$) for an additional 240 min compared to those in the plasma. Another illustration is the oral delivery of 5-fluorouracil (5-FU), which was explored in terms of the drug release profile into the gut from hyaluronic acid-coated chitosan nanoparticles (Fricker et al. 2010; Yoo and Park 2001; KoopaeiM et al. 2014). The 5-FU release profile indicates that the medicine was shielded from discharge in the stomach and small intestine as it travelled from the stomach to the colon. Additionally, the high concentration of drugs at the tumour site might be able to improve antitumor effectiveness while lowering systemic toxicity. Experimental studies show that targeting and bioavailability of drug by chitosan nanocarrier can be improved by chemical modification of nanoparticles.

13.4.1.2 Cellulose Based Nanocarrier

In drug delivery systems, cellulose and its derivatives are frequently utilized as nano carriers, primarily to change the solubility and gelation of the pharmaceuticals that result in the control drug release profile (Sun et al. 2019). Release of repaglinide orally (an anti-hyperglycemic-RPG) was investigated by using cellulose nanocrystals and chitosan nanoparticles (Elseoud et al. 2018). Chitosan nanoparticles had a mean size distribution of 197 nm, whereas hybrid chitosan nanoparticles and oxidized cellulose nanocrystals containing repaglinide had mean diameters of 251–310 nm. Because of the drug's hydrogen bonds with the cellulose nanocrystals, the drug was continuously released. In contrast to nanoparticles derived from native cellulose nanocrystals, those based on oxidized cellulose nanocrystals had a reduced release profile. Acyclovir was used as the drug molecule to study the drug release into the nasal mucosa utilizing nanoparticles of hydroxypropyl methylcellulose, cationic hydroxyethyl cellulose, methylcellulose, and sodium carboxymethyl cellulose (Hansen et al. 2015). The polymers' suitability as excipients for nasal release applications was also evaluated with regard to the frequency of their ciliary beats and their infusion into the tissue system of the nasal passages. It was observed that, thermally induced viscosity was increased by polymer grafted copolymer with the cellulose derivatives. Additionally, the permeability of acyclovir into the nasal mucosa was seen to be enhanced when combined with cationic hydroxyethyl cellulose. As evaluated by ciliary beat frequency, negative effects on tissues were not observed all the cellulose derivatives (Elseoud et al. 2018; Hansen et al. 2015).

13.4.1.3 Liposomes

Liposomes are spherical vesicles made up of phospholipids and steroids that are generally between 50 and 450 nm in size. Phospholipids are generallyrecognized as safe components and thus minimize the chance of potential side effects. Liposomes are used for the transportation of various molecules in the cosmetics and pharmaceutical industry and one of the most extensively studied methods for delivering drugs. Liposomes are thought to be a superior medication carrier because of how close their membrane structure is to that of the cell membrane. They also make it easier for drugs to be incorporated into them (Bozzuto and Molinari 2015). Experimental studies have shown that liposomes make drug molecule stable, enhance their biodistribution, are also biocompatible and biodegradable and can be used for both hydrophilic and hydrophobic drugs. Additionally, the in vivo stability or the targeting ligand can be increased by functionalization of the liposome surface with 'stealth' material, which also allow liposomes to be delivered preferentially. Liposomes have been effectively employed as effective carriers for several medications, including antivirals, antineoplastics, antibacterial, insulin, and plasmid DNA, thanks to their multifunctional properties. Table 13.2 lists a few of the chosen biomedical uses for liposome nanocarriers.

Table 13.2 Liposomes biomedical application

Liposome	Composition	Drug application	References
Hydrogenated soya, phosphatidylcholine and distearoylphosphatidylglycerol (DSPG)	Amphotericin B	Aspergillus fumigatus	Takemoto et al. (2004)
Dipalmitoyl-phosphatidylcholine (DPPC),cholesterol and dimethylammonium ethane carbamoyl cholesterol (DC-chol)	Benzyl penicillin	Staphylococcus aureus	Kim and Jones (2004)
Stearylamine (SA) and dicetyl phosphate	Zidovudine	HIV	Kaur et al. (2008)
Liposome	Daunorubicin and doxorubicin	Breast cancer	Park (2002)
DC-Chol liposome	Plasmid DNA	Gene transfer insubcutaneous tumor	Whitemore et al. (2001)

13.4.1.4 Dendrimers

Dendrimers are three-dimensional globular formations that are well defined, monodisperse, and extremely divergent. These structures provide ideal drug delivery systems because their surface may be readily and precisely functionalized (Whitemore et al. 2001; Kesharwani et al. 2015). There are two methods for producing dendrimers: the first is a divergent technique, where the synthesis begins in the core and is subsequently extended outward. The other one is convergentwhich starts from the synthesis of outside of the dendrimer followed by extended to words the inside core (Madaan et al. 2014). The presence of amine group in dendrimers limits its clinical use, because the amine groups are either cationic or positively charged, making them toxic. Normally, modified dendrimers are used in order to reduced or eliminate the toxicity. Dendrimers are categorised based on their functionalization moiety such aspolyamidoamine (PAMAM), polyethyleneimine (PEI), polyethylene glycol (PEG), melamine, polypropyleneimine(PPI), poly L-glutamic acid (PG). The mechanisms are used to load drugs into dendrimers: electrostatic interaction, simple encapsulation, and covalent conjugationare some of the mechanisms via which drug is loaded in the dendrimers (Cheng et al. 2008).Basically, dendrimers delivered the drug via two different path ways, (i) in vivo release of drug from dendrimer is possible in presence of suitable enzyme which can cleave the drug dendrimer's covalent bonding; (ii) change in physical environment like pH, temperatureetc.,will also lead to the release of drug (Tripathy and Das 2013). Dendrimers have successfully applied in ocular, oral, transdermal, pulmonary and targeteddrug delivery (Cheng et al. 2008). Few applications of dendrimers in biomedicals are given in Table 13.3.

Table 13.3 Biomedical application of dendrimers

Dendrimer	Composition	Drug application	Refs.
PAMAM (polyamidoamine)	Nadifloxacin, prulifloxacin	Various bacteria	Madaan et al. (2014), Cheng et al. (2008)
PAMAM (polyamidoamine)	Propranolol	Hypertension	D'Emanuele et al. (2004)
Polylysine dendrimer	VivaGel (SPL7013 Gel)	HIV, HSV and sexuallytransmitted infections	Rupp et al. (2007)
PAMAM	5-Fluorouracil	Tumor	Zhuo et al. (1999)
Dendrimer	Isotope of boron (10B)	Cancer	Hawthorne (1993)
Poly(L-glutamic acid),polyamidoamine andpoly(ethyleneimine)	Folic acid	Breast cancer	Kukowska-Latallo et al. (2005)

13.4.1.5 Polymeric Micelles

Amphiphilic copolymers are used to produce polymeric micelles, which self-assemble in aqueous solutions to form a core shell structure. Drugs that are hydrophobic can be loaded into the micelle's hydrophobic core, but the hydrophilic shell renders the system water soluble and remains stable the core. To prevent rapid renal excretion, polymeric micelles typically have a size under 100 nm and a narrow dispersion. Given that these drugs can be accommodated by the micelles' internal core structure, they have a significant potential for hydrophobic drug delivery. As a result, drug stability and bioavailability are increased (Xu et al. 2013). Formation of micelles was affected by many factorssuch asconcentration of amphiphiles, size of the hydrophobic chain in the system of the molecule, temperature, and solvent (Bhushan et al. 2016). The crucial micelle concentration is the lowest concentration at which micelle assembly creation begins (Kulthe et al. 2012). The amphiphilic moleculesremain independent at lower concentration. Drugs can be loaded onto polymeric micelles via the solvent evaporation technique, direct dissolving process, and dialysis procedure (Bhushan et al. 2016; Kulthe et al. 2012).The drug targeting properties of polymericmicelles can be established by different mechanism of action. A few examples of this are the clustering of monoclonal antibodies to the micelle crown, the complexing of a particular ligand to the micelle surface, or enhanced penetrability and holding effect stimuli.

13.4.1.6 Protein and Polysaccharides Nanoparticles

Natural biopolymers such as proteins and polysaccharides are derived from a variety of biological sources, including microbes, animals, plants, and marine sources (Balaji et al. 2017). The majority of protein-derived nanoparticles are biodegradable, metabolizable, and functionalized for attachment to certain drugs and other targeted ligands. Bovine and human serum albumin, as well as gliadin, are examples of water-soluble proteins that can be used to make protein nanoparticles (Bassas-Galia et al. 2017). To deliver the drug at exact cells and tissues, the protein nanoparticles are combined with targeting ligand via chemical alteration of the nanoparticle (Balaji et al. 2017; Bassas-Galia et al. 2017). Likewise, the polysaccharides are composedof monosaccharides bonded through*O*-glycosidic link. The significant disadvantage of polysaccharide nanoparticles for drug administration is their oxidative destruction at high temperatures (beyond their melting point), which is frequently required in industrial procedures. The polysaccharides' inability to be used in any way since they are all soluble in water is another issue (Lohcharoenkal et al. 2014). However, various technique has been established the stability of the polysaccharide chains must be ensured.

13.4.2 Inorganic Nanoparticle as Drug Carrier

Various inorganic nanoparticles, such as metal nanoparticles, Mesoporoussilicananoparticles, quantum dots, and nanotubes possesses unique properties which used for differentbiomedical applications. Some of them are discussed below.

13.4.2.1 Metal Nanoparticle

In recent years, biomedical application of metal nanoparticle in target/sustained drug delivery, biosensors, bio-imaging and photo ablation therapy has been growing drastically (McNamara and Tofail 2017). Further, functionalization and modification of these nanoparticle improve their binding capacity with drugs, other ligands and antibodies, thus making them more promising candidate for biomedical applications (Kudr et al. 2017). Most extensive studies have been carried out with some metal nano particle like gold, iron, silver and copper for their ability as drug carrier. In numerous cases, the stability of polymeric nanoparticles has been increased by covalently bounded them with goldnanoparticles (Aryal et al. 2009).

13.4.2.2 Mesoporous Silica Nanoparticle

Over the last two-decade, application of mesoporous silica for drug delivery have been studied extensively. The diverse properties of mesoporous silica such as tenable pore size, high surface area, monodispersity and diverse functionalization makes them more suitable for drug delivery system. A diverse range of drugs such as paclitaxel, telmisartan, camptothecin, chlorambucil and doxorubicin have been either loaded in mesoporous silica or covalently grafter to it. In order to get enhance drug delivery capacity, functionalized silica nanoparticles are usually used. Forexample, silica nanoparticle functionalized with galactose or mannose shows higher cytotoxicity and efficient target profile to cancer cellsthan unfunctionalized ones (Bharti et al. 2015; Baeza et al. 2016).

13.4.2.3 Quantum Dot

The quantum dots are coreshellnanocrystalscontaining interface amongvarious semiconductor. Usually, the size of quantum dots ranges from 2 to 10 nmin diameterand increases to 5–20 nm after polymer encapsulation (Choi et al. 2007a, b). Hydrophilic therapeuticagents can be immobilized via covalentor non-covalent bondsontothe hydrophilic side of the amphiphilic polymer. This fully designed nano structure will identify the diseased cells and treat it. As Quantum dot nanoparticle emit detectable signals, it helps for real-time monitoring of its trajectory (Qi and Gao 2008).

13.5 Clinical Development and Approved Nano Medicines

13.5.1 Cancer Therapy

Cancer can be viewed as a sickness of many diseases because it is a very complex biological phenomenon. Many patients' lives have been saved by the medicine that is now used to treat cancer. Still, the treatment's severe side effects, which affect every part of the body because of the non-specificity of chemotherapeutics, are also responsible for a huge number of fatalities. Among the qualities of cancerous cells is their ability for replication, which allows them to spread quickly and uncontrolled (Hanahan and Weinberg 2011). The primary goal of the available chemotherapy is to prevent all rapid cell growth. The drawback of this therapy is that it also destroys other quickly dividing cells in the body, such as intestinal epithelial cells and hair follicles, which might have life-altering side effects for the patient (Baudino 2015). Another technique for the delivery of chemotherapeutic drugs is via micelles and liposomes. Moreover, micelles are fantastic how in overcoming the hydrophobic core of insoluble medications and making them soluble shell that is hydrophilic. Suppose the micelle's surface is PEGylated once more through passive diffusion. In that case, it improves the nanocarriers' capacity to penetrate the fenestrated vasculature of malignancies and inflammatory tissue, a greater drug concentration in tumors due to improved medication delivery. There are numerous polymeric micelles on the market now that contain anticancer drugs. The NK105, NK012, NC-6004, NK911, SP1049C and NC-6004 studies are examples of clinical trials and Genexol-PM is one such system (Varela-Moreira et al. 2017).Patients with breast cancer are permitted to take (paclitaxel) there are other types of nanoparticles as well, carbon nanotubes are one of the newest systems that has demonstrated potential when used to treat cancer (Varela-Moreira et al. 2017; Rahamathulla et al. 2021). Carbon nanotubes (CNTs) that fit into the category of a type of carbon with a cylinder-shaped structure that deepens on several sheets in cylinder-shaped concentric circlesfor instance, single-walled carbon nanotubes (SWCNTs) and multiwalled carbon nanotubes (MWCNTs) (Rahamathulla et al. 2021).Because the hollow interior of carbon nanotubes is extremely hydrophobic, drugs that are not soluble in water can be simply added. The expansive area makes it possible to functionalize the outside of objects, particularly for a specific cancer receptor and contrast agents (Hahn et al. 2011).

Last but not least, the anticancer effects of the spherical compound Buckminsterfullerene C60 and its compounds are studied. Fullerene C60 is a potent collector of reactive oxygen species due to its capability to bind up to six electrons (Usenko et al. 2008). Fullerene nanocrystals (Nano-C60) can boost the cytotoxicity of chemotherapeutic medicines, hence future research into a Nano-C60 adjuvant chemotherapy is possible (Zhang et al. 2008). Further research conducted by Prylutska and Skivka (2015) using the Fullerene C60 and Doxorubicin combination showed that the tumour volumes of the treated rats (C60 + Dox) were 1.4 times lower than those of the control group. Additionally, the manner in which the C60 + Dox complex affects tumour cells and its implications on immune regulation.

13.5.2 Diagnostic Testing

Despite extensive study, it is currently impossible for clinical applications to utilize nanoparticles for diagnostic testing (Mitchell et al. 2021; Kolluru and Rizvi 2013). The limitations of fluorescent light hamper diagnostic testing indicators, including color and fluorescence that disappear after usage. Researchers now have a means to get around the problems with matching and limiting the usage of dyes because of a bleeding effect thanks to fluorescent nanoparticles (Wolfbeis 2015).Quantum dots, which can be produced in a large variety of distinctly diverse hues, were discovered, which was a huge advance. They have an adjustable emission spectrum, photo-stability in the visible light spectrum, and an absorption spectrum that ranges from UV wavelengths to high quantum yield. The location of a particular particle inside the spectrum is determined by the size of the nanodot. The nanodot's size dictates where a specific particle is located in the spectrum (Wolfbeis 2015). The advantages of quantum dots are numerous as they may connected to biomolecules that can spend a lot of time investigating a biological system while applying white light they become agitated (Datta and Jaitawat 2006). Recent advances in theranostic nanoparticles, which employed for diagnosis and therapy have drawn a lot of interest. There has been a lot of interest in recent developments in theranostic nanoparticles, which are used for both diagnosis and treatment. Janib et al. (2010) has been used demonstrated that a variety of nanoparticle types, such as surfactants and dendrimers, include drug conjugate aggregates (vesicles and micelles), carbon, and core–shell nanotubes. It is possible to track the path and the positioning of these nanoparticles at the targeted area in addition to the drug taking steps to evaluate therapeutic response by combining a drug and an imaging agent in one ingenious distinct formulation (Bhojani et al. 2010). If the human immunodeficiency virus (HIV) or acquired immune deficiency syndrome (AIDS) is not treated, the patient's immune system will be almost completely destroyed, and they will die (Bhojani et al. 2010; Vishnu and Aboulafia 2015). Treatment was intentionally developed for this sickness when it started, with most patients taking 30 to 40 pills daily. The previous ten years, therapeutic advances have decreased the need for pills, daily countdown to a select few (Mamo et al. 2010; Haberer et al. 2011; Khalil et al. 2011; Crabtree-Ramirez et al. 2010).Research has proven how to create polymeric nanoparticles that transport antiretroviral (ARV) medicines to increase the effectiveness of this therapy (Mamo et al. 2010; Haberer et al. 2011). In order to prevent HIV infections, this nanotechnology based can also be utilized in conjunction with vaccinations (Mamo et al. 2010; Khalil et al. 2011). Antiretroviral medicines used to treat HIV can be classified according to the stages of the HIV life cycle to which they respond best. To avoid the emergence of tolerance and to actively prevent the progression of HIV, highly effective antiretroviral therapy (HAART), a mixture of several drugs is administered. Using nanotechnology has greatly facilitated the delivery of antiretroviral medications and in increasing compliance. When administered orally or through other non-parental methods, antiretroviral medications mucosal epithelium barrier must be permeable (suppository and patches, etc.) (Crabtree-Ramirez et al. 2010).

13.5.3 Nutraceuticals Delivery

Observable health benefits can be found in nutraceuticals, which are standardized components. They are frequently eaten in tandem with several allopathic treatments to enhance health benefits and lower the risk of particular chronic illnesses (Das et al. 2012). As with any other medicines, food matrix interactions, water solubility, degradation, and epithelial permeability all affect the bioavailability of oral nutraceuticals, and subsequently their effectiveness (Gonçalves et al. 2018). Most supplements are made of lipophilic compounds, such as polyunsaturated lipids, phytochemicals, and fat-soluble vitamins (vitamin E, A, K, and D). Extensive research has been focused on enhancing the nutraceuticals' dissolving processes utilizing combinations with nanoparticles, and nanotechnology once again and offers all-encompassing supports (Crabtree-Ramirez et al. 2010; Das et al. 2012; Gonçalves et al. 2018; Acosta 2009; McClements 2015; Din et al. 2017; Ravichandran 2010). The most well-known and well-researched nutraceutical is curcumin (diferuloylmethane), which has anti-inflammatory, antioxidant, antiapoptotic, and antiangiogenic properties. Curcumin's oral bioavailability was found to be 9 times greater when taken without piperine, an absorption booster (Shaikh et al. 2009). When compared to curcumin powder, a different study utilising colloidal nanoparticles of curcumin called theracurmin shown inhibitory effects against alcohol intoxication as well as a 40 times more surface area under the a 27-fold higher AUCarea under the curve in healthy human volunteers compared to rats (Sasaki et al. 2011).

The Vitis vinifera, Labrusca, and Muscadine grapes are the most prevalent sources of resveratrol, a notable non-flavonoid polyphenol present in many plants. It is universally acknowledged for its anti-inflammatory, anti-cancer, and antioxidant potential (Summerlin et al. 2015). Despite its low solubility and low bioavailability, the body rapidly metabolizes resveratrol and eliminates it (Kapetanovic and Muzzio 2011). The more prevalent and biologically active of the geometric isomers of resveratrol, trans-resveratrol, is photosensitive and transforms into cis-resveratrol when exposed to light. Extensive studies have focused on resveratrol nano-formulations that enhance the pharmacokinetic profile and bioavailability. Bonfanti Santos, polymeric nanoparticles, and zein-based nanoparticles are a few of them. Recent studies have concentrated on zein-based nanoparticles, nano-emulsions, liposomes, cyclodextrins, and dual nano-encapsulation methods. The neuroprotective properties of resveratrol were recently evaluated by developing solid lipid nanoparticles coated with apolipoprotein E (apoE) for low-density lipoprotein (LDL) receptor recognition on the blood–brain barrier (Sessa et al. 2013; Penalva et al. 2015; Neves et al. 2016).

13.5.4 Clinical Approvals and Market Status of Nano Medicines

Market-available nanomedicines come in a variety of forms and are based on the most cutting-edge liposome technology. The other commercially available nanomedicines are based on polymeric micelles, polymeric prodrugs, and nanoparticles. 41 nanocarrier-based formulations are being tested at various stages of clinical development and have gone from the lab to the patient's bedside. Many additional formulations based on nanocarriers are now under preclinical research. Unsurprisingly, to avoid the additional manufacturing, regulatory, financial, and polydispersity challenges associated with active, ligand-receptor-based targeting, the majority of these nanomedicines (37) rely on passive targeting via the increased permeability and retention effect. Only 4 actively targeting nanocarriers, 3 of which use TfR as their target and one of which uses prostate-specific membrane antigen (PSMA) are now being developed clinically. Drugs created at the nanoscale have several physical and biological benefits. These benefits may then result in enhanced therapeutic effectiveness and decreased toxicity. A bigger number of medicines are being researched for a variety of disorders in clinical settings, despite the fact that only about 50 nanotherapeutics have previously reached clinical use (Svenson 2012; Caster et al. 2017). The development of nanomedicines due to incorporating nanotechnology in medicine is changing how diseases are treated. This is especially true for nanomedicines that use lipid nanoparticle (LNP) drug delivery systems because more than the FDA has already approved 10 pharmaceuticals that employ LNPs to deliver treatments to illness sites. Most of these nanomedicines are cancer therapy formulations that function better and have lesser toxicity than the "free" drug (Allen and Cullis 2013). We now have a strong grasp of the criteria for the effective clinical translation of LNP systems for the delivery of small molecules as a result of the clinical success of LNP-based drug delivery systems. Size ranges of 100 nm or below, extremely effective encapsulation methods, a low surface charge, reliable, scalable production procedures, and appropriate product stability are examples of translational parameters (Cullis et al. 1989; Akinc et al. 2019). Table 13.4 gives the details of some nanomaterial based drug, its market status and involvement of active ingredient.

13.6 Advantages of Nanomaterial for Drug Delivery

In general, drugs are taken either through injection or via orally. In this process, the drug circulates all over the patient's body which may causes harmful side effect. Additionally, after oral administration, some of the drugs are poorly absorbed due to their vulnerability across theintestinal epithelium. As a result, high dose of drug is required in conventional drug delivery system to make up the bio-availability of drug

Table 13.4 Nanomaterial based drug, active molecules, disease target and market status

Nanomaterial based drug delivery	Drug product	Active ingredient	Status	Disease target	Company/ Manufacturar	FDA approved date/clinical trial status	References
Liposomes	Doxil_/Caelyx	Pegylated doxorubicin	Marketed	Metastatic breast and ovarian cancers	Janssen Pharmaceuticals/ Orthobiotech, Schering-Plough	November 1995	Svenson (2012), Caster et al. (2017), Pillai (2014)
	Myocet	Liposome-encapsulated doxorubicin	Marketed (EU)	Metastatic breast and ovarian cancers	Cephalon Inc. (EU)/ Sopherion therapeutics (US, CAN)/Elan Pharmaceuticals/ Sopherion Therapeutics	2000 approved in Europe and Canada	
	DaunoXome_	Liposome-encapsulated daunorubicin	Marketed	HIV-associated	Galen Ltd	April 1996	
	EZN-2208	Multi-arm mPEG-SN38 conjugate	Phase 2 NCT01036113b NCT00931840b	Metastatic breast cancer	Enzon Pharmaceuticals Inc	–	Svenson (2012), Cullis et al. (1989)
Polymer based	Genexol-PM	Paclitaxel-loaded polymeric micelle	Marketed	Breast cancer	Samyang pharmaceuticals	Marketed in Europe, Korea	Svenson (2012), Cullis et al. (1989), Akinc et al. (2019), Pillai (2014)

(continued)

Table 13.4 (continued)

Nanomaterial based drug delivery	Drug product	Active ingredient	Status	Disease target	Company/ Manufacturar	FDA approved date/clinical trial status	References
	NK911	mPEG-poly(aspartic acid)-doxorubicin conjugate	Phase 1	Solid tumors	Nippon Kayaku Co., Ltd.	–	Svenson (2012), Pillai (2014)
	XMT1001	Polyacetal camptothecin conjugate	Phase 1 NCT00455052b	SCLC and NSCLC	Mersana Therapeutics	–	
	SPI-077	Stealth liposomal cisplatin	Phase 2 NCT00004083b	Platinum-sensitive ovarian cancer	New York University School of Medicine/Alza	–	
Legend-receptor based	MBP-426	Transferrin-targeted nanoparticles with oxaliplatin	Phase 2 NCT00964080a	Metastatic gastric, Cancer, gastro-esophageal junction	Mebiopharm Co., Ltd	–	Svenson (2012), Caster et al. (2017), Allen and Cullis (2013), Cullis et al. (1989), Akinc et al. (2019)
	CALAA-01	Transferrin-beta-cyclodextrin polymer nanoparticle complexed with siRNA	Phase 1 NCT00689065a	Tumors	Calando Pharmaceuticals, Inc		

inside the body. By using nanoparticles as drug carrier, one can able to tune the parameters such as biodistribution of drugs, solubility, diffusivity, toxicity and pharmacokinetics of the drug molecule inside body (Zhang et al. 2008, 2014). Thus, nanoparticles offer substantial advantages over the conventional drug delivery in terms of high specificity, high stability, efficient drug carrying capacity and controlled drug release. It also has the ability to deliver both hydrophilic and hydrophobic molecules inside the body. These properties of nano carrier offer improved drug bioavailability and less dosing frequency. Nanoparticles possess unique physicochemical properties like functionalizable structure, high reactivity, large surface area to mass ratio and manageable and ultra-small size. They can easily enter through the biological membranes, tissues and organs, whereas large size particle of conventional medicine cannot cross these barriers easily (Yao et al. 2020). The most significant features of nanoparticle drug delivery system is their target specific delivery ability. By this they only affect the diseased tissue, which results in increase drug concentration at target site, improve its efficacy and also reduce the unwanted side effect caused by drug molecule. Due to these advantages, targeted nanocarriers are being highly focused in therapeutic and research. Overall, using nanoparticles as a drug carrier have the following advantages.

- Controlled and target specific release of the therapeutic agent during the transportation which increase the drug efficiency and reduces the side effects.
- Drug can be encapsulated in to the system without any chemical reaction.
- There is no drug wastage, and thus increased the drug bioavailability at target site for prolonged time period and lower the frequency of drug intake.
- By tuning the ligand of nanocarriers, the solubility of poorly water-soluble drugs can easily increase.
- Additionally, it also extends the half-life of drug circulation by decreasing immunogenicity.

13.7 Toxicity and Hazards of Nanoparticles

If nanotechnology is to be utilized in nanomedicine, safety and toxicological risks must be given full attention to resolve. The difference between the doses indicated for clinical efficacy and the levels that result in undesirable pharmaceutical side effects is known as a therapeutic ratio/index. However, these specific compositions also need for a toxicological evaluation. That's also particularly true when medications are delivered via nanoparticles. Particles are intentionally injected into the environment and human body for a variety of purposes, some of which anticipate major improvements in healthcare (Abeer 2012). When toxicologists argued that new science, practises, and protocols are required, opinions began to shift (Krewski et al. 2020). Unlike bulk materials, which are produced for their distinctive features, nanoparticles. These special characteristics need to be looked into from a toxicological perspective because surface contact with bodily tissue is crucial in determining particle response. When nanoparticles are chosen for their distinct reactive properties, it stands to reason that these properties would also affect how dangerous these particles are. Although there may be several issues related to the usage of these nanoparticles that can be found using existing testing and methodologies in medication and device review, these tests cannot be relied upon to identify all potential dangers, both biological and non-biological in nature.

As compared to micron-sized particles, nanoparticles are fundamentally different in their physico-chemical composition, which may enable them to modify body distribution, cross the blood–brain barrier, and activate blood coagulation pathways. Studies into the (pharmaco) dynamics and dispersion of nanoparticles should get further attention in light of these properties. What is currently lacking is a fundamental knowledge of how nanoparticles behave biologically in terms of in vivo dispersion at the organ and cellular level. Diseased people are primarily affected by combustion-derivednanoparticles in populations exposed to the environment. Typical pre-clinical testing is nearly typically conducted on healthy animals and volunteers, which means that particle concerns may not be discovered until much later. Utilizing nanoparticles as drug delivery systems may lessen the toxicity of the medicine that is added. It is thus impossible to differentiate between the toxicity of pharmaceuticals and that of nanoparticles. When using slow- or non-degradable particles to deliver medications, this is essential since they may build up and remain eventually, the drug delivery site will produce persistent inflammatory reactions (Abeer 2012; Krewski et al. 2020).

13.7.1 Nanoparticle Toxicity Evidence

With PM10 particulate matter with a size below 10 mm, the strongest database on the toxicity of nanoparticles has been created wherein nanoparticle manufacturing has shown to be an influential driving factor for research (Borm et al. 2006). It is crucial to talk about these data in the belief that it would help shed light on the toxicity of synthetic nanoparticles. Several sources support the hypothesis that combustion-derived nanoparticles have a significant role in causing the negative consequences of ambient particle air pollution or PM10. The concept that PM10 contains a component has grown as a result of the assumption that a sizable portion of PM10's mass is non-toxic, with combustion-derived nanoparticles appearing to be a plausible option really because many PM10 particles are believed to be non-toxic, inflammation has developed, and combustion-derived nanoparticle appears to be a potential possibility. Since they make up the majority of particles, nanoparticles may be significant because they have a large surface area per mass while being tiny. According to particle toxicology, a larger particle surface corresponds to a greater level of toxicity for hazardous particles in general. The claim that nanoparticles in PM10 are significant contributors to negative effects is supported by significant data from toxicology and limited information from epidemiology sources (Borm et al. 2006; Pope et al. 2004).

Respiratory disease respondents to indicate fatalities, hospitalizations, and deaths from cardiovascular and respiratory disease are all indicators of the harmful impact of PM on human health (Pope et al. 2004). These negative consequences are linked together by inflammation, making nanoparticles capacity to elicit inflammation one of their key characteristics. It is unclear which nanoparticles effects require lower exposures may have an influence on the consequences of pulmonary inflammation as a precursor than those that cause inflammation. Additionally, there is a chance that pulmonary inflammation will alter membrane permeability, which could then affect the likelihood that particles will spread outside the lung. Some nanoparticles may also have the capacity to directly influence cardiovascular disease. Following the intake of diesel exhaust particles, vascular function was hampered (Hoet et al. 2004; Ehsanifar et al. 2021). Several data are still sparse, and not all nanoparticle investigations have detected a significant transfer of substance from the lung to the blood. Very little translocation has occurred in several studies that are important to understand the kinematics of how ambient air nanoparticles are cleaned in order to realize the potential for adverse effects of inhaling them (Kreyling et al. 2004; Takenaka et al. 2006).

13.7.2 Toxicological Effects of Nanoparticles

Nanoparticles have several extremely unique characteristics that are crucial for the future design and toxicity testing of artificial nanomaterials. The impacts of tiny particles are only quantitatively different in some cases (Fourches et al. 2011). In this situation, nanoparticles might have similar effects to "conventional" nanoparticles (such as lung cancer or inflammation), but they might be more effective due to their large surface area. Nevertheless, nanoparticles may result in new effects that have not before been observed with larger particles (e.g., mitochondrial damage, platelet aggregation, cardiovascular effects). Ligand-coated based nanoparticles researched and used to deliver drugs or for molecular imaging. The ability of nanoparticles to penetrate tissue without impairing tissue function has significantly enhanced. Anionic nanoparticles are generally non-toxic in nature, in contrast to cationic nanoparticles, which have been shown to cause hemolysis and blood coagulation when present, such as Au and polystyrene (Fourches et al. 2011; Tang et al. 2012).Additionally, drug-loaded nanoparticles have been utilized to lengthen half-lives or better understand difficulties, and they have demonstrated which features of the particles need to be changed to permit delivery while remaining biocompatible (Tang et al. 2012).

Identification of hazards is the standard procedure for assessing the safety of healthcare goods, its advised is to include testing based on the intended use and risk classification. Inhalation risks may be associated with some designed nanoparticles that become airborne, while continuous exposures may result from using nanoparticles in cosmetics (Fourches et al. 2011; Tang et al. 2012). Parenteral administration requires careful consideration of systemic distribution, interactions with blood components, and kinetics when employing synthetic nanoparticles as instruments to deliver drugs to specific target. Each nanoparticle composition needs to be separately investigated using the appropriate techniques, paying close attention to their route of entry. Empty particles' possible harmful (toxic) consequences should also be considered in this context (Tang et al. 2012). A number of fundamental factors must be taken into account when developing testing processes. Although nanoparticles have an effects as "conventional" nanoparticles (such as inflammation and lung cancer), their increased surface area may make them more effective. Additionally, nanoparticles may have brand-new effects that were not previously observed with bulk substances or larger particles. The main issue at hand is whether the testing and categorization procedures used today are sufficient or appropriate. These will be able to spot specific hazardous consequences, as shown by the studies that have already been released. However, it is possible that not all dangers will be found, necessitating more targeted testing. Additionally, nanotechnology encourages the blending of technologies; for instance, similar materials may be used in the automobile and health sciences industries. Data interchange between sectors is advised to promote the manufacturing and marketing of safe nanomaterials (Fourches et al. 2011; Gnach et al. 2015).

13.8 Challenges and Future Scope

The application of nanotechnology to cancer therapy has brought about a new era in cancer treatment. Both organic and inorganic are among the nanoparticles are already used in cancervarious types of treatments. Nanoparticles based drug delivery systems are superior to customary pharmaceuticals in linked with better biocompatibility, pharmacokinetics, stability, and disease targeting, while also contributing in important contribution to lowering systemic toxicity and overcoming resistance to drugs. Due to these benefits, Nanoparticlesbased medications can be broadly used in radiation, targeted therapy, and chemotherapy, gene therapy and hyperthermia. Better platforms are provided by nano material based drug delivery systems for combination treatment, it aids in overcoming drug resistance pathways, including over expression of the tumour microenvironment under hypoxia, impaired apoptotic pathway, and efflux transporter. According to different drugs resistance mechanisms, MDR can be overcome by using loaded nanoparticles that have different targeting agents paired with cytotoxic agents.

As it developed, other types of nanoparticles have display superior delivery qualities and sparked additional interest attention. Additional research in biological traits of identifying certain cancers will result in more focused in the directions of research for these medicines. Additionally, creating hybrid nanoparticles that are designing nanoparticles with targeting moieties that more specifically target cancer cells makes sense for cancer therapy more investigation particularly, the relationships between nanoparticles complicated, as is the immune system (Sailor and Park 2012). In addition, a variety of considerations must be considered to ensure proper execution of a particular drug release and its practical uses in human bodies. For instance, to obtain an acceptable encapsulation effect the nanocomposites with multi-stimulatory responsive characteristics may be produced as effectively adapted to complicated biological systems. The nanoparticles surface, composition, size, and form are all the elements that influence how nanoparticles interact with the defense mechanism. Although classical immunotherapy has shown greater efficacy than nanovaccines and synthetic APCs, the clinical efficacy of this therapy is still inadequate, and further research into the safety and tolerance of these novel techniques is required. Nanotechnology will undoubtedly continue to progress, and in the near future, non-toxic, biocompatible, and hemo-compatible drug delivery systems will successfully transport therapeutic agents to afflicted locations in human bodies to treat cancer. Additionally, creating nanoparticles laden with immunomodulatory factors may increase the potency of immunotherapy vaccines. Therefore, for drug design and exploration, a deeper understanding of nanotechnology and additional research into the interaction between nanoparticles in drug delivery systems and target disease immunity are required.

References

Abeer S (2012) Future medicine: nanomedicine. Jimsa 25(3):187–192

Acosta E (2009) Bioavailability of nanoparticles in nutrient and nutraceutical delivery. Curr Opin Colloid Interface Sci 14(1):3–15

Akinc A, Maier MA, Manoharan M, Fitzgerald K, Jayaraman M, Barros S, Ansell S, Du X, Hope MJ, Madden TD, Mui BL (2019) The Onpattro story and the clinical translation of nanomedicines containing nucleic acid-based drugs. Nat Nanotechnol 14(12):1084–1087

Allen TM, Cullis PR (2013) Liposomal drug delivery systems: from concept to clinical applications. Adv Drug Deliv Rev 65(1):36–48

Aryal S, Pilla S, Gong S (2009) Multifunctional nano-micellesformed by amphiphilic gold-polycaprolactone- methoxy poly-(ethylene glycol) (Au-PCL-MPEG) nanoparticles for potentialdrug delivery applications. J Nanosci Nanotechnol 9:5701–5708

Baeza A, Manzano M, Colilla M, Vallet-Regí M (2016) Recent advances in mesoporous silica nanoparticles for antitumor therapy: our contribution. Biomater Sci 4(5):803–813

Baker Jr, JR (2009) Dendrimer-based nanoparticles for cancer therapy. Hematol Am Soc Hematol Educ Program 708–719

Balaji AB, Pakalapati H, Khalid M, Walvekar R, Siddiqui H (2017) Natural andsynthetic biocompatible and biodegradable polymers. In: Shimpi NG (ed) Biodegradable and biocompatible polymer composites:processing, properties and applications. Woodhead Publishing series incomposites science and engineering. Woodhead Publishing, Duxford, pp 3–32

Bassas-Galia M, Follonier S, Pusnik M, Zinn M (2017) Natural polymers: a sourceof inspiration. In: Bioresorbable polymers for biomedical applications. Elsevier, New York, pp 31–64

Baudino TA (2015) Targeted cancer therapy: the next generation of cancer treatment. Curr Drug Discov Technol 12(1):3–20

Bharti C, Nagaich U, Pal AK, Gulati N (2015) Mesoporous silica nanoparticles in target drug delivery system: a review. Int J Pharm Investig 5(3):124

Bhojani MS, Van Dort M, Rehemtulla A, Ross BD (2010) Targeted imaging and therapy of brain cancer using theranostic nanoparticles. Mol Pharm 7(6):1921–1929

Bhushan SP, Vladimir PT (2016) Targeted drug delivery systems: strategies and challenges. In: Devarajan PV, Jain S (eds) Targeted drug delivery: concepts and design. Springer, Switzerland, pp 3–38

Borm PJA, Robbins D, Haubold S et al (2006) The potential risks of nanomaterials: a review carried out for ECETOC. Part Fiber Toxicol 3:11

Bozzuto G, Molinari A (2015) Liposomes as nanomedical devices. Int J Nanomedicine 10:975–999

Caster JM, Patel AN, Zhang T, Wang A (2017) Investigational nanomedicines in 2016: a review of nanotherapeutics currently undergoing clinical trials. Wiley Interdiscip Rev Nanomed Nanobiotechnol 9(1):e1416

Cheng Y, Xu Z, Ma M, Xu T (2008) Dendrimers as drug carriers: applications indifferent routes of drug administration. J Pharm Sci 97(1):123–143

Choi AO, Cho SJ, Desbarats J, Lovric J, Maysinger D (2007a) Quantum dot-induced cell death involves Fas upregulation andlipid peroxidation in human neuroblastoma cells. J Nanobiotechnol 5:1–4

Choi HS, Liu W, Misra P, Tanaka E, Zimmer JP, Ipe BI, Bawendi MG, Frangion JV (2007b) Renal clearance of quantumdots. Nat Biotechnol 25(10):1165–1170

Chung EJ, Cheng Y, Morshed R, Nord K, Han Y, Wegscheid ML, Auffinger B, Wainwright DA, Lesniak MS, Tirrell MV (2014) Fibrin-binding, peptide amphiphile micelles for targeting glioblastoma. Biomater 35(4):1249–1256

Crabtree-Ramirez B, Villasís-Keever A, Galindo-Fraga A, del Río C, Sierra-Madero J (2010) Effectiveness of highly active antiretroviral therapy (HAART) among HIV-infected patients in Mexico. AIDS Res Hum Retroviruses 26(4):373–378

Cullis PR, Mayer LD, Bally MB, Madden TD, Hope MJ (1989) Generating and loading of liposomal systems for drug-delivery applications. Adv Drug Deliv Rev 3(3):267–282

Das L, Bhaumik E, Raychaudhuri U, Chakraborty R (2012) Role of nutraceuticals in human health. J Food Sci Technol 49(2):173–183

Datta R, Jaitawat S (2006) Nanotechnology - the new frontier of medicine. Med J Armed Forces India 62(3):263–268

Davis ME, Zuckerman JE, Choi CHJ, Seligson D, Tolcher A, Alabi CA, Yen Y, Heidel JD, Ribas A (2010) Evidence of RNAi in humans from systemically administered siRNA via targeted nanoparticles. Nature 464(7291):1067–1070

Din FU, Aman W, Ullah I, Qureshi OS, Mustapha O, Shafique S, Zeb A (2017) Effective use of nanocarriers as drug delivery systems for the treatment of selected tumors. Int J Nanomed12:7291

Ding CZ, Li ZB (2017) A review of drug release mechanisms from nanocarriersystems. Mater Sci Eng C 76:1440–1453

D'Emanuele A, Jevprasesphant R, Penny J, Atwood D (2004) Theuse of a dendrimer-propranolol prodrug to bypass efflux transportersand oral bioavailability. J Control Release 95(3):447–453

Ehsanifar M, Banihashemian SS, Ehsanifar M (2021) Exposure to air pollution nanoparticles: Oxidative stress and neuroinfl ammation. J Biomed Res Environ Sci 2(10):964–967

Elseoud WSA, Hassan ML, Sabaa MW, Basha M, Hassan EA, Fadel SM (2018) Chitosan nanoparticles/cellulose nanocrystals nanocomposites as acarrier system for the controlled release of repaglinide. Int J Biol Macromol 111:604–613

Fais S, O'Driscoll L, Borras FE, Buzas E, Camussi G, Cappello F, Carvalho J, Da Silva AC, Del Portillo H, El Andaloussi S, Trcek TF (2016) Evidence-based clinical use of nanoscale extracellular vesicles in nanomedicine. ACS Nano 10(4):3886–3899

Farhood H, Gao X, Son K, Lazo JS, Huang L, Barsoum J, Bottega R, Epand RM (1994) Cationic liposomes for direct gene transfer in therapy of cancer and other diseases a. Ann N Y Acad Sci 716(1):23–35

Farokhzad OC, Cheng J, Teply BA, Sherifi I, Jon S, Kantoff PW, Richie JP, Langer R (2006) Targeted nanoparticle-aptamer bioconjugates for cancer chemotherapy in vivo. Proc Natl Acad Sci 103(16):6315–6320

Felgner PL, Gadek TR, Holm M, Roman R, Chan HW, Wenz M, Northrop JP, Ringold GM, Danielsen M (1987) Lipofection: a highly efficient, lipid-mediated DNA-transfection procedure. Proc Natl Acad Sci 84(21):7413–7417

Fourches D, Pu D, Tropsha A (2011) Exploring quantitative nanostructure-activity relationships (QNAR) modeling as a tool for predicting biological effects of manufactured nanoparticles. Comb Chem High Throughput Screen 14(3):217–225

Fricker G, Kromp T, Wendel A, Blume A, Zirkel J, Rebmann H, Setzer C, Quinkert R-O, Martin F, Muller-Goymann C (2010) Phospholipidsand lipid-based formulations in oral drug delivery. Pharm Res 27(8):1469–1486

Gnach A, Lipinski T, Bednarkiewicz A, Rybka J, Capobianco JA (2015) Upconverting nanoparticles: assessing the toxicity. Chem Soc Rev 44(6):1561–1584

Gonçalves RF, Martins JT, Duarte CM, Vicente AA, Pinheiro AC (2018) Advances in nutraceutical delivery systems: from formulation design for bioavailability enhancement to efficacy and safety evaluation. Trends Food Sci Technol 78:270–291

Haberer JE, Cook A, Walker AS, Ngambi M, Ferrier A, Mulenga V, Kityo C, Thomason M, Kabamba D, Chintu C, Gibb DM (2011) Excellent adherence to antiretrovirals in HIV+ Zambian children is compromised by disrupted routine, HIV nondisclosure, and paradoxical income effects. PloS One 6(4):e18505

Hahn MA, Singh AK, Sharma P, Brown SC, Moudgil BM (2011) Nanoparticles as contrast agents for in-vivo bioimaging: current status and future perspectives. Anal Bioanal Chem 399(1):3–27

Hainfeld JF, Slatkin DN, Smilowitz HM (2004) The use of gold nanoparticles to enhance radiotherapy in mice. Phys Med Biol 49(18):309–315

Hanahan D, Weinberg RA (2011) Hallmarks of cancer: the next generation. Cell 144:646–674

Hansen K, Kim G, Desai KG, Patel H, Olsen KF, Curtis-Fisk J, Tocce E, Jordan S, Schwendeman SP (2015) Feasibility investigation of cellulose polymersfor mucoadhesive nasal drug delivery applications. Mol Pharm 12:2732–2741

Hawthorne MF (1993) The role of chemistry in the development ofboron neutron capture therapy of cancer. Angew Chem Int Ed 32(7):950–984

Hirsch LR, Stafford RJ, Bankson JA, Sershen SR, Rivera B, Price RE, Hazle JD, Halas NJ, West JL (2003) Nanoshell-mediated near-infrared thermal therapy of tumors under magnetic resonance guidance. Proc Natl Acad Sci 100(23):13549–13554

Hoet PH, Brüske-Hohlfeld I, Salata OV (2004) Nanoparticles–known and unknown health risks. J Nanobiotechnol 2(1):1–15

Holowka E, Bhatia SK (2016) Drug delivery. Springer, New York

Hrkach J, Von Hoff D, Mukkaram Ali M, Andrianova E, Auer J, Campbell T et al (2012) Preclinical development and clinical translation of a PSMAtargeted docetaxel nanoparticle with a differentiated pharmacological profile. Sci Transl Med 4(128):28ra139

Janib SM, Moses AS, MacKay JA (2010) Imaging and drug delivery using theranostic nanoparticles. Adv Drug Deliv Rev 62(11):1052–1063

Kapetanovic IM, Muzzio M et al (2011) Pharmacokinetics, oral bioavailability, and metabolic profile of resveratrol and its dimethylether analog, pterostilbene, in rats. Cancer Chemother Pharmacol 68:593–601

Kaur CD, Nahar M, Jain NK (2008) Lymphatic targeting ofzidovudine using surface-engineered liposomes. J Drug Target 16(10):798–805

Keith DP, Cui H (2014) Fabrication of drug delivery systems using self-assembled peptide nanostructures. In: Micro and nanofabrication using self-assembled biological nanostructures. Elsevier, pp 91–111

Kesharwani P, Xie L, Banerjee S, Mao G, Padhye S, Sarkar FH, Iyer AK (2015) Hyaluronic acid-conjugated polyamidoamine dendrimers for targeteddelivery of 3, 4-difluorobenzylidene curcumin to CD44 overexpressingpancreatic cancer cells. Colloids Surf B 136:413–423

Khalil N, Carraro E, Cótica LF, Mainardes RM (2011) Potential of polymeric nanoparticles in AIDS treatment and prevention. Expert Opin Drug Deliv 8(1):95–112

Kim HJ, Jones MN (2004) The delivery of benzyl penicillin toStaphylococcus aureus biofilms by use of liposomes. J Liposome Res 14(3–4):123–139

Kolluru LP, Rizvi SAA et al (2013) Formulation development of albumin based theragnostic nanoparticles as a potential tumor targeting and delivery system. J Drug Target 21:77–86

KoopaeiM N, KhoshayandM R, Mostafavi SH, Amini M, Khorramizadeh MR, Tehrani MJ, Atyabi F, Dinarvand R (2014) Docetaxel loaded PEG-PLGA nanoparticles: optimized drugloading, in-vitro cytotoxicity and in-vivo antitumor effect. Iran J Pharm Res 13(3):819–833
Krewski D, Andersen ME, Tyshenko MG, Krishnan K, Hartung T, Boekelheide K, Wambaugh JF, Jones D, Whelan M, Thomas R, Yauk C (2020) Toxicity testing in the 21st century: progress in the past decade and future perspectives. Arch Toxicol 94(1):1–58
Kreyling WG, Semmler M, Möller W (2004) Dosimetry and toxicology of ultrafine particles. J Aerosol Med 17:140–152
Kudr J, Haddad Y, Richtera L, Heger Z, Cernak M, Adam V, Zitka O (2017) Magnetic nanoparticles: from design and synthesis to real world applications. Nanomaterials 7:243
Kukowska-Latallo JF, Candido KA, Cao Z, Nigavekar SS, Majoros IJ, Thomas TP, Balogh LP, Khan MK, Baker JR (2005) Nanoparticle targeting of anticancer drug improvestherapeutic response in animal model of human epithelial cancer. Cancer Res 65(12):5317–5324
Kulthe SS, Choudhari YM, Inamdar NN, Mourya V (2012) Polymeric micelles:authoritative aspects for drug delivery. Design Monomers Polym 15:465–521
Lee JE, Lee N, Kim H, Kim J, Choi SH, Kim JH, Kim T, Song IC, Park SP (2010) Uniform mesoporous dye-dopedsilica nanoparticles decorated with multiple magnetite nano crystals for simultaneous enhanced magnetic resonance imaging, fluorescence imaging, and drug delivery. J Am Chem Soc 132:552–557
Lin R, Cheetham AG, Zhang P, Lin YA, Cui H (2013) Supramolecular filaments containing a fixed 41% paclitaxel loading. Chem Commun 49(43):4968–4970
Liu S, Yang S, Ho PC (2018) Intranasal administration of carbamazepine-loadedcarboxymethyl chitosan nanoparticles for drug delivery to the brain. Asian J Pharm Sci 13:72–81
Lohcharoenkal W, Wang L, Chen YC, Rojanasakul Y (2014) Protein nanoparticlesas drug delivery carriers for cancer therapy. BioMed Res Int 2014:180549
Lombardo D, Kiselev MA, Caccamo MT (2019) Smart nanoparticles for drug delivery application: development of versatile nanocarrier platforms inbiotechnology and nanomedicine. J Nanomater 2019:3702518
Lu H, Wang J, Wang T, Zhong J, Bao Y, Hao H (2016) Recent progress on nano-structures for drug delivery applications. J Nanomater 2016:5762431
Madaan K, Kumar S, Poonia N, Lather V, Pandita D (2014) Dendrimers in drugdelivery and targeting: drug-dendrimer interactions and toxicity issues. J Pharm Bioallied Sci 6(3):139–150
Mamo T, Moseman EA, Kolishetti N, Salvador-Morales C, Shi J, Kuritzkes DR, Langer R, Andrian UV, Farokhzad OC (2010) Emerging nanotechnology approaches for HIV/AIDS treatment and prevention. Nanomedicine 5(2):269–285
Mattos BD, Rojas OJ, Magalhaes WLE (2017) Biogenic silica nanoparticlesloaded with neem bark extract as green, slow-release biocide. J Clean Prod 142:4206–4213
McClements DJ (2015) Nanoscale nutrient delivery systems for food applications: improving bioactive dispersibility, stability, and bioavailability. J Food Sci 80(7):N1602–N1611
McNamara K, Tofail SA (2017) Nanoparticles in biomedical applications. Adv Phys 2:54–88
Meng J, Fan J, Galiana G, Branca RT, Clasen PL, Ma S, Zhou J, Leuschner C, Kumar CSSR, Hormes J, Otiti T (2009) LHRH-functionalized superparamagnetic iron oxide nanoparticles for breast cancer targeting and contrast enhancement in MRI. Mater Sci Eng C 29(4):1467–1479
Mirza AZ, Siddiqui FA (2014) Nanomedicine and drug delivery: a mini review. Int Nano Lett 4:1–7
Mitchell MJ, Billingsley MM, Haley RM, Wechsler ME, Peppas NA, Langer R (2021) Engineering precision nanoparticles for drug delivery. Nat Rev Drug Discov 20(2):101–124
Mottaghitalab F, Farokhi M, Fatahi Y, Atyabi F, Dinarvand R (2019) New insights into designing hybrid nanoparticles for lung cancer: diagnosis and treatment. J Control Release 295:250–267
Mudshinge SR, Deore AB, Patil S, Bhalgat CM (2011) Nanoparticles: Emerging carriers for drug delivery. Saudi Pharm J 19(3):129–141
Nagpal K, Singh SK, Mishra DN (2010) Chitosan nanoparticles: a promising system in novel drug delivery. Chem Pharm Bull 58(11):1423–1430

Naresh V, Lee N (2021) A review on biosensors and recent development of nanostructured materials-enabled biosensors. Sensors 21(4):1109
Neves AR, Queiroz JF, Reis S (2016) Brain-targeted delivery of resveratrol using solid lipid nanoparticles functionalized with apolipoprotein E. J Nanobiotechnol 14:27
Park JW (2002) Liposome-based drug delivery in breast cancertreatment. Breast Cancer Res 4(3):95–99
Patra JK, Das G, Fraceto LF, Campos EVR, Rodriguez-Torres MDP, Acosta-Torres LS, Diaz-Torres LA, Grillo R, Swamy MK, Sharma S, Habtemariam S (2018) Nano based drug delivery systems: recent developments and future prospects. J Nanobiotechnol 16(1):1–33
Penalva R, Esparza I, Larraneta E, González-Navarro CJ, Gamazo C, Irache JM (2015) Zein-based nanoparticles improve the oral bioavailability of resveratrol and its anti-inflammatory effects in a mouse model of endotoxic shock. J Agric Food Chem 63(23):5603–5611
Pillai G (2014) Nanomedicines for cancer therapy: an update of FDA approved and those under various stages of development. SOJ Pharm Pharm Sci 1(2):13
Pope CA, Burnett RT, Thurston GD, Thun MJ, Calle EE, Krewski D, Godleski JJ (2004) Cardiovascular mortality and long-term exposure to particulate air pollution: epidemiological evidence of genearal pathophysiological pathways of disease. Circulation 109:71–77
Prylutska SV, Skivka LM et al (2015) Complex of C60 fullerene with doxorubicin as a promising agent in antitumor therapy. Nanosc Res Lett 10(1):499
Puri A, Loomis K, Smith B, Lee JH, Yavlovich A, Heldman E, Blumenthal R (2009) Lipid-based nanoparticles as pharmaceutical drug carriers: from concepts to clinic. Crit Rev Ther Drug Carrier Syst 26(6):523–580
Qi L, Gao X (2008) Emerging application of quantum dots for drugdelivery and therapy. Expert Opin Drug Deliv 5(3):63–67
Rabbani M, Hoque ME, Mahbub ZB (2020) Nanosensors in biomedical and environmental applications: perspectives and prospects. In: Nanofabrication for smart nanosensor applications, pp 163–186
Ragelle H, Danhier F, Préat V, Langer R, Anderson DG, (2017) Nanoparticle-based drug delivery systems: a commercial and regulatory outlook as the field matures. Expert Opin Drug Deliv 14(7):851–864
Rahamathulla M, Bhosale RR, Osmani RA, Mahima KC, Johnson AP, Hani U, Ghazwani M, Begum MY, Alshehri S, Ghoneim MM, Shakeel F (2021) Carbon nanotubes: current perspectives on diverse applications in targeted drug delivery and therapies. Materials 14(21):6707
Ravichandran R (2010) Nanotechnology applications in food and food processing: innovative green approaches, opportunities and uncertainties for global market. International J Green Nanotech: Phys Chem 1(2):P72–P96
Rupp R, Rosenthal SL, Stanberry LR (2007) VivaGel (SPL7013Gel): a candidate dendrimer–microbicide for the prevention of HIVand HSV infection. Int J Nanomed 2(4):561–566
Sahu T, Ratre YK, Chauhan S, Bhaskar LVKS, Nair MP, Verma HK (2021) Nanotechnology based drug delivery system: current strategies and emerging therapeutic potential for medical science. J Drug Deliv Sci Technol 63:102487
Sailor MJ, Park JH (2012) Hybrid nanoparticles for detection and treatment of cancer. Adv Mater 24(28):3779–3802
Salim M, Minamikawa H, Sugimura A, Hashim R (2014) Amphiphilicdesigner nano-carriers for controlled release: from drug deliveryto diagnostics. MedChemComm 5(11):1602–1618
Sasaki H, Sunagawa Y, Takahashi K, Imaizumi A, Fukuda H, Hashimoto T, Wada H, Katanasaka Y, Kakeya H, Fujita M, Hasegawa K, Morimoto T (2011) Innovative preparation of curcumin for improved oral bioavailability. Biol Pharm Bull 34(5):660–665
Sessa M, Balestrieri ML, Ferrari G, Servillo L, Castaldo D, D'Onofrio N, Donsì F, Tsao R (2013) Bioavailability of encapsulated resveratrol into nanoemulsion-based delivery systems. Food Chem 147:42–50

Shaikh J, Ankola DD, Beniwal V, Singh D, Ravi Kumar MNV (2009) Nanoparticle encapsulation improves oral bioavailability of curcumin by at least 9-fold when compared to curcumin administered with piperine as absorption enhancer. Eur J Pharm Sci 37(3–4):223–230

Sharma A, Madhunapantula SV, Robertson GP (2012) Toxicological considerations when creating nanoparticle-based drugs and drug delivery systems. Expert Opin Drug Metabolism Toxicol 8(1):47–69

Siepmann F, Herrmann S, Winter G, Siepmann J (2008) A novel mathematicalmodel quantifying drug release from lipid implants. J Control Release 128:233–240

Sowjanya K (2015) A review on current advancements in nanotechnology. Res Rev J Med Health Sci 4(3)

Sultana A, Zare M, Thomas V, Kumar TS, Ramakrishna S (2022) Nano-based drug delivery systems: conventional drug delivery routes, recent developments and future prospects. Medicine Drug Discov 100134

Summerlin N, Soo E, Thakur S, Qu Z, Jambhrunkar S, Popat A (2015) Resveratrol nanoformulations: challenges and opportunities. Int J Pharm 479(2):282–290

Sun B, Zhang M, Shen J, He Z, Fatehi P, Ni Y (2019) Applications of cellulose-basedmaterials in sustained drug delivery systems. Curr Med Chem 26(14):2485–2501

Svenson S (2012) Clinical translation of nanomedicines. Curr Opin Solid State Mater Sci 16(6):287–294

Takemoto K, Yamamoto Y, Ueda Y, Sumita Y, Yoshida K, Niki Y (2004) Comparative studies on the efficacy of Am Bisomeand Fungizone in a mouse model of disseminated aspergillosis. J Antimicrob Chemother 53(2):311–317

Takenaka S, Karge E, Kreyling WG, Lentner B, Möller W, Behnke-Semmler M, Jennen L, Walch A, Michalke B, Schramel P, Heyder J, Schulz H (2006) Distribution pattern of inhaled ultrafine gold nanoparticles in the rat lung. Inhalation Toxicol 18:733–740

Tang F, Li L, Chen D (2012) Mesoporous silica nanoparticles: synthesis, biocompatibility and drug delivery. Adv Mater 24(12):1504–1534

Tran MA, Watts RJ, Robertson GP (2009) Use of liposomes as drug delivery vehicles for treatment of melanoma. Pigment Cell Melanoma Res 22(4):388–399

Tripathy S, Das M (2013) Dendrimers and their applications as novel drugdelivery carriers. J Appl Pharm Sci 3(9):142–149

Usenko CY, Harper SL, Tanguay RL (2008) Fullerene C60 exposure elicits an oxidative stress response in embryonic zebrafish. Toxicol Appl Pharmacol 229(1):44–55

Varela-Moreira A, Shi Y, Fens MH, Lammers T, Hennink WE, Schiffelers RM (2017) Clinical application of polymeric micelles for the treatment of cancer. Mater Chem Front 1(8):1485–1501

Vishnu P, Aboulafia DM (2015) Haematological manifestations of human immune deficiency virus infection. Br J Haematol 171(5):695–709

Wang H, Agarwal P, Zhao G, Ji G, Jewell CM, Fisher JP et al (2018) Overcoming ovarian cancer drug resistance with a cold responsive nanomaterial. ACS Cent Sci 4:567–581

Whitemore M, Li S, Huang L (2001) Liposome vectors for in vivogene delivery. Curr Protoc Hum 12.8.1–12.8.9

Wolfbeis OS (2015) An overview of nanoparticles commonly used in fluorescent bioimaging. Chem Soc Rev 44(14):4743–4468

Xu W, Ling P, Zhang T (2013) Polymeric micelles, a promising drug deliverysystem to enhance bioavailability of poorly water-soluble drugs. J Drugdeliv 2013:340315

Yan L, Chen X (2014) 7-Nanomaterials for drug delivery. In: Nanocrystalline materials, 2nd edn. Elsevier, pp 221–268

Yao Y, Zhou Y, Liu L, Xu Y, Chen Q, Wang Y, Wu S, Deng Y, Zhang J, Shao A (2020) Nanoparticle-based drug delivery in cancer therapy and its role in overcoming drug resistance. Front Mol Biosci 7:193

Yoo HS, Park TG (2001) Biodegradable polymeric micellescomposed of doxorubicin conjugated PLGA–PEG blockcopolymer. J Control Release 70(1–2):63–70

Zhang S, Guo N, Wan G, Zhang T, Li C, Wang Y et al (2019) PH and redox dual-responsive nanoparticles based on disulfide-containing poly(βamino ester) for combining chemotherapy and COX-2 inhibitor to overcome drug resistance in breast cancer. J Nanobiotechnol 17:109

Zhang X, Huang Y, Li S (2014) Nanomicellar carriers for targeted delivery of anticancer agents. Ther Deliv 5(1):53–68

Zhang L, Gu FX, Chan JM, Wang AZ, Langer RS, Farokhzad OC (2008) Nanoparticles in medicine: therapeutic applications and developments. Clin Pharm Therap 83(5):761–769

Zhou J, Kroll AV, Holay M, Fang RH, Zhang L (2020) Biomimetic nanotechnology toward personalized vaccines. Adv Mater 32(13):1901255

Zhuo RX, Du B, Lu ZR (1999) In vitro release of 5-fluorouracilwith cyclic core dendritic polymer. J Contr Rel 57(3):249–257

Chapter 14
Role of Nanotechnology in Medicine: Opportunities and Challenges

Ikechukwu P. Ejidike, Oluwaferanmi Ogunleye, Mercy O. Bamigboye, Obiamaka M. Ejidike, Athar Ata, Michael O. Eze, Hadley S. Clayton, Victor U. Nwankwo, and Johnson O. Fatokun

Abstract Most of the buzz surrounding the word "nanotechnology" has been on over the last 15–20 years. The application of nanotechnology in diverse fields is based on two motives. Firstly, an increasing surface/volume ratio, nanoparticles become more active. Secondly, a smaller particle size results in greater quantum confinement which is responsible for the changes in electron mobility, effective mass, relative dielectric constant, and optical properties. Application of nanotechnology in medicine (nanomedicine) includes Nanodiagnosis, Nano therapy (controlling drug

I. P. Ejidike (✉) · O. Ogunleye
Department of Chemical Sciences, Faculty of Natural, Applied and Health Sciences, Anchor University, Lagos, Nigeria
e-mail: iejidike@aul.edu.ng

I. P. Ejidike · H. S. Clayton
Department of Chemistry, College of Science, Engineering and Technology, University of South Africa, Florida Campus, South Africa

I. P. Ejidike · A. Ata · M. O. Eze
Department of Chemistry, Faculty of Science, The University of Winnipeg, Winnipeg, MB, Canada

M. O. Bamigboye
Department of Industrial Chemistry, Faculty of Physical Sciences, University of Ilorin, Ilorin, Nigeria

O. M. Ejidike
Department of Geology, Faculty of Science, University of Benin, Benin, Nigeria

A. Ata
Department of Chemistry, College of Sciences and Arts, Michigan Technological University, MI, USA

V. U. Nwankwo
Department of Physics, Faculty of Natural, Applied and Health Sciences, Anchor University, Lagos, Nigeria

J. O. Fatokun
Department of Mathematical Sciences, Faculty of Natural, Applied and Health Sciences, Anchor University, Lagos, Nigeria

M. P. Shah et al. (eds.), *Biogenic Nanomaterials for Environmental Sustainability: Principles, Practices, and Opportunities*, Environmental Science and Engineering,
https://doi.org/10.1007/978-3-031-45956-6_14

delivery), and Remedial (renovated) medicine. Nano diagnosis is the evolution of nanodevices and image devices to detect and analyze diseased conditions, cellular malfunction, and unmoral cells. Nano therapy assigns effective nano-systems by performing a more effective therapy while minimizing side effects while renovated medicine aims to repair or reproduce deteriorated tissues and organs using nanotechnology materials. Although nanotechnology is a rapidly growing field, availability of product is far away from reach, this is because of various hurdles at different stages of development. Major hurdles needed to be overcomed include toxicity, production cost, environmental hazards, and accessibility to unreachable far-off areas. The obstacles to growth if overcome can cause ground-breaking modifications in the field of health care and medicine. It is hoped that this chapter will provide updated information useful for material scientists, biotechnologists, pharmacologists, clinicians, researchers in the health sector, or scholars of the allied fields, who have an interest in drug discovery and development, nano-diagnosis, nanobiotechnology, and nanomedicine.

Keywords Nanotechnology · Nanomaterials · Nanoparticles · Nanomedicine · Medicine · Nanodiagnosis

14.1 Introduction

Nanoscience is the study of a solid body of theory that can be used to construct nanoscale technology. It involves the study of matter at the nanoscale, as well as methods for manipulating it (Ramsden 2005). The word ‘nano’ in nanotechnology refers to a billionth (1×10^{-9}) (Jackson et al. 2017). Nanotechnology can be characterized as the study and control of matter at scales between 1 to 100 nanometers (Hulla et al. 2015). Nanotechnology studies the science and technology of objects at the nanoscale, which differ significantly in properties from those of their constituent material at the macroscopic or even microscopic scale (Sengupta and Sarkar 2015). The essence of nanotechnology is size and control since it involves designing and fabricating structures, materials, and systems at nanometer dimensions (Ramsden 2005).

Nanomedicine can be described as the repair and control of biological systems in the human body using instruments created and designed according to nanotechnology level and principles (Radwan 2018). It entails using engineered nanodevices and nanostructures to diagnose, prevent, treat, and control disease in human biological systems at the molecular level. Natural or man-made nanoparticles (NPs) are materials with at least two dimensions between 1 and 100 nm (Khan et al. 2019). Manmade or engineered nanoparticles (ENPs) can either be metal-based or carbon-based. Fullerenes and carbon nanotubes are two types of carbon-based materials, whereas metals, metal oxides, and quantum dots are metal-based materials (Peralta-Videa et al. 2011). Nanomaterials are organic, inorganic, or organometallic material with physical, chemical, and electrical properties that changes as a result of the material's size and shape (Hochella et al. 2019). They are materials with an internal nanoscale surface structure with at least one dimension between 1 and 100 nm (Mullard 2017). The mixture of two or more known nanomaterials is called nanocomposites. This may include any combinations of metal-based, nanofibers, carbon-based, or organic-based nanowires, and ceramic, metal, or polymer bulk materials (Khan 2020).

Nanobiotechnology or bionanotechnology refers to the study of materials and processes at the nanometer scale that is focused on biomimetic, biological, and nano-technological devices used to monitor or manipulate biological procedures like biochips (Ramsden 2005).

14.1.1 History of Nanotechnology

Richard Zsigmondy, the Nobel Laureate in Chemistry in 1925, was the first to propose the idea of a "nanometer." He was the first to use the microscope to determine the size of particles such as gold colloids, and he invented the term nanometer to describe particle size (Hulla et al. 2015). On the evening of 29th December 1959, under the title '**There's plenty of room at the bottom** physicist Richard Feynman outlined the fascinating opportunities that would open up if scientists could learn to manipulate single atoms, molecules and improve the efficiency of instruments such as electron microscopes (Toumey 2009). The possibilities of nano etching texts, storing and retrieving data in an atom-size code are described by the phrase "plenty of room.“ He imagined using electron beams to etch lines a few atoms deep, basically foreshadowing the invention of electron-beam lithography, which is still used to manufacture silicon chips today. Using a scanning tunneling microscope, he suggested manipulating individual atoms to create new small structures with very different properties, which has now been achieved (Jackson et al. 2017). He also envisioned electronic miniaturization, the challenges of miniaturization, a swallowable mechanical surgeon, and a system of "a billion tiny factories" cooperating (Toumey 2009).

Norio Taniguchi, a Japanese physicist, was the first to use the term "nanotechnology" to characterize semiconductor practices in form of a nanometer, almost after 15 years Feynman's lecture. Nanotechnology, he said, was described as the processing, consolidation, separation, and deformation of materials by a single atom or molecule (Hulla et al. 2015). The golden age of nanotechnology initiated in the 1980s, with the inventions of scanning tunneling microscopy (STM) and atomic force microscopy by Binnig and Rohrer (2009). Kroto et al. discovered Fullerene C_{60} in 1985. Following that, in 1986, Eric Drexler from the Massachusetts Institute of Technology followed the foot-step of Feynman's "There is Plenty of Room at the Bottom" and Taniguchi's term nanotechnology in his book, "Engines of Creation: The Coming Era of Nanotechnology." (Hulla et al. 2015).

Another Japanese scientist, Saumio Iijima, advanced nanotechnology in 1991 when he created carbon nanotubes (Hulla et al. 2015; Sumio 1991). By the year 2000, the U.S. government had launched the National Nanotechnology Initiative

(NNI). This was a visionary research and development program for nanotechnology-based investments coordinated by 16 different US departments and independent organizations), which paved the way for advancements in nanotechnology research and development (Binnig and Rohrer 2009; Miyazaki and Islam 2007).

14.1.2 Classification of Nanomaterials

Nanomaterials are classified based on structure, shape, size, chemical synthesis as well as a source of origin.

14.1.2.1 Carbon Nanomaterials

Carbon nanomaterials are nanomaterials that contain carbon atoms. Carbon nanomaterials can be made in a variety of sizes, including hollow tubes and spheres (Kumar and Kumbhat 2016). Because of their particular physical properties, nanotubes have some distinct advantages over other drug delivery and diagnostic devices (Nikalje 2015). Carbon nanomaterials also include graphene, carbon nanotubes, carbon onions, carbon nanofibers, carbon black, and fullerenes.

14.1.2.2 Inorganic Nanomaterials

Metals and metal oxides, also known as metal and metal oxide nanomaterials or inorganic nanomaterials, can be used to make NMs. Silver (Ag) and Gold (Au) nanomaterials (major biological use), and metal oxides-based nanomaterials such as zinc oxide (ZnO) and titanium dioxide (TiO_2) nanomaterials are some of these NMs (Khan 2020). Metallic nanoparticles have been used in drug delivery, especially for cancer treatment, as well as biosensors (Nikalje 2015).

14.1.2.3 Quantum Dots

Quantum dots have a diameter of 2–10 nm and are made of semiconductors that can change the color of light, such as cadmium selenide (Sharma et al. 2020). Quantum dots absorb white light and reemit it at a certain wavelength a few nanoseconds later. QDs have been reported to show a long blood circulation time, exhibit fluoresce for a number of months in vivo, and are photo-bleaching resistant (Ballou et al. 2005). These features have made Quantum dots (QDs) a subject of high priority in molecular profiling and cellular imaging of pathological tissue specimens from cancer patients for diagnosis of disease types and phases, prognosis forecast, and handling strategy direction (Rahman et al. 2022).

14.1.2.4 Organic Nanomaterials

Organic matter is common in these NMs, with no carbon or inorganic-based nanomaterials. One important characteristics of organic nanomaterials is the possession of noncovalent bonds (weak- easily broken). These organic materials can easily be modified into various nanomaterial shapes such as liposomes, polymers, micelles, and dendrimers (Khan 2020). Due to their small scale, liposomes have been extensively investigated for targeted drug delivery (50–200 nm). They are biocompatible, flexible, and effective at entrapment. It can be used to deliver genes, proteins, and peptides both passively and actively via the circulatory system. Dendrimers (those with a diameter of less than a millimeter) are useful in the circulatory system for the regulated delivery of bioactive material, the targeted delivery of bioactive particles to macrophages, and the targeted delivery of bioactive particles to the liver (Nikalje 2015).

14.1.2.5 Nanocomposites

Nanocomposites are made up of one type of nanomaterial combined with another type of nanomaterial. The nanomaterials may be combined with other forms of nanowires, nanofibers, or larger-scale materials. Any mixture of metal-based, carbon-based, or organic-based nanowires, nanofibers, and ceramic, metal, or polymer bulk materials may be used in these nanocomposites (Khan 2020).

14.1.3 Classification Based on the Source of Origin

14.1.3.1 Natural Nanomaterial

Natural nanomaterials are present in biological organisms such as bacteria or plants, as well as in human activities. Natural nanomaterials are abundant in the hydrosphere, atmosphere, lithosphere, and biosphere, making their production a simple process. They are supplementary nanomaterials released to the atmosphere as a result of mining activities, forest fires, transportation and factory emissions, and urban and anthropogenic processes, as well as nanomaterials naturally formed and present in rivers, holding ponds, groundwater oceans, wastewater treatment plants, lakes, rocks, soils, landfills, magma, or lava, as well as microbial organisms. These all have an effect on the Earth system as a whole (Hochella et al. 2019). Biocompatible and biodegradable natural materials facilitate cell adhesion, migration, and differentiation (Mishra 2013).

14.1.3.2 Synthetic or Chemical-Based Nanomaterial

The synthetic method, which allows for the development of nanomaterials through biological, physical, chemical, or hybrid methods, is the most widely used method for making nanomaterials. One of the benefits of synthetically generated nanomaterials is the ability to make a large number of nanomaterials in various sizes and shapes. Other important feature of the synthetic method is the precision with which diverse chemicals and reagents can be connected or conjugated with nanomaterials. Among synthetically engineered nanomaterials, the main concern is whether current knowledge is sufficient to predict their efficiency. Furthermore, they behave differently in different environments than natural nanomaterials. For various biological applications, various sources of nanomaterials are currently developed (Bundschuh et al. 2018). Manmade nanoparticles (engineered nanoparticles, ENPs) can be carbon-based or metal-based. Fullerenes and carbon nanotubes are two types of carbon-based materials, while metals, metal oxides, and quantum dots are metal-based materials (Peralta-Videa et al. 2011). Among the most produced and used metal-based ENPs are zinc oxide (nZnO), silver (nAg), copper oxide (nCuO), cerium oxide ($nCeO_2$), titanium dioxide ($nTiO_2$), and gold (nAu) NPs. Other NPs like nMnO, nFe_3O_4, nCoO, and $nCoFe_2O_4$ are also widely used (Al-Whaibi and Mohamed 2015).

Synthetic materials are commonly used in tissue engineering owing to their strong mechanical strength, their simple molding characteristics, their relatively easy processing, and their ability to control dissolution and degradation. While synthetic materials are biocompatible, they lack natural sites for cell adhesion, necessitating the modification of surface properties such as topography, structure, and hydrophobicity to achieve a good cell-material interaction (Mishra 2013).

14.2 Synthesis of Nanomaterial

14.2.1 Synthesis of Nanomaterial

Nanomaterials are synthesized in a variety of ways, depending on the form and nature of the nanomaterial. In a broad sense, the two most common methods for synthesizing nanomaterials are "top-down" and "bottom-up." Bulk materials are reduced to nanomaterials in the top-down process, while nanomaterials are synthesized from the elementary level in the bottom-up method. Chemical vapor deposition, gas phase method, hydrothermal synthesis, microwave synthesis, solvothermal synthesis, thermal decomposition, templating process, combustion method, pulsed laser ablation, and traditional Sol–Gel method are some of the methods used to synthesize nano-materials (Khan 2020).

14.2.1.1 Production of Nanomaterials by Top-Down Method

This approach entails breaking or slicing the bulk material into smaller particles (nano-sized material) using physical routes like crushing, grinding, and milling. In general, this approach is ineffective for producing uniformly formed nanomaterials, and it is extremely difficult to obtain very small nanoparticles, even when using a lot of energy. The process wastes material and is constrained by the resolution of the resources available. The lack of surface structure, which has a direct effect on the physical properties and surface chemistry of n, is one of the method's major challenges. The process wastes material and is constrained by the resolution of the resources available. The lack of surface structure, which has a direct effect on the physical characteristics and surface chemistry of nanomaterials, is one of the method's major challenges. Furthermore, the processed shapes suffer significant crystallographic loss as a result of this system (Khan 2020). The different methods of lithographic techniques, the hydrothermal process, and pulsed laser ablation are examples of this type of approach.

Hydrothermal Method

The hydrothermal method is classically performed in a pressurized container called an "Autoclave" allowing temperature and pressure to be well-ordered and regulated. The temperature at the boiling point of water can be elevated during nanomaterial synthesis, allowing the vapour to become saturated (Yu et al. 2003). This process has been widely used in the development of various nanoparticles. The benefit of this approach is that it can be used to regulate reaction temperature, strain, solvent properties, solution composition, and additives to monitor material size, particle morphology, crystalline phase, and surface chemistry (Caep et al. 2004).

Method of Thermal Decomposition and Pulsed Laser Ablation

Decomposing metal alkoxides, heat, electricity, or salts can be utilized during doped metals preparation. Also, the nanomaterials properties are extremely influenced by the precursor concentrations rate in the reactions flow and the environment in which they are formed. TiO_2 nanoparticles with a diameter of less than 30 nm have been confirmed to be synthesized using the titanium alkoxide thermal decomposition at 1200°C (Kim et al. 2005). In another report, TiO_2 nanoparticles with a diameter (3–8 nm) were manufactured by pulsed laser ablation technique (Liang et al. 2004). The solution combustion method have been used for doped anatase TiO_2 nanoparticles synthesis (Nagaveni et al. 2004a). However, the high cost, low yield, and difficulty in controlling the structure and morphology of the nanomaterials are all drawbacks of this process.

14.2.1.2 Production of Nanomaterials by Bottom-Up Method

Materials are prepared atom-by-atom or molecule-by-molecule (self-assembling) to produce a large number of materials in this process. This process is more commonly used to make the majority of nanomaterials. This method can generate nanomaterials that are uniform in size, shape, and distribution. It precisely regulates the chemical synthesis process to prevent unwanted particle growth. This approach is critical in the development and processing of nanomaterials with improved particle size distribution and morphology (Khan 2020). Another significant advantage is that it is an environmentally friendly and cost-effective method for producing nanoparticles (much less expensive) (Khan 2020). There are many approaches for synthesizing nanomaterials using bottom-up techniques like combustion synthesis (Nagaveni et al. 2004b), gas-phase methods (Jones and Chalker 2003), sol–gel processing and microwave synthesis, described below.

Chemical Vapor Deposition Method

High-performance thin nano-films have been fabricated via the chemical vapor deposition (CVD) route. The substrate is treated with volatile precursors that act on the substrate surface to create the anticipated films in this process. Gas flow through the reaction chamber typically eradicatesvolatile by-products. The consistency of the deposited materials on the surface is influenced by several factors, including reaction rate, temperature, and precursor volume (Kim et al. 2005). A report has shown that Sn (Sengupta and Sarkar 2015) a doped TiO_2 nanoparticle films were fashioned by the CVD technique (Cao et al. 2004). Another doped TiO_2 nanoparticle was manufactured by the CVD pattern where TiO_2 is crystallized into rutile structures subject to the type and number of cations available in the chemical reactions (Gracia et al. 2004). The benefit of this process is that the nanofilm glaze is consistent, but it has some drawbacks, comprising the advanced temperatures needed for chemical reactions and the trouble of scaling up (Sudarshan 2003).

The Solvothermal Method

The solvothermal method has similarity to the hydrothermal technique, with the exclusionthat it does not use water as a solvent. Remarkably, when organic solvents or chemicals with high boiling points are used, this process is more efficient in the synthesis of nanomaterials with good distribution. Additionally, compared to the hydrothermal system, this method permits for more accurate control of material size and shape. With or without the addition of surfactants, this procedure produces nanomaterials or nanorods (Khan 2020).

Templating Method

The templating method is a method for generating materials with identical morphologies. The processing of nanomaterials using the templating scheme has recently become extremely common. By combining morphological characteristics with reactive deposition, this approach allows for the preparation of a large number of new materials with a consistent and regulated morphology by simply modifying the morphology of the template materials. Several models have been developed in recent years to synthesize various nanomaterials (Kim et al. 1797; Iwasaki et al. 2004). This approach has some drawbacks, such as complicated synthetic procedures where models must be extracted, usually by calcination, thus, raising the engineering costs and upsurges the risk of contamination (Bavykin et al. 2006).

Combustion Method

Rapid heating of a solution containing redox groups is used in the combustion process (Nagaveni et al. 2004b). This approach produces highly crystalline nanoparticles with wide surface areas. To make the crystalline materials, the temperature is raised to about 650 °C for 1–2 min. during processing.

Gas Phase Method

This method is good to produce thin film because it can be performed chemically or physically. Nanomaterials are produced in the gas phase due to chemical reaction or decomposition of a precursor (Jones and Chalker 2003; Lee et al. 2017). Electron beam evaporation is the method of producing TiO_2 thin films by heating the TiO_2 material with an electron beam and thus, producing electrons. There are several benefits of using E-beam evaporation for the deposition of TiO_2 over CVD, this include smoothness and better conductivity (Krol et al. 1999).

Microwave Radiation Method

Microwave radiation can also be used to make nanomaterials, and this process has many advantages, including not requiring long periods of high-temperature calcination and being a simple method of producing crystalline nanomaterials (Corradi et al. 2005). Consequently, by combining hydrothermal and microwave methods, high-quality rutile rods can be generated, and TiO_2 hollow open-ended nanotubes can be made by reacting anatase and rutile crystals in a solution of NaOH. (Wu et al. 2005).

Conventional Sol–Gel Method

The ability to impregnate or co-precipitate nanomaterials, that can be used to incorporate dopants, is one of the advantages of this method. Multiple oxide materials have been synthesized using the sol–gel process (Fernández-García et al. 2004), this approach also allows for greater monitoring of texture formation, chemical reaction, and morphological properties of solid materials. The ability to scale up with high-quality nanomaterials is a big benefit for the Sol–Gel technique (Kolen'ko et al. 2005). The polymerization and hydrolysis reactions of the precursors yield a colloidal suspension in the Sol–Gel technique (Pierre 2020). Besides that, any factor that influences one or both reactions are likely to alter the gel-forming properties, and these factors are termed to as Sol–Gel technique factors. Type of solution, water content, acid or base content, and various types of precursors, precursor concentration, and temperature are all factors to consider. The composition of the initial gel shape is influenced by these influences. The wet gel can then be matured in a diverse solvent after this stage. Aging is the period between the formation of a gel and its drying (Chen and Mao 2007).

14.2.2 Application of Nanotechnology in Medicine

The application of nanotechnology to different areas of medicine is known as nanomedicine. It is described as the process of repairing and controlling the biological systems of the human body using nanotechnology-based instruments and principles (Radwan 2018). Nanomaterials and nanoelectronic biosensors are used in nanomedicine (Farjadian et al. 2019). Nanomedicine is a multidisciplinary field of research. For a simple project, physicists, mathematicians, engineers, chemists, biologists, and users such as orthopedic surgeons would all be required (Radwan 2018). Nanomedicine involves at least three major fields:

- Nano-diagnosis,
- Nano-therapy (controlling drug delivery) and
- Remedial (renovated) medicine.

Nano diagnosis is the evolution of nanodevices and image devices to detect and interpret diseased conditions, cellular failure, and unmoral cells (Radwan 2018).The Nano therapy assigns effectivenano-systems. To perform a more effective therapy while minimizing side effects, drugs are transferred and delivered exclusively to cells or influenced areas (Radwan 2018; Jin et al. 2020; Filipponi and Sutherland 2011).Renovated medicine aims to use nanotechnology materials to restore or reproduce deteriorated tissues and organs. Tissue engineering uses artificially stimulated cells, which could revolutionize organ transplantation and artificial implant placement (Nikalje 2015). Nanoscale devices can interact easily with biomolecules on both the surface and inside of cells due to their small size. Drugs can be delivered directly to diseased cells in the body using nanoparticles.

14.2.3 Medical Diagnostics

In the field of diagnosis, nanotechnology has made significant progress (nano-diagnostics). It involves using a deliberately built system in at least one dimension on the nanoscale to diagnose events that occur at that size. Nano-diagnostics can aid in early detection and improved sensitivity. It may also be used in combination with other diagnostic and imaging technologies that are more traditional (Alhibshi et al. 2020). They can be innovative and authoritative tool for the cancer detection system. The traditional diagnostic procedures are incapable to detect tumors in their primitive stage and more inaccurate in distinguishing the benign from the malignant stage. Therefore, in the conventional methods, innovative nanoparticles (NPs) are proficient of yielding discerning imaging of affected areas (Patel et al. 2015).

This technology is facilitating the creation of platforms for the identification of single nucleotide polymorphisms and other disease biomarkers that are not detected using traditional methods (Alhibshi et al. 2020). For example, quantum dots, a kind of semiconducting nanocrystal, may improve biological imaging for medical diagnostics (Maclurcan 2005; Mathuria 2009; Gardner 2015). They emit a broad range of bright colors when exposed to ultraviolet radiation, which can be used to detect and distinguish particular types of cells and biological activities. These crystals have up to 1000 times better optical detection than traditional dyes used in many biological experiments, such as MRIs, which provide considerably more information (Ibrahim 2015). Nanotechnology is being investigated for use in the production of portable point-of-care diagnostic kits, which will allow for the quick and precise diagnosis of multiple diseases at the same time (Yen et al. 2015).Some nanomaterials have been used in medical diagnostics and these activities can be summarized as follows.

Single-walled Carbon nanotubes (SWNT) was used for the monitoring of blood nitric oxide level in inflammatory diseases employing fluorescent signal (Iverson et al. 2013). Nanoflares have been utilized to enable live cell detection of intracellular mRNA concerning the principle of fluorescence (Halo et al. 2014; Kwong et al. 2013). Target-specific magnetic nanoparticles allow for constant bloodstream monitoring of glioblastoma multiform microvesicles (Shao et al. 2012). Detection and isolation of circulating tumor cells by the principle of laser microdissection (LMD) were accomplished using anti-EpCAM antibody-coated silicon nanowires overlaid with polydimethylsiloxane to create a nano Velcro chip (Lu et al. 2013). In an experiment based on the Surface Enhanced Raman Spectroscopy theory (SERS), Silver nanorod array substrate has been used for the separation and identification of biological agents such as bacteria and viruses in blood, urine, saliva, and food using on-chip technology (Negri and Dluhy 2013).

Gold nanoparticles coated with influenza A has been used for the deletion of influenza virus in samples, based on the principle of dynamic light scattering (DLS) (Driskell et al. 2011), while monoclonal antihemagglutinin antibody-modified gold nanoparticles (mAb) were reported to have been used to detect influenza A virus in the blood, based on the principle of colorimetric immunosensing (Liu et al. 2015).

In-vitro diagnostic procedure for detecting nanomolar myeloperoxidase concentrations (MPO). A low-cost, fast-detecting immunofluorescence sensor known as μQLIDA (microfabricated Quantum dot-linked immunodiagnostic assay) has been used to detect nanomolar concentrations of nanomolar myeloperoxidase concentrations (MPO) (Rica and Stevens 2012; Yu et al. 2015).Iron-oxide magnetic nanoparticles coated with peptide (polydopamine) which uses near-infrared laser irradiation have been reported to detect cancerous cell clusters during photothermal cancer therapy and magnetic resonance imaging (MRI) (Montalti et al. 2015; Wu et al. 2015; Braas et al. 2012). Gold nanoparticle-based molecular diagnostic platform, used as a nanosensor was approved by the FDA for use in a genetic test for warfarin sensitivity (Lefferts et al. 2010).

14.2.4 Clinical Therapy and Drug Delivery Systems

Nanoparticles are ideal carriers for delivering anticancer drugs and other therapeutic drugs at the site of target with optimal proficiency and minimal collateral damage to bordering healthy tissues (Patel et al. 2015), implying that nanotechnology can be used to develop new forms of therapy and drug delivery (2006). Drugs can be distributed to specific areas in the body with greater precision using nanotechnology through self-assembling nano-scale vesicles with long-lasting encapsulation key (Gardner 2015; Mamo et al. 2010). The therapeutic focus has shifted away from the cell itself and toward intracellular molecular targets (Patel et al. 2015). The drugs are designed such that the active ingredient permeates cell membranes more effectively, resulting in a lower dosage requirement (Shetty 2010; Gardner and Dhai 2014). For example, micelles acquired from block co-polymers, are used for drug encapsulation. They are important for the transportation of small drug molecules to the target location. Similarly, nano-electromechanical systems have been used for the active drug release (Nikalje 2015). Iron nanoparticles or gold shells are finding a lot of use in cancer care. As a result, nanotechnology is revolutionizing drug delivery by resolving obstacles such as drug toxicity, sustaining drug release in the body, and improving bioavailability. Because of its potential to cause a reduction in drug administration time and drugs side effects minimization; patient compliance could progress, thus contributing to the successful management and treatment of infection (Gardner 2015). A targeted medicine decreases drug consumption and treatment expenditures, making the handling of patients cost-effective.

Biocompatible packaging materials may be used to introduce gene-silencing small interfering RNAs, gene-encoding DNAs, or recombinant proteins into cells. Liposomes, bacterial toxins, and viral NPs have commonly used packaging scaffolds, but they are also degraded and cleared from circulation before reaching the potential target site (Patel et al. 2015). Recent advances in reducible polymers have gotten a lot of attention because they can be molecularly programmed using sensors that can respond to variations in ion concentrations in the microenvironment and thus distinguish between intracellular and extracellular sites (Klein and Wagner 2014).

14.2.5 Tissue Growth and Regenerative Medicine

Tissue regenerative medicine research aims to create implants or scaffolds that can transport medications, growth factors, and hormones for tissue reconstruction. They deliver bioactive molecules over time to promote cell survival, infiltration, and proliferation in tissue engineering. Total tissue regeneration and functional rehabilitation are expected, as a result of this treatment modality. The use of carbon nanotubes, nanowires, and nanoparticles improves the structure of the extracellular matrix (Patel et al. 2015).

14.2.5.1 Biomimetic Hydrogels

To speed up bone regeneration, biomimetic hydrogels are used as a regulated biomolecule delivery system for growth factors (Fong and Wood 2006; Oupický and Li 2014; Lienemann et al. 2012). It has been discovered that in the case of bone regeneration, the scaffold is known to facilitate osteoblast proliferation (Mishra 2013). Cells, signals (biochemical factors), and scaffolds are considered to be the three major "pillars" that sustain bone tissue engineering (Mishra 2013). These three essential components, osteogenesis, osteoinduction, and osteoconduction, all play important roles in the processes that promote bone development.

14.2.5.2 Nano-filled Composites

As compared to traditional composite microparticles, nano-filled composites have greater compressibility, tensile strength, and flexure strength.

14.2.5.3 Chondroitin Sulfate Nanoparticles

Chondroitin sulfate nanoparticles (CSnps) within the scaffold of chitin blended with poly(butylene succinate) was used for skin repair in wounds (Koushki et al. 2015; Deepthi et al. 2014). Since it is biocompatible, biodegradable, and forms a porous substrate for improved nutrient exchange, it has a superior aesthetic value.

14.2.5.4 Polyethylene Glycol-Based Hydrogel Scaffold

In myocardial infarction, polyethylene glycol-based hydrogel scaffolds help in the transplanted cardiac cells retention and growth (Grover et al. 2014).

14.2.5.5 Graphene Oxide Film

Human Adipose-Derived Stem Cells' Adhesion and Osteogenic Differentiation Are Stimulated by a Glass Slide Covered with the Graphene Oxide film (Lyu et al. 2015).

14.2.6 Bioseparation

Magnetic nanoparticles owing to their distinctive size and physical properties are increasingly beingusedin analytical biochemistry, medicine, and biotechnology, to immobilize proteins, enzymes, and other bioactive agents. Since these particles are superparamagnetic, they can be labelled to the biomaterial of interest and removed from a matrix using a magnetic field without agglomerating (Mishra 2013; Cao et al. 2004). As such, proteins could be immobilized on magnetic nanoparticles for use in biological applications, because when proteins are immobilized directly on the surface of the particles, environmental factors such as pH, temperature, protease, and immunological responses might have an important/significant impact on their stability and operation.

14.3 Opportunities and Challenges of Nanotechnology

14.3.1 Opportunities

The nanotechnology revolution has presented scientists and engineers with new scientific insights and technological advances which have enabled them to begin to fill in a previously missing piece of our understanding of the Earth system (Hochella et al. 2019).

14.3.1.1 Antibiotic Resistance

The use of nanoparticles in combination therapy has helped to reduce antibiotic resistance (Nikalje 2015). By meddling with various proteins that are involved in antibiotic resistance or pharmacological pathways of drugs, zinc oxide (ZnO) nanoparticles may reduce antibiotic resistance and enhance the antibacterial action of Ciprofloxacin against microorganisms (Nikalje 2015). The integration of antibiotics into nanoparticles such as hydro-dispersible magnetite nanoparticles (HMNPs) allows for improved tissue penetration, and the drugs can be released at a regulated and predetermined pace for a long enough time to reach the target location. That could potentially increase the therapeutic index and the efficacy of treatment against Gram-negative (Ellis et al. 2013; Bolocan et al. 2015; Keskinbora and Jameel 2018).

Antibiotics such as Ciprofloxacin and Kanamycin were prepared with Nanoparticles of crystalline size (8–16 nm) in self-assembling principles, resulting in more potent antibiotics (Hasanova et al. 2015). The conjugation of gold nanoparticles (AuNPs) with antibiotics like streptomycin and kanamycin results in a type of these antibiotics that has a lower minimum inhibitory concentration (MIC), a higher bactericidal ability, and is more effective in remote areas where proper storage conditions are unavailable (Keskinbora and Jameel 2018; Saha et al. 2007).

14.3.1.2 Human Implantable Nanodevices (HINDs)

Nanosensors are built based on optical fibers or minute electrodes, in which the physiologically relevant molecules concentrations in the blood and other biofluids and tissues can be quantified. The heartbeat, blood pressure, and blood oxygenation of patients in intensive care are all monitored with this implantable sensor (Ramsden 2005; Keskinbora and Jameel 2018). The powerful computing devices enabled by nanotechnology are used for more sophisticated design recognition-based analysis of the data collected wirelessly by those sensors. The working ideologies of these sensors are well known. Example include the glucose sensor, already in use commercially as off-line devices. Miniaturization of the sensor, as well as its power supply and data transmission capability, is needed for implanting these sensors in the body. If such sensors are successfully created, they can be used not only in hospitals but also in everyday situations where a person is thought to be at risk of succumbing to any abnormality by himself or his physician. The sensors must be able to do the following for this strategy to be truly functional. To be truly effective, the sensors must be able to detect the early stages of the disease, which would necessitate developments in medical knowledge especially the detection of novel biomarkers for various physiological states (Ramsden 2005).

14.3.2 Challenges for Nanotechnology

Nanotechnology is a swiftly growing field, product availability remains a long way off due to numerous barriers at various stages of growth. The following hurdles to progress, once solved, will result in revolutionary changes in the field of health care and medicine.

14.3.2.1 Inadequate Knowledge of Nanoparticle Components

Nanostructures come in a variety of shapes, sizes, and compositions. The physicochemical phenomena of these NPs both in vitro and in vivo are not well known. As a result, finding the appropriate nanomaterial for the intended action is critical (Patel et al. 2015). Polyethylenimine(PEI) has been identified as a good cargo for intracellular nucleic acid targeting. Despite this, it is known as a potent cytotoxic mediator. Methods have been invented to mitigate its toxicity by low molecular weight PEI crosslinking to dithiodipropionic acid di (*N*-succinimidyl ester) and thus improving its availability and proficiency (Patel et al. 2015; Klein and Wagner 2014).

14.3.2.2 Problem of Standardization in Model Systems and Test Assay

To expound the chemical, physical, and biological behavior of nanomaterials accurately, the use of in vivo model must be in place and this is lacking. It is also challenging to verify the effects of nanoparticle interactions with cells because the results differed between different cell types, even though the test assay conditions were the same (Patel et al. 2015).

14.3.2.3 There is No Uniformity in Nanotoxicity

Different structures, proportions, and shapes of nanomaterials can be toxic to different types of cells under different exposure conditions. The structure, size, form, charge, aggregation, coating, and solubility of nanoparticles affect the target cell and target moieties for toxicity (Patel et al. 2015). Carbon nanotubes are cytotoxic to human T-cells at 400 μg/ml, 3.06 μg/cm (Jackson et al. 2017) on alveolar macrophages whereas, at 3.8 μg/ml, exposed cell cultures showed no cytotoxicity (Klein and Wagner 2014). Also, human keratinocytes exposed to the insoluble single-walled carbon nanotubes revealed apoptosis and oxidative stress. CdTe quantum dots was observed to have exhibited genotoxic and cytotoxic effects on HUVECs. Quantum dots-induced DNA damage time stress may be associated with reactive oxygen species (ROS) generation (Wang et al. 2010; Abdal Dayem et al. 2017).

14.3.2.4 Incomplete Understanding of the Impact on the Biological System

In terms of cellular or tissue toxicity, and genotoxic or carcinogenic effects, the impact on health and safety concerns is still unknown. These compounds are small enough to be inhaled, and the fragments build up in the alveoli of the lungs, causing inflammation or carcinogenesis. This is a major problem as employees may be exposed to occupational hazards (Patel et al. 2015).

14.3.2.5 Inadequate Analytical Tools

With the fact that nanotechnology deals with nanoscale structures, new analytical techniques are required for accurately determining the surface charge, particle size, surface chemistry, aggregation state, crystalline state, and size distribution of nanomaterials. Technological advancements in metrology are needed to forecast the action of nanoparticles/ nanomaterials in biological media (Patel et al. 2015).

14.3.2.6 Lack of In-Vivo Monitoring Systems

The lack of a substantial infrastructure for *in-vivo* nanomedicine research, the failure to track many probes, and the need for patients to be admitted for analysis are all major factors that prevent biological activities from being optimized (Patel et al. 2015; Klein and Wagner 2014).

14.3.2.7 Lack of Standard Synthesis Procedure

Nanomaterials are synthesized via a variety of synthetic reagents, many of which are toxic or carcinogens. It is necessary to create an effective synthetic pathway that is devoidof the use of hazardous contaminants. High purity and improved biocompatible nanoparticles are permeated by careful use of synthetic materials and compliance with safety standards (Patel et al. 2015; Klein and Wagner 2014).

14.3.2.8 Inadequate Standardized Safety Guidelines

It is difficult to define a specific safety guideline for a specific nanoparticle due to the complex nature of nanomedicine and its multiform toxicity. Empirical research and rigorous pre-clinical tests are needed to envisage a safety protocol (Patel et al. 2015; Klein and Wagner 2014).

14.4 Conclusion

Nanotechnology provides the capability to build large products that are extremely influential. Nanotechnology has a promising future, its integration with other technologies and the consequent appearance of complex and innovative hybrid technologies. Nanomaterials have distinct physicochemical and biological properties. These factors including surface-to-volume ratio; gravitational forces; the proportion of atoms that come in contact with surrounding objects; electromagnetic forces; quantum mechanics laws; the strong Brownian motion of nanoparticles has major consequences and importance for the successful generation of well-defined nanomaterials, and this can greatly influence their interactions with biomolecules and cells. Nanomedicine has aided drug absorption into cancerous cells, protected drugs from degradation in the body, and increased specificity when attacking the diseased cell with minimal collateral damage to the surrounding tissue. Nanoparticles also can be used to create excellent single-walled carbon; images of tumor sites. Nanotubes have been used as high-efficiency biomolecule delivery transporters into cells. The toxicity, environmental hazards, manufacturing expense, and connectivity to the unreachable in far-flung areas are the main obstacles that are yet to be overcome.

Acknowledgements The authors gratefully acknowledge the support received from the National Research Foundation, South Africa (grant no: 120790), The University of Winnipeg, Winnipeg, Canada, and Anchor University, Lagos, Nigeria.

References

Abdal Dayem A, Hossain MK, Lee SB, Kim K, Saha SK, Yang GM, Choi HY, Cho SG (2017) The role of reactive oxygen species (ROS) in the biological activities of metallic nanoparticles. Int J Mol Sci 18:120

Al-Whaibi MH, Mohammad F, Mohamed H (2015) Nanotechnology and plant sciences: Nanoparticles and their impact on plants. Springer International Publishing, Switzerland

Alhibshi AH, Alamoudi WA, Farooq RK (2020) Applications of nanomaterials in neurological diseases, neuronal differentiation, neuronal protection, and neurotoxicity. In: Khan FA (ed.) Applications of Nanomaterials in Human Health 1st (Ed.), pp 83–124

Alqathami M, Blencowe A, Yeo UJ, Franich R, Doran S, Qiao G, Geso M (2013) Enhancement of radiation effects by bismuth oxide nanoparticles for kilovoltage x-ray beams: A dosimetric study using a novel multi-compartment 3D radiochromic dosimeter. J Phys Conference Series 444, 012025

Baac HW, Ok JG, Maxwell A, Lee KT, Chen YC, Hart AJ, Xu Z, Yoon E, Guo LJ (2012) Carbon-nanotube optoacoustic lens for focused ultrasound generation and high-precision targeted therapy. Sci Rep 2:989

Ballou B, Ernst LA, Waggoner AS (2005) Fluorescence imaging of tumors *in vivo*. Current Medical Chemistry 12:795–805

Bavykin DV, Friedrich JM, Walsh FC (2006) Protonated titanates and TiO2 nanostructured materials: Synthesis, properties, and applications. Adv Mater 18:2807–2824

Bobadilla AD, Samuel ELG, Tour JM, Seminario JM (2013) Calculating the hydrodynamic volume of poly(ethylene oxylated) single-walled carbon nanotubes and hydrophilic carbon clusters. J Phys Chem B 117:343–354

Bolocan A, Mihaiescu DE, Andronescu E, Voicu G, Grumezescu AM, Ficai A, Vasile BŞ, Bleotu C, Chifiriuc MC, Pop CS (2015) Biocompatible hydrodispersible magnetite nanoparticles used as antibiotic drug carriers. Rom J Morphol Embryol 56:365–370

Braas D, Ahler E, Tam B, Nathanson D, Riedinger M, Benz MR, Smith KB, Eilber FC, Witte ON, Tap WD, Wu H, Christofk HR (2012) Metabolomics strategy reveals subpopulation of liposarcomas sensitive to gemcitabine treatment. Cancer Discov 2:1109–1117

Bundschuh M, Filser J, Lüderwald S, McKee MS, Metreveli G, Schaumann GE, Schulz R, Wagner S (2018) Nanoparticles in the environment: where do we come from, where do we go to? Environ Sci Eur 30:6

Caep O, Huisman CL, Reller AR (2004) Photoinduced reactivity of titanium dioxide. Prog Solid State Chem 32:33–177

Cao Y, Yang W, Zhang W, Liu G, Yue P (2004) Improved photocatalytic activity of Sn^{4+} doped TiO_2 nanoparticulate films prepared by plasma-enhanced chemical vapor deposition. New J Chem 28:218–222

Charafeddine RA, Makdisi J, Schairer D, O'Rourke BP, Diaz-Valencia JD, Chouake J, Kutner A, Krausz A, Adler B, Nacharaju P, Liang H, Mukherjee S, Friedman JM, Friedman A, Nosanchuk

JD, Sharp DJ (2015) Fidgetin-Like 2: A microtubule-based regulator of wound healing. J Invest Dermatol 135:2309–2318

Chen X, Mao SS (2007) Titanium dioxide nanomaterials: Synthesis, properties, modifications and applications. Chem Rev 107:2891–2959

Cooper DR, Bekah D, Nadeau JL (2014) Gold nanoparticles and their alternatives for radiation therapy enhancement. Front Chem 2:1–13

Corradi AB, Bondioli F, Focher B, Ferrari AM, Grippo C, Mariani E, Villa C (2005) Conventional and microwave-hydrothermal synthesis of TiO_2 nanopowders. J Am Ceram Soc 88:2639–2641

Davis ME (2009) The first targeted delivery of siRNA in humans via a self-assembling, cyclodextrin polymer-based nanoparticle: From concept to clinic. Mol Pharm 6:659–668

Deepthi S, Sidhy Viha CV, Thitirat C, Furuike T, Tamura H, Jayakumar R (2014) Fabrication of chitin/poly(butylene succinate)/chondroitin sulfate nanoparticles ternary composite hydrogel scaffold for skin tissue engineering. Polymers (Basel) 6, 2974–298

Driskell JD, Jones CA, Tompkins SM, Tripp RA (2011) One-step assay for detecting influenza virus using dynamic light scattering and gold nanoparticles. Analyst 136:3083–3090

Ellis CE, Vulesevic B, Suuronen E, Yeung T, Seeberger K, Korbutt GS (2013) Bioengineering a highly vascularized matrix for the ectopic transplantation of islets. Islets 5:216–225

Farjadian F, Ghasemi A, Gohari O, Roointan A, Karimi M, Hamblin MR (2019) Nanopharmaceuticals and nanomedicines currently on the market: challenges and opportunities. Nanomedicine (lond) 14:93–126

Fernández-García M, Martínez-Arias A, Hanson JC, Rodriguez JA (2004) Nanostructured oxides in chemistry: characterization and properties. Chem Rev 104:4063–4104

Filipponi L, Sutherland D (2011) Introduction to Nanoscience and Nanotechnologies.Interdiscip. Nanosci. Center. Aarhus Univ. Denmark. 2–29

Fong J, Wood F (2006) Nanocrystalline silver dressings in wound management: A review. Int J Nanomed 1:441–449

Gardner J (2015) Nanotechnology in medicine and healthcare : Possibilities, progress and problems. South African Journal of Bioethics and Law. 8:50–53

Gardner J, Dhai A (2014) Nanotechnology and Water: Ethical and Regulatory Considerations. In Mishra AK (ed.) Application of Nanotechnology in Water Research. 1st Ed., Wiley, pp 3–19

Gracia F, Holgado JP, Caballero A, Gonzalez-Elipe AR (2004) Structural, optical, and photoelectrochemical properties of Mn+-TiO2 model thin film photocatalysts. Journal of Physical Chemistry B. 108, 17466–17476

Grover GN, Rao N, Christman KL (2014) Myocardial matrix-polyethylene glycol hybrid hydrogels for tissue engineering. Nanotechnology 25:014011

Gu Z, Dang TT, Ma M, Tang BC, Cheng H, Jiang S, Dong Y, Zhang Y, Anderson DG (2013) Glucose-responsive microgels integrated with enzyme nanocapsules for closed-loop insulin delivery. ACS Nano 7:6758–6766

Halo TL, McMahon KM, Angeloni NL, Xu Y, Wang W, Chinen AB, Malin D, Strekalova E, Cryns VL, Cheng C, Mirkin CA (2014) NanoFlares for the detection, isolation, and culture of live tumor cells from human blood. The Proceedings of the National Academy of Sciences U. S. A.111, 17104–17109

Hasanova UA, Ramazanov MA, Maharramov AM, Hajiyeva SF, Parfyonova YV, Eyvazova GM, Hajiyeva FV, Guliyeva NA, Veliyeva SB (2015) The improvement of antimicrobial activity of kanamycin and ciprofloxacin antibiotics coupled with biocompatible magnetite nanoparticles and characterization of their structure. J Nanotechnol Eng Med 6, 041006

Hochella Jr MF, Mogk DW, Ranville J, Allen IC, Luther GW, Marr LC, McGrail BP, Murayama M, Qafoku NP (2019) Natural, incidental, and engineered nanomaterials and their impacts on the Earth system. Science. 363, eaau8299

Hu CMJ, Fang RH, Copp J, Luk BT, Zhang L (2013) A biomimetic nanosponge that absorbs pore-forming toxins. Nat Nanotechnol 8:336–340

Hulla JE, Sahu SC, Hayes AW (2015) Nanotechnology : History and Future. 34:1318–1321

Ibrahim HMM (2015) Green synthesis and characterization of silver nanoparticles using banana peel extract and their antimicrobial activity against representative microorganisms. Journal of Radiation Research and Applied Sciences. 8:265–275

Iverson NM, Barone PW, Shandell M, Trudel LJ, Sen S, Sen F, Ivanov V, Atolia E, Farias E, McNicholas TP, Reuel N, Parry NMA, Wogan GN, Strano MS (2013) In vivo biosensing via tissue-localizable near-infrared-fluorescent single-walled carbon nanotubes. Nat Nanotechnol 8:873–880

Iwasaki M, Davis SA, Mann (2004) Spongelike macroporous TiO_2 monoliths prepared from starch gel template. J Sol-Gel Sci Technol 32, 99–105

Jackson T, Patani B, Ekpa D (2017) Nanotechnology in diagnosis: A review. Advances in Nanoparticles. 6:93–102

Jeevanandam J, Barhoum A, Chan YS, Dufresne A, Danquah MK (2018) Review on nanoparticles and nanostructured materials: history, sources, toxicity, and regulations. Beilstein J Nanotechnol 9:1050–1074

Jin C, Wang K, Oppong-Gyebi A, Hu J (2020) Application of nanotechnology in cancer diagnosis and therapy - A mini-review. International Journal of Medical Science 17:2964–2973

Joachim E, Kim ID, Jin Y, Kim K, Lee JK, Choi H (2014) Gelatin nanoparticles enhance the neuroprotective effects of intranasally administered osteopontin in rat ischemic stroke model. Drug Delivery and Translational Research. 4, 395–399

Jones AC, Chalker PR (2003) Some recent developments in the chemical vapour deposition of electroceramic oxides. J Phys D Appl Phys 36:R80

Kanasty R, Dorkin JR, Vegas A, Anderson D (2013) Delivery materials for siRNA therapeutics. Nat Mater 12:967–977

Keskinbora KH, Jameel MA (2018) Nanotechnology applications and approaches in medicine: A review. Journal of Nanoscience & Nanotechnology Research. 2:6

Khan I, Saeed K, Khan I (2019) Nanoparticles: Properties, applications, and toxicities. Arab J Chem 12:908–931

Khan FA (2020) Applications of nanomaterials in human health. Springer Nat. Singapore Pte Ltd.

Kim CS, Nakaso K, Xia B, Okuyama K, Shimada S (2005) A new observation on the phase transformation of TiO_2 nanoparticles produced by a CVD method. Aerosol Sci Technol 39:104–112

Kim J, Lee JW, Lee TG, Nam SW, Han J (2005) Nanostructured titania membranes with improved thermal stability. J Mater Sci 40, 1797–1799

Klein PM, Wagner E (2014) Bioreducible polycations as shuttles for therapeutic nucleic acid and protein transfection. Antioxid Redox Signal 21:804–817

Kolen'ko YV, Kovnir KA, Gavrilov AI, Garshev AV, Meskin PE, Churagulov BR, Bouchard M, Colbeau-Justin C, Lebedev OI, Van Tendeloo G, Yoshimura M (2005) Structural, textural, and electronic properties of a nanosized mesoporous ZnxTi1-xO_2-x solid solution prepared by a supercritical drying route. J Phys Chem B. 109, 20303

Koushki N, Katbab AA, Tavassoli H, Jahanbakhsh A, Majidi M, Bonakdar S (2015) A new injectable biphasic hydrogel based on partially hydrolyzed polyacrylamide and nanohydroxyapatite as scaffold for osteochondral regeneration. RSC Advanc 5, 9089–9096

van de Krol R, Goossens A, Meulenkamp EA (1999) In Situ X-Ray Diffraction of Lithium Intercalation in Nanostructured and Thin Film Anatase TiO_2. J Electrochem Soc 146:3150

Kumar A, Patel A, Duvalsaint L, Desai M, Marks ED (2014) Thymosin β4 coated nanofiber scaffolds for the repair of damaged cardiac tissue. Journal of Nanobiotechnology. 12:10

Kumar Teli M, Mutalik S, Rajanikant GK (2010) Nanotechnologygoing small means aiming big and Nanomedicine. Curr Pharm Des 16:1882–1892

Kumar N, Kumbhat S (2016) Carbon-Based Nanomaterials. In: Essentials in Nanoscience and Nanotechnology. Hoboken, NJ, U.S.A. John Wiley & Sons, Inc. pp 189–236

Kwong GA, Von Maltzahn G, Murugappan G, Abudayyeh O, Mo S, Papayannopoulos IA, Sverdlov DY, Liu SB, Warren AD, Popov Y, Schuppan D, Bhatia SN (2013) Mass-encoded synthetic biomarkers for multiplexed urinary monitoring of disease. Nat Biotechnol 31:63–70

Lee EJ, Huh BK, Kim SN, Lee JY, Park CG, Mikos AG, Choy YB (2017) Application of materials as medical devices with localized drug delivery capabilities for enhanced wound repair. Progress in Materials Science. 89,392–410

Lefferts JA, Schwab MC, Dandamudi UB, Lee HK, Lewis LD, Tsongalis GJ (2010) Warfarin genotyping using three different platforms. American Journal of Translational Research. 2:441–446

Liang C, Shimizu Y, Sasaki T, Koshizaki N (2004) Synthesis, characterization, and phase stability of ultrafine TiO_2 nanoparticles by pulsed laser ablation in liquid media. J Mater Res 19:1551–1557

Lienemann PS, Lutolf MP, Ehrbar M (2012) Biomimetic hydrogels for controlled biomolecule delivery to augment bone regeneration. Adv Drug Deliv Rev 64:1078–1089

Liu Y, Zhang L, Wei W, Zhao H, Zhou Z, Zhang Y, Liu S (2015) Colorimetric detection of influenza A virus using antibody-functionalized gold nanoparticles. Analyst 140:3989–3995

Lu Y-T, Zhao L, Shen Q, Garcia MA, Wu D, Hou S, Song M, Xu X, OuYang W-H, OuYang WW-L, Lichterman J, Luo Z, Xuan X, Huang J, Chung LWK, Rettig M, Tseng H-R, Shao C, Posadas EM (2013) NanoVelcro chip for CTC enumeration in prostate cancer patients. Methods 64:144–152

Lyu CQ, Lu JY, Cao CH, Luo D, Fu YX, He YS, Zou DR (2015) Induction of osteogenic differentiation of human adipose-derived stem cells by a novel self-supporting graphene hydrogel film and the possible underlying mechanism. ACS Appl Mater Interfaces 7:20245–20254

Ma Y, Liu H, Wu J, Yuan L, Wang Y, Du X, Wang R, Marwa PW, Petlulu P, Chen X, Zhang H (2019) The adverse health effects of bisphenol A and related toxicity mechanisms. Environ Res 176:108575

Maclurcan DC (2005) Nanotechnology and developing countries Part 1: What possibilities? AzONano-Online Journal of Nanotechnology. 1:1–10

Mamo T, Moseman EA, Kolishetti N, Salvador-Morales C, Shi J, Kuritzkes DR, Langer R, Andrian UV, Farokhzad OC (2010) Emerging nanotechnology approaches for HIV/AIDS treatment and prevention. Nanomedicine (lond) 5:269–285

Mathuria JP (2009) Nanoparticles in tuberculosis diagnosis, treatment and prevention: A hope for future. Dig J Nanomater Biostruct 4:309–312

Mishra AK (2013) Nanomedicine for drug delivery and therapeutics. John Wiley & Sons, Inc., Scrivener Publishing

Mita, M., Burris, H., LoRusso, P., Hart, L., Eisenberg, P., Mita, A., Low, S., Summa, J., Berk, G.& Sachdev, J. A phase 1 study of BIND-014, a PSMA-targeted nanoparticle containing docetaxel, administered to patients with refractory solid tumors on a weekly schedule. *Cancer Research.* **74**, CT210 (2014).

Miyazaki K, Islam N (2007) Nanotechnology systems of innovation-An analysis of industry and academia research activities. Technovation 27:661–675

Montalti M, Cantelli A, Battistelli G (2015) Nanodiamonds and silicon quantum dots: Ultrastable and biocompatible luminescent nanoprobes for long-term bioimaging. Chem Soc Rev 44:4853–4921

Mullard A (2017) 2016 FDA drug approvals. Nat Rev Drug Discovery 16:73–76

Nagaveni K, Hegde MS, Ravishankar N, Subbanna GN, Madras G (2004b) Synthesis and structure of nanocrystalline TiO_2 with lower band gap showing high photocatalytic activity. Langmuir 20:2900–2907

Nagaveni, K., Hegde, M. S. & Madras, G. Structure and photocatalytic activity of $Ti_{1-x}\,M_xO_{2\pm\delta}$(M = W, V, Ce, Zr, Fe, and Cu) synthesized by solution combustion method. *Journal of Physical Chemistry B.***108**, 20204–20212 (2004a).

Nanostructures for drug delivery (2009) Ochekpe, N. A., Olorunfemi, P. O. & Ngwuluka., N. C. Nanotechnology and drug delivery Part 2. Trop J Pharm Res 8:275–287

Negri P, Dluhy RA (2013) Ag nanorod based surface-enhanced Raman spectroscopy applied to bioanalytical sensing. J Biophotonics 6:20–35

Nikalje AP (2015) Nanotechnology and its applications in medicine. Med Chem 5:081–089

Oh B, Lee CH (2013) Nanofiber for cardiovascular tissue engineering. Expert Opin Drug Deliv 10:1565–1582

Oupický D, Li J (2014) Bioreducible polycations in nucleic acid delivery: Past, present, and future trends. Macromol Biosci 14:908–922

Patel S, Nanda R, Sahoo S (2015) Nanotechnology in healthcare: Applications and challenges. Medicinal chemistry (Shāriqah (United Arab Emirates)). 5, 528–533

Peralta-Videa JR, Zhao L, Lopez-Moreno ML, de la Rosa G, Hong J, Gardea-Torresdey JL (2011) Nanomaterials and the environment: A review for the biennium 2008–2010. J Hazard Mater 186:1–15

Pierre AC (2020) Introduction to Sol-Gel Processing. Introduction to Sol-Gel Processing.Springer Nature Switzerland AG, pp 165–208. https://doi.org/10.1007/978-3-030-38144-8

Radwan FAA (2018) Nanotechnology and medicine. Materials Science and Nanotechnology. 2:10–11

Rahman MM, Islam F, Afsana Mim S, Khan MS, Islam MR, Haque MA, Mitra S, Emran TB, Rauf A (2022) Multifunctional therapeutic approach of nanomedicines against inflammation in cancer and aging. J Nanomater Article ID 4217529, 19 p

Ramsden JJ (2005) What is nanotechnology? Nanotechnology Perceptions. 1:3–17

De La Rica R, Stevens MM (2012) Plasmonic ELISA for the ultrasensitive detection of disease biomarkers with the naked eye. Nat Nanotechnol 7:821–824

Ryan JJ, Bateman HR, Stover A, Gomez G, Norton SK, Zhao W, Schwartz LB, Lenk R, Kepley CL (2007) Fullerene Nanomaterials Inhibit the Allergic Response. J Immunol 179:665–672

Saha B, Bhattacharya J, Mukherjee A, Ghosh AK, Santra CR, Dasgupta AK, Karmakar P (2007) In vitro structural and functional evaluation of gold nanoparticles conjugated antibiotics. Nanoscale Res Lett 2:614

Sengupta A, Sarkar CK (2006) Scientific Committee on Emerging and Newly Identified Health Risks (SCENIHR). The appropriateness of existing methodologies to assess the potential risks associated with engineered and adventitious products of nanotechnologies. European Commission Health & Consumer Protection Directorate-General. 246, 1–79 (2006).

Sengupta, A. & Sarkar, C. K. Introduction to nano: Basics to nanoscience and nanotechnology. *Springer*, London.Pp. 241 (2015).

Shao H, Chung J, Balaj L, Charest A, Bigner DD, Carter BS, Hochberg FH, Breakefield XO, Weissleder R, Lee H (2012) Protein typing of circulating microvesicles allows real-time monitoring of glioblastoma therapy. Nat Med 18:1835–1840

Sharma M, Easha P, Tapasvi G, Reetika R (2020) Nanomaterials in Diagnostic Tools and Devices. In: Nanomaterials in biomedical diagnosis. Elsevier, Pp. 57–83

Shetty P (2010) Nanotechnology for health: Facts and figures. London Science and Development Network

Shoffstall AJ, Atkins KT, Groynom RE, Varley ME, Everhart LM, Lashof-Sullivan MM, Martyn-Dow B, Butler RS, Ustin JS, Lavik EB (2012) Intravenous hemostatic nanoparticles increase survival following blunt trauma injury. Biomacromol 13:3850–3857

Sudarshan TS (2003) In Coated powders - new horizons and applications, Advances in Surface Treatment: Research & Applications (ASTRA), Proceedings of the International Conference, Hyderabad, India, 412–422

Sumio I (1991) Helical microtubules of graphitic carbon. Nature 354:56–58

Tian F, Zhang RH, Wang X (2018) A coupled ocean physics-biology modeling study on tropical instability wave-induced chlorophyll impacts in the pacific. Journal of Geophysical Research Ocean. 123:5160–5179

Toumey C (2009) Plenty of room, plenty of history. Nat Nanotechnol 4, 783–784

Verma A, Kumar N, Malviya R, Sharma PK (2014) Emerging trends in noninvasive insulin delivery. J Pharmaceutics 378048

Wang L, Zhang J, Zheng Y, Yang J, Zhang Q, Zhu X (2010) Bioeffects of CdTe quantum dots on human umbilical vein endothelial cells. J Nanosci Nanotechnol 10, 8591–8596

Wu X, Jiang Q-Z, Ma Z-F, Fu M, Shangguan W-F (2005) Synthesis of titania nanotubes by microwave irradiation. Solid State Commun 136:513–517

Wu M, Zhang D, Zeng Y, Wu L, Liu X, Liu J (2015) Nanocluster of superparamagnetic iron oxide nanoparticles coated with poly (dopamine) for magnetic field-targeting, highly sensitive MRI and photothermal cancer therapy. Nanotechnology 26:115102

Xiao Y, Gao X, Taratula O, Treado S, Urbas A, Holbrook RD, Cavicchi RE, Avedisian CT, Mitra S, Savla R, Wagner PD, Srivastava S, He H (2009) Anti-HER2 IgY antibody-functionalized single-walled carbon nanotubes for detection and selective destruction of breast cancer cells. BMC Cancer 9:351

Yen C-W, de Puig H, Tam JO, Gómez-Márquez J, Bosch I, Hamad-Schifferli K, Gehrke L (2015) Multicolored silver nanoparticles for multiplexed disease diagnostics: Distinguishing dengue, yellow fever, and Ebola viruses. Lab Chip 15:1638–1641

Yoon HJ, Lee K, Zhang Z, Pham TM, Nagrath S (2011) Nanoassembly of graphene oxide for circulating tumor cell isolation. In: 15th International Conference on Miniaturized Systems for Chemistry and Life Sciences 2011, MicroTAS 2011. 2–6, 1098–1100

Yu C, Kim GB, Clark PM, Zubkov L, Papazoglou ES, Noh M (2015) A microfabricated quantum dot-linked immuno-diagnostic assay (μQLIDA) with an electrohydrodynamic mixing element. Sens Actuators, B Chem 209:722–728

Yu J-G, Yu JC, Cheng B, Hark SK, Iu K (2003) The effect of F^--doping and temperature on the structural and textural evolution of mesoporous TiO_2 powders. J Solid State Chem 174:372–380

van Kortryk TS (2018) The doctoral students of Richard Feynman. History and Philosophy of Physics. https://doi.org/10.1063/PT.5.9100

Chapter 15
Smart Targeted-Nanocarriers for Cancer Therapeutics

Débora Ferreira, Renata Silva, Laura Figueiredo, and Ligia Rodrigues

Abstract The main hindrance in modern nanomedicine includes drug non-specific tissue biodistribution. This is particularly important for cancer, which is the second leading cause of death worldwide. Conventional chemotherapy is a beneficial therapy, however serious adverse events, as well as multidrug resistance, hampers the success of this treatment option. Innumerous research efforts have been conducted over the years aiming to create novel and more effective treatments, including the development of novel smart drug delivery systems (DDS) as an alternative to chemotherapy. Smart DDS should specifically target pathologic tissues with reduced dosage frequency in a spatially controlled manner, minimizing off-target effects of the active therapeutic agents on healthy tissues. Nanoparticle (NP) surfaces functionalization with targeting moieties specific to cell surface components, that are exclusive or upregulated in tumor tissues, has been a good strategy to enhance NP targeting. This chapter highlights the most recent advances in NP targeting over the last years. Ligand-NP conjugation chemistries are covered along with the various moieties' classes, including aptamers, peptides, small molecules, antibodies and/or proteins. Each ligand class exhibits advantages, drawbacks and unique features that are discussed in detail. The current challenges and future research scope in the field are also presented.

Abbreviations

ASGP	asialoglycoprotein
BBB	blood-brain barrier
CD44	cluster determinant 44
CPMV	cowpea mosaic virus
CuAAC	Copper-catalyzed azide-alkyne cycloaddition

D. Ferreira · R. Silva · L. Figueiredo · L. Rodrigues (✉)
CEB—Centre of Biological Engineering, University of Minho, 4710-057 Braga, Portugal
e-mail: lrmr@deb.uminho.pt

LABBELS—Associate Laboratory, Braga, Guimarães, Portugal

M. P. Shah et al. (eds.), *Biogenic Nanomaterials for Environmental Sustainability: Principles, Practices, and Opportunities*, Environmental Science and Engineering,
https://doi.org/10.1007/978-3-031-45956-6_15

DDS	drug delivery systems
DNA	single-stranded deoxyribonucleic acid
Dox	doxorubicin
dsDNA	double stranded DNA
EGFR	epidermal growth factor receptor
EphA2	ephrin receptor A2
EPR	enhanced permeability and retention
FA	folic acid
Fc	fragment crystallizable
Fab	fragment antibody binding
FDA	Food and Drug Administration
HA	hyaluronic acid
HAS	human serum albumin
HER2	human epidermal growth factor receptor 2
K_d	dissociation constant
Lf	lactoferrin
mAbs	monoclonal antibodies
MUC1	mucin-1
NPs	nanoparticles
OBOC	one-bead one-compound
PAM	peptide amphiphile micelles
PEG	polyethylene glycol
PLGA	poly(lactic-co-glycolic acid)
PSMA	prostate-specific membrane antigen
RGD	arginine-glycine-aspartic acid
RNA	ribonucleic acid
RRM2	M2 subunit of ribonucleotide reductase
scFv	single-chain variable fragment
SELEX	Systematic Evolution of Binding by Exponential Enrichment
siRNA	small interference RNA
SPIO	super paramagnetic iron oxide
ssDNA	single-stranded deoxyribonucleic acid
SSTR2	somatostatin receptor 2
Tf	transferrin
VEGFR	vascular endothelial growth factor receptor

15.1 Nanoparticles

Since the last century, when the Nobel laureate Richard Feynman introduced the concept of nanotechnology, the field has fully expanded into an active research area, undergoing revolutionary developments (Fig. 15.1) (Cypriyana et al. 2021). Nanotechnology have gained relevance as a tailor platform of shape, size, design, and chemical, physical and surface properties of materials at a nanoscale level. Nanoparticles (NPs), in particular, are clusters of atoms with a size of 100 nm or less in at least one dimension (Cypriyana et al. 2021; Selmani et al. 2022). Undetectable to the human eye, they display unique characteristics in comparison to the bulk materials from which they originate, being composed by three distinct layers: the surface, the shell layer and the core portion (Selmani et al. 2022).

Overtime, NPs were organized by different classifications, considering, for example, the core source material and chemical composition (Selmani et al. 2022; Mitchell et al. 2021; Jeevanandam et al. 2018). NPs can be labelled as inorganic when synthesized from gold, iron and silica, offering unique qualities for their application in diagnostic, imaging and photothermal therapies (Mitchell et al. 2021; Jeevanandam et al. 2018). Organic NPs, such as liposomes, micelles, dendrimers and polymeric-derived NPs, present a highly biocompatible profile with improved drug loading capacity, colloidal stability and tunable properties (Jeevanandam et al. 2018). NPs are commonly subdivided in a third group, the carbon-based, that has attracted interest due to their strength, structure, electron affinity and versatility (Selmani et al. 2022). Every time any of these NPs are combined with a larger or bulk-type material, such as any form of metal, ceramic or polymer, they are designated as composites (Jeevanandam et al. 2018).

NPs increased prominence is due to their small size, stability and their specifically tailored physicochemical properties, such as shape, wettability, hydrophilicity and surface charge. Beyond this, one of the most attractive NPs features is their surface area to volume ratio, since these entities offer, with a very limited volume, an enormous surface area for transport, interaction and chemical reactions with biological systems (Selmani et al. 2022; Jeevanandam et al. 2018; Nikzamir et al. 2021). Moreover, NPs can improve bioavailability, stability and solubility of encapsulated cargos, enhance the plasma half-life and also the circulation time (Mitchell et al. 2021; Greish et al. 2014). Nevertheless, the cornerstone of traditional NPs application is the enhanced permeability and retention (EPR) effect. The EPR effect exploitation takes advantage from the blood vasculature leakiness and defective lymphatic drainage

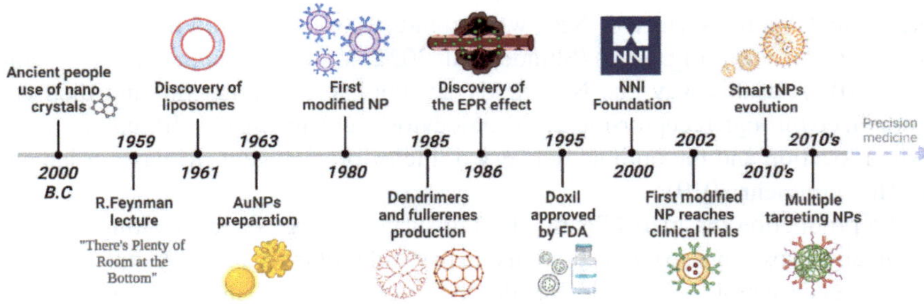

Fig. 15.1 Chronological landmarks of nanoparticle development. AuNPs: gold nanoparticles; EPR: enhanced permeability and retention effect; FDA: Food and Drug administration; NNI: national nanotechnology initiative; NPs: nanoparticles

system of tissues undergoing certain pathologies. As a result, NPs accumulate at the tumor site and can access distant and close cellular populations due to the improved permeability (Greish et al. 2014; Shinde et al. 2022). Overall, NPs offer an increased efficacy while retaining a safe profile and reduced toxicity (Selmani et al. 2022; Mitchell et al. 2021; Greish et al. 2014).

Owing to all these exceptional characteristics, NPs gained increased interest and a great investment as therapeutic, diagnostic and delivery agents in biomedicine applications, resulting in several Food and Drug Administration (FDA) approved nanomedicines for inflammatory-related, hematological and oncological pathologies (Mitchell et al. 2021; Nikzamir et al. 2021).

Notwithstanding the promising outcomes, the number of nanomedicine strategies approved are critically below the projections outlined in 2000 by the National Nanotechnology Initiative (Mitchell et al. 2021). The productivity has been tempered by requirements like biocompatibility, non-toxicity and non-immunogenicity and by defensive elimination mechanisms such as the host immune system and liver or kidneys clearance. Additionally, the NPs efficacy is dependent on factors like cellular cycle, intracellular location and particularly by cellular binding and internalization actions. Given the lack of cellular precision of the NPs, active agents are released early, without criteria and extravasate to organs, such as liver and spleen (Greish et al. 2014; Shinde et al. 2022). In addition, the NPs accumulation in pathological tissues also leads to long-term retention which hampers further penetration and significantly limits the therapeutic outcome (Shinde et al. 2022). Some researchers have also started to question the reliability of the EPR effect, since factors such as tumor location and type, blood perfusion, vascular architecture and pore dimensions are highly heterogeneous. Moreover, the EPR effect is well-characterized in primary environments, however the knowledge on secondary and metastatic sites is premature and limited (Greish et al. 2014; Shinde et al. 2022).

Considering this, the field is now withdrawing from these passive techniques and researchers are engineering NPs to target specific sites of interest, mitigate adverse side effects on the surrounding tissue and enhance the therapeutic effect (Mitchell et al. 2021; Shinde et al. 2022). Therefore, while passive targeting relies on NP's intrinsic properties and the EPR effect, an active targeting approach conjugates ligands on the surface of NPs, which binds to overexpressed receptors on the pathological tissues (Fig. 15.2) (Shinde et al. 2022; Molavipordanjani and Hosseinimehr 2019). In this way, the NPs firstly accumulate on tumor site and then are internalized through receptor-mediated endocytosis, enhancing the NPs uptake at the desired location and the optimal delivery of the cytotoxic agent (Molavipordanjani and Hosseinimehr 2019).

The production of smart NPs by altering surface charge or inactivating reactive chemical groups can also reduce toxicity, in particular of inorganic NPs, and stabilize their presence in biological fluids. Besides the biocompatibility increment, modifying NPs can possibly overcome conventional delivery limitations, such as low bioavailability, rapid clearance, and several biological hindrances like the blood brain barrier (BBB). In particular, smart NPs can modulate the host immune response and expand

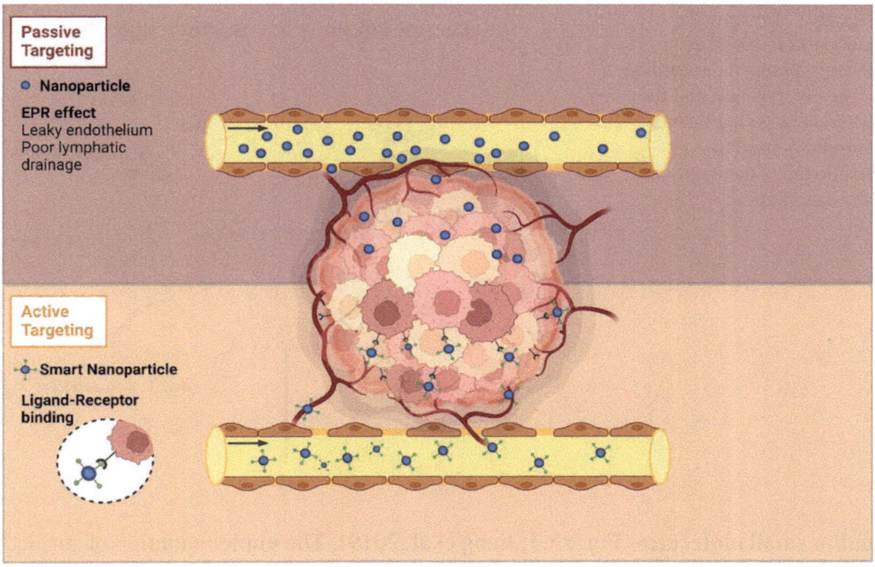

Fig. 15.2 Schematic representation of active and passive targeting at tumor site. EPR: enhanced permeability and retention effect

their plasma half-life, by promoting an escape of the macrophages from the reticuloendothelial system (Barbero et al. 2017). On another hand, this ability can be exploited to promote immune responses against antigens for the next generation of vaccines (Cappellano et al. 2019). Undoubtedly, as mentioned above, the most influential asset is the high precision of these smart NPs dwindling the uncontrollable release of agents and enhancing their uptake.

In this chapter, we describe a wide range of ligands that have been used in smart NPs and highlight the distinct conjugation strategies available to graft them onto the NP's surface. In addition, we provide a brief state-of-the-art on the existing smart NPs on clinical trials, as well as discuss possible future challenges and next steps.

15.2 Surface Ligands

As previously mentioned, NPs undergo passive targeting, preferentially accumulating in cancer tissues. However, the outcome of the EPR effect is uncertain, and NPs perse have no specificity for cancer cells.

Therefore, to achieve active targeting, it could be necessary to use NPs functionalized with recognition moieties, such as antibodies, aptamers, peptides, proteins,

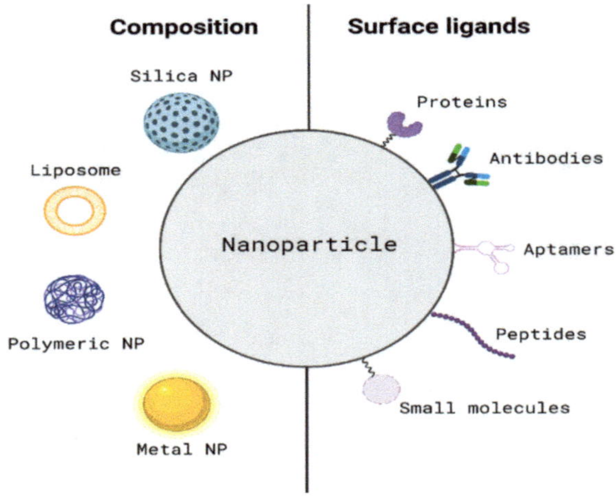

Fig. 15.3 Design of a nanoparticle-based drug delivery platform, according to the composition and the different targeting ligands commonly used to functionalize nanoparticles. NP: nanoparticle

and/or small molecules (Fig. 15.3; Jiang et al. 2019). The implementation of surface-functionalized nanocarriers will improve tumor selectivity, reducing off-target effects and toxicity, thus increasing the therapeutic efficiency of DDS (Zhao et al. 2020).

The current section covers various classes of moieties, highlighting their advantages and drawbacks, showing the most recent literature where smart DDS were applied.

15.2.1 Proteins

Proteins are large biomolecules formed by the reaction of polymerization by condensation of amino acid residues (Silva et al. 2018). There are many naturally occurring proteins, with three-dimensional structures that present a sharp affinity for specific endogenous targets and, for this reason, have the potential to be used for targeted delivery purposes (Sousa et al. 2019).

Transferrin (Tf) is a polypeptide glycoprotein found in blood, with approximately 80 kDa, responsible for iron transport in the body (Salahpour Anarjan 2019; Mao et al. 2021; Yoo et al. 2019). Tf-conjugated NPs have been explored due to the overexpression of Tf receptors in tumor cells and tissues, including breast, lung, and brain cancer. Therefore, by using Tf as a targeting moiety against cancer cells, the entry of NPs will be facilitated by the receptor-mediated endocytosis pathway (Salahpour Anarjan 2019). In a recent study, Tf was combined with a protein-lipid hybrid NP loaded with two different drugs to achieve active targeting in non-small cells lung cancer (Mao et al. 2021). Authors noticed that NPs conjugated with this glycoprotein had a higher uptake efficiency than unconjugated NPs.

Similarly, lactoferrin (Lf) is also a low molecular iron-binding glycoprotein belonging to the Tf family that is present in many tissues and secretory biological fluids (Silva et al. 2018; Sousa et al. 2019; Kapoor et al. 2019). This protein is associated with several physiological functions, such as anticancer effect, anti-inflammatory properties and, of course, regulation of iron absorption (Sousa et al. 2019; Santos-Pereira et al. 2022). Moreover, Lf receptors have been reported to be highly expressed in BBB-representative cells (Huang et al. 2007). Due to this evidence, in 2017, a group of researchers used Lf to functionalize poly (lactic-co-glycolic acid) (PLGA)- polyethylene glycol (PEG) NPs loaded with shikonin to improve and facilitate the transport of this drug through the BBB up to glioma cells via receptor-mediated pathways (Li et al. 2018).

Furthermore, synthetic proteins, such as affibodies, can also be conjugated to NPs to allow a targeted delivery. Their main advantage is that they do not compete with proteins of natural origin that are extremely abundant (Sousa et al. 2019). Affibodies are engineered high-affinity synthetic proteins that can be obtained by ribosomal or phage display methodologies. These proteins have low immunogenicity and lower molecular weight than antibodies (around 8 kDa) (Shipunova et al. 2021). Recently, Shipunova et al. (2021) successfully used the anti-human epidermal growth factor receptor 2 (HER2) affibody to functionalize PLGA-based NPs to achieve targeted delivery and thus obtain selective interaction with HER2-overexpressing cancer cells. In addition, Zhang and co-workers (Zhang et al. 2020) also took advantage of an affibody against HER2 to achieve targeted delivery to HER2-positive breast cancer cells.

Although non-antibody proteins, like the ones mentioned above, are seen as attractive ligands, they have some limitations, including their large size, which causes the increase of the nanocarrier size and the possibility of immune system activation (Sousa et al. 2019).

15.2.2 Antibodies and Antibodies Fragments

An antibody, also known as immunoglobulin, is a Y-shaped glycoprotein expressed on the surface of B lymphocytes used in the recognition and neutralization of external and unknown substances, like viruses or bacteria (Sousa et al. 2019; Marques et al. 2020). This large protein, around 150 kDa and 10 nm can be abundantly found in body fluids and is composed of four polypeptide chains: two identical heavy chains connected between them and linked to two identical light chains by disulfide bonds (Sousa et al. 2019; Marques et al. 2020). So far, there are five known classes of antibodies, also known as isotypes: IgA, IgE, IgD, IgG, and IgM, which can be structurally distinguished by their heavy chains and their distinct functions (Marques et al. 2020).

Antibodies can be synthesized through different strategies, namely the use of transgenic mice and in vitro selection technologies, like phage or ribosome display technologies (Sousa et al. 2019; Marques et al. 2020). Generally, antibodies differ from animal to animal, and can be differentiated into three categories according to human and non-human parts in their constitution: humanized antibodies, fully human antibodies and chimeric antibodies (Sousa et al. 2019; Salahpour Anarjan 2019).

The antibody-cell interaction is a result of the antibody binding to the antigen present on the surface of the target cell, through a weak interaction between complementary three-dimensional structures (Salahpour Anarjan 2019). The binding of the antibody to its complementary antigen serves as an inducing mechanism for suppression of protein expression and interference with the ligand-receptor binding, for example (Sousa et al. 2019).

Because of their strong affinity, with a dissociation constant (K_d) in the sub-nanomolar range (Zhao et al. 2020), and the high specificity of the antibody-antigen interaction, this type of ligand as a whole, or its fragments, is widely used (Zhao et al. 2020; Silva et al. 2018). Furthermore, the antibodies can be used to functionalize NPs as a strategy to overcome challenges present when using non-targeted NPs. A directional conjugation leading to an ideal orientation should be accomplished, leaving the fragment crystallizable (Fc) region linked to the NP surface and the fragment antibody-binding (Fab) section free to interact properly with the antigen (Silva et al. 2018; Marques et al. 2020).

One of the exceptional features of antibodies is that they, among other ligands, can be used both for therapeutic and targeting purposes. Nevertheless, despite the high specificity and target affinity of antibodies, these entities hold some intrinsic disadvantages (Sousa et al. 2019; Yoo et al. 2019; Friedman et al. 2013). Their structure, as previously stated, restricts their correct orientation at the NPs surface and their large size makes the penetration of the functionalized nanocarriers challenging, hampering the conjugation with NPs since it restricts the antibody surface density (Friedman et al. 2013). In addition, specific monoclonal antibodies (mAbs) have high manufacturing costs, as their production requires the use of large quantities of mammalian cells and undergoes extensive purification processes (Silva et al. 2018; Sousa et al. 2019). Besides, mAbs and antibody fragments may present immunogenicity problems, which can be a major obstacle for their usein clinical applications (Silva et al. 2018). Other challenges that may arise when developing antibody-NP conjugates include the conjugation itself, which must be specific and efficient, not disrupting the specificity of the antibodies; the circulation time may be also affected by the linker used to build the conjugate (Marques et al. 2020).

The first evidence of the so-called active targeting was observed using an antibody as a ligand, in 1980, where small liposomes were functionalized with mAbs through covalent binding (Leserman et al. 1980). In addition, the use of mAbs for disease therapeutic dates back to 1986 when the first mAb (Orthoclone OKT3; Janssen-Cilag) was approved by the FDA as a medicine in the treatment of acute transplant rejection. It was later, in 1997, that mAbs entered the field of oncology as drugs for anti-tumor therapy, through the FDA approval of Rituximab (an anti-CD20 monoclonal antibody) (Arruebo et al. 2009).

More recently, Comparetti et al. (2020) developed multi-walled carbon nanotubes loaded with the widely used chemotherapeutic agent paclitaxel, and surface-functionalized with antibodies targeting the prostate-specific membrane antigen

(PSMA), a cell surface marker found at high levels in serum when metastatic cells are present. It was found that these functionalized carbon nanotubes induced a greater internalization and higher early apoptosis when compared with non-functionalized NPs (containing only paclitaxel), demonstrating a successful uptake by prostate cancer cells. Moreover, Liszbinski et al. (2009) produced gold NPs loaded with 5-fluorouracil and functionalized with anti-epidermal growth factor receptors (EGFR) antibodies, commonly upregulated in a wide range of tumors. The authors pointed out that the functionalization of gold NPs with mAbs improved the cytotoxicity and the apoptosis rate in cells as the antibody directs NPs towards target cells.

Antibody fragments, such as Fab or single-chain variable fragment (scFv), can overcome some of the mAbs limitations, for example, the risk of inactivation when applying full antibody. A Fab can be obtained by enzymatic cleavage using papain or pepsin combined with a reducing agent and it is composed of one constant and one variable portion of the heavy and the light chain (Marques et al. 2020; Seidu et al. 2022). High-affinity scFv is a fusion of the variable regions of the heavy and light chains of immunoglobulins, connected with a peptide linker, and can be screened and isolated using, for instance, in vitro selection technologies (Marques et al. 2020).

The use of such fragments offers several advantages, such as their lower molecular weight, thus allowing a greater penetration in tumor tissues and, consequently, greater efficiency. However, this property and the lack of Fc region may lead to a higher clearance and therefore to less tumor accumulation. This drawback can be circumvented by, for example, conjugating these fragments with PEG moieties. Moreover, as opposed to full-length antibodies, antibody fragments do not trigger Fc-mediated cytotoxic immune mechanisms that may lead to the development of major adverse effects (Marques et al. 2020).

A popular antibody is the anti-HER2, an antibody that aims to target the HER2 receptor, a receptor highly expressed by cancer cells. In a recent study, HER2-targeted NPs were developed using anti-HER2 Fab fragments for active targeted drug delivery to breast cancer cells. It was found that NPs functionalized with Fab fragments exhibited higher tumor accumulation than when the intact antibody was used, thus proving the potential of antibody fragments to overcome the disadvantage of the higher molecular weight of the antibodies that limits their targeting application (Duan et al. 2018).

15.2.3 Aptamers

Aptamers, first introduced in 1990 (Ellington and Szostak 1990; Tuerk and Gold 1990), are peptide or oligonucleotide molecules that can bind to various targets, such as cells, bacteria, proteins, and small organic or inorganic molecules (Byun 2021; Fu and Xiang 2020).

Nucleic acid aptamers, a class of aptamers, are short single-stranded deoxyribonucleic acid (ssDNA) or ribonucleic acid (RNA) structures, usually 20 to 80 nucleotide residues long, that can present multiple secondary structures and further fold into single three-dimensional structures (Fu and Xiang 2020). These relatively small molecules (about 15 kDa) bind with high affinity and specificity to their targets through spatial complementarity mediated by hydrogen bonding, Van der Waals forces, and electrostatic/ionic interaction (Byun 2021; Fu and Xiang 2020). The aptamer-target binding affinity can be characterized by their K_d, which typically lies between pico- to nanomolar (Byun 2021).

Systematic Evolution of Binding by Exponential Enrichment (SELEX) is an iterative in vitro method that through the application of repeated cycles with steps such as binding, washing and amplification allows the obtention of aptamers from random oligonucleotide libraries (Silva et al. 2018; Byun 2021; Ferreira et al. 2021). This process is usually divided in three steps: incubation of the library with the desired target; partitioning (discarding the unbound sequences followed by the recovery/collection of the bound sequences); and amplification by polymerase chain reaction (PCR) (Fu and Xiang 2020). Subsequent studies have led to the development of several variations of conventional SELEX, such as counter SELEX or cell-SELEX (Sousa et al. 2019). For example, cell-SELEX allows the production of aptamers with the ability to bind to target proteins on the surface of living cells that are used as targets during the process. This allows the screening of aptamers that are specific to a given cell line, without any prior knowledge of biomarkers of this target cell (Ferreira et al. 2021).

Aptamers exhibit several advantages over other types of commonly used ligands. Firstly, they are more versatile, since they can be developed, unlike other moieties, against numerous targets and have low-to-no inherent immunogenicity and therefore present no problems of immune rejection. Moreover, the aptamers small molecular weight provides higher permeability and easiness to penetrate tissues, they have a long shelf-life and are not temperature sensitive, even if denaturation takes place this can be easily reversed, returning to their original three-dimensional structure. Finally, because they are chemically synthesized aptamers, they exhibit low batch-to-batch variation and are easier and cheaper to produce than antibodies (Zhao et al. 2020; Fu and Xiang 2020).

Despite the aptamers assets as ligands, this strategy is limited by rapid renal excretion, mainly associated with their low molecular weight and susceptibility to degradation by nucleases. To avoid degradation of nucleases during in vivo applications, that is particularly common when using RNA aptamers, mirror imaging aptamers (Spiegelmers) can be produced and the phosphate-sugar backbone can be modified by replacing the 2'-hydroxyl with 2'-fluorine, 2'-amino or 2'-methoxy, chemically modifying bases, among other strategies (Byun 2021; Fu and Xiang 2020). The conjugation of aptamers with high molecular mass molecules (e.g., PEG and cholesterol) to the 5′ or the 3′ ends of the RNA aptamer, works in two ways as it is a method that increases the renal retention time and, at the same time, decreases degradation by nucleases (Fu and Xiang 2020).

The 1990 studies published by Ellington and Szostak (Ellington and Szostak 1990) and Tuerk and Gold (Tuerk and Gold 1990), in which RNA aptamers were identified, signaled the 'birth of aptamers' and their use as ligands. Since then, and due to the ability of aptamers to bind with specificity and affinity to the tumor biomarkers for which they are specifically selected, they have been selected against cell surface receptors that are overexpressed in tumors (Fu and Xiang 2020; Pereira et al. 2022). For example, Tsogtbaatar et al. (2014) identified the OS-7.9 ssDNA aptamer, which has been shown to bind specifically to an osteosarcoma cell line. Moreover, Ferreira and co-workers (2017) using cell-SELEX, identified two ssDNA aptamer candidates that specifically bind to a triple negative breast cancer cell line (MDA-MB-231).

Additionally, aptamers can be easily modified with functional groups at their ends and conjugated with chemotherapeutics agents, therapeutic RNAs, and NPs, to make a targeted delivery to tumorigenic cells. The resulting NP-aptamer conjugates do not have an increase in size compared to free NPs, thus being able to easily penetrate the tumor tissue (Fu and Xiang 2020).

The first proof-of-concept study, in which a conjugated aptamer-NP was tested to ensure an active drug delivery, was carried out in 2004. The authors synthesized poly (lactic acid)-PEG copolymer NPs and covalently bound the specific RNA aptamer targeting an overexpressed protein in the prostate cancer epithelium (Farokhzad et al. 2004). Other examples include the work by Peng et al. (2004) that report the use of gold NPs conjugated with a deoxyribonucleic acid (DNA) aptamer that effectively targets the EGFR variant III, and by Xiao and co-workers (Xiao et al. 2022) that designed NPs with an EGFR aptamer to specifically target the rare malignant bone tumor cells called chordomas cells that highly express this receptor.

Despite the potential of aptamers, to date, no cancer therapy takes advantage of conjugated NP-aptamers approved by the FDA or other agencies, and neither are any conjugated NP-aptamers in clinical trials (Sousa et al. 2019).

15.2.4 Peptides

Peptides are molecules that consist of chains of amino acids (usually 2 to 50) connected by chemical bonds called peptide bonds. These moieties can be seen as a compromise between large molecules, such as antibodies, and small molecules, offering advantages over the ligands herein mentioned (Jiang et al. 2019; Yoo et al. 2019). Their relatively small size allows a high-density decoration of the NPs (Silva et al. 2018; Yoo et al. 2019), and due to the abundance of functional groups in peptides, they are easily manipulated and modified (Jiang et al. 2019). Furthermore, peptide-based ligands have less immunogenicity, present easy and low-cost production compared to protein-based ligands for instance, and have much higher binding specificity and affinity than small molecule ligands (Jiang et al. 2019; Silva et al. 2018; Yoo et al. 2019).

Due to the facility to chemically modify this class of ligands, these can be used to generate various types of decorated nanocarriers. Peptides can be conjugated with drugs, grafted onto NPs and even lead to the generation of self-assembled nanocarriers. These moieties can be categorized into two classes: cell-penetrating peptides that are non-specific, and cell-targeting peptides with receptor specificity (Zhao et al. 2020; Marques et al. 2020). Such strategies take advantage of the functional groups present on the cysteine and lysine residues, as well as the C- and N- terminus of peptides (Zhao et al. 2020).

Peptides that are used in active DDS are usually identified through large-scale screening of peptide libraries (Jiang et al. 2019). Phage display is a widely used method, consisting of a cyclic selection process, in which a peptide library (up to 10^{11} different sequences) is displayed on the surface of bacteriophages (Silva et al. 2018; Rodrigues et al. 2017; Ferreira and Martins 2017). This randomized library can be incubated with an already known purified target molecule or with specific cell types whose receptors are unknown, which allows the identification of peptides that exhibit affinity for the given target (Jiang et al. 2019; Zhao et al. 2020; Sousa et al. 2019). This screening tool can be applied to in vitro and in vivo conditions (Sousa et al. 2019). In addition, peptides can also be identified using chemical libraries, with the one-bead one-compound (OBOC) library being the most prevalent, where each bead displays a different peptide. Unlike phage display libraries, which contain only natural L-amino acid sequences with simple configurations, OBOC libraries grant more structural possibilities and offer the opportunity to employ non-natural amino acids, however, they can only be used for in vitro screenings (Jiang et al. 2019).

Several peptides have been identified using phage display (Ferreira et al. 2019; Silva et al. 2016; Nobrega et al. 2016; Pereira et al. 2021). For instance, Ferreira and co-workers (Ferreira et al. 2019) identified and characterized peptides specific to the colorectal cancer cell line RKO. A synthetic peptide, RKO pep, demonstrated an excellent binding ability to several colorectal cancer cell lines, showing no affinity for non-cancerous cells.

The most well-known peptides are the arginine-glycine-aspartic acid (RGD) peptides, that rely on the sequence of these three amino acids (Silva et al. 2018). Given that, RGD peptides have a high capacity to selectively bind to cell surface receptors, often overexpressed on the surface of tumor cells, known as integrins, namely avβ3 and avβ5, which are frequently used as tumor-targeting ligands, both in diagnosis and therapy (Zhao et al. 2020; Silva et al. 2018). For instance, Gong et al. (2014) designed RGD peptide self-assembled NPs to obtain active targeting towards liver cancer cells. The authors found that NPs carrying Dox showed greater cellular uptake than free Dox.

Despite their many advantages, the application of peptides for targeted drug delivery is hampered by their sensitivity and predisposition to enzymatic microenvironments, such as blood circulation and cleavage. In order to overcome proteolysis, peptide cyclization can be achieved through the formation of chemical bonds, as well as the modification of the peptide sequence through the introduction of unnatural amino acids, as is the case of the D-amino acids (Jiang et al. 2019).

15.2.5 Small Molecules

Small molecules are small entities that have shown great potential as target moieties (Zhao et al. 2020; Silva et al. 2018). These molecules, such as vitamins and sugars, were among the first ligands to be explored due to their simple chemistry, thus easy to combine with other entities, such as NPs (Zhao et al. 2020). They present low production costs, ease of manufacturability and modification, low immunogenicity, improved stability and low molecular weight (Jiang et al. 2019; Zhao et al. 2020; Sousa et al. 2019). Furthermore, small molecules exhibit very high selectivity towards their receptors, making them appealing tools for targeting cells which express their targets (Sousa et al. 2019).

One of the most widely employed small molecules is folic acid (FA), a water-soluble vitamin, whose receptors are highly expressed in tumor cells compared to non-tumorigenic cells. Thus, FA-modified NPs can selectively target folate receptors on the cell membrane of tumor cells. For example, Ren et al. (2021) formulated gold NPs that carried paclitaxel and were functionalized with FA as an active DDS to selectively target cancer cells that overexpress the folatereceptor (Ren et al. 2021).

Somatostatin receptor 2 (SSTR2) is one of the subtypes of the somatostatin receptor and has been abundantly detected in tumor cells, such as leukemic cells. Because of this, somatostatin and its analogues have been identified as promising targeting moieties for the treatment of leukemia. Abdellatif and co-workers (Abdellatif et al. 2020) formulated an NPs-octreotide conjugate thought the functionalization of amino-PEGylated quantum dots with a somatostatin analogue (octreotide), able to target blood cells overexpressing SSTR2 (Abdellatif et al. 2020).

Other small molecules that display specific and tight binding to uniquely expressed or overexpressed receptors in cancer can also be used as surface ligands to target cancer cells. For example, sigma receptors are membrane proteins that are overexpressed in some human tumors to which a few benzamide derivatives bind with high affinity. Bearing this in mind, Jalilian et al. (2021) developed lipid NPs containing docetaxel and functionalized with a low molecular weight benzamide derivative, anisamide, to be applied in prostate cancer therapy, whose cells are sigma receptor positive.

15.2.5.1 Polysaccharides

Hyaluronic acid (HA) is a natural anionic polysaccharide that is composed of alternating repeats of D-glucuronic acid and N-acetyl-D-glucosamine (Silva et al. 2018; Jia et al. 2022). This linear mucopolysaccharide is a large molecule and is a extracellular matrix major component (Yoo et al. 2019). Given its composition with carboxyl and hydroxyl groups, HA can undergo unlimited modifications, for easier conjugation with nanocarriers, (Jia et al. 2022). Other than that, HA is biocompatible, biodegradable, non-immunogenic and non-toxic (Meng et al. 2022). As a ligand, HA presents some advantages, namely the fact that it can be used as a hydrophilic backbone polymer of NPs and at the same time as a targeting moiety (Yoo et al. 2019). Moreover, the hydrophilic nature of HA allows lower adsorption and permeation of proteins, avoiding the formation of a protein corona, thus favoring the circulation and stability of the HA-engineered DDS invivo (Jia et al. 2022). However, NPs functionalized with HA tend to accumulate in the liver, due to the high expression of the receptors of this polysaccharide in the hepatic tissue (Yoo et al. 2019).

HA binds to the cluster determinant 44 (CD44) receptor, which is a transmembrane glycoprotein that holds potential for use as a target in active targeting therapy. CD44 receptor has low expression in healthy cells yet, in contrast, is overexpressed on the surface of various cancer cells (Silva et al. 2018; Jia et al. 2022). Therefore, due to its excellent characteristics and the ability to be used in active targeting, HA and its derivatives have been used as targeting ligands in nanocarrier systems for the delivery of various therapeutic agents (Silva et al. 2018; Meng et al. 2022).

Recently, Meng et al. [52] reported the use of HA as a targeting moiety. In this study, HA-coated liposomes loaded with shikon in were developed to achieve an active targeting to triple negative breast cancer cells that overexpress the CD44 receptor. In the present study, the authors observed that both the internalization ability and retention time were significantly higher for the HA functionalized liposomes.

15.2.5.2 Sugar Molecules

Sugar molecules, such as glucose, galactose, and mannose, are relatively small molecules and represent another class of target molecules of interest to be applied in DDS. These carbohydrates have the ability to selectively recognize cell surface receptors, like lectins (Silva et al. 2018; Sousa et al. 2019). As the cell surface is rich in carbohydrates bound to both glycolipids and glycoproteins, sugar molecules present a great potential to be used as recognition ligands (Silva et al. 2018).

Lectin-like receptors, such as mannose, are carbohydrate-responsive receptors (Zhao et al. 2020). Mannose is a highly expressed sugar in cells of the immune system and, as so, has the potential to be used to functionalize DDS (Sousa et al. 2019). Moreover, mannose receptors have recently been found to be abundantly present on the cell surface of the MDA-MB-231 cell line, and for that reason, Li et al.[54]have functionalized branched gold NPs with a modified mannoside molecule as an active targeting strategy against triple negative breast cancer cells.

Galactose is a simple sugar molecule that has the ability to target asialo glycoprotein (ASGP) receptors, which are abundantly expressed on liver cell surfaces and in many cancer cells. ASGP receptors can recognize both glycoproteins and glycoconjugates that have galactose termination, so for example galactose functionalized nanocarriers can be used to target the liver (Zheng et al. 2018). Therefore, Zheng et al. (2018) designed a lactobionic acid functionalized nanocarrier (with galactose in its constitution) as an approach for hepatocellular carcinoma treatment. Furthermore, galactose in its monosaccharide form can be directly conjugated to nanocarriers and used as a tumor-targeting moiety (Xia et al. 2019).

In addition, mannan, a natural polysaccharide with repeated carbohydrate units produced by yeast, can be recognized by mannose receptors. Therefore, mannan-decorated NPs have been used as protein vaccine carriers for targeting dendritic cells via mannose-binding lectin-mediated active transportation (Xu et al. 2022).

15.3 Strategies for Conjugating Functional Ligands into NPs

Targeting a moiety without losing its functionality after attachment to a NP is the main goal of a targeted ligand conjugation. For instance, the binding of targeting ligands to NPs without considering the recognition site can shield the functional groups leading to decreased targeting abilities. Therefore, a deep understanding on the chemical structure and nature of both ligand and NP is of utmost relevance. The functionalization process can take place through covalent and non-covalent binding of a moiety onto the surface of NPs. This process of covalently or non-covalently connecting ligands to NPs to create novel functionalized NPs encompasses various chemical, as well as molecular biology procedures. The following sections survey various conjugation chemistries used to link the aforementioned targeting ligands into a wide range of functional NPs.

15.3.1 Covalent Conjugation

Covalent conjugation strategies are the most usual binding methods between atoms. The reaction of one functional group with another leads to the establishment of a covalent bond through the sharing of electrons between atoms (Sunasee and Narain 2014). Some of the most dominant covalent reactions that are used in the conjugation of the above-listed targeting ligands onto various types of NPs include amide linkage (carboxylic acid and amine reaction), Schiff base linkage (aldehyde reaction with amine or hydrazide) and reactions known as click-chemistry (Molavipordanjani and Hosseinimehr 2019).

Amide linkage of moieties and NP's surface occurs through the condensation of carboxylic acid and amine reactive groups. Theoretically, targeting ligands can possess amine functional groups and NPs can have carboxylic groups and vice versa; great part of the published reports state that targeting ligands hold amine groups (Molavipordanjani and Hosseinimehr 2019). As activation of carboxylic acid reactive groups by turning it to an active ester form is required before the reaction, different reagents have been used to synthetize active ester forms from the carboxylic acid, including 1-ethyl-3- (3-dimethylaminopropyl) carbodiimide (EDC), N-hydroxysuccinimide (NHS) and N-Hydroxysulfosuccinimide (sulfo-NHS). Tan and collaborators (Tan et al. 2016) described a simple strategy to detect leukemia cells, where carboxyl-labelled fluorescent silica NPs were conjugated with the amine-modified Sgc8 aptamer, which binds tightly to T acute lymphocytic leukemia cells, through amide coupling between carboxyl and amine reactive groups. Moreover, a broad range of peptides were described to be functionalized onto the surface of NPs using this amide linkage strategy (Su et al. 2017; Okur et al. 2016; Gao et al. 2015). In particular, Gao and co-workers (Gao et al. 2015) performed the successful anchoring of the cell-penetrating peptide RLW onto the surface of the carboxyl units of PEG-poly (e-caprolactone) NPs for lung cancer targeting.

The reaction between an aldehyde and an amine produces a Schiff base linkage (Molavipordanjani and Hosseinimehr 2019). As, normally, targeting ligands do not possess aldehyde functional groups, mild oxidation of such molecules can provide such reactive groups. The reaction of this reactive group with hydrazide or amine-functionalized NPs can lead to the successful decoration of NPs with this modified targeting ligand (Joshi et al. 2013; Kumar et al. 2008). Aldehyde-functionalized NPs can also react directly with functional moieties containing hydrazide or amine reactive groups (Yu et al. 2010; Brunel et al. 2010). In a particular study, the viral NP cowpea mosaic virus (CPMV) was covalently modified to yield a benzaldehyde-labeled CPMV (Brunel et al. 2010). The hydrazide group on the vascular endothelial growth factor receptor (VEGFR-1) specific peptide, F56, reacted specifically with the benzaldehydes on CPMV and the F56-modified CPMV recognized VEGFR-1 on endothelial cells and VEGFR-1-expressing tumor xenografts in mice.

Click reactions can provide a broad spectrum of ligand-functionalized NPs. Copper-catalyzed azide-alkyne cycloaddition (CuAAC) is recognized as one of the best click reactions to date and one of its main features includes the ability to proceed under mild reaction conditions in aqueous solutions with high yield (Molavipordanjani and Hosseinimehr 2019; Yu et al. 2012). For instance, Chen and collaborators (Chen et al. 2018) described a high-efficient capping of azide-modified mesoporous silica NPs by alkyne-labelled double stranded DNA (dsDNA) containing thrombin aptamer through CuAAC. The developed system was successfully applied to detect thrombin in serum samples with high selectivity. In another report, it was demonstrated the fruitful incorporation of the alkyne-containing cyclic tumor-targeting peptide LyP-1, which binds to p32 mitochondrial protein-overexpressing tumor cells, to azide-PEG-iron oxide NPs (Maltzahn et al. 2008). These LyP-1 "click" NPs were able to stably circulate in vivo and specifically bind to p32-expressing cells in tumors.

The thiol-ene click reaction of thiol reactive groups with maleimide and acrylic derivatives can enable the linking of various targeting moieties onto NP's surface via thioether linkage (Molavipordanjani and Hosseinimehr 2019). Maleimide-functionalized polyethylene glycol-functionalized nanographene oxide (PEGylated n-GO) was successfully conjugated with mucin-1 receptor immunoglobulin G antibody (anti-MUC1 IgG) by taking advantage of the thiol-ene reaction between the thiol reactive groups of cysteines residues in anti-MUC1 and maleimide groups in the NP's surface (Rubio et al. 2015). Breast cancer targeting was confirmed in vitro using various techniques, including flow cytometry and confocal laser scanning microscopy. Liu and collaborators (Liu et al. 2017) also reported the functionalization of colloidal NPs with N- (3,4-dihydroxyphenethyl) acrylamide. Using this thiol-ene strategy, the thiol-modified Sgc8 aptamer was conjugated onto the colloidal NP's surface. The resulting Sgc8-nanoconjugate demonstrated an excellent target binding capability. Thiol-ene "click" chemistry was also employed with success for the modification of different NPs with diverse peptides (Choi et al. 2017; Kuang et al. 2016).

Diels–alder "click" chemistry includes a cycloaddition process between an electron-poor dienophile (alkene) and an electron-rich diene to generate a cyclohexene derivative (Molavipordanjani and Hosseinimehr 2019; Sunasee and Narain 2014). In one particular study, Logie and co-workers (Logie et al. 2017) reported the development of novel polymeric micelles targeted with the incorporation of the anti-HER2 fragment antibody, Fab 73 J. The terminal furan groups of polymeric micelles enabled the facile antibody Fab 73 J conjugation via Diels–Alder chemistry for targeted delivery.

15.3.2 Non-Covalent Conjugation

In addition to the broad range of covalent reactions that can be applied to conjugate targeting ligands onto NPs, non-covalent bonds are commonly used as well. Non-covalent interactions rely profoundly on physical interactions through intermolecular non-covalent forces, encompassing affinity interactions, electrostatic, hydrophobic and hydrogen bonding and hybridization methods (Sunasee and Narain 2014).

Various bioconjugation approaches use biological recognition motifs, being the strongest known non-covalent interaction biotin- (strept)avidin a classical example (Friedman et al. 2013). An RNA–DNA hybrid A10-3-J1 biotinylated aptamer, which recognizes the PSMA extracellular domain was incorporated onto streptavidin-modified super paramagnetic iron oxide (SPIO) NPs (Leach et al. 2016). This SPIO-Apt bioconjugate showed improved specificity to $PSMA^{-}$positive prostate cancer cells. In one recent study, Cathcart et al. (2022) incorporated a biotinylated peptide (ISV4), which decreases MMP-14-mediated cell invasion, into polysaccharide-based NPs via the biotin-avidin binding. ISV4-conjugated NPs competently prevented MMP-14 mediated cell invasion leading to an enhanced uptake in comparison to control peptides. In another novel report, Marshall and collaborators (Marshall et al. 2022) developed a biotin-modified epithelial cell adhesion molecule (EpCAM) antibody targeted biomimetic coated nanocarrier, which successfully accomplished the delivery of a radiotherapeutic agent to breast cancer cells.

Base-pairing hybridization has emerged as powerful tool for the generation of NP-aptamer bioconjugates (Javier et al. 2008; Lee et al. 2011). Recently, Smith and collaborators (Smith et al. 2018) described the design of peptide amphiphile micelles (PAM) co-assembled with a DNA oligonucleotide (anti-tail amphiphile) to form A/PAMs. A human B cell leukemia and lymphoma targeting DNA aptamer containing a tail sequence complementary to the anti-tail amphiphile was annealed to the A/PAM forming the aptamer ~ A/PAMs, which was found to be stable in biofluids and able of selective binding.

Over the years, researchers from different fields of science, including chemistry, biochemistry, nanotechnology and biotechnology, have largely relied in traditional covalent and non-covalent bioconjugation chemistries for the generation of novel bioconjugates for application in cancer diagnosis and/or treatment. While great part of the bioconjugation strategies relies in the robust approach of covalent chemistry, non-covalent strategies have also led to excellent results due to its ability to generate complex structures from simple building blocks. As these bioconjugation methodologies are active areas of research, the design of novel powerful covalent, as well as non-covalent reactions to expand the toolkit of bioconjugation approaches will unquestionably continue to be developed in the next years.

15.4 Overview of Smart Nanoparticles on Clinical Context

Currently, NPs are seen as the forefront of health research, granting investment from several companies. Researchers explore the NPs capacity to perform high precision tasks, such as safely encapsulate several cytotoxic agents and effectively delivery it to the intended site by overcoming physiological barriers. Moreover, these new technologies not only rely on the EPR effect, that allows NPs accumulation on tumor sites, but are improved with regard to their targeting capacity through the use of NPs conjugated with engineered ligands.

As cancer remains one of the leading causes of death in the world, there is a demand to design more effective and safer strategies for both therapeutic and diagnostic applications. Therefore, the majority of articles published about targeted drug delivery aims to treat solid malignant neoplasms. Up to date, 17 ligand-targeted NPs with oncological therapeutic ends have entered clinical trials (Table 15.1). The proposed nanomedicines include lipid and polymer-based NPs, as well as a retroviral vector and bacterially derived minicells, all ranging from 20 to 400 nm. Moreover, these proposed strategies rely on the application of antibodies, proteins and small molecules.

The first generation of smart NPs were engineered with antibodies. Two nanomedicines, SGT-53 and SGT-94, display at their surface a scFv that targets the Tf receptor present on tumor cells (Leung et al. 2021; Siefker-Radtke et al. 2016). SGT-53, a liposome loaded with DNA plasmid encoding a wild type p53 tumor suppressor protein, was particularly developed as a coadjuvant agent of chemo and radiotherapy. Recently, a phase II trial (NCT02340156) employed, simultaneously, SGT-53 and temozolomide, a chemotherapy agent, for the treatment of recurrent glioblastomas (Leung et al. 2021). Similarly, the liposomal nanodelivery complex SGT-94 carries a gene encoding the tumor suppressor protein RB94. The phase I trial (NCT01517464) concluded in 2017 proved that systemically delivered SGT-94 was well tolerated and induced clinical activity against a group of patients with solid tumors, while showing evidence of selective tumor targeting (Siefker-Radtke et al. 2016).

HER2, paramount in hormonal tumors, such as breast cancer, activates tumorigenic cells, stimulating their proliferation. The in vitro and in vivo findings of several researchers allowed the development of MM-302, HER2-targeted antibody-liposomes conjugates loaded with Dox applied in the treatment of HER2-positive advanced breast cancers. While the phase I trial (NCT01304797) showed promising

Table 15.1 Overview of engineered nanoparticles for cancer treatment undergoing clinical trials. BCL2L12: Bcl-2-like protein; EGFR: epidermal growth factor receptor; EphA2: Ephrin receptor A2; Fab: fragment antigen-binding; FRα: folate receptor α; HER2: Human epidermal growth factor receptor 2; MYH14: myosin-14; PSMA: prostate specific membrane antigen; RRM2: M2 subunit of ribonucleotide reductase; scFv: single-chain variable fragment

Surface modification	Type of NPs	Cargo	Name	Application	Phase	Refs.
Antibodies						
Anti-EGFR (Cetuximab fab)	Liposomes	Doxorubicin	Anti-EGFR ILs-DOX	Solid tumors	I	Mamot et al. (2012)
Anti-HER2	Liposomes	Doxorubicin	MM-302	Breast cancer	I	Munster et al. (2018)
Anti-transferrin receptor scFv	Liposomes	p53 plasmid DNA	SGT-53	Pancreatic cancer and glioblastomas	II	Leung et al. (2021)
Anti-transferrin receptor scFv	Liposomes	RB94 plasmid DNA	SGT-94	Solid tumors	I	Siefker-Radtke et al. (2016)
Anti-EGFR (Cetuximab)	Bacterially derived mini-cell	Paclitaxel	Erbitux-EDVsPAC	Solid tumors	II	Solomon et al. (2015)
Anti-EphA2 scFv	Liposomes	Docetaxel	MM-310	Solid tumors	I	Ernstoff et al. (2018)
Anti-EGFR (Cetuximab fab)	Liposomes	Doxorubicin	C225-ILS-Dox	Glioblastoma	I	Kasenda et al. (2022)
Single domain antibody	Liposomes	Melanoma antigens and IFNγ	Lipovaxin-MM	Melanoma	I	Gargett et al. (2018)
Anti-MYH14 fab (GAH)	Liposomes	Doxorubicin	MCC-465	Metastatic stomach cancer	I	Ura et al. (2004)
Proteins						
Transferrin	Liposomes	Oxaliplatin	MBP-426	Metastatic gastric and esophageal adenocarcinoma	I-II	Alavi et al. (2022)

(continued)

Table 15.1 (continued)

Surface modification	Type of NPs	Cargo	Name	Application	Phase	Refs.
Glutathione	Liposomes	Doxorubicin	2B3-101	Solid tumors and high-grade gliomas	I-IIa	Brandsma et al. (2014)
Transferrin	Cyclo-dextrin polymer based	RRM2 siRNA	CALAA-01	Solid tumors	Ia-Ib	Zuckerman et al. (2014)
Human serum albumin	Polymeric	Paclitaxel	TENPA	Advanced solid tumors	I	Chung et al. (2020)
Small molecules						
PSMA-targeting ligand	Polylactic-acid based	Docetaxel	BIND-014	Metastatic castration-resistant prostate cancer	II	Autio et al. (2018)
Collagen-binding motif	MLV-based retroviral vector	Dominant negative cyclin-G1 DNA construct	Rexin-G	Pancreatic adenocarcinoma, osteosarcoma, soft tissue sarcoma	I-II	Gordon et al. (2006), Chawla et al. (2009), Chawla et al. (2019)
Folic acid analogs	Silica-encased (CDots)	Exatecan	ELU001	FRα-overexpressing tumors	I-II	Ma et al. (2022)
Somatostatin analogs	Polymeric	Cetuximab	Qdots-SST	Colorectal cancer	I	Hafez Abdellatif et al. (2018)

efficacy and safety, MM-302 has recently missed the endpoint of the phase II trial (Munster et al. 2018).

EGFR overexpressing tumors can be targeted via engraftment of Fab fragments of an anti-EGFR monoclonal antibody, such as cetuximab. Cetuximab, first approved by FDA in 2004, is the active targeting moiety in 2 active trials: Anti-EGFR ILs-Dox and C225-ILS-Dox (NCT01702129 and NCT03603379). Both exploit the application of liposomes as NPs and Dox as an active cargo and are part of phase I clinical trials. Anti-EGFR ILs-Dox were administered in 26 patients with EGFR-overexpressing solid tumors from which one showed complete response, another a partial response and ten individuals exhibited stable disease for up until a year (Mamot et al. 2012). C225-ILS-Dox preliminary phase I trial data was recently published which demonstrated that these anti-EGFR immunoliposomes can target EGFR-amplified glioblastomas, successfully delivering Dox. However, they were not able to cross the intact BBB (Kasenda et al. 2022). In addition, Erbitux-EVDs PAC, currently undergoing a phase II clinical trial, targets EGFR by modifying a new nanocarrier platform, the bacterially derived mini cells. These large and promising NPs were produced by mutants that inactivated the genes responsible for cellular division, were loaded with paclitaxel and surface engineered with bispecific antibodies that recognize an O-polysaccharide component of the EFGR receptor. This approach presented promising in vivo data, showing a reverse multidrug resistance effect in several solid tumors. The phase I results granted the safety and determined the optimal dosage to pursue phase II trials (Solomon et al. 2015).

Antibodies are still the chosen ligand of three other clinical trials, MM-310, lipovaxin-MM and MCC-465, all relying on liposomes as NPs. Aiming to treat a range of solid tumors, MM-310 displays an Ephrin receptor A2 (EphA2)-targeting antibody at the NPs surface, delivering the cytotoxic drug Docetaxel with higher specificity. However, in 2019, the phase I study (NCT03076372) was discontinued due to the cumulative toxicity observed in patients (Ernstoff et al. 2018). Lipovaxin-MM is a lipid-based vaccine that does not directly target malignant melanoma cells but delivers melanoma antigens to dendritic cells that then activate tumor-specific cytotoxic T cells. The targeting of dendritic cells is achieved via coupling of a domain antibody (DMS5000). Twelve patients with metastatic cutaneous melanoma were selected for a phase I study (NCT01052142) that promoted, with high tolerance, a partial response and stable disease in one and two individuals, respectively (Gargett et al. 2018). MCC-465 conjugates a Dox-loaded pegylated liposome with a Fab fragment of the human GAH selected antibody. After an enthusiastic in vivo outcome, with an increased efficacy and diminished cytotoxicity, this approach enrolled in a phase I trial that promoted the disease (metastatic stomach cancer) stability in 10 of 18 patients (Ura et al. 2004). However, since its conclusion, no updates were published, and it remains uncertain the reasons for the discontinuity of the work.

Besides antibodies, that represent more than half of the existent proposals, researchers also explored the engraftment of proteins and small molecules.

MBP-426 is an oxaliplatin-loaded liposome conjugated to the Tf protein developed to target gastric, gastroesophageal, and esophageal adenocarcinomas as a second line of treatment. Phase Ib trial (NCT00964080) data shows that thrombocytopenia was the main dose-limiting toxicity event, but MBP-426 still showed a favorable safety profile and the recommended dose of 170 mg/m^2 was calculated for the next steps (Alavi et al. 2022). Another clinical trial relies on Tf to promote selective cellular internalization; however, it resorts to another type of NP and is the vehicle of a nucleic acid. CALAA-01 is a polymeric NP that induces the knockdown of the M2 subunit of ribonucleotide reductase (RRM2) through small interference (siRNA) technology. Phase Ia-Ib trials (NCT00689065) concluded that RNAi was able to occur in humans, affecting mRNA and protein levels, due to the systemic delivery of siRNA contained in CALAA-01 (Zuckerman et al. 2014).

Two more proteins engrafted onto NPs joined in clinical trials: glutathione and human serum albumin (HAS), for 2B3-101 and TENPA nanomedicines, respectively. Applied for solid tumors and glioma therapy, 2B3-101 is a Dox-loaded liposome with glutathione as a targeting ligand, aiming to cross the BBB via glutathione transporters. 2B3-101 entered a phase I/IIA trial (NCT01386580) to determine its safety as a single agent or in combination with trastuzumab (Brandsma et al. 2014). TENPA, on the other hand, is a polymeric NP engineered with HAS and loaded with paclitaxel. In 2016, enrolled on a phase I clinical trial (NCT02979392), however preliminary results are not available (Chung et al. 2020).

BIND-014 is a polymeric NP studied in patients with metastatic castration resistant prostate cancer. This NP is decorated with a small molecule, S,S-2-[3-[5-amino-1-carboxypentyl]-ureido]-pentanedioic acid (ACUPA), that specifically binds to prostate-specific membrane antigen (PSMA) and allows an effective delivery of Docetaxel. In a phase II clinical trial (NCT01812746), promising outcomes determined that the median progression-free survival of 42 patients was 9.9 months, while granting the nanomedicine safety (Autio et al. 2018).

Very recently, two other NPs displaying small molecules as ligands have entered clinical trials: Qdots-SST and ELU001. Qdots-SST were created to target the somatostatin receptors commonly found in colorectal cancer cells. The oral polymeric NPs decorated with somatostatin analogue and loaded with cetuximab were approved to enter a phase I clinical trial in 2018 (NCT03774680) (Hafez Abdellatif et al. 2018). Nevertheless, the current status of the study in unknown. Moreover, ELU001 is a new chemical entity described as a CDot drug conjugate that combines payloads of exatecans and FA analogs as targeting moieties. In 2021, it became the first engineered NP to employ FA tested in clinical conditions (NCT05001282), focusing on adults with advanced folate receptor alpha (FRα) overexpressing tumors (Ma et al. 2022).

Retroviral vectors efficiently integrate their payload in the DNA of the target cells, however these NPs are hindered by the lack of tissue specificity. Rexin-G, developed for the treatment of osteosarcoma, sarcoma, and pancreatic tumors, aims to solve this limitation by engrafting a high-affinity collagen binding motif on the surface of a retroviral vector. This popular nanomedicine has already enrolled in five different clinical trials (NCT00505713, NCT00504998, NCT00505271, NCT00572130 and NCT00121745). In a phase I/II trial, carried out in the Philippines, Rexin-G was well tolerated, exhibited anti-tumor activity against metastatic pancreatic cancer and did not induce organ damage (Gordon et al. 2006). In the USA, phase I/II trials for the treatment of all three abovementioned types of cancer successfully established a correlation between Rexin-G administration and overall survival, while no dose-limiting toxicity was discovered, as well as no organ related toxicity, antibody response and off-target transfection (Chawla et al. 2009, 2019). Moreover, in all clinical trials, Rexin-G also improved physiological conditions, namely liver function, blood chemistry and wound healing. Based on these outcomes, Rexin-G gained orphan drug status in the U.S.A. for the treatment of all three proposed pathologies.

15.5 Challenges and Future Steps

Although several engineered NPs have been formulated, evaluated, and submitted to in vivo studies, only a small part has progressed into clinical trials and none of them have yet been approved. To date, there are still many hurdles hindering the therapeutic and clinical applications, including the expensiveness and the toxicity of smart NPs, the reproducibility of large-scale production, and the complexity of biological barriers and environment.

A major drawback for the success of smart NPs in terms of enhanced tissue specificity, cellular uptake and ligand-receptor interactions is the NPs toxicity. In order to overcome this limitation, researchers must define appropriate physicochemical properties, including size, shape, surface charge and rigidity, and should not underestimate the optimization of surface ligand density (Kiio and Park 2021). This is a crucial aspect, since the engrafted moieties need to guarantee the tumor uptake and consequent accumulation, but not exceeding the saturation point. Another major bottleneck is the large-scale production that commonly results in aggregation and alters NPs characteristics and stability. Given that the production of the existent NPs and the conjugation techniques to obtain functionalized NPs were outlined for laboratorial scale, improvements to these methodologies to apply them on a large scale and the implementation of innovative low-energy-input methods are devised (Zhao et al. 2022). All this development pipeline, including the NPs design and the ligand, particularly antibodies, production and conjugation, are very costly processes. Therefore, pursuing these strategies remains an economically unsustainable investment, associated with increased financial risks and potential clinical translation fail (Cheng et al. 2012).

Despite the attractiveness of exploiting smart NPs as DDS, the inherent complexity and dynamic of biological environments hampers these nanomedicines' effectiveness. Broadly, smart DDS have to circulate through the body, identify the targeted tissue, accumulate in tumor-site, penetrate into deeper regions, be internalized into the cells, and finally release the active agents. In order to achieve the therapeutic goal, the first step is to study the best administration route to improve the biodistribution and face the need for multiple dosing, already reported in NPs application (Xu et al. 2021). Then, smart NPs must escape from lymphatic drainage and penetrate two barriers, the tumoral, and possibly the BBB (if the intention is to reach the brain). Afterwards, at tumor site, smart NPs face the heterogeneity of the EPR effect and the surface variations between tumor types and disease stages. More importantly, it is essential to understand tumor cells interaction and behavior since the therapeutic outcomes relies on pharmacokinetic, tissue distribution, tumor accumulation, and penetration. However, these aspects are highly heterogeneous between cellular cultures, animal model species and humans, culminating in a translational gap between preclinical studies and clinical trials. In addition, the development pipeline does not consider the heterogeneity amongst patients by not promoting studies on the NPs interaction with stratified patient populations (Mitchell et al. 2021; Seidu et al. 2022).

Therefore, in order to keep pace with pre-clinical studies, several topics must be addressed. First and foremost, researchers must exploit the influence of tumor size, stage, location, and grade and standardize aspects of clinical research, such as patient screening, drug selection, and combination with existent therapies. In addition, new strategies must be employed to improve the NPs performance, either therapeutically or diagnostically, such as the development of fusogenic NPs and NPs multifunctionalization (Friedman et al. 2013; Seidu et al. 2022). With the purpose of decreasing adverse side effects, resistance and immune responses, authors pursued the production of hybrid NPs. These fusogenic NPs are constructed from at least two distinct classes of NPs taking advantage from their different properties, overcoming limitations of single-component NPs, and gathering multiple functionalities (Wu et al. 2015; Gao et al. 2017). Following the same line of thought, NPs multifunctionalization, either by multifunctional targeting or chimeric targeting approach, is now trending in the field. Both strategies attach more than one type of ligands on the NPs surface in order to target distinct individual targets, but while multifunctional targeting relies on ligands from the same class, chimeric targeting uses moieties across categories (Kluza et al. 2012; Ko et al. 2011). Both strategies aim to expand multiple-targeting specificity, enhance cellular uptake and increase the benefit against pathological tissues while sustaining the healthy ones.

As a result, given all the advantages of using smart NPs in cancer therapies and the extensive work that researchers all over the worldare developing to overcome the disadvantages and problems that still exist, it can be expected that in the next couple of years more functionalized NPs will enter clinical trials and that eventually some of these will be approved for use in cancer therapy.

References

Abdellatif AAH, Hennig R, Pollinger K, Tawfeek HM, Bouazzaoui A, Goepferich A (2020) Fluorescent Nanoparticles Coated with a Somatostatin Analogue Target Blood Monocyte for Efficient Leukaemia Treatment. Pharm Res 37:217. https://doi.org/10.1007/s11095-020-02938-1

Abdellatif AA, Abdelhafez WA, Sarhan HA (2018) Somatostatin Decorated Quantum Dots for Targeting of Somatostatin Receptors., Iran. J. Pharm. Res. IJPR. 17 (2018) 513–524. http://www.ncbi.nlm.nih.gov/pubmed/29881409

Alavi N, Rezaei M, Maghami P, Fanipakdel A, Avan A (2022) Nanocarrier System for Increasing the Therapeutic Efficacy of Oxaliplatin. Curr Cancer Drug Targets 22:361–372. https://doi.org/10.2174/1568009622666220120115140

Arruebo M, Valladares M, González-Fernández Á (2009) Antibody-Conjugated Nanoparticles for Biomedical Applications. J Nanomater 2009:1–24. https://doi.org/10.1155/2009/439389

Autio KA, Dreicer R, Anderson J, Garcia JA, Alva A, Hart LL, Milowsky MI, Posadas EM, Ryan CJ, Graf RP, Dittamore R, Schreiber NA, Summa JM, Youssoufian H, Morris MJ, Scher HI (2018) Safety and Efficacy of BIND-014, a Docetaxel Nanoparticle Targeting Prostate-Specific Membrane Antigen for Patients With Metastatic Castration-Resistant Prostate Cancer. JAMA Oncol 4:1344. https://doi.org/10.1001/jamaoncol.2018.2168

Barbero F, Russo L, Vitali M, Piella J, Salvo I, Borrajo ML, Busquets-Fité M, Grandori R, Bastús NG, Casals E, Puntes V (2017) Formation of the Protein Corona: The Interface between Nanoparticles and the Immune System. Semin Immunol 34:52–60. https://doi.org/10.1016/j.smim.2017.10.001

Brandsma D, Kerklaan BM, Diéras V, Altintas S, Anders CK, Ballester MA, Gelderblom H, Soetekouw PM, Gladdines W, Lonnqvist F, Jager A (2014) Phase 1/2A Study of Glutathione Pegylated Liposomal Doxorubicin (2B3–101) in Patients with Brain Metastases (Bm) from Solid Tumors or Recurrent High Grade Gliomas (Hgg), Ann. Oncol. 25 (2014) iv157. https://doi.org/10.1093/annonc/mdu331.32.

Brunel FM, Lewis JD, Destito G, Steinmetz NF, Manchester M, Stuhlmann H, Dawson PE (2010) Hydrazone Ligation Strategy to Assemble Multifunctional Viral Nanoparticles for Cell Imaging and Tumor Targeting. Nano Lett 10:1093–1097. https://doi.org/10.1021/nl1002526

Byun J (2021) Recent Progress and Opportunities for Nucleic Acid Aptamers. Life 11:193. https://doi.org/10.3390/life11030193

Cappellano G, Comi C, Chiocchetti A, Dianzani U (2019) Exploiting PLGA-Based Biocompatible Nanoparticles for Next-Generation Tolerogenic Vaccines against Autoimmune Disease. Int J Mol Sci 20:204. https://doi.org/10.3390/ijms20010204

Cathcart J, Suarato G, Li W, Cao J, Meng Y (2022) Peptide-Functionalized Nanoparticles for the Targeted Delivery of Cytotoxins to MMP-14-Expressing Cancer Cells. Biophysica 2:203–220. https://doi.org/10.3390/biophysica2030021

Chawla SP, Bruckner H, Morse MA, Assudani N, Hall FL, Gordon EM (2019) A Phase I-II Study Using Rexin-G Tumor-Targeted Retrovector Encoding a Dominant-Negative Cyclin G1 Inhibitor for Advanced Pancreatic Cancer. Mol. Ther. Oncolytics. 12:56–67. https://doi.org/10.1016/j.omto.2018.12.005

Chawla SP, Chua VS, Fernandez L, Quon D, Saralou A, Blackwelder WC, Hall FL, Gordon EM (2009) Phase I/II and Phase II Studies of Targeted Gene Delivery In Vivo: Intravenous Rexin-G for Chemotherapy-resistant Sarcoma and Osteosarcoma. Mol Ther 17:1651–1657. https://doi.org/10.1038/mt.2009.126

Chen Z, Sun M, Luo F, Xu K, Lin Z, Zhang L (2018) Stimulus-response click chemistry based aptamer-functionalized mesoporous silica nanoparticles for fluorescence detection of thrombin. Talanta 178:563–568. https://doi.org/10.1016/j.talanta.2017.09.043

Cheng Z, Al Zaki A, Hui JZ, Muzykantov VR, Tsourkas A (2012) Multifunctional nanoparticles: cost versus benefit of adding targeting and imaging capabilities., Science. 338 (2012) 903–10. https://doi.org/10.1126/science.1226338

Choi J, Rustique E, Henry M, Guidetti M, Josserand V, Sancey L, Boutet J, Coll J-L (2017) Targeting tumors with cyclic RGD-conjugated lipid nanoparticles loaded with an IR780 NIR dye: In vitro and in vivo evaluation. Int J Pharm 532:677–685. https://doi.org/10.1016/j.ijpharm.2017.03.007

Chung H-J, Kim H-J, Hong S-T (2020) Tumor-specific delivery of a paclitaxel-loading HSA-haemin nanoparticle for cancer treatment, Nanomedicine Nanotechnology. Biol Med 23:102089. https://doi.org/10.1016/j.nano.2019.102089

Comparetti EJ, Romagnoli GG, Gorgulho CM, de Albuquerque Pedrosa V, Kaneno R (2020) Anti-PSMA monoclonal antibody increases the toxicity of paclitaxel carried by carbon nanotubes, Mater. Sci. Eng. C. 116 (2020) 111254. https://doi.org/10.1016/j.msec.2020.111254

Cypriyana J, Angalene LA, Samrot AV, Kumar S, Ponniah P, Chakravarthi S (2021) Overview on toxicity of nanoparticles, it's mechanism, models used in toxicity studies and disposal methods – a review, Biocatal. Agric Biotechnol 36 (2021) 102117. https://doi.org/10.1016/j.bcab.2021.102117.

Duan D, Wang A, Ni L, Zhang L, Yan X, Jiang Y, Mu H, Wu Z, Sun K, Li Y (2018) Trastuzumab- and Fab′ fragment-modified curcumin PEG-PLGA nanoparticles: preparation and evaluation in vitro and in vivo. Int J Nanomedicine 13:1831–1840. https://doi.org/10.2147/IJN.S153795

Ellington AD, Szostak JW (1990) In vitro selection of RNA molecules that bind specific ligands. Nature 346:818–822. https://doi.org/10.1038/346818a0

Ernstoff MS, Ma WW, Tsai FY, Munster PN, Zhang T, Kamoun W, Pipas JM, Chen S, Santillana S, Askoxylakis V (2018) A phase 1 study evaluating the safety, pharmacology and preliminary activity of MM-310 in patients with solid tumors., J. Clin. Oncol. 36 (2018) TPS2604–TPS2604. https://doi.org/10.1200/JCO.2018.36.15_suppl.TPS2604

Farokhzad OC, Jon S, Khademhosseini A, Tran T-NT, LaVan DA, Langer R (2004) Nanoparticle-Aptamer Bioconjugates. Cancer Res 64:7668–7672. https://doi.org/10.1158/0008-5472.CAN-04-2550

Ferreira D, Barbosa J, Sousa DA, Silva C, Melo LDR, Avci-Adali M, Wendel HP, Rodrigues LR (2021) Selection of aptamers against triple negative breast cancer cells using high throughput sequencing. Sci Rep 11:8614. https://doi.org/10.1038/s41598-021-87998-y

Ferreira D, Silva AP, Nobrega FL, Martins IM, Barbosa-Matos C, Granja S, Martins SF, Baltazar F, Rodrigues LR (2019) Rational Identification of a Colorectal Cancer Targeting Peptide through Phage Display. Sci Rep 9:3958. https://doi.org/10.1038/s41598-019-40562-1

Ferreira D, Martins IM (2017) Artificial virus particles, in: Bioinspired Mater. Med. Appl., Elsevier, 2017: pp. 427–450. https://doi.org/10.1016/B978-0-08-100741-9.00015-2

Friedman AD, Claypool SE, Liu R (2013) The smart targeting of nanoparticles. Curr Pharm Des 19:6315–6329

Fu Z, Xiang J (2020) Aptamers, the Nucleic Acid Antibodies, in Cancer Therapy. Int J Mol Sci 21:2793. https://doi.org/10.3390/ijms21082793

Gao F, Zhang J, Fu C, Xie X, Peng F, You J, Tang H, Wang Z, Li P, Chen J (2017) iRGD-modified lipid-polymer hybrid nanoparticles loaded with isoliquiritigenin to enhance anti-breast cancer effect and tumor-targeting ability. Int J Nanomedicine 12:4147–4162. https://doi.org/10.2147/IJN.S134148

Gao H, Zhang Q, Yang Y, Jiang X, He Q (2015) Tumor homing cell penetrating peptide decorated nanoparticles used for enhancing tumor targeting delivery and therapy. Int J Pharm 478:240–250. https://doi.org/10.1016/j.ijpharm.2014.11.029

Gargett T, Abbas MN, Rolan P, Price JD, Gosling KM, Ferrante A, Ruszkiewicz A, Atmosukarto IIC, Altin J, Parish CR, Brown MP (2018) Phase I trial of Lipovaxin-MM, a novel dendritic cell-targeted liposomal vaccine for malignant melanoma. Cancer Immunol Immunother 67:1461–1472. https://doi.org/10.1007/s00262-018-2207-z

Gong Z, Liu X, Zhou B, Wang G, Guan X, Xu Y, Zhang J, Hong Z, Cao J, Sun X, Gao Z, Lu H, Pan X, Bai J (2021) Tumor acidic microenvironment-induced drug release of RGD peptide nanoparticles for cellular uptake and cancer therapy. Colloids Surfaces B Biointerfaces. 202:111673. https://doi.org/10.1016/j.colsurfb.2021.111673

Gordon EM, Lopez FF, Cornelio GH, Lorenzo CC, Levy JP, Reed RA, Liu L, Bruckner HW, Hall FL (2006) Pathotropic nanoparticles for cancer gene therapy Rexin-G IV: three-year clinical experience., Int. J. Oncol. 29 :1053–64

Huang R, Ke W, Qu Y, Zhu J, Pei Y, Jiang C (2007) Characterization of lactoferrin receptor in brain endothelial capillary cells and mouse brain. J Biomed Sci 14:121–128. https://doi.org/10.1007/s11373-006-9121-7

Jalilian M, Derakhshandeh K, Kurd M, Lashani H (2021) Targeting Solid Lipid Nanoparticles with Anisamide for Docetaxel Delivery to Prostate Cancer: Preparation, Optimization, and In-vitro Evaluation., Iran. J. Pharm. Res. IJPR. 20 327–338. https://doi.org/10.22037/ijpr.2020.113436.14302.

Javier DJ, Nitin N, Levy M, Ellington A, Richards-Kortum R (2008) Aptamer-Targeted Gold Nanoparticles As Molecular-Specific Contrast Agents for Reflectance Imaging. Bioconjug Chem 19:1309–1312. https://doi.org/10.1021/bc8001248

Jeevanandam J, Barhoum A, Chan YS, Dufresne A, Danquah MK (2018) Review on nanoparticles and nanostructured materials: history, sources, toxicity and regulations. Beilstein J Nanotechnol 9:1050–1074. https://doi.org/10.3762/bjnano.9.98

Jia Y, Chen S, Wang C, Sun T, Yang L (2022) Hyaluronic acid-based nano drug delivery systems for breast cancer treatment: Recent advances, Front. Bioeng. Biotechnol. 10 (2022). https://doi.org/10.3389/fbioe.2022.990145.

Jiang Z, Guan J, Qian J, Zhan C (2019) Peptide ligand-mediated targeted drug delivery of nanomedicines. Biomater Sci 7:461–471. https://doi.org/10.1039/C8BM01340C

Joshi PP, Yoon SJ, Hardin WG, Emelianov S, Sokolov KV (2013) Conjugation of Antibodies to Gold Nanorods through Fc Portion: Synthesis and Molecular Specific Imaging. Bioconjug Chem 24:878–888. https://doi.org/10.1021/bc3004815

Kapoor D, Bhatt S, Kumar M, Maheshwari R, Tekade RK (2019) Ligands for Targeted Drug Delivery and Applications, in: Basic Fundam. Drug Deliv., Elsevier, pp. 307–342. https://doi.org/10.1016/B978-0-12-817909-3.00008-X

Kasenda B, König D, Manni M, Ritschard R, Duthaler U, Bartoszek E, Bärenwaldt A, Deuster S, Hutter G, Cordier D, Mariani L, Hench J, Frank S, Krähenbühl S, Zippelius A, Rochlitz C, Mamot C, Wicki A, Läubli H (2022) Targeting immunoliposomes to EGFR-positive glioblastoma. ESMO Open. 7:100365. https://doi.org/10.1016/j.esmoop.2021.100365

Kiio TM, Park S (2021) Physical properties of nanoparticles do matter. J Pharm Investig 51:35–51. https://doi.org/10.1007/s40005-020-00504-w

Kluza E, Jacobs I, Hectors SJCG, Mayo KH, Griffioen AW, Strijkers GJ, Nicolay K (2012) Dual-targeting of αvβ3 and galectin-1 improves the specificity of paramagnetic/fluorescent liposomes to tumor endothelium in vivo. J Control Release 158:207–214. https://doi.org/10.1016/j.jconrel.2011.10.032

Ko HY, Choi K-J, Lee CH, Kim S (2011) A multimodal nanoparticle-based cancer imaging probe simultaneously targeting nucleolin, integrin αvβ3 and tenascin-C proteins. Biomaterials 32:1130–1138. https://doi.org/10.1016/j.biomaterials.2010.10.034

Kuang Y, Jiang X, Zhang Y, Lu Y, Ma H, Guo Y, Zhang Y, An S, Li J, Liu L, Wu Y, Liang J, Jiang C (2016) Dual Functional Peptide-Driven Nanoparticles for Highly Efficient Glioma-Targeting and Drug Codelivery. Mol Pharm 13:1599–1607. https://doi.org/10.1021/acs.molpharmaceut.6b00051

Kumar S, Aaron J, Sokolov K (2008) Directional conjugation of antibodies to nanoparticles for synthesis of multiplexed optical contrast agents with both delivery and targeting moieties. Nat Protoc 3:314–320. https://doi.org/10.1038/nprot.2008.1

Leach JC, Wang A, Ye K, Jin S (2016) A RNA-DNA Hybrid Aptamer for Nanoparticle-Based Prostate Tumor Targeted Drug Delivery. Int. J. Mol. Sci. 17 (2016) 380. https://doi.org/10.3390/ijms17030380.

Lee I-H, An S, Yu MK, Kwon H-K, Im S-H, Jon S (2011) Targeted chemoimmunotherapy using drug-loaded aptamer–dendrimer bioconjugates. J Control Release 155:435–441. https://doi.org/10.1016/j.jconrel.2011.05.025

Leserman LD, Barbet J, Kourilsky F, Weinstein JN (1980) Targeting to cells of fluorescent liposomes covalently coupled with monoclonal antibody or protein A. Nature 288:602–604. https://doi.org/10.1038/288602a0

Leung CP, Barve MA, Wu M-S, Pirollo KF, Strauss JF, Liao W-C, Yang S-H, Nunan RA, Adams J, Harford JB, Chang EH (2021) A phase II trial combining tumor-targeting TP53 gene therapy with gemcitabine/nab-paclitaxel as a second-line treatment for metastatic pancreatic cancer. J Clin Oncol 39:4139–4139. https://doi.org/10.1200/JCO.2021.39.15_suppl.4139

Li H, Tong Y, Bai L, Ye L, Zhong L, Duan X, Zhu Y (2018) Lactoferrin functionalized PEG-PLGA nanoparticles of shikonin for brain targeting therapy of glioma. Int J Biol Macromol 107:204–211. https://doi.org/10.1016/j.ijbiomac.2017.08.155

Lin H-C, Hsu K-F, Lai C-L, Wu T-C, Chen H-F, Lai C-H (2020) Mannoside-Modified Branched Gold Nanoparticles for Photothermal Therapy to MDA-MB-231 Cells. Molecules 25:1853. https://doi.org/10.3390/molecules25081853

Liszbinski RB, Romagnoli GG, Gorgulho CM, Basso CR, Pedrosa VA, Kaneno R (2020) Anti-EGFR-Coated Gold Nanoparticles In Vitro Carry 5-Fluorouracil to Colorectal Cancer Cells. Materials (basel). 13:375. https://doi.org/10.3390/ma13020375

Liu Y, Hou W, Sun H, Cui C, Zhang L, Jiang Y, Wu Y, Wang Y, Li J, Sumerlin BS, Liu Q, Tan W (2017) Thiol–ene click chemistry: a biocompatible way for orthogonal bioconjugation of colloidal nanoparticles. Chem Sci 8:6182–6187. https://doi.org/10.1039/C7SC01447C

Logie J, Ganesh AN, Aman AM, Al-Awar RS, Shoichet MS (2017) Preclinical evaluation of taxane-binding peptide-modified polymeric micelles loaded with docetaxel in an orthotopic breast cancer mouse model. Biomaterials 123:39–47. https://doi.org/10.1016/j.biomaterials.2017.01.026

Ma WW, Tolcher AW, Strauss JF, Bekaii-Saab TS, Zhao Y, Perez CA, Hamilton EP, Adams GP, Reddick C, Bayever E (2022) ELU-FRα-1: A study to evaluate ELU001 in patients with solid tumors that overexpress folate receptor alpha (FRα)., J. Clin. Oncol. 40 (2022) TPS3158–TPS3158. https://doi.org/10.1200/JCO.2022.40.16_suppl.TPS3158

von Maltzahn G, Ren Y, Park J-H, Min D-H, Kotamraju VR, Jayakumar J, Fogal V, Sailor MJ, Ruoslahti E, Bhatia SN (2008) In Vivo Tumor Cell Targeting with "Click" Nanoparticles. Bioconjug Chem 19:1570–1578. https://doi.org/10.1021/bc800077y

Mamot C, Ritschard R, Wicki A, Stehle G, Dieterle T, Bubendorf L, Hilker C, Deuster S, Herrmann R, Rochlitz C (2012) Tolerability, safety, pharmacokinetics, and efficacy of doxorubicin-loaded anti-EGFR immunoliposomes in advanced solid tumours: a phase 1 dose-escalation study. Lancet Oncol 13:1234–1241. https://doi.org/10.1016/S1470-2045(12)70476-X

Mao K, Zhang W, Yu L, Yu Y, Liu H, Zhang X (2021) Transferrin-Decorated Protein-Lipid Hybrid Nanoparticle Efficiently Delivers Cisplatin and Docetaxel for Targeted Lung Cancer Treatment. Drug Des. Devel. Ther. 15:3475–3486. https://doi.org/10.2147/DDDT.S296253

Marques AC, Costa PJ, Velho S, Amaral MH (2020) Functionalizing nanoparticles with cancer-targeting antibodies: A comparison of strategies. J Control Release 320:180–200. https://doi.org/10.1016/j.jconrel.2020.01.035

Marshall SK, Panrak Y, Makchuchit N, Jaroenpakdee P, Saelim B, Taweesap M, Pachana V (2022) Anti-EpCAM Functionalized I-131 Radiolabeled Biomimetic Nanocarrier Sodium/Iodide-Symporter-Mediated Breast-Cancer Treatment. Bioengineering 9:294. https://doi.org/10.3390/bioengineering9070294

Meng L, Ren J, Liu Z, Zhao Y (2022) Hyaluronic acid-coated shikonin liposomes for the treatment of triple-negative breast cancer via targeting tumor cells and amplification of oxidative stress. J. Drug Deliv. Sci. Technol. 70:103193. https://doi.org/10.1016/j.jddst.2022.103193

Mitchell MJ, Billingsley MM, Haley RM, Wechsler ME, Peppas NA, Langer R (2021) Engineering precision nanoparticles for drug delivery. Nat Rev Drug Discov 20:101–124. https://doi.org/10.1038/s41573-020-0090-8

Molavipordanjani S, Hosseinimehr SJ (2019) Strategies for Conjugation of Biomolecules to Nanoparticles as Tumor Targeting Agents. Curr Pharm Des 25:3917–3926. https://doi.org/10.2174/1381612825666190903154847

Munster P, Krop IE, LoRusso P, Ma C, Siegel BA, Shields AF, Molnár I, Wickham TJ, Reynolds J, Campbell K, Hendriks BS, Adiwijaya BS, Geretti E, Moyo V, Miller KD (2018) Safety and pharmacokinetics of MM-302, a HER2-targeted antibody–liposomal doxorubicin conjugate, in patients with advanced HER2-positive breast cancer: a phase 1 dose-escalation study. Br J Cancer 119:1086–1093. https://doi.org/10.1038/s41416-018-0235-2

Nehoff H, Parayath NN, Domanovitch L, Taurin S, Greish K (2014) Nanomedicine for drug targeting: strategies beyond the enhanced permeability and retention effect. Int J Nanomedicine 2539. https://doi.org/10.2147/IJN.S47129

Nikzamir M, Akbarzadeh A, Panahi Y (2021) An overview on nanoparticles used in biomedicine and their cytotoxicity. J. Drug Deliv. Sci. Technol. 61:102316. https://doi.org/10.1016/j.jddst.2020.102316

Nobrega FL, Ferreira D, Martins IM, Suarez-Diez M, Azeredo J, Kluskens LD, Rodrigues LR (2016) Screening and characterization of novel specific peptides targeting MDA-MB-231 claudin-low breast carcinoma by computer-aided phage display methodologies. BMC Cancer 16:881. https://doi.org/10.1186/s12885-016-2937-2

Okur AC, Erkoc P, Kizilel S (2016) Targeting cancer cells via tumor-homing peptide CREKA functional PEG nanoparticles. Colloids Surfaces B Biointerfaces. 147:191–200. https://doi.org/10.1016/j.colsurfb.2016.08.005

Peng L, Liang Y, Zhong X, Liang Z, Tian Y, Li S, Liang J, Wang R, Zhong Y, Shi Y, Zhang X (2020) Aptamer-Conjugated Gold Nanoparticles Targeting Epidermal Growth Factor Receptor Variant III for the Treatment of Glioblastoma. Int J Nanomedicine 15:1363–1372. https://doi.org/10.2147/IJN.S238206

Pereira AC, Ferreira D, Santos-Pereira C, Vieira TF, Sousa SF, Sales G, Rodrigues LR (2021) Selection of a new peptide homing SK-BR-3 breast cancer cells. Chem Biol Drug Des 97:893–903. https://doi.org/10.1111/cbdd.13816

Pereira A, Sousa D, Pina A, Ferreira D, Santos-Pereira C, Melo LD, Sales MGF, Sousa S, Rodrigues LR (2022) Identification of novel aptamers targeting Cathepsin B-overexpressing prostate cancer cells. Mol Syst Des Eng. https://doi.org/10.1039/D2ME00022A

Ren B, Cai Z-C, Zhao X-J, Li L-S, Zhao M-X (2021) Evaluation of the Biological Activity of Folic Acid-Modified Paclitaxel-Loaded Gold Nanoparticles. Int J Nanomedicine 16:7023–7033. https://doi.org/10.2147/IJN.S322856

Rodrigues JL, Ferreira D, Rodrigues LR (2017) Synthetic biology strategies towards the development of new bioinspired technologies for medical applications, in: Bioinspired Mater. Med. Appl., Elsevier, 451–497. https://doi.org/10.1016/B978-0-08-100741-9.00016-4.

Rubio N, Mei K-C, Klippstein R, Costa PM, Hodgins N, Wang JT-W, Festy F, Abbate V, Hider RC, Chan KLA, Al-Jamal KT (2015) Solvent-Free Click-Mechanochemistry for the Preparation of Cancer Cell Targeting Graphene Oxide. ACS Appl Mater Interfaces 7:18920–18923. https://doi.org/10.1021/acsami.5b06250

Salahpour Anarjan F (2019) Active targeting drug delivery nanocarriers: Ligands, Nano-Structures & Nano-Objects. 19 100370. https://doi.org/10.1016/j.nanoso.2019.100370

Santos-Pereira C, Andrés MT, Fierro JF, Rodrigues LR, Côrte-Real M (2022) A review on lactoferrin as a proton pump inhibitor. Int J Biol Macromol 202:309–317. https://doi.org/10.1016/j.ijbiomac.2022.01.075

Seidu TA, Kutoka PT, Asante DO, Farooq MA, Alolga RN, Bo W (2022) Functionalization of Nanoparticulate Drug Delivery Systems and Its Influence in Cancer Therapy. Pharmaceutics 14:1113. https://doi.org/10.3390/pharmaceutics14051113

Selmani A, Kovačević D, Bohinc K (2022) Nanoparticles: From synthesis to applications and beyond. Adv Colloid Interface Sci 303:102640. https://doi.org/10.1016/j.cis.2022.102640

Shinde VR, Revi N, Murugappan S, Singh SP, Rengan AK (2022) Enhanced permeability and retention effect: A key facilitator for solid tumor targeting by nanoparticles. Photodiagnosis Photodyn Ther 39:102915. https://doi.org/10.1016/j.pdpdt.2022.102915

Shipunova VO, Sogomonyan AS, Zelepukin IV, Nikitin MP, Deyev SM (2021) PLGA Nanoparticles Decorated with Anti-HER2 Affibody for Targeted Delivery and Photoinduced Cell Death. Molecules 26:3955. https://doi.org/10.3390/molecules26133955

Siefker-Radtke A, Zhang X, Guo CC, Shen Y, Pirollo KF, Sabir S, Leung C, Leong-Wu C, Ling C-M, Chang EH, Millikan RE, Benedict WF (2016) A Phase l Study of a Tumor-targeted Systemic Nanodelivery System, SGT-94, in Genitourinary Cancers. Mol Ther 24:1484–1491. https://doi.org/10.1038/mt.2016.118

Silva AL, Moura LI, Carreira B, Conniot J, Matos AI, Peres C, Sainz V, Silva LC, Gaspar RS, Florindo HF (2018) Functional Moieties for Intracellular Traffic of Nanomaterials, in: Biomed.

Appl. Funct. Nanomater., Elsevier, pp 399–448. https://doi.org/10.1016/B978-0-323-50878-0.00014-8

Silva VL, Ferreira D, Nobrega FL, Martins IM, Kluskens LD, Rodrigues LR (2016) Selection of novel peptides homing the 4T1 cell line: Exploring alternative targets for triple negative breast cancer, PLoS One. 11. https://doi.org/10.1371/journal.pone.0161290.

Smith JD, Cardwell LN, Porciani D, Nguyen JA, Zhang R, Gallazzi F, Tata RR, Burke DH, Daniels MA, Ulery BD (2018) Aptamer-displaying peptide amphiphile micelles as a cell-targeted delivery vehicle of peptide cargoes. Phys Biol 15:065006. https://doi.org/10.1088/1478-3975/aadb68

Solomon BJ, Desai J, Rosenthal M, McArthur GA, Pattison ST, Pattison SL, MacDiarmid J, Brahmbhatt H, Scott AM (2015) A First-Time-In-Human Phase I Clinical Trial of Bispecific Antibody-Targeted. Paclitaxel-Packaged Bacterial Minicells, PLoS One. 10:e0144559. https://doi.org/10.1371/journal.pone.0144559

Sousa D, Ferreira D, Rodrigues JL, Rodrigues LR (2019) Nanotechnology in Targeted Drug Delivery and Therapeutics. Appl. Target. Nano Drugs Deliv. Syst Elsevier, pp 357–409. https://doi.org/10.1016/B978-0-12-814029-1.00014-4

Su C-W, Yen C-S, Chiang C-S, Hsu C-H, Chen S-Y (2017) Multistage Continuous Targeting with Quantitatively Controlled Peptides on Chitosan-Lipid Nanoparticles with Multicore-Shell Nanoarchitecture for Enhanced Orally Administrated Anticancer In Vitro and In Vivo. Macromol Biosci 17:1600260. https://doi.org/10.1002/mabi.201600260

Sunasee R, Narain R (2014) Covalent and Noncovalent Bioconjugation Strategies, in: Chem. Bioconjugates, John Wiley & Sons, Inc., Hoboken, NJ, USA, 2014: pp. 1–75. https://doi.org/10.1002/9781118775882.ch1.

Tan J, Yang N, Hu Z, Su J, Zhong J, Yang Y, Yu Y, Zhu J, Xue D, Huang Y, Lai Z, Huang Y, Lu X, Zhao Y (2016) Aptamer-Functionalized Fluorescent Silica Nanoparticles for Highly Sensitive Detection of Leukemia Cells. Nanoscale Res Lett 11:298. https://doi.org/10.1186/s11671-016-1512-8

Tsogtbaatar K, Sousa DA, Ferreira D, Tevlek A, Aydın HM, Çelik E, Rodrigues L (2021) In vitro selection of DNA aptamers against human osteosarcoma. Invest New Drugs. https://doi.org/10.1007/s10637-021-01161-y

Tuerk C, Gold L (1990) Systematic evolution of ligands by exponential enrichment: RNA ligands to bacteriophage T4 DNA polymerase, Science (80). 249 (1990) 505–510. https://doi.org/10.1126/science.2200121.

Ura T, Shirao K, Matsumura Y, Hamaguchi T, Miyata Y, Okura H, Chin K, Yamao T, Kannami A, Takahashi K (2004) Clinical trial of MCC-465, a doxorubicin encapsulated in PEG-immunoliposome, in patients with metastatic stomach cancer. Drug Deliv. Syst. 19:439–444. https://doi.org/10.2745/dds.19.439

Wu B, Yu P, Cui C, Wu M, Zhang Y, Liu L, Wang C-X, Zhuo R-X, Huang S-W (2015) Folate-containing reduction-sensitive lipid–polymer hybrid nanoparticles for targeted delivery of doxorubicin. Biomater Sci 3:655–664. https://doi.org/10.1039/C4BM00462K

Xia Y, Zhong J, Zhao M, Tang Y, Han N, Hua L, Xu T, Wang C, Zhu B (2019) Galactose-modified selenium nanoparticles for targeted delivery of doxorubicin to hepatocellular carcinoma. Drug Deliv 26:1–11. https://doi.org/10.1080/10717544.2018.1556359

Xiao D, Huang Y, Huang S, Zhuang J, Chen P, Wang Y, Zhang L (2022) Targeted delivery of cancer drug paclitaxel to chordomas tumor cells via an RNA nanoparticle harboring an EGFR aptamer. Colloids Surfaces B Biointerfaces. 212:112366. https://doi.org/10.1016/j.colsurfb.2022.112366

Xu Y, Ma S, Zhao J, Chen H, Si X, Huang Z, Yu Z, Song W, Tang Z, Chen X (2022) Mannan-decorated pathogen-like polymeric nanoparticles as nanovaccine carriers for eliciting superior anticancer immunity. Biomaterials 284:121489. https://doi.org/10.1016/j.biomaterials.2022.121489

Xu L, Faruqu FN, Lim YM, Lim KY, Liam-Or R, Walters AA, Lavender P, Fear D, Wells CM, Wang JT, Al-Jamal KT (2021) Exosome-mediated RNAi of PAK4 prolongs survival of pancreatic cancer mouse model after loco-regional treatment, Biomaterials. 264 (2021) 120369. https://doi.org/10.1016/j.biomaterials.2020.120369

Yoo J, Park C, Yi G, Lee D, Koo H (2019) Active Targeting Strategies Using Biological Ligands for Nanoparticle Drug Delivery Systems. Cancers (basel). 11:640. https://doi.org/10.3390/cancers11050640

Yu D-H, Lu Q, Xie J, Fang C, Chen H-Z (2010) Peptide-conjugated biodegradable nanoparticles as a carrier to target paclitaxel to tumor neovasculature. Biomaterials 31:2278–2292. https://doi.org/10.1016/j.biomaterials.2009.11.047

Yu MK, Park J, S. Jon (2012) Targeting Strategies for Multifunctional Nanoparticles in Cancer Imaging and Therapy, Theranostics. 2. https://doi.org/10.7150/thno.3463

Zhang C, Zhang F, Han M, Wang X, Du J, Zhang H, Li W (2020) Co-delivery of 5-fluorodeoxyuridine and doxorubicin via gold nanoparticle equipped with affibody-DNA hybrid strands for targeted synergistic chemotherapy of HER2 overexpressing breast cancer. Sci Rep 10:22015. https://doi.org/10.1038/s41598-020-79125-0

Zhao Z, Ukidve A, Kim J, Mitragotri S (2020) Targeting Strategies for Tissue-Specific Drug Delivery. Cell 181:151–167. https://doi.org/10.1016/j.cell.2020.02.001

Zhao B, Chen S, Hong Y, Jia L, Zhou Y, He X, Wang Y, Tian Z, Yang Z, Gao D (2022) Research Progress of Conjugated Nanomedicine for Cancer Treatment., Pharmaceutics. 14 (2022). https://doi.org/10.3390/pharmaceutics14071522

Zheng G, Zhao R, Xu A, Shen Z, Chen X, Shao J (2018) Co-delivery of sorafenib and siVEGF based on mesoporous silica nanoparticles for ASGPR mediated targeted HCC therapy. Eur J Pharm Sci 111:492–502. https://doi.org/10.1016/j.ejps.2017.10.036

Zuckerman JE, Gritli I, Tolcher A, Heidel JD, Lim D, Morgan R, Chmielowski B, Ribas A, Davis ME, Yen Y (2014) Correlating animal and human phase Ia/Ib clinical data with CALAA-01, a targeted, polymer-based nanoparticle containing siRNA. Proc Natl Acad Sci 111:11449–11454. https://doi.org/10.1073/pnas.1411393111

Chapter 16
Biogenic Nanomaterials as a Catalyst for Photocatalytic Dye Degradation

Parvindar M. Sah, Smita Ganpat Gite, Ratnamala Sonawane, and Rajesh W. Raut

Abstract The demand for dyes is increasing across a variety of industries, including food, rubber, leather, and plastics. Due to their low cost, stability, and performance, people prefer synthetic and semi-synthetic dyes over natural dyes. However, as these dyes are resistant to biological oxidation they are therefore harmful to the environment. More than 1 lakh dyes are available commercially, of which around 10–15% end up in the environment without being treated and it causes adverse effects on the environment. Nowadays these problems are tackled with the help of nanotechnology. The nanoparticles act as a catalyst for the degradation of dyes in the presence of sunlight. In the process of photocatalysis irradiation of light generates electron–hole pair that reacts with water molecules which produces hydroxyl radicals. If the dye reacts with hydroxyl radical it starts decomposing. At the end of the process, H_2O and CO_2 get released. the biogenic nanoparticle used in these process such as TiO_2, ZnO, Fe_2O_3, Ag NPs, AgO NPs, Au NPs etc. This chapter discusses photocatalytic dye degradation using biogenic nanomaterials.

Keywords Photocatalysts · nanomaterials · Advanced oxidation processes · Dyes · Nanomaterials · Nanocatalysts · Green nanocatalyst

---

P. M. Sah · R. W. Raut (✉)
Department of Botany, The Institute of Science, Dr. Homi Bhabha State University, 15 Madam Cama Road, Mumbai, Maharashtra 400032, India
e-mail: rajesh.w.raut@gmail.com

S. G. Gite · R. Sonawane
Department of Chemistry, The Institute of Science, Dr. Homi Bhabha State University, 15 Madam Cama Road, Mumbai, Maharashtra 400032, India

M. P. Shah et al. (eds.), *Biogenic Nanomaterials for Environmental Sustainability: Principles, Practices, and Opportunities*, Environmental Science and Engineering,
https://doi.org/10.1007/978-3-031-45956-6_16

16.1 Introduction of Biogenic Nanomaterials as a Catalyst for Photocatalytic Dye Degradation

In the textile industry, dyes are one of the largest groups of organic compounds that are widely used. Although natural and synthetic dye products have played an instrumental role in making our world spectacular, their abandoned discharge over the last decade has contributed to non-aesthetic pollution which has led to ecosystem destruction (Kumar 2021). There are many different types of dyes, including azo, basic, acidic, anthraquinone, and metal complexes. Among the most significant synthetic organic compounds released by industries such as plastic, paper, food, tanneries, pharmaceuticals, cosmetics, and textiles are dyes (Wu et al. 2008). Out of the industries mentioned, the textile industry produces the most wastewater containing high concentrations of dyes in the range of 10–200 mg L^{-1}. In this industry, 15–20% of consumer paint is estimated to enter the sewage system. Colour content in dye adsorbs at a wavelength of 350–700 nm (visible light region) and reflects sunlight entering polluted water, which hinders photosynthesis and interferes with aquatic species development (Chokalingam et al. 2019). In humans and animals, these organic pollutants can cause skin irritation, blood disorders, liver and kidney damage, mutagenic and carcinogenic effects, and poisoning of the central nervous system. The removal of such carcinogenic compounds from water is therefore critical from a biological and environmental perspective. The degradation of such compounds results in a less toxic product. This is because the compounds are highly soluble in water and other solvents with high stability. Conventional methods for removing toxic dyes include activated carbon sorption, flocculation, ultrafiltration, and chemical, photochemical, electrochemical, and biological methods. Chemically stable dyes and dye pollutants often resist degradation by physicochemical processes. Despite being cost-effective, biodegradation methods are slow and are not adequate to degrade dyes because they are harmful to microorganisms (Anjana et al. 2021). These methods, while perhaps effective, may also generate new compounds that need further treatment.

In recent years, Nanocatalysts has become an alternative to conventional wastewater treatment methods for efficient degradation and adsorption of toxic dyes.

Nano biotechnology is the intersection of nanotechnology and biotechnology, which focuses on building, improving, and using nanoscale structures for advanced biotechnology (Bandala et al. 2020). The application of nanotechnology in remediation involves the use of reactive nanomaterials for the transformation and detoxification of chemicals through chemical reduction or catalysis. Metal nanoparticles have finite sizes and large surface-to-volume ratios, making them excellent catalysts. Dye degradation by chemical reduction is thermodynamically favorable, but kinetically unfavorable. By reducing the activating energy at the nanoscale, gold, silver and copper provide an alternative path for the reaction, thus reducing the kinetic barrier, making the reaction thermodynamically and kinetically feasible (Kumar 2021).

In the past, nanomaterials were synthesized using a variety of physical and chemical techniques. Traditionally, nanoparticles are produced using capping agents, toxic solvents, harsh chemicals, and other additives, which limit their use in biomedical and clinical applications. In most cases, these methods require a significant amount of energy and capital. Researchers focused on developing an eco-friendly protocol for chemical synthesis so that solvents and toxic reagents are avoided, waste by-products are avoided, and renewable, affordable, biodegradable resources are used, which is a need of the industry (Bandala et al. 2020). By synthesizing at physiological temperatures and pH levels at relatively low costs, bio-based approaches eliminate harsh processing conditions. There is no hazardous waste generated by these biological methods, and the products usually do not need to be purified. A biological method reduces metal ions by using natural reducing agents found in prokaryotes and plant extracts (Ahluwalia et al. 2016). In comparison to conventional production, green synthesis improves the steric stability of nanoparticles by preventing aggregation. It also eliminates the problem of using sodium borohydride as a reduction agent because it is corrosive and flammable (Kulkarni and Bhanage 2014).

Recently, bacterial species such as *Pseudomonas aeruginosa,Rhizopus oryzae,Zoogloearamigera,* and many others have been explored for their ability to produce metallic nanoparticles (Srivastava 2019). Silver, selenium, titanium dioxide, and gold (metal) nanoparticles (AuNPs) have been synthesized using different bacterial strains, including *Bacillus amyloliquefaciens, B. clausii,* and *Azoarcus species* (Nadaf and Kanase 2019). In comparison with other microorganisms, fungi are considered to be a better resource for producing AuNPs due to their high yields, short synthesis time, and high ion concentration tolerance (Dhillon et al. 2012).

AuNPs synthesized by *Trichoderma viridae* were found to be effective catalysts for the complete reduction of 4-nitrophenol in water. Meanwhile, (Narayanan and Park 2015) found that the AuNPs intracellularly synthesized by *Flammulinavelutipes* could be used as a heterogeneous catalyst for reducing organic pollutants, such as methylene blue and 4-nitrophenol. Silver nanoparticles (AgNPs) have been prepared from silver salt by infusing potato (*Solanum tuberosum*) tubers that are readily available in common markets and cheaper than other vegetables. Glucose, fructose, polyphenols, tannin, Gallic acid, amino acids, citric acid, and alkaloids are active chemical constituents that form stable metal nanoparticles (Edison et al. 2016). Nanoparticles synthesized with plant extracts are more stable over a prolonged period. They do not require stabilizing agents. This is due to the presence of integral components that serve as both capping and stabilizing agents. According to (Bogireddy et al. 2016), AgNPs were employed as a catalyst to reduce/degrade phenol red, methyl orange, methylene blue, 4-nitrophenol, and dark blue. Another study reported the green synthesis of AuNPs using polysaccharides, including *Acacia nilotica*leaf extract, xanthan gum, and gellan gum. Green-synthesized AuNPs containing *Salmaliamalabarica*gum were shown to be efficient catalysts for the reduction of Methylene blue (MB) and Crystal red (CR) dyes (Ganapuram et al. 2015). The biological synthesis of nanoparticles provides a feasible and non-toxic method for removing aromatic pollutants from the environment. Hence these synthesis processes can be used to replace both chemical and other physical methods used in industry for the large-scale production of nanoparticles (NPs) that are used for dye reduction or degradation.

Today, developed and developing countries share problems related to pollution and scarcity of energy due to climate change, rapid population growth, and new 21st-century problems such as COVID-19. The second problem is that global energy demand is expected to increase by 35% (pessimistic variant) or 45–60% (optimistic variant) by 2030. According to the International Energy Agency (IEA), in 2000, 205 million barrels of oil were consumed (Tvaronavičienė et al. 2020); however, this number increased in 2030 to 334 million barrels. This is a troublesome worldwide issue with a need for renewable and environmentally sound solutions. The two major issues facing our planet are pollution (air pollution, and water pollution) and an increase in the cost of energy supply. Without a solution for these critical problems, people's well-being will be seriously compromised. In addition, if the world faces an increasing shortage of energy resources in the future, it will be hard to meet demand with the current methods. Compared with all these issues, pollution is one of the largest problems which the WHO estimates kills 9 million people every year compared with war violence statistics of 50%. What are the solutions to all? Has any groundbreaking treatments and methods been known? One possible solution is using a photocatalytic technique. Tahir et al. (2021) discussed the need for a renewable energy resource that can meet our needs while being ecologically safe, low-cost, and long-lasting. In the energy sector, photocatalytic techniques can meet all of our needs by providing us with a clean hydrogen production solution. And on top of that, it's environmentally safe and free from side effects.

16.2 Dye Generated by Various Industries

Dye is a synthetic aromatic compound that is used to develop color and sometimes to change the color of substances. A wide variety of industrial sectors use dyes for various purposes, including coloring fibers and cloths, biological and biochemical stains, food and cosmetics, color photography, electronics and lasers, solar cells, display panels, pigments in modern paint, printer inks, and leather products, among others. The majority of dyes are composed of organic compounds with covalently attached unsaturated or saturated functional groups, such as chromophore and auxochrome groups. They are responsible for absorbing light in the visible zone, approximately (350–750 nm), where they only show colour.

The chromophore group is an electron acceptor which is responsible for the colorization of the dye while the auxosome represents the electron donor which has the capacity of enhancing the color of the dye. Dye is considered to be soluble in solvents occurring naturally and synthetic in nature. All the color compounds are not considered to be a dye. Dye is applied for various purposes such as it is used in art supplies, beverages, glass wax, construction, biomedicine, soap, and plastic products. Safranin and crystal violet are the most commonly used dye for the microbiological experiment to identify the two bacterial strains Gram-positive and Gram-negative bacteria. A dye has excellent shine power, suitable color, and a good resistant ability to the action of light, water, soap, detergents, sweat, and other chemical substances during the washing or dry-cleaning process (Kishor et al. 2019).

Classification of dyes on the basis of their physical properties and application is given in Table 16.1.

Table 16.1 Types of dye

Class	Principal substrates	Method of application	References
Acid dye	Nylon, wool, silk, paper, inks and leather	Usually from neutral to acidic dye baths	Kishor et al. (2019)
Basic dye	Paper, polyacrylonitrile, modified nylon, polyester, and inks	Applied from acidic dye baths	Kishor et al. (2019)
Direct dye	Cotton, rayon, paper, leather, and nylon	Applied from neutral or slightly alkaline baths containing additional electrolyte	
Disperse dye	Polyester, polyamide, acetate, acrylic, and plastics	Fine aqueous dispersions often applied by high temperature/ pressure or lower emperature carrier methods; dye may be padded on cloth and baked on or Thermo fixed	
Reactive	Cotton, wool, silk, and nylon	Reactive site on dye reacts with a functional group on fibre	
Vat dyes	Cotton, rayon and Wool	Water-insoluble dyes solubilized by reducing with sodium hydrogen sulfite,	
Azoic components and composition	Cotton, rayon, cellulose acetate and polyester	Fibre impregnated with coupling component and treated with a solution of stabilized diazonium salt	
Fluorescent brighteners	Soaps and detergents and all fibres, oils, paints and plastics	From solution, dispersion or suspension in a mass	
Mordent	Wool, leather and anodized aluminium	Applied in conjunction with Cr salts	
Oxidation bases	Hair, fur and cotton	Aromatic amines and phenols oxidized on the substrate	
Sulphur	Cotton and rayon	Aromatic substrate vatted with sodium sulphide and reoxidized to insoluble sulphur-containing products on fibre	

16.3 Drawbacks

Synthetic dyes are not all harmful, for instance, fiber-reactive dyes bind covalently to fibres such as cotton, rayon, and soy. There is no need to use harmful mordants on them. The toxicity of a dye is determined by its structure rather than its dyeing process. There is a great deal of impact caused by dyes containing heavy metals and dyes that cause cancer. Low-impact dyes are not those that cause allergic reactions. Dyes that have passed eco-standards such as a blue sign, Global Organic Textile Standard (GOTS), OekoTex 100, and Cradle to Cradle have been thoroughly tested for toxicity and should be chosen overdyes those that have not. As a result of oral ingestion and inhalation, acute toxicity is mainly associated with skin irritation and sensitization. This is mainly triggered by reactive dyes for cotton and viscose, some acid dyes for polyamide fibers, and dispersible dyes for polyester, polyamide, and acetate rayon.

16.4 Biogenic Versus Conventional Nanomaterials

In Table 16.2, the major advantages and drawbacks of traditional and biogenic nanomaterials are outlined. A major concern in recent years has been the unknown toxic effects of conventional nanomaterials after they are released into the environment. These effects have been demonstrated to negatively affect plants, soil biodiversity, or even consumers of agricultural products that are exposed to conventional nanomaterials. A biogenic nanomaterial's large surface/volume ratio provides many reactive sites and requires lower NM loads. This reduces the need for water treatment and minimizes waste generation, and they cause less pollution. Green, eco-friendly, and biogenic syntheses have precedence over traditional methods that often use hazardous and organic materials and solutions that can cause cytotoxicity, carcinogenicity, and environmental toxicity. Biological processes are easy to scale up, low cost, and environmentally friendly for synthesizing nanoparticles. Agricultural crop waste and food industry waste are inexpensive sources of active ingredients for NM synthesis. However, the potentially toxic nature of BNMs after release into the environment is less understood. Several reports have found that even naturally occurring nanomaterials—generally considered part of the soil, and therefore not toxic—may have toxic properties that have not been fully assessed. Several studies have documented BNMs' toxic evolution (Wypij et al. 2020), but none have used photocatalytic materials in other applications.

It has been suggested that biogenic-derived capping agents may stabilize BNM behavior in a physiological environment, but they also found that the mechanisms underlying that stabilization and the role of biomolecules and amino acid moieties in BNMs remain unclear. The lack of information on the toxicity of BNMs with photocatalytic activity is also a significant knowledge gap that requires attention when BNM use is considered for environmental applications such as water treatment. In

Table 16.2 Major advantages and drawbacks of conventional and biogenic compound

Materials type	Major advantage	Major drawbacks
Conventional NMs	Smaller and simpler in chemical composition Reliable sizes, shapes, surface characteristics, and photocatalytic ability Standardized synthetic procedure Known production costs and yields Replicable production conditions Full-scale application feasible	It is unknown what will happen to this substance after it has been released into the environment Highly reactive and unstable Toxicity is known and concerning unknown effects on human health and ecosystems life cycle is relatively known
Biogenic NMs	Synthesis process that is environmentally friendly Avoid the use of toxic/ organic solvents waste feedstock can be used as raw material BNMs stabilized by capping agents Recyclable biogenic source	After releasing into the environment, unknown fate/toxicity larger and chemically more complex Variable synthetic procedure Unknown production mechanisms High variability of size, shapes, surface characteristics, photocatalytic ability, and capping agents life cycle unknown

addition to the fact that the nature of molecules covering biogenic materials remains unclear, the wide variety of biogenic sources that may be used to produce BNMs highlights the need for further investigation.

As compared to conventional NMs, biogenic NMs are generally larger and chemically more complex due to the composition of their stabilizing agents.There are a large number of chemically different stabilizers attached to BNM surfaces, ranging from proteins and cellulose to small organic molecules (e.g., flavones). The production costs and ecological gains of BNMs make them far superior to synthetic NMs once stabilized and using one polymer or other chemicals.

Nevertheless, Navia-Mendoza et al. (2021) have found that a better, fairer comparison needs to be developed between the performance of conventional and biogenic NMs to determine the best option and the Considering one material over another based on certain conditions.In particular, very few studies have reported using BNMs with photocatalytic capabilities to degrade organic contaminants or inactivate pathogens in water. However, there is no mention in the available studies of how the BNMs compare with their conventional counterparts or whether it is possible to compare their performance with other studies that used conventional NMs a very interesting direction for further research because of the lack of information about a systematic, fair comparison of the performance of conventional and biogenic NMs as photocatalysts for water treatment.

The variety of particle sizes, shapes, surface characteristics, and photocatalytic abilities of conventional nanomaterials in the different studies currently available provide interesting opportunities for comparison with the variety of particle sizes, shapes, capping agents, and biogenic origins of BNMs and their feasibility for use in water treatment. Once the basic comparison is completed, other highly interesting options will emerge, such as comparing the production costs, environmental footprint, reuse potential, environmental pathways, toxicity, and side effects that may lead to a real assessment of the advantages and disadvantages of each material.

The European Union's Scientific Committee on Emerging and Newly Identified Health Risks (SCENIHR) defines nanomaterials as objects whose extension lies between 1 and 100 nm in all three dimensions. Material extending in just two dimensions on the nm scale is termed nanotube or particulate objects, whereas material showing just one dimension under 100 nm is termed Nano pellets. Materials within the nanometer range have been known and produced for many decades. These materials are classified according to production methods into synthetic nanomaterials obtained through physical or chemical synthesis and biogenic nanomaterials (BNMs) obtained using biological synthetic approaches. The natures of BNMs can be very different from biomolecules, metals, and metal oxide micelles. Therefore, synthesizing and characterizing these BNMs can be very challenging and it is not uncommon to obtain different physical–chemical data on the same BNM. Herein, biogenic nanomaterials in the classical three-dimensional definition are discussed and their production using biogenic synthetic processes is considered (Navia-Mendoza et al. 2021).

16.5 The General Process for the Synthesis of Biogenic Nanoparticles

16.6 Gold Nanoparticles

Figure 16.1 Demonstrates the synthesis of various biogenic metal nanoparticles. These days, numerous research works have been done in connection with gold nanoparticles and have attracted increasing interest in their unique properties. It is evident that their morphology-dependent optical, chemical and electronic properties offer promising applications in many areas, including catalysis. In organic chemistry, photocatalysis involves the use of solar light to drive physicochemical reactions. Gold nanoparticles can be biosynthesized using a minimal amount of resources and are regarded as a more environmentally friendly and cost-effective alternative to chemical synthesis. Recently, extracellular biosynthesis methods employing various plant extracts, including *Cinnamomum zeylanicum* leaf, *Zingiber officinale* root, Aloe vera,*Plukenetiavolubilis* oil, Rambutan peel, sugar beet pulp, edible mushroom, and *Lantana camara* berry have emerged as non-hazardous alternatives to chemical synthesis procedures for the bulk production of nanoparticles. *A. nigra* leaves synthesis gold nanoparticle showed very efficient photocatalytic degradation of methyl orange (83.25%) and Rhodamine B (87.64%), respectively in the presence of direct sunlight. In another study by Desai et al. (2018), green synthesis of photo luminescent gold nanoparticles using Kokum fruit *(Garcinia indica Choisy)* extract with enhanced photocatalytic properties of gold nanoparticles for the degradation of methylene violet dye in the presence of visible light (89.17%,) was slightly faster than UV light (86.02%,). By using an environmentally friendly aqueous extract of *Angelica gigasstem*, (Kumar et al. 2018) fabricated gold nanoparticles that had significant applications in environmental remediation. In this study, the synthesized gold nanoparticles exhibited photocatalytic degradation activity on eosin Y (83%) and malachite green (65%) dyes under UV irradiation (Vinay et al. 2020).

16.7 Silver Nanoparticles

One of the most exciting aspects of modern nanoscience and nanotechnology is the fabrication of silver nanoparticles from scratch. By exposing silver nanoparticles to light under the UV–visible region, they exhibit remarkable photocatalytic activity. Patil and Sastry (1997) reported green synthesized spheres of silver nanoparticles using onion *(Allium cepa)* extract that varies in particle size from 50 to 100 nm have an excellent The degradation of dyes such as methylene blue, methyl red, eosin yellow, and safranin has been achieved by this catalyst.

The biosynthesized silver nanoparticles reported by Vanaja et al. (2014) were spherical, showed SPR at 414 nm, and also exhibited photocatalytic degradation of eosin Y (67%) and malachite green (64%), respectively. The rapid microwave-assisted fabrication of silver nanoparticles using leaf extract (aq) of *Mussaenda glabrata* was studied by Abbasi et al. (2016). A spherical shape of silver nanoparticles with an average size of 51.32 nm was found in the SPR peak at 415 nm, indicating that rhodamine degradation was lower at this wavelength. ($k = 0.4464\ min^{-1}$) and methyl orange ($k = 0.7910\ m^{-1}$) with respect to the synthesized gold nanoparticles.

16.8 Iron and Iron Oxide Nanoparticles

It has been reported that synthesized Fe_3O_4NPs with an average size of 54.5 ± 24.6 nm exhibited photocatalytic degradation activity of organic dyes such as MB ($k = 0.0105475\ min^{-1}$), CR ($k = 0.0043240\ min^{-1}$), and Methyl orange (MO) ($k = 0.0028930\ min^{-1}$), efficiently. Desalegn et al. (2019) synthesized iron nanoparticles (FeNPs) using green, Oolong, and black tea extracts (caffeine/polyphenols) as reducing and capping agents, instead of using environmentally toxic sodium borohydride. They concluded that the most effective method of degrading malachite green dye was FeNPs synthesized with green tea extracts. Recently, Fe_3O_4 nanoparticles were prepared from *Cynara cardunculus* leaf extract, presenting a novel, environmentally benign, and eco-friendly synthetic route for producing iron oxide nanoparticles (Ruíz-Baltazar et al. 2019). They synthesized Fe_3O_4 NPs that showed a catalytic degradation activity against methylene blue dye. Another study by Tahir et al. (2020) successfully synthesized FeNPs via an eco-friendly method using aqueous leaf extract of *Plantago majo*r and evaluated their ability to remove dyes (Wu et al. 2008).

16.9 Mechanism for Degradation of Organic Dyes Using Nanoparticles

A photocatalytic material creates electron–hole pairs when exposed to solar radiation. The photocatalytic activity of phytochemicals-modified metal nanoparticles has been studied extensively because of its potential use in sterilization, sanitation, and remediation under exposure to solar/visible/ultraviolet. Furthermore, it can also be used as a cleaning agent, an anti-soiling agent, an antifungal, an antibacterial, and an antiviral agent. It is also used for air purification and deodorization, as well as wastewater treatment. It has been extensively studied how gold, silver, and iron/iron oxide nanoparticles can be used to remediate organic dyes and treat wastewater (Kumar 2021). The mechanism of catalysis in the presence of $NaBH_4$ and metal nanoparticles (MNPs)/Au/Ag/FeNPs can be described as an electron transfer process from donor $NaBH_4$ to an acceptor dye. MNPs acts as an electron relay and trigger the movement of the electron from the BH_4 - ion (donor B_2H_4/BH_4^-) to the organic dye (acceptor) and thus cause a reduction of the dye. The BH_4 − ion was simultaneously adsorbed on the surface of NPs, and thus, electron transfer occurs from the BH_4^- ion to the organic dye through NPs. Solar energy is an interesting aspect of photocatalytic technologies. Solar photocatalysis has become an active area of research in which sunlight is the source of illumination to perform various photocatalytic reactions to different kinds of dyes. In the literature, research on photocatalysis reveals that photocatalytic activity can be strongly dependent on the crystallographic structure, morphology, and particle size (Ramkumar et al. 2017).

The photocatalytic mechanism of nanomaterials for dye degradation is schematically illustrated in Fig. 16.2. The process of photocatalysis is also explained by indirect photocatalysis, which is thermodynamically feasible but less common. In this approach, the excitons formed do not react directly with the dye molecules. Instead, they react with the exciton-trapping molecules adsorbed on the catalyst surface to form free radicals, which then interact with the dye molecules. During the photocatalytic process, the absorption of photons by the Nanocatalysts leads to the excitation of electrons from the valence band (VB) to the conduction band (CB), generating electron (e−)/hole (h+) pairs. The holes can react with the H_2O molecule and transform to hydroxyl radical (·OH) as shown in Fig. 16.2. Although direct photocatalysis is thermodynamically feasible, indirect photocatalysis is another way to explain photocatalysis. As a result of this approach, the excitons formed do not react directly with the dye molecules. As a result, the dye molecules react with free radicals that are

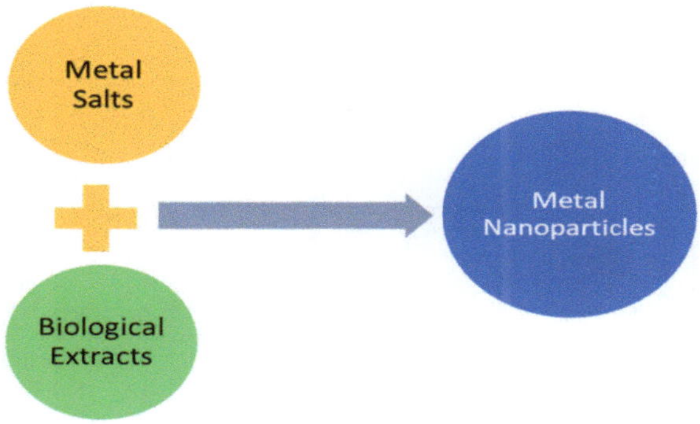

Fig. 16.1 Biogenic synthesis of nanomaterials

formed by their reaction with the exciton-trapping molecules adsorbed on the surface of the catalyst. During the photocatalytic process, the absorption of photons by the Nanocatalysts leads to the excitations of electrons from the valence band (VB) to the conduction band (CB), generating electron (e^-)/hole (h^+) pairs. The holes can react with the H_2O molecule and transform to hydroxyl radical (·OH) and hydronium ions (H^+), as depicted in Eq. (16.1); the holes may also react with the surface-adsorbed hydroxide ions to form hydroxyl radicals as in Eq. (16.2). The excited electrons react with oxygen molecules and hydronium ions to generate hydrogen peroxide, which can further dissociate in the presence of oxygen to form hydroxyl ions, as depicted in Eqs. (16.3)–(16.5). After transforming the active species, the organic dye could be effectively reacted with H^+. The electrons in the conduction band are captured by oxygen molecules dissolved in the suspension and in the valence band captured by OH^- or H_2O species absorbed This produces hydroxyl radicals (·OH) on the catalyst surface. Bhattacharjee and Ahmaruzzaman (2015) report that these hydroxyl radicals also oxidize chemicals to produce small inorganic molecules (Bhattacharjee and Ahmaruzzaman 2015). Additionally, MNPs may cause the colorless final product to desorb due to their large surface area.

The photocatalytic degradation process may be represented by the following reaction:

$$MNPs + h \rightarrow h^{++}e^{-}(photogeneration of exciton\tilde{s}) \quad (1)$$

$$H_2O + h^+ \rightarrow \cdot H^- + H^+ \quad (2)$$

$$OH^- + h^+ \rightarrow \cdot OH \quad (3)$$

$$e^- + O_2 \rightarrow \cdot O_2 \quad (4)$$

$$O_2^- + H^+ \rightarrow \cdot OOH/H_2O_2 \quad (5)$$

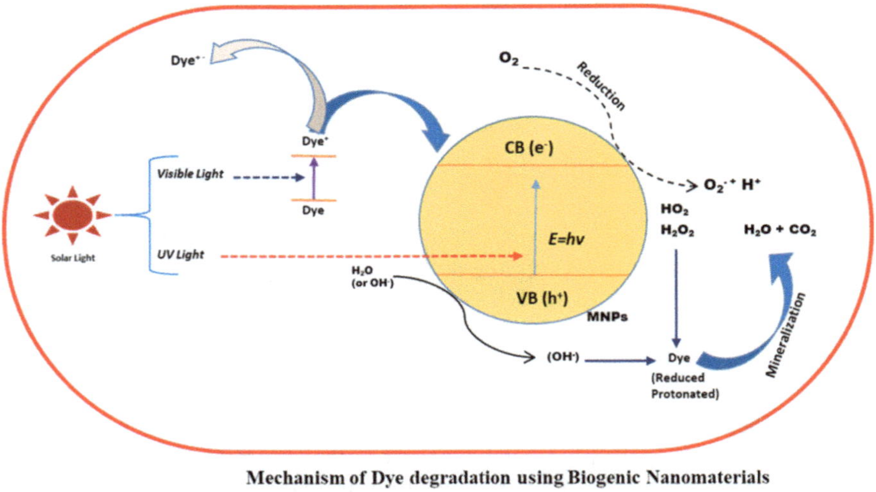

Fig. 16.2 Mechanism of dye degradation using biogenic nanomaterials

$$\text{Dye} + \text{h}\nu \rightarrow \text{Dye}* \tag{6}$$

$$\text{Dye}* + O_2^- \rightarrow \text{(unstable intermediate)} \rightarrow \text{degradation products} \left(CO_2^+ H_2O\right) \tag{7}$$

16.10 Photocatalytic Application Biogenic Nanomaterials: Degradation of Organic Contaminant

As the world experiences shortages of clean water, it becomes even more critical to find efficient ways to treat wastewater. Groundwater, surface water, and drinking water all contain organic contaminants. The environment, animals, and humans are all affected by these contaminants. Organic contaminants must be removed from wastewater with low-cost, high-efficiency treatments. Current water treatment technologies are able to adsorb or concentrate pollutants from one phase to another but are not able to destroy or eliminate pollutants completely. Filtration, sedimentation, and chemical oxidation are other methods of treating water, but they lack the capability to completely remove pollutants, are costly, and take a long time to complete. It is necessary to develop new methods in order to improve the quality of water. One such method is photocatalysis, which can be used for solar water splitting and purifying pollutants in the air and water. Fujishima et al. (2007) studied heterogeneous photocatalysis in 1972 and discovered the photochemical splitting of water in the presence of a titanium dioxide (TiO_2) catalyst.

Photolysis uses chemical compounds and generates radicals that accomplish photochemical reactions due to the absorption of radiation energy that elevates them to an excited state. The radiation is solar energy from mercury lamps. When water splits, hydroxyl compounds are generated, which further react with pollutants. Metallic salts are used as catalysts to accelerate the process that generates the oxidation method.

16.11 Organic Effluent

Since it cannot decompose in aqueous solutions, its presence in sewage, seawater, and sludge presents significant environmental and health risks. A large portion of global warming can be attributed to volatile organic compounds (VOCs). In addition to being carcinogenic and toxic, VOCs are also mutagenic, which means they are resistant to antibiotics. Organic matter (NOM) creates a more harmful effect since it reacts with chloramines, which are toxic and carcinogenic to water.

16.12 Rhodamine B: Triphenylmethane

Rhodamine B dyes are the most common type of Rhodamine B used in textiles, photographic processing, printing, and cosmetics. It remains in the environment for a long time because it has resistance to bacterial attacks. Rhodamine B is a common phosphorescent dye that consists of four N-ethyl groups flanked by xanthene rings (Tahir et al. 2020). It is worth pointing out that relatively few studies have been done to explore the photocatalytic activity of BNMs, perhaps because photocatalysis involves the surface of the materials, which may pose a significant impediment for BNMs containing capping agents. As a result of interaction with radiation of the appropriate wavelength, charge carriers are formed in the photocatalyst, which triggers photocatalytic processes. In the BNM synthesis process, layers of proteins, amino acids, or other chemical structures may reduce their potential for photocatalytic applications because they participate in redox processes or control particle size.

Despite this, no study was found in which the authors included a control that performed the synthesis without the plant extract to demonstrate the importance of the biogenic material. It is fundamental to demonstrate the significance of the synthetic process that proper quality control/assurance is included in the experimental design. Biogenic synthesis has also been reported using alternative methods. Gopinath et al. (2017) investigated the use of electrochemically active biofilms to narrow the band gap energy in Degussa P25 TiO_2. Shkrob et al. (2004) found that adding the Ti_4^+ salt solution to proteins, lipids, polyphenols, or carbohydrates shaped the Ti_4^+ ions into specific patterns that eventually led to the formation of anatase-TiO_2 crystals. These authors, however, failed to provide any experimental evidence or reference to back up their claim in their paper. According to another study (Gautam et al. 2019) Ag nanoparticles were deposited into ZnO through a deposition/precipitation procedure involving fennel seed extract. An important research avenue worth exploring is the systematic phytochemical study of plant extracts used for biogenetic nanomaterial synthesis.

To significantly reduce the perception of empiricism caused by the lack of proper characterization in several of the currently available studies, it is necessary to identify biogenic sources that are successfully used to synthesize NMs and determine the extract partition and identify the compounds that are major players in the synthetic process.

A majority of the organic compounds tested were dyed, with a few other pollutants (e.g. 4-nitrophenol and chlorpyrifos) also included. The lack of information about the efficacy of BNMs for photocatalytic degradation of organic contaminants is another significant knowledge gap, even though it is clear that dyes are particularly easy to follow during discoloration processes because a UV–Vis spectrophotometer is required for analysis. Many organic contaminants other than dyes have been identified as public health concerns due to their biological activity, persistence, or bioaccumulation. The recent study by Gautam et al. (2019) suggesting the use of BNMs for removing drug compounds from water is a good example of this. In sewage water, they used biogenic manganese oxides and bio-palladium (bio-Pd) to remove recalcitrant pharmaceuticals, and in contrast mediums, they used bio-Pd via catalytic reduction. Photocatalytic degradation was not reported for any of the BNMs, but they were used as adsorbents, sequestrants, oxidants/reducers, catalysts, or electro catalysts. Research avenues that need to be explored include testing BNMs' ability to produce photocatalytic processes using either UV or visible radiation to remove antibiotics, pharmaceuticals, pesticides, and poly-fluoroalkyl substances (PFAS), as well as identifying reaction mechanisms, by-products, reaction kinetics, or mineralization (Rodriguez-Narvaez et al. 2017).

16.13 Photocatalytic Application Biogenic Nanomaterials: Inactivation of Microorganism

Antimicrobial agents have been developed in the past using some non-biogenic nanomaterials. For wound dressings, silver nanoparticles are considered potent antibacterial agents, while iron nanoparticles have been reported to be great antimicrobial agents. A particle's antibacterial activity is determined by its size and shape, but mostly by its ability to produce ROS, which is responsible for the oxidation of cellular components in all cases. BNMs have been widely studied as antibacterial, antifungal, and even anticancer agents (Chatzimitakos and Stalikas 2016). It is noticeable that AgNPs or their derivatives are the BNM group that is most commonly tested for antibacterial activity against a wide variety of Gram-positive and Gram-negative species.

Other BNMs, such as copper, titanium, zinc, iron, and manganese, have also been reported to produce oxidative stress, metal ion release, and non-oxidative stress, which are the most common mechanisms for antibacterial activity (Singh et al. 2018).

16.14 Types of Nanomaterials Used as Catalysts for Photocatalytic Dye Degradation

16.14.1 Catalysts in AOPs for Removal of Dye from the Food and Chemicals Industry

In recent years the use of catalysts has been increased for the degradation of organic and inorganic pollutants released in the environment. Many catalysts are used for dye degradation, including metals, metal oxides, semiconductors, metal–organic frameworks (MOF), even quantum dots, and dendrimers with magnetic core (Garrido-Cardenas et al. 2020).

For their synthesis, a variety of strategies have been developed, including precipitation, flame hydrolysis, microwave radiation, impregnation, sol–gel, hydrothermal, solvothermal, chemical vapour deposition, electrodeposition, and electrospinning (Azimi 2013). As a semiconductor material, titanium dioxide (TiO_2) is widely used for photocatalysis. Its abundance, chemical stability, good optical transparency, high refractive index, non-toxic nature, and low cost make it the catalyst of choice for decomposing dyes into mineralized products (Opinion 2010). By adding TiO_2 under UV irradiation, the photodegradation process of Brilliant Blue was analysed.

Nevertheless, more recent studies have shown that TiO_2 can be harmful to human health, leading researchers to explore alternatives to TiO_2 in AOPs. Furthermore, the simplest compounds were mineralized within a shorter period of time with relatively high doses. Other semiconductors, such as ZnO, CdS, WO_3, and Fe_2O_3, have also been used in studies, but to a lesser extent since they are heavy metals. By thermal decomposition of the zirconium hydroxide-urea complex, King et al. (2018) developed and characterized an N-doped ZrO_2photocatalyst (N-ZrO_2). In a study using amaranth dye under visible or UV light irradiation, this photocatalyst showed an improved catalytic activity (decolourization of 67.2%) than pure ZrO_2. In addition, UVA light at 3.5 Wm2 was demonstrated to be a highly effective alternative to eliminate the amaranth dye completely without generating toxic compounds. Doping with transition metals has been used to extend the adsorption band to the visible region, as they exhibit excellent photocatalytic activity (Castro-León et al. 2020). Meanwhile, photocatalytic processes are increasingly being combined with electrochemical advanced oxidation processes (EAOPs) to improve degradation performance through reactive species formation. The disadvantage of electro-oxidation, however, is that in certain cases, recalcitrant by-products could develop. As a result of the release of metal particles or microparticles and the production of by-products that are more toxic than the primary pollutant, this process can cause poisoning effects. Since the objective is not to achieve complete mineralization, it is necessary to evaluate the toxicity of the treated effluent. In order to improve the efficiency of advanced oxidative processes, it is necessary to use new economical and environmentally-friendly methods, such as green catalysts, or nanocatalyst, such as nanoparticles, nanotubes, Nano spheres, and nanofibers.

16.15 Nanocatalysts

In recent decades, nanometer-sized materials (<100 nm) have been considered promising options, as they can adopt different shapes, sizes, and structures, and improve the catalytic activity, selectivity, and stability of the system (Aoudjit et al. 2018). Unlike catalysts, nanocatalyst, due to their nano size, have a larger surface area, allowing for better surface–volume interaction and, therefore, greater diffusion of reagents and products.

It is sometimes necessary to spread these nanoparticles on support, couple them with an adsorbent, or dope them with metals or adsorbents with high porosities in order to increase the efficiency of the process. As far as support materials go, zeolites, activated carbon, silicon films, and hybrid materials are the most commonly usedIn order to ensure the stability of the nanoparticles, their synthesis must be carried out correctly, since, when they bind ineffectively to the support, they are easily detached after the first use and washing (Aoudjit et al. 2018).

In addition to low sludge production, they are also able to be reused through regeneration methods integrated into the system. In aqueous solutions, there are a variety of nanocatalyst available, including electro catalysts, Fenton-based catalysts, and antimicrobial catalysts. As an example, Thakur et al. (2017) synthesized gelatin-Zr (IV) phosphate nanocomposites (GT/ZPNC) to degrade Fast Green dye, achieving an 89.91% degradation rate over 5 h; even antimicrobial activity against E. coli was demonstrated (Korkmaz et al. 2004).

Sol–gel auto-combustion was used to synthesize cadmium vanadate nanostructures with a 67% photocatalytic degradation of erythrosine. As a result, Li et al. (2020) produced, for the first time, Gd2ZnMnO6/ZnO nanocomposites (GZMO/ZnO NC) by using a sol–gel auto combustion technique using grape or coffee syrup as green fuel, resulting in a product with a forbidden band of around 3.27 eV that was effective at removing dyes. According to this study, the photocatalytic activity of the catalyst was superior to that of the control group by 73.8% towards anionic dye degradation (erythrosine).

In contrast to spherical nanostructures, nanocatalyst with complex morphologies, such as cubes, prisms, stars, or rods, exhibit enhanced adsorption due to their multipolar resonances. Ion exchange catalyzed photo-Fenton reaction in the aqueous phase for the degradation of tartrazine by Palas et al. (2017). During tests, both nanotubes effectively degrade the food dye, even tartrazine mineralization is achieved with nanotubes containing 0.70 wt.%. Darwish et al. (2016) synthesized PVP-coated cadmium sulphide (CdS) nanoparticles using polyol with ethylene glycol, a facile method that decreased photo corrosion. As a result of rapid electron–hole pair's generation by photoexcitation, CdS exhibits better photocatalytic functions than TiO_2. A 21.92 eV optical energy band gap enabled it to degrade the food dye Allura Red AC by 96.99% in six minutes, resulting in excellent photocatalytic efficiency.

16.16 Metal Oxide-Based Nanocatalysts

A wide range of hazardous pollutants can be removed from wastewater using oxide-based nanocatalyst, which are inorganic catalysts that are generally prepared with metals or non-metals. High Brunauer–Emmett–Teller (BET) surface areas, low solubility, minimal environmental impact, and absence of secondary pollutants distinguish them. There are three generations of metal oxide nanocatalyst: the first generation consists of a single component, such as TiO2 and ZnO; the second generation contains multiple components suspended in suspension, such as WO_3/NiWO4, BiOI/ $ZnTiO_2$. Finally, the third generation consists of photocatalysts immobilized on solid substrates (FTO/WO_3 -ZnO, Steel/TiO_2 -WO_3). Nanocatalysts have three natural states, anatase, rutile, and brookite, with anatase being the best nano-photocatalyst for degrading anionic and cationic dyes, and dyes with different chromophore groups (Navia-Mendoza et al. 2021). A To improve the surface area and structural porosity of TiO_2 anatase, and to achieve higher photocatalytic performance without using support material. This research obtained excellent results, exceeding the efficiency of other TiO_2 catalysts, in 15 min, degrading 95.6% of Amaranth dye (30 ppm) using 0.3 g of the catalyst under UV irradiation. Furthermore, this nanocatalyst showed good stability for 5 cycle. Lamba et al. (2017) synthesized and characterized heterostructures of Ag_2O-decorated ZnO nanowires using the ultrasound-assisted precipitation method. Through photocatalytic reaction processes, photogenerated electron/hole pairs actively participate in the chemical reactions for the generation of superoxide radicals and oxidizing agents, which are responsible for the degradation of the dye. The photocatalytic activity was measured by the degradation of Amaranth dye under visible light irradiation, exhibiting a photocatalytic yield of about 94% in 30 min. These heterostructures showed a synergistic effect, which was more efficient than pure ZnO nanowires and Ag_2ONPs separately, even presenting higher activity under visible light irradiation than commercially available photocatalysts, such as Merck ZnO, TiO_2-P25, and TiO_2-PC-50 (Aoudjitet al. 2018).

on the other hand, evaluated the feasibility of using PVDF-TrFE photocatalytic membranes with immobilized TiO_2 nanoparticles as a catalyst to degrade tartrazine in a solar photo reactor. A high removal performance of the pollutant (78%) was obtained, but a second use under the same experimental conditions showed a reduction of approximately 10% in degradation efficiency. A major reason for the loss of efficiency is that not all nanoparticles are effectively bound to the polymeric matrix. They tend to detach after the first use and subsequent washing, and they are also affected by a large amount of dye retained on the membrane surface. Both of these factors become the major limitations of immobilized systems because specific photocatalytic active sites decrease, resulting in an overall reduction in efficiency. Alcantara-Cobos et al. (2020) prepared zeolite–ZnO nanoparticles (Ze–nano ZnO) for a coupled adsorption–photocatalysis process. The coupling of pure ZnO nanoparticles and nanoZnO nanoparticles produced lower degradation percentages; nevertheless, nannanoZnO is more expensive because it is more difficult to separate, making the process more expensive. As a final step, Türkylmaz et al. (2017) synthesized and analysedZnO nanostructures doped with Ni, Mn, Fe, and Ag, resulting in a tartrazine degradation mechanism. Maximum Metal–organic frameworks (MOFs) are defined as compounds formed by the interaction of metal ions or metal clusters (secondary building units, SBUs) and organic ligands ideal (usually carboxylic acid or nitrogen-containing ligands) for the fabrication of high-performance multifunctional compounds. They are characterized by very good surface areas and high porosity (porous crystalline materials), as well as by dispersed active sites and functionalities within this porosity. Due to their ability to act as chemical Nano reactors, they can synthesize and stabilize a catalytically active species that would otherwise be difficult to access. The band gap and the light energy differ in photodegradation processes. Using MOFs in this case, the electrons in the valence band gain energy when the light source emits more energy than the forbidden band's energy, and are excited to the conduction band, creating a positive hole in the valence band, which in turn generates electrons and holes undergoing reduction and oxidation half-reactions. The photocatalytic degradation mechanism, where MOFs considered potential photocatalysis degradation of 98.2% was achieved with Ni/ZnO in 60 min. As a result of multivalent ions coexisting in the ZnO host and their highly agglomerated structures, Mn and Fe did not give good results.

16.17 Metals Organic Frameworks Synthesized with Nanocomposite

MOFs (metal–organic frameworks) are compounds fabricated by the interaction of metal ions or metal clusters (secondary building units, SBUs) with organic ligands (usually carboxylic acids or nitrogen-containing ligands) for the fabrication of high performance multifunctional compounds. They have very good surface areas and high porosity (porous crystalline materials), as well as dispersed active sites and functionalities within the porosity. By acting as chemical nanoreactors, they are capable of synthesizing and stabilizing catalytically active species that would otherwise be difficult to obtain (Alcantara-Cobos et al. 2020). Photodegradation processes involve a difference between the energy of light and the band gap. As a result, MOFs work by generating positive holes in the valence band when the light source produces more energy than the forbidden band's energy. The electrons gain energy in the valence band and are excited into the conduction band, thus generating reduction and oxidation half-reactions involving electrons and holes, respectively. As a potential photocatalyst, MOFs behaving as semiconductors could be used in this degradation mechanism. For example, iron-based MOF materials, such as MIL-88A, MIL-53, and MIL-68, are presented as potential photocatalytic tools for dye degradation. Zhang et al. (2018) developed for the first time a Fe nanoparticle-containing MOF, MIL-101 as a support to immobilize ferrocene (Fc) chemically. There have been few studies examining the use of Fe-based catalysts to activate Oxone in amaranth degradation (Zhang et al. 2018) demonstrated that Oxone catalyzed by MIL-101-NH2 can completely degrade amaranth dye within 30 min of visible light irradiation in Fe nanoparticle-based MOF.

However, the micro porosity of the MOFs prevents the complete adsorption of larger dye molecules due to the difficulty of entering the internal pores or voids of the MOFs. In addition, there would be a selectivity of charge and/or size, since MOF-based nanocatalyst would be designed for a specific type of dye, which is unfavorable for the treatment of wastewater with complex dye mixtures with this method (Balu et al. 2020).

16.18 Nanomaterials Materials Synthesized by Green Technologies

The nanomaterials synthesized by using chemical and physical methods are considered to be efficient systems but not all the nanomaterials are economical and environmentally friendly, because the nanomaterials synthesized by conventional routes are hazardous and volatile which lead to the formation of secondary pollutants. Therefore it is very important to find out new and efficient methods, which have high catalytic activity, are easy to handle, and are most eco-friendly in the environment. The development and application of green chemistry concepts are the most recent alternative studies of nanomaterials synthesized by green technologies or biogenic nanoparticles (BNPs). Nowadays biogenic nanomaterials are considered to be a new tool for photocatalytic water remediation. With their large surface area, ion exchange mechanism, unique morphology, high biocatalytic reactivity, efficient regeneration, and colloidal properties, these biogenic nanomaterials can be separated from effluents by gravity filtration, sedimentation, or coagulation-flocculation. Several studies have demonstrated the efficacy of these processes in removing various food contaminants. Deepika et al. (2017) identified bioactive compounds in the bark and leaves of *Ailanthus excelsa*Roxb to remove synthetic food dyes. The researchers observed that *A. excelsa* bark extract exhibits better photocatalytic activity by degrading 55% of sunset yellow dye FCF. In the same way, Hassanien et al. (2019) synthesized selenium BNPs from drumstick leaf extracts and evaluated their photocatalytic activity against sunset yellow FCF (for coloring food) under solar and UV illumination. A dye degradation rate of 83.8 and 76.6% was achieved in the two reactions. (Navia-Mendoza et al. 2021b) proposed the synthesis of silver BNPs (AgNP) by using

Viburnum opulus L. fruit extract, which acts as a reducing agent of silver ions and stabilizer of the obtained nanoparticles.

16.19 Advanced Method for Dye Degradation Using Biogenic Nanoparticle

The latest type of nanoparticle such as CuO nanoparticles are synthesized from Ficus Carica fruit extract are able degradation of toxic hazardous Alizarian Yellow R dye. For synthesized of nanoparticle phytochemical present in fruit extract are responsible. These biologically synthesized nanoparticle almost 89.71% dye degradation.

In most of type of nanoparticles are synthesized from single metal ions and used in dye degradation but recently bimetallic metal oxide nanoparticle such as $ZnOSnO_2$ are used for photocatalytic degradation of dye. The bimetallic nanoparticle synthesized from plant *Sutherlandia frutescens (S. frutescens)* shows degradation of Methylene Blue and it also capable for degradation of antibiotic sulfisoxazole and sulfamethoxazole. The highest 88% efficiency of methylene blue degradation was achieved in bimetallic nanoparticle.

Other than for the full-scale application of BNMs. The study suggests using living cells to produce compounds of interest avoids extraction and purification steps, ensures structural integrity, and allows continuous cell culture and downstream processing. However, controlling BNMs properties remains a major challenge because while some biomolecules such as glycolipids and proteins have been shown to control the size and shape of the nanomaterials, a better understanding of the biochemical pathways and biomolecules involved remains a pending research problem. Scalability is also heavily influenced by the nature of the biogenic process. According to Ramkumar et al. (2017), extracellular biogenic processes have the greatest potential for scaling up as compared to intracellular biogenic processes, since the biological entity used to produce BNMs is usually recyclable. Separating the produced BNMs will be easier than with intracellular biogenic processes, which prevents disrupting/killing the biological entity by tearing it apart. In order to determine if a biogenic process is suited to scale up to industrial production, these authors identified the following basic conditions: (i) the availability and abundance of raw materials; (ii) well-defined process variables that can be automatically controlled; (iii) raw materials, products, and by-products that are environmentally friendly; and (iv) processes that are economically feasible.

The nanotechnology field has experienced a significant lack of straightforward and reproducible synthetic protocols to prepare large amounts of nanomaterials for industrial applications. Due to slow production rates, as well as a lack of specific biological strains that are required for their synthesis, the full-scale application of BNMs may be at risk, as well as a lack of control and ability to fine tune the reaction kinetics and continue flow synthesis, which may lead to consistent, reproducible, and scalable production (Trojanowicz 2020). They can identify this lack of information as another highly significant knowledge gap that is worth exploring.

16.20 Conclusion and Prospects

The use of biogenic nanomaterials promote the photocatalytic degradation of organic contaminants and inactivates pathogenic microorganisms in water. The BNMs have higher potential environmental applications for the treatment of wastewater. A detailed analysis of BNMs for photocatalytic applications was conducted in this chapter, as well as a detailed comparison of the advantages and disadvantages of BNMs with conventional nanomaterials. As well as analyzing the main challenges facing BNMs in full-scale production and application, a topic for which very little research has been conducted.

In most cases, bottom-up approaches were used for BNMs synthesis, while top-down approaches were used for conventional NMs. A large number of studies reported different advantages and drawbacks for extracellular and intracellular processes in the bottom-up approach, which should be kept in mind for those interested in getting involved in BNMs preparation and testing. As compared with extracellular BNMs synthesis, intracellular BNMs synthesis is the most suitable method for producing small, uniform nanoparticles. However, in both cases, significant challenges were observed along with a need for additional information on the biological and biochemical mechanisms involved. Using this information.They can increase production yields and homogenize the material quality at the end of the process.

BNMs were mostly characterized using the same conventional characterization techniques reported for conventional NMs. The varied nature of many BNMs as well as the widely accepted capping layer generation on their surface necessitated additional specific characterization approaches. BNMs have been evaluated mainly for the photocatalytic degradation of dyes in synthetic water rather than for the photocatalytic degradation of organic contaminants.

In addition to a lack of control over the synthesis processes used to generate the BNMs tested as photocatalysts, some apparent flaws in preparation methods might result in artifacts and overstatement of the photocatalytic capabilities of the tested materials. A need for accurate quality assurance/quality control (QA/QC) has been identified for many studies already reported in the literature in order to generate comparable results that allow fair comparisons among BNMs and their performance against conventional NMs. BNMs have been studied extensively for their antimicrobial activity, but using them to inactivate microorganism's photo catalytically has more limitations than their capacity for photocatalytic degradation or their presence in the water. There was a lack of understanding of the mechanisms of either antimicrobial or photocatalytic activity, which was related to the role played by biogenic materials and/or the types of components in BNMs.

Lastly, the challenges of engineering BNMs for engineered use were revised.It is found that, at the current state of the art, there are still a lot of significant challenges that need to be addressed, including achieving uniformity in the material characteristics, improving material characterization, and enhancing BNMs purification processes, as well as standardizing synthetic processes, production costs, and yields. BNMs production scaling up and final quality standardization has been considered and reported in very few studies, which limits their ability to be applied on a large scale.

References

Abbasi E, Milani M, Aval SF et al (2016) Silver nanoparticles: Synthesis methods, bio-applications and properties. Crit Rev Microbiol 42:173–180. https://doi.org/10.3109/1040841X.2014.912200

Al-Dhabi NA, Mohammed Ghilan AK, Arasu MV (2018) Characterization of silver nanomaterials derived from marine Streptomyces sp. Al-Dhabi-87 and its in vitro application against multidrug resistant and extended-spectrum beta-lactamase clinical pathogens. Nanomaterials 8:1–13. https://doi.org/10.3390/nano8050279

Alcantara-Cobos A, Gutiérrez-Segura E, Solache-Ríos M et al (2020) Tartrazine removal by ZnO nanoparticles and a zeolite-ZnO nanoparticles composite and the phytotoxicity of ZnO nanoparticles. Microporous Mesoporous Mater 302. https://doi.org/10.1016/j.micromeso.2020.110212

Anjana VN, Joseph M, Francis S et al (2021) Microwave assisted green synthesis of silver nanoparticles for optical, catalytic, biological and electrochemical applications. Artif Cells, Nanomedicine Biotechnol 49:438–449. https://doi.org/10.1080/21691401.2021.1925678

Aoudjit L, Martins PM, Madjene F et al (2018) Photocatalytic reusable membranes for the effective degradation of tartrazine with a solar photoreactor. J Hazard Mater 344:408–416. https://doi.org/10.1016/j.jhazmat.2017.10.053

Azimi S (2013) Sol-Gel Synthesis and Structural Characterization of Nano-Thiamine Hydrochloride Structure. ISRN Nanotechnol 2013:1–4. https://doi.org/10.1155/2013/815071

Balu S, Chen YL, Yang TCK et al (2020) Effect of ultrasound-induced hydroxylation and exfoliation on P90–TiO2/g-C3N4 hybrids with enhanced optoelectronic properties for visible-light photocatalysis and electrochemical sensing. Ceram Int 46:18002–18018. https://doi.org/10.1016/j.ceramint.2020.04.115

Bandala ER, Stanisic D, Tasic L (2020) Biogenic nanomaterials for photocatalytic degradation and water disinfection: A review. Environ Sci Water Res Technol 6:3195–3213. https://doi.org/10.1039/d0ew00705f

Castro-León G, Baquero-Quinteros E, Loor BG et al (2020) Waste to catalyst: Synthesis of catalysts from sewage sludge of the mining, steel, and petroleum industries. Sustain 12:1–16. https://doi.org/10.3390/su12239849

Chatzimitakos TG, Stalikas CD (2016) Qualitative Alterations of Bacterial Metabolome after Exposure to Metal Nanoparticles with Bactericidal Properties: A Comprehensive Workflow Based on 1H NMR, UHPLC-HRMS, and Metabolic Databases. J Proteome Res 15:3322–3330. https://doi.org/10.1021/acs.jproteome.6b00489

Chokkalingam M, Jahan Rupa E, Huo Y et al (2019) Photocatalytic degradation of industrial dyes using Ag and Au nanoparticles synthesized from Angelica gigas ribbed stem extracts. Optik (stuttg) 185:1213–1219. https://doi.org/10.1016/j.ijleo.2019.04.065

Darwish M, Mohammadi A, Assi N (2016) Microwave-assisted polyol synthesis and characterization of pvp-capped cds nanoparticles for the photocatalytic degradation of tartrazine. Mater Res Bull 74:387–396. https://doi.org/10.1016/j.materresbull.2015.11.002

Deepika S, Harishkumar R, Dinesh M et al (2017) Photocatalytic degradation of synthetic food dye, sunset yellow FCF (FD&C yellow no. 6) by Ailanthus excelsaRoxb. possessing antioxidant and cytotoxic activity. J PhotochemPhotobiol B Biol 177:44–55. https://doi.org/10.1016/j.jphotobiol.2017.10.015

Desai MP, Sangaokar GM, Pawar KD (2018) Kokum fruit mediated biogenic gold nanoparticles with photoluminescent, photocatalytic and antioxidant activities. Process Biochem 70:188–197. https://doi.org/10.1016/j.procbio.2018.03.027

Desalegn B, Megharaj M, Chen Z, Naidu R (2019) Green synthesis of zero valent iron nanoparticle using mango peel extract and surface characterization using XPS and GC-MS. Heliyon 5:e01750. https://doi.org/10.1016/j.heliyon.2019.e01750

Dhillon GS, Brar SK, Kaur S, Verma M (2012) Green approach for nanoparticle biosynthesis by fungi: Current trends and applications. Crit Rev Biotechnol 32:49–73. https://doi.org/10.3109/07388551.2010.550568

de Ruíz-Baltazar Á, J, Reyes-López SY, Mondragón-Sánchez M de L, et al (2019) Eco-friendly synthesis of Fe3O4 nanoparticles: Evaluation of their catalytic activity in methylene blue degradation by kinetic adsorption models. Results Phys 12:989–995. https://doi.org/10.1016/j.rinp.2018.12.037

Espenti CS, Rao KSVK, Rao KM (2016) Bio-synthesis and characterization of silver nanoparticles using Terminalia chebula leaf extract and evaluation of its antimicrobial potential. Mater Lett 174:129–133. https://doi.org/10.1016/j.matlet.2016.03.106

Fujishima A, Zhang X, Tryk DA (2007) Heterogeneous photocatalysis: From water photolysis to applications in environmental cleanup. Int J Hydrogen Energy 32:2664–2672. https://doi.org/10.1016/j.ijhydene.2006.09.009

Ganapuram BR, Alle M, Dadigala R et al (2015) Catalytic reduction of methylene blue and Congo red dyes using green synthesized gold nanoparticles capped by salmaliamalabarica gum. Int Nano Lett 5:215–222. https://doi.org/10.1007/s40089-015-0158-3

Gautam PK, Singh A, Misra K et al (2019) Synthesis and applications of biogenic nanomaterials in drinking and wastewater treatment. J Environ Manage 231:734–748. https://doi.org/10.1016/j.jenvman.2018.10.104

Gopinath V, Priyadarshini S, Loke MF et al (2017) Biogenic synthesis, characterization of antibacterial silver nanoparticles and its cell cytotoxicity. Arab J Chem 10:1107–1117. https://doi.org/10.1016/j.arabjc.2015.11.011

Hassanien R, Abed-Elmageed AAI, Husein DZ (2019) Eco-Friendly Approach to Synthesize Selenium Nanoparticles: Photocatalytic Degradation of Sunset Yellow Azo Dye and Anticancer Activity. ChemistrySelect 4:9018–9026. https://doi.org/10.1002/slct.201901267

Joshi R, Satyal P, Setzer W (2016) Himalayan Aromatic Medicinal Plants: A Review of their Ethnopharmacology, Volatile Phytochemistry, and Biological Activities. Medicines 3:6. https://doi.org/10.3390/medicines3010006

Karkmaz M, Puzenat E, Guillard C, Herrmann JM (2004) Photocatalytic degradation of the alimentary azo dye amaranth: Mineralization of the azo group to nitrogen. Appl Catal B Environ 51:183–194. https://doi.org/10.1016/j.apcatb.2004.02.009

King G, Soliz JR, Gordon WO (2018) Local Structure of Zr(OH)4 and the Effect of Calcination Temperature from X-ray Pair Distribution Function Analysis. Inorg Chem 57:2797–2803. https://doi.org/10.1021/acs.inorgchem.7b03137

Koul B, Poonia AK, Yadav D, Jin JO (2021) Microbe-mediated biosynthesis of nanoparticles: Applications and future prospects. Biomolecules 11. https://doi.org/10.3390/biom11060886

Kulkarni A, a, Bhanage BM, (2014) Anand A. Kulkarni and Bhalchandra M. Bhanage *. ACS Sustain Chem Eng 2:1007–1013

Kumar B (2021) Green synthesis of gold, silver, and iron nanoparticles for the degradation of organic pollutants in wastewater. J Compos Sci 5. https://doi.org/10.3390/jcs5080219

Lamba R, Umar A, Mehta SK, Kansal SK (2017) Enhanced visible light driven photocatalytic application of Ag2O decorated ZnO nanorods heterostructures. Sep Purif Technol 183:341–349. https://doi.org/10.1016/j.seppur.2017.03.070
Li J, Diao X, Xiao Y et al (2020) Effect of surfactant SDS on the morphology and photocatalytic performance of Zn2GeO4nanorods. Mater Res Express 7:85005. https://doi.org/10.1088/2053-1591/abaac7
Majeed A, Ullah W, Anwar AW et al (2018) Cost-effective biosynthesis of silver nanoparticles using different organs of plants and their antimicrobial applications: A review. Mater Technol 33:313–320. https://doi.org/10.1080/10667857.2015.1108065
Nadaf NY, Kanase SS (2019) Biosynthesis of gold nanoparticles by Bacillus marisflavi and its potential in catalytic dye degradation. Arab J Chem 12:4806–4814. https://doi.org/10.1016/j.arabjc.2016.09.020
Narayanan KB, Park HH (2015) Homogeneous catalytic activity of gold nanoparticles synthesized using turnip (Brassica rapa L.) leaf extract in the reductive degradation of cationic azo dye. Korean J Chem Eng 32:1273–1277. https://doi.org/10.1007/s11814-014-0321-y
Navia-Mendoza JM, Filho OAE, Zambrano-Intriago LA et al (2021) Advances in the application of nanocatalysts in photocatalytic processes for the treatment of food dyes: A review. Sustain 13. https://doi.org/10.3390/su132111676
Opinion S (2010) Scientific Opinion on the re-evaluation of Brown HT (E 155) as a food additive. EFSA J 8:1–31. https://doi.org/10.2903/j.efsa.2010.1536
Palas B, Ersöz G, Atalay S (2017) Photo Fenton-like oxidation of Tartrazine under visible and UV light irradiation in the presence of LaCuO3 perovskite catalyst. Process Saf Environ Prot 111:270–282. https://doi.org/10.1016/j.psep.2017.07.022
Patil V, Sastry M (1997) Electrostatically controlled diffusion of carboxylic acid derivatized Q-state CdS nanoparticles in thermally evaporated fatty amine films. J Chem Soc - Faraday Trans 93:4347–4353. https://doi.org/10.1039/a704355d
Ramkumar R, Balasubramani G, Raja RK et al (2017) Lantana camara Linn root extract-mediated gold nanoparticles and their in vitro antioxidant and cytotoxic potentials. Artif Cells, Nanomedicine Biotechnol 45:748–757. https://doi.org/10.1080/21691401.2016.1276923
Rodriguez-Narvaez OM, Peralta-Hernandez JM, Goonetilleke A, Bandala ER (2017) Treatment technologies for emerging contaminants in water: A review. Chem Eng J 323:361–380. https://doi.org/10.1016/j.cej.2017.04.106
Shkrob IA, Sauer MC, Gozstola D (2004) Efficient, rapid one-electron photooxidation of chemisorbed polyhydroxyl alcohols and carbohydrates by TiO 2 nanoparticles in an aqueous solution. J Phys Chem B 108:12512–12517
Singh P, Garg A, Pandit S et al (2018) Antimicrobial Effects of Biogenic Nanoparticles. Nanomaterials 8:1–19. https://doi.org/10.3390/nano8121009
Singh PK, Singh RL (2017) Bio-removal of Azo Dyes: A Review. Int J Appl Sci Biotechnol 5:108–126. https://doi.org/10.3126/ijasbt.v5i2.16881
Srivastava SK (2019) Microbial Membrane-Supported Catalysts: A Paradigm Shift in Clean Energy and Greener Production. ACS Sustain Chem Eng 7:19321–19331. https://doi.org/10.1021/acssuschemeng.9b05084
Tahir MB, Rafique M, Rafique MS, et al (2020) Photocatalytic nanomaterials for degradation of organic pollutants and heavy metals. Elsevier Inc.
Tahir MB, Sohaib M, Sagir M, Rafique M (2021) Role of Nanotechnology in Photocatalysis. Encycl Smart Mater 578–589. https://doi.org/10.1016/B978-0-12-815732-9.00006-1
Thakur M, Sharma G, Ahamad T et al (2017) Efficient photocatalytic degradation of toxic dyes from aqueous environment using gelatin-Zr(IV) phosphate nanocomposite and its antimicrobial activity. Colloids Surfaces B Biointerfaces 157:456–463. https://doi.org/10.1016/j.colsurfb.2017.06.018
Trojanowicz M (2020) Flow chemistry in contemporary chemical sciences: A real variety of its applications. Molecules 25. https://doi.org/10.3390/molecules25061434

Vanaja M, Paulkumar K, Baburaja M et al (2014) Degradation of methylene blue using biologically synthesized silver nanoparticles. Bioinorg Chem Appl 2014. https://doi.org/10.1155/2014/742346

Vinay SP, Udayabhanu NG et al (2020) Hydrothermal synthesis of gold nanoparticles using spider cobweb as novel biomaterial: Application to photocatalytic. Chem Phys Lett 748:137402. https://doi.org/10.1016/j.cplett.2020.137402

Wu W, He Q, Jiang C (2008) Magnetic iron oxide nanoparticles: Synthesis and surface functionalization strategies. Nanoscale Res Lett 3:397–415. https://doi.org/10.1007/s11671-008-9174-9

Wypij M, Jedrzejewski T, Ostrowski M et al (2020) Biogenic silver nanoparticles: Assessment of their cytotoxicity, genotoxicity and study of capping proteins. Molecules 25:1–19. https://doi.org/10.3390/molecules25133022

Chapter 17
Biogenic Metal Based Nanomaterials as Antimicrobial Agents

Samuel S. Mgiba, Vimbai Mhuka, Nomso C. Hintsho-Mbita, and Nomvano Mketo

Abstract Recently, there has been a significant interest on biogenic nanomaterials as part of nanotechnology which is a new branch of science that has a potential to solve many problems in medical sciences, drug delivery, electronics, agriculture, etc. This can be attributed to simple, eco-friendly and low-cost synthetic routes of these biogenic nanomaterials amongst other reasons. Greener nanomaterial synthesis includes the use biogenic materials such as extracts from plants to produce biogenic nanomaterials that have been employed for different used including as an antimicrobial agent in biomedical application. These biogenic nanomaterials include copper, silver, gold and titanium. Properties such as surface and size of the biogenic nanomaterials makes them very dependable for antimicrobial activities. This chapter focuses on the current trends on synthetic routes of different biogenic nanomaterials and their antimicrobial activity. The properties of the nanomaterials in relation to their antimicrobial activity are also correlated on this chapter.

Keywords Antimicrobial agents · Biogenic nanomaterials · Green synthesis

S. S. Mgiba · V. Mhuka · N. Mketo (✉)
Department of Chemistry, College of Science, Engineering and Technology (CSET), University of South Africa, Florida Science Campus, Johannesburg 1709, South Africa
e-mail: nomvano.mketo@gmail.com; mketon@unisa.ac.za

S. S. Mgiba
Department of Maths, Science and Technology Education, Faculty of Humanities, University of Limpopo, Polokwane 0727, South Africa

N. C. Hintsho-Mbita
Department of Chemistry, Faculty of Science and Agriculture, University of Limpopo, Polokwane 0727, South Africa

M. P. Shah et al. (eds.), *Biogenic Nanomaterials for Environmental Sustainability: Principles, Practices, and Opportunities*, Environmental Science and Engineering,
https://doi.org/10.1007/978-3-031-45956-6_17

17.1 Introduction

Nanoscience is one important scientific research branch that has gained much attention over the past 20 years. This research branch is important in field such as computer industry, sensors, drug delivery, nanobots, medical application, etc. (Phanjom et al. 2012). In recent times, the global economy has been burdened by infectious diseases such human immunodeficiency virus (HIV), H_5N_1, chronic obstructive dieses (COPD) which may cause death to millions of people worldwide. Antibiotics are the first preference for the treatment of these kind of dieses because it inhibits the presence of microbes (Yu et al. 2009). Different types of nanomaterials have been applied previously for medical applications. Copper, Zinc, Silver, Gold, Magnesium, Titanium, etc. has been used as antimicrobial agents. One important property about nanoparticles in drug delivery as antimicrobial agent is the surface and volume ratio, their chemical and physical composition (Rai et al. 2009). There has been development of high- microbial infections and antibiotic resistance which is considered as the biggest threat to global heath by World Health Organisation (WHO) that can also lead medical conditions and/or even death (Davis et al. 2018). For example, pathogenic bacteria has gained resistance towards chemically synthesized antibiotics hence the need of novel antimicrobials agents such as biogenic nanomaterials which can be effective on a wide range of bacteria (Mahmud et al. 2022).

There are commonly used chemical synthetic routes for nanomaterials which includes chemical reduction (Zhang et al. 2021), electrochemical techniques (Li et al. 2013) and photochemical reaction (Sakamoto et al. 2009). Physical methods have been employed also for the synthesis of metallic nanomaterials, these includes melt-mixing, laser ablation, physical vapour deposition and sputtering (Singh et al. 2018). Toxic compounds are produced by some nanoparticles when certain chemical synthetic routes are followed. The toxic compounds can be adsorbed on the surface of the nanoparticles and can lead to unexpected medical complications on humans (Khan and Jamil Khan 2017). The chemical synthetic routes of some nanoparticles use high energy, pressure and temperature. They are costly and uses toxic compounds such as thiophenol (Khorrami et al. 2018).

Due to the aforementioned challenges, different synthetic routes have been developed to circumvent the challenges by decorating the conventional antimicrobial agents. The use of biogenic materials is one green synthetic routes currently followed in order to minimize the bottlenecks and to produce nanoparticles that are more stable with more control of the size and shapes (Ingle et al. 2014). The decoration of the surface of the nanoparticles with biological molecules makes them biocompatible when compared with chemically prepared nanoparticles (Sharma et al. 2019). Biogenic materials are natural products such as vitamins, amino acids, polyphenol, biopolymers and surfactants (Bandala et al. 2020). The production of biogenic nanomaterials makes them compatible with biomedical application since no toxic compounds are involved. Furthermore, biogenic synthesis of nanomaterials is deemed eco-friendly, cost effective and rapid (Suganthy et al. 2018). Synthesized biogenic nanomaterials tend to have high catalytic reactivity (Li et al. 2011) and biogenic materials can be used as reducing, capping, and stabilizing agents during the synthesis of nanomaterials (Shaheen and Ahmad 2020). Biogenic approach towards the synthesis is also preferred due to majority of enzymes that abide in ambient conditions of different pressure, pH and temperature (Li et al. 2011). The biogenic materials also produces a substantial amount of proteins which has an effect in metal ions reduction thus allowing the control of morphology of the nanomaterials produced (Patil et al. 2021).

17.2 Synthesis of Biogenic Nanomaterials

The synthesis of nanomaterials follows two classical approaches, i.e. the top-down approach and the bottom-approach (Gwo et al. 2016). Although the two methods follow different approaches, the final product of desired characteristics is obtained. The top-down approach involves the grinding of bulk materials to small pieces with the view of yielding fine nanoparticles in large quantity. In essence, nanomaterials are removed from bulk materials. Surface imperfection and damage of nanomaterials are some of the disadvantages this approach possesses (Thakkar et al. 2010). The bottom-up approach involves the assembling of atoms to generate nanomaterials. Sol–gel processing, self-assembly of polymer or monomer molecules and chemical vapour deposition are some of the processes that this approach follows. This approach does not produce homogeneous nanoparticles at most, thus broad particle size distribution with average size varying from one batch to another. This approach is mostly used for industrial purposes (Decarolis et al. 2020).

Biological synthesis of nanomaterials follows the bottom-up approach which uses unicellular to multicellular biological products such as algae, fungi and plant products. Fungi, bacteria, yeast, etc., are metal ion permissive and they able to prosper on environmental conditions. These enable them to convert metals into their respective metal ions (Li et al. 2011). This spectacle of reducing metals into their respective metal ions is through redox reaction. Figure 17.1, shows a general representation of biogenic nanomaterials synthesis. The process involves metal being brought together with biogenic materials and reduced to form their respective metal ions. (Khandel et al. 2018). There are three main elements involved in the production of biogenic nanomaterials, this are solvent, environmentally friendly reducing agent and also a stabilizing agent. During the synthesis, different parameters play an important role on the reaction. Incubation time, pH and temperature have an effect on the synthesized nanomaterials. For example, the average size of the nanomaterials are reduced at lower temperatures whilst at higher temperature the rate of synthesis is increased (Gericke and Pinches 2006).

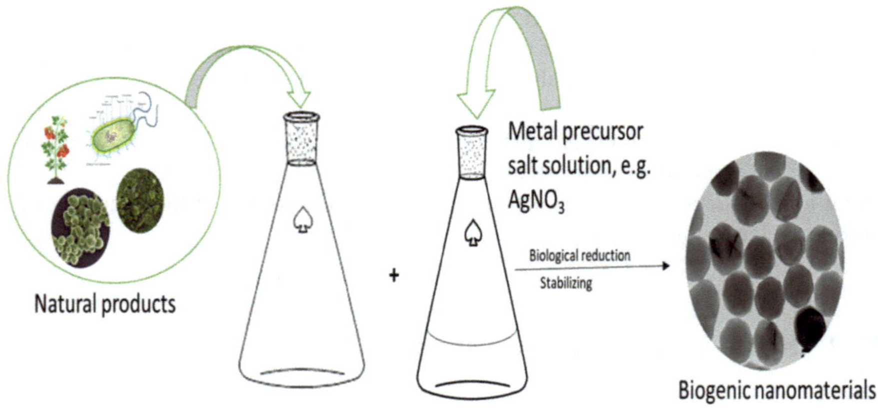

Fig. 17.1 General representation for biogenic nanomaterials synthesis

17.3 Biogenic Nanomaterials as Anti-microbial Agents and Their Morphology

Biogenic nanomaterials exhibit a range of biocidal activities against Gram-positive and Gram-negative bacterium and eukaryotes due to the presence of peptidoglycan layer which acts as barrier for the prevention of nanoparticles. Biogenic nanomaterials functions by releasing metal ions, interaction with cell membrane this resulting on the cell membrane being damaged and gaps being formed on the cell membrane leading to fragmentation of cell membrane (Rajkuberan et al. 2015). Nanoparticles such as silver nanoparticles ţend to interfere with thiol group of the bacterial cell causing inhibition of the respiratory chain reaction andcell division (Rajkuberan et al. 2017). Figure 17.2 shows some other antimicrobial activities of biogenic nanomaterials which includes enzyme deactivation, DNA damage etc. The production of reactive oxygen species (ROS) is caused by the bactericidal activity of nanoparticles which leads to cell wall damage and the collapse of internal proteins or DNA or RNA (Karakoti et al. 2006).

17.3.1 Silver Nanoparticles

Antibacterial activity of silver nanoparticles has been previously explored on multidrug delivery resistance bacteria. These includes, *Escherichia coli, Streptococcus pyogenes, Pseudomonas aeruginosa.* The silver nanoparticles also showed antimicrobial activity of antibiotics such as *penicillin G, amoxicillin, vancomycin, clindamycin and Escherichia coli* (Vu et al. 2018). Different synthetic routes have been followed for silver nanoparticles production. These includes heat decomposition in organic solvents, reduction of silver ions in a aqueous solution, photo reduction,

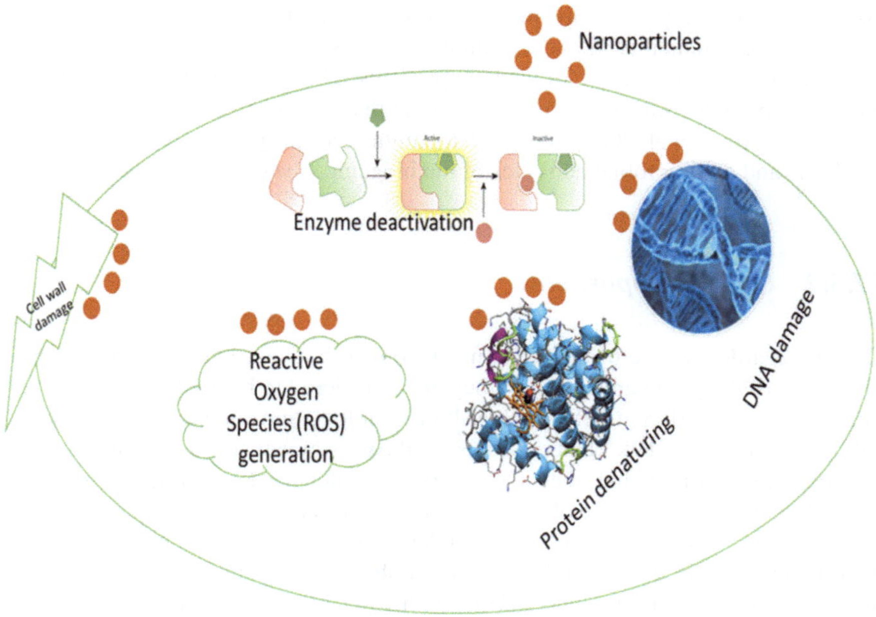

Fig. 17.2 General antimicrobial mechanism for biogenic nanomaterials

chemical reduction in reverse micelles (Pileni 2000). Most of these synthetic routes uses poisonous substances and some are costly, thus there's a need to develop an environmentally synthetic routes, i.e., biological approaches to nanoparticles formation. Biogenic nanomaterials are synthesized using different biological extracts which includes plants extracts, fungus, plant biomass, etc., due to rapid growth rate and increased protein expression, bacteria, plants extracts and fungi are the most used candidates used as biological products for biogenic nanomaterials (Khambhaty et al. 2009).

A study by Ahmed and co-workers focused on the preparation of biogenic nanomaterials by mixing *Skimmia laureola* which is a leaf extract together with silver nitrate at a ratio of 3:7. The prepared nanomaterials were confirmed using XRD and SEM analysis and then evaluated for antibacterial activity towards five bacterial strains i.e. *Escherichia coli*, *Klebsiella. pneumoia*, *Staphylococcus aureus*, *Pseudomonas aeuruginose* and *Proteus vulgaris*. The leaf extract served as reducing and capping agent of silver nitrate nanoparticles. The antimicrobial activity of the nanoparticles shows maximum growth inhibition affect against *Staphylococcus Aureus* strain with low affect against the *Escherichia Coli* strain. The mechanism for the whole process of antimicrobial activity is yet to be studied (Jamil et al. 2015). Another green synthesis of silver nanoparticles involved the use of plant product in a form of root hair extract. *Escherichia coli* and *Candida Albicans* were the bacterial strain which were evaluated against the biogenic nanomaterial. It showed 100%

potency for the strain with only 40 μg/mL concentration of the silver nanoparticles. The study also finds that the zone inhibition against the strains increased with the increase on the concentration of the nanoparticles. The antimicrobial mechanism of these nanoparticles was found to be consistent that of Amro and co-workers (Oves, et al. 2017; Amro et al. 2000). Table 17.1, summarizes projects embarked on for the synthesis and antimicrobial activities of the biogenic silver nanoparticles.

17.3.2 Gold Nanoparticles

Gold is regarded as one of the rarest metals on earth. Due to distinct photoelectric properties of the gold nanoparticles, they are considered to be noble metal nanoparticles. It used in different fields such as biotechnology and chemical spheres (Balasubramanian et al. 2019). Plants extracts are the most widely used biogenic material as a capping and reducing agents for different nanoparticles synthesis due to bio capability and less labour involved for extraction (Li et al. 2016).

Plants leave extracts were used by Sanderam and co-workers to prepare the gold nanoparticles. This green synthetic method showed a good antibacterial activity against *Escherichia* and *Bacillus subtilis* with 100 μg/mL concentration. The bacterial strains were chosen based on their association on previously with food borne disease, urinary tract infection and other illness. The mechanism for the antibacterial activity is summarized on Table 17.2 (Sunderam et al. 2019).

Vegetable waste was also used for the synthesis of biogenic nanoparticles which includes fresh skin of vegetables, fresh and damaged leaves. *Klebsiella sp* and *Staphylococcus sp* are the two pathogenic bacteria that were assessed for antimicrobial activity using a well diffusion method. The antibacterial activity was evaluated at four different concentration of gold nanoparticles which showed an improved inhibition as the concentration of the nanoparticles increased (Mythili et al. 2018). This study paved a way for other researchers to explore biological waste for the synthesis of gold nanoparticles. Waste water melon was also used to synthesis the gold nanoparticles and evaluated against two bacterial strains, *Escherichia. coli* and *Staphylococcus epidermidis* (Chums-ard et al. 2019). *Ganoderma lucidum* mushroom was used for the synthesis of biogenic gold nanoparticles. *Bacillus subtilis, Staphylococcus.aureus, Bacillus cereus, Escherichia coli* and *Pseudomonas aeruginosa* are the bacterial strains that it was evaluated against through the agar disk diffusion method. The gold nanoparticles showed a multi-shaped structure including hexagonal, pentagonal, rhombus, spherical and Nano rods shape. The concentration ranges of the gold nanoparticles (20–160 μg/mL). The nanoparticles did show some antibacterial activity but not all strains that were tested. Currently no particular reason could be deduced in order to substantiate the phenomena (Nguyen et al. 2022). Table 17.2, give summary of other studies on biogenic gold nanoparticles.

Table 17.1 Summary of studies on biogenic silver nanoparticles

Biological source	Bacterial strains	Antimicrobial mechanism	Concentration (µg/mL)	Morphology of NP's	Size of NP's (nm)	Refs.
Basidiomycetous fungi	*Bacillus subtilis, Staphylococcus aureus, Micrococcus luteus, Bacillus cereus, Bacillus megaterium, Escherichia coli, Enterobacter aerogens, Klebsiella pneumoniae, Salmonella typhimurium, Proteus vulgaris, Pseudomonas aeruginosa* and *Salmonella paratyphi*	Attacks respiratory chain, cell division leading to cell death	113.12	Spherical to round	15–20	Gudikandula et al. (2017)
Leaves	. *Staphylococcus aureus, K. Klebsiella pneumonia, Pseudomonas aeruginosa* and *Escherichia. coli*	Silver nanoparticles penetrate the cell wall of bacteria, causing deterioration in plasma membrane, which leads to bacterial cell death	5–10	Spherical shape	40	Jamil et al. (2015)
Leaf extract	*Escherichia coli, Pseudomonas aeruginosa, Staphylococcus aureus* and *Bacillus subtilis*	Anchoring or penetrating the bacterial cell wall and subsequently modulating cellular signalling by dephosphorylating putative key peptide substrates on tyrosine residues	193.3	Spherical	10–33	Ghaedi et al. (2015)

(continued)

Table 17.1 (continued)

Biological source	Bacterial strains	Antimicrobial mechanism	Concentration (μg/mL)	Morphology of NP's	Size of NP's (nm)	Refs.
Plant extract,	*Staphylococcus aureus, Escherichia coli, Klebsiella Pneumonia* and *Bacillus subtilis*	-	40	Spherical	20- 50	Lakshmanan et al. (2018)
Aqueous rhizome	*Bacillus subtilis, Bacillus cereus* and *Staphylococcus aureus*	Inhibited the bacterial growth at the log phase which is otherwise the active phase when the bacterial cells show exponential growth	92.48 and 69.44	Spherical	31.83	Nakkala et al. (2014)
Green tea extract	*Staphylococcus aureus, Escherichia coli, Pseudomonas aeruginosa* and *Salmonella enterica*	–	7–60	Spherical	34.68 ± 4.95	Rolim, et al. (2018)
Fungus strain	*Candida albicans, Pseudomonas aeruginosa, Bacillus subtilis, Staphylococcus aureus* and *Escherichia coli*	Yet to be explained	50 and 200	Approximately spherical	4–55	Ma et al. (2017)

(continued)

Table 17.1 (continued)

Biological source	Bacterial strains	Antimicrobial mechanism	Concentration (μg/mL)	Morphology of NP's	Size of NP's (nm)	Refs.
Root hair extract	*Escherichia coli* and. *Candida albicans*	Nanoparticles could attach to the bacterial cell wall or releases free radicals that can inhibit the growth of bacteria. Nanoparticles damage the cell membranes of E. coli and C albicans; the formation of pits or grooves in the membranes of later organisms and destruction of whole cells in case of former microbe	40	Spherical shape	15–40	Oves et al. (2017)
Leaf extract	. *Staphylococcus aureus*	Nanoparticles releases lipopolysaccharides and membrane proteins, generation of free radicals responsible for the damage of membrane and dissipation of the proton motive force resulting in the collapse of the membrane potential	40	–	30–50	Abdel-Aziz et al. (2014)

(continued)

Table 17.1 (continued)

Biological source	Bacterial strains	Antimicrobial mechanism	Concentration (μg/mL)	Morphology of NP's	Size of NP's (nm)	Refs.
Macroalgae	*Staphylococcus aureus, Bacillus cereus, S typhimurium,Escherichia coli, L. Listeria monocytogenes, Pseudomonas aeruginosa* and *Klebsiella*	The suggested mechanism is that after nanopartcles penetration into the bacteria can inactivate their enzymes, generate hydrogen peroxide and cause bacterial cell death	–	Quasi-spheres	17.6	Salari et al. (2016)
Olive leaf extract	*Staphylococcus aureus (S. aureus), Pseudomonas aeruginosa (P. aeruginosa)* and *Escherichia coli (E. coli)*	Nanoparticles may attach to the surface of the bacterial cell membrane via interacting with sulfur containing proteins (Feng et al., 2000), disturbing permeability and respiration functions of the cell resulting in cell death	30–70	Mostly spherical	20–25	Khalil et al. (2014)

(continued)

Table 17.1 (continued)

Biological source	Bacterial strains	Antimicrobial mechanism	Concentration (μg/mL)	Morphology of NP's	Size of NP's (nm)	Refs.
Leaf extract	*Escherichia coli*	Due to the cytoplasmic membrane disorganization and the consequent leakage of various biomolecules such as amino acids, protein and carbohydrates	–	Spherical	70	Raja et al. (2017)
Ginger rhizomes extract	*Vibrio anguillarum*, *Vibrio alginolyticus*, *Aeromonas punctata*, *Vibrio parahaemolyticus*, *Vibrio splendidus* and *Vibrio harveyi*	AgNPs have been found to attach to the surface of the bacterial membrane, cause structural changes of the membrane, and finally lead to death of the microbe cells	0.8–50	Spherical	20–80	Yang et al. (2017)

(continued)

Table 17.1 (continued)

Biological source	Bacterial strains	Antimicrobial mechanism	Concentration (μg/mL)	Morphology of NP's	Size of NP's (nm)	Refs.
Marine algae	*Staphylococcus aureus and Proteus mirabilis*	When silver nanoparticles enter the bacterial cell, it forms a low molecular weight region in the centre of the bacteria to which the bacteria conglomerates thus protecting the DNA from the silver ions. The nanoparticles preferably attack the respiratory chain cell division finally leading to cell death	5–15	Almost spherical	5–25	Kathiraven et al. (2015)
Keratinase	*Escherichia coli*	Through the interaction of AgNPs with sulphur and phosphorus containing biomolecules in the bacterial cell, the particles enter into the cell, where cell-killing is initiated through the attack of the respiratory chain and cell division	150	Spherical	5–30	Lateef et al. (2015)

Table 17.2 Summary of biogenic gold nanoparticles

Biological source	Bacterial strain	Antimicrobial mechanism	Concentration of gold nanoparticles (µg/mL)	Size of nanoparticles (nm)	Morphology of nanoparticles	Refs.
Mushroom	*Pseudomonas aeruginosa and Escherichia coli, B. subtilis, S. aureus,* and *B. cereus*	-	20–160	15–40	Cubic structure	Nguyen et al. (2022)
Vegetable waste	*Klebsiella sp.* and *Staphylococcus sp.*	Growth inhibition of pathogens	25, 50, 75 and 100	10–70	Spherical and triangle shapes	Mythili et al. (2018)
Leaves extract	*Escherichia coli* and *Bacillus subtilis*	Damage membrane, cell wall, ribosomes and mitochondria. Further inhibits the thiol groups in bacterial cells	20 and 40	40	Spherical	Sunderam et al. (2019)
Seaweed	*Aeromonas. hydrophila*	Increased protein by the gold nanoparticles by charging the permeability of the membrane	100	16	Triangular, spherical in structure,	Vinosha et al. (2019)
Leaf extract	*Streptococcus pyogenes, Staphylococcus aureus, Escherichia coli and Klebsiella pneumonia)*	The inside the microbe could be due to interaction of thiols and phosphorus group of the DNA. The nano particles facilitates the production of ROS which lead to protein and DNA destruction	100	8–37	Spherical	Balasubramanian et al. (2019)
Root extract	*Bacillus subtilis, Staphylococcus aureus, and Escherichia coli*	No antibacterial activity due to low concentration	7.5–120	20 ± 3.2	Spherical	Doan (2020)

(continued)

Table 17.2 (continued)

Biological source	Bacterial strain	Antimicrobial mechanism	Concentration of gold nanoparticles (μg/mL)	Size of nanoparticles (nm)	Morphology of nanoparticles	Refs.
Bacterium	*Escherichia coli and Staphylococcus aureus)*	Nanoparticles bind to the DNA and stops the DNA transcription in the bacteria	100	43.75	Spherical, triangular and irregular	Li et al. (2016)
Watermelon	*Escherichia coli and Staphylococcus epidermis*	Nanoparticles inflict cellular damage which leads to the death of the bacteria	–	200–500	Hexagonal plate-like	Chumsard et al. (2019)

17.3.3 Metal Oxide Nanomaterials

Metal oxides has also attracted attention from researchers as nanoparticles. Nanoparticles like zinc oxide exhibits some wide range of significant applications including as antimicrobial agents. Biogenic zinc oxide nanoparticles are also synthesized from different biological species such as fungi and bacteria with common substrate being plants extracts (Sharma et al. 2020). The plant extracts of waste pees which are the extracts of *Prunus domestica L* for the synthesis of metallic oxide form. FTIR, XRD and SEM were used to confirm the synthesised nanoparticles and their antimicrobial activity evaluated thereafter. Four bacterial strains were used for evaluation with different concentration of the biogenic zinc oxide nanoparticles and the study reported high effectivity towards *Staphylococcus aureus* (Ajmal and Saraswat 2017). Another plant extract used for biogenic synthesis is a leaf. Hexagonal structured nanoparticles with diameter of 11–25 nm were synthesized and also evaluated for their antimicrobial activity on six pathogens using an agar diffusion method. It was found that as the concentration of nanoparticles increases, the inhibition was also increased. The mechanism for the antimicrobial activity was not discussed on the study (Suresh et al. 2018). Magnesium oxide (Essien et al. 2020), titanium oxide (Anbumani et al. 2022) and Iron oxide (Ezealigo et al. 2021) nanoparticles are some of the metal oxides that were synthesized using biogenic materials and evaluated for their antimicrobial activities. Some of the metal oxide nanoparticles synthesized through this process, paved a way for a significant clinical process.

17.4 Conclusion

The green synthesis of metal and metal oxide nanoparticles using biogenic material has proven to be a safe and environmentally friendly method when compared to other methods (physical and chemical methods). These biogenic extracts act as reducing and capping agents for the respective nanoparticles. The application of the biogenic nanoparticles as antimicrobial agents shows a wide range of effects including inhibition and bacterial cell damage which can be a breakthrough on subjects such as MDR and toxicity. Furthermore, optimization of the parameters such as temperature and pH control the shape and the size of the nanoparticles which on turn has an effect on antimicrobial activity. This method is the cost-effective method as it does not use many chemicals and less hazardous method for nanomaterials synthesis.

References

Abdel-Aziz MS, Shaheen MS, El-Nekeety AA, Abdel-Wahhab MA (2014) Antioxidant and antibacterial activity of silver nanoparticles biosynthesized using Chenopodium murale leaf extract. J Saudi Chem Soc 18(4):356–363. https://doi.org/10.1016/j.jscs.2013.09.011

Ajmal N, Saraswat K (2017) Advances in bio reasearch In-vitro Antibacterial and Antioxidant activities of zinc oxide nanoparticles synthesized using Prunus domestica L. (Plum) agro-. Adv. Biores. 8(May):39–46. https://doi.org/10.15515/abr.0976-4585.8.3.3946

Amro NA, Kotra LP, Wadu-Mesthrige K, Bulychev A, Mobashery S, Liu GY (2000) High-resolution atomic force microscopy studies of the Escherichia coli outer membrane: structural basis for permeability. Langmuir 16(6):2789–2796. https://doi.org/10.1021/la991013x

Anbumani D et al (2022) Green synthesis and antimicrobial efficacy of titanium dioxide nanoparticles using Luffa acutangula leaf extract. J. King Saud Univ. - Sci. 34(3):101896. https://doi.org/10.1016/j.jksus.2022.101896

Balasubramanian S, Kala SM, Pushparaj TL (2019) Biogenic synthesis of gold nanoparticles using Jasminum auriculatum leaf extract and their catalytic, antimicrobial and anticancer activities," J. Drug Deliv. Sci. Technol., vol. 57, no. May 2019, p. 101620. https://doi.org/10.1016/j.jddst.2020.101620.

Bandala ER, Stanisic D, Tasic L (2020) Biogenic nanomaterials for photocatalytic degradation and water disinfection: A review. Environ Sci Water Res Technol 6(12):3195–3213. https://doi.org/10.1039/d0ew00705f

C. Rajkuberan, K. Sudha, G. Sathishkumar, and S. Sivaramakrishnan, "Spectrochimica Acta Part A : Molecular and Biomolecular Spectroscopy Antibacterial and cytotoxic potential of silver nanoparticles synthesized using latex of Calotropis gigantea L .," vol. 136, pp. 924–930, 2015, doi: https://doi.org/10.1016/j.saa.2014.09.115.

Chums-ard W, Fawcett D, Fung CC, Poinern GEJ (2019) Biogenic synthesis of gold nanoparticles from waste watermelon and their antibacterial activity against Escherichia coli and Staphylococcus epidermidis. Int. J. Res. Med. Sci. 7(7):2499. https://doi.org/10.18203/2320-6012.ijrms20192874

D. Decarolis, Y. Odarchenko, J. J. Herbert, C. Qiu, and A. M. Beale, "Identification of the key steps in the self-assembly of homogeneous gold metal nanoparticles produced using inverse micelles †," pp. 18824–18834, 2020, doi: https://doi.org/10.1039/c9cp03473k.

Davis M, Whittaker A, Lindgren M, Djerf-Pierre M, Manderson L, Flowers P (2018) Understanding media publics and the antimicrobial resistance crisis. Glob Public Health 13(9):1158–1168. https://doi.org/10.1080/17441692.2017.1336248

Doan VD et al (2020) Biosynthesis of Silver and Gold Nanoparticles Using Aqueous Extract of Codonopsis pilosula Roots for Antibacterial and Catalytic Applications," *J. Nanomater.*, vol 2020. https://doi.org/10.1155/2020/8492016.

Essien ER, Atasie VN, Okeafor AO, Nwude DO (2020) Biogenic synthesis of magnesium oxide nanoparticles using Manihot esculenta (Crantz) leaf extract. Int. Nano Lett. 10(1):43–48. https://doi.org/10.1007/s40089-019-00290-w

Ezealigo US, Ezealigo BN, Aisida SO, Ezema FI (2021) Iron oxide nanoparticles in biological systems: Antibacterial and toxicology perspective. JCIS Open, vol. 4, no. October, p. 100027. https://doi.org/10.1016/j.jciso.2021.100027

Gericke M, Pinches A (2006) Biological synthesis of metal nanoparticles. Hydrometallurgy 83(1–4):132–140. https://doi.org/10.1016/j.hydromet.2006.03.019

Ghaedi M, Yousefinejad M, Safarpoor M, Khafri HZ, Purkait MK (2015) Journal of Industrial and Engineering Chemistry Rosmarinus officinalis leaf extract mediated green synthesis of silver nanoparticles and investigation of its antimicrobial properties. J Ind Eng Chem 31:167–172. https://doi.org/10.1016/j.jiec.2015.06.020

Gudikandula K, Vadapally P, Singara Charya MA (2017) Biogenic synthesis of silver nanoparticles from white rot fungi: their characterization and antibacterial studies. OpenNano, vol. 2, no. June, pp. 64–78. https://doi.org/10.1016/j.onano.2017.07.002.

Ingle AP, Duran N, Rai M (2014) Bioactivity, mechanism of action, and cytotoxicity of copper-based nanoparticles: A review. Appl Microbiol Biotechnol 98(3):1001–1009. https://doi.org/10.1007/s00253-013-5422-8

Jamil M, Murtaza G, Mehmood A (2015) Green synthesis of silver nanoparticles using leaves extract of Skimmia laureola : Characterization and antibacterial activity. Mater Lett 153:10–13. https://doi.org/10.1016/j.matlet.2015.03.143

Karakoti AS, Hench LL, Seal S (2006) The role of surfaces on NM toxicity. Jom 58(7):77–82

Kathiraven T, Sundaramanickam A, Shanmugam N, Balasubramanian T (2015) Green synthesis of silver nanoparticles using marine algae Caulerpa racemosa and their antibacterial activity against some human pathogens. Appl Nanosci 5(4):499–504. https://doi.org/10.1007/s13204-014-0341-2

Khalil MMH, Ismail EH, El-Baghdady KZ, Mohamed D (2014) Green synthesis of silver nanoparticles using olive leaf extract and its antibacterial activity. Arab J Chem 7(6):1131–1139. https://doi.org/10.1016/j.arabjc.2013.04.007

Khambhaty Y, Mody K, Basha S, Jha B (2009) Kinetics, equilibrium and thermodynamic studies on biosorption of hexavalent chromium by dead fungal biomass of marine Aspergillus niger. Chem Eng J 145(3):489–495. https://doi.org/10.1016/j.cej.2008.05.002

Khan NT, Jamil Khan M (2017) Biogenic Nanoparticles: An Introduction to What They Are and How They Are Produced. Int. J. Biotechnol. Bioeng., vol. 3, no. 3, pp. 66–70. https://doi.org/10.25141/2475-3432-2017-3.0066.

Khorrami S, Zarrabi A, Khaleghi M, Danaei M, Mozafari MR (2018) Selective cytotoxicity of green synthesized silver nanoparticles against the MCF-7 tumor cell line and their enhanced antioxidant and antimicrobial properties. Int J Nanomedicine 13:8013–8024. https://doi.org/10.2147/IJN.S189295

Lakshmanan G, Sathiyaseelan A, Kalaichelvan PT, Murugesan K (2018) ScienceDirect Plant-mediated synthesis of silver nanoparticles using fruit extract of Cleome viscosa L.: Assessment of their antibacterial and anticancer activity. Karbala Int. J. Mod. Sci. 4(1):61–68. https://doi.org/10.1016/j.kijoms.2017.10.007

Lateef A, Adelere IA, Gueguim-Kana EB, Asafa TB, Beukes LS (2015) Green synthesis of silver nanoparticles using keratinase obtained from a strain of Bacillus safensis LAU 13. Int. Nano Lett. 5(1):29–35. https://doi.org/10.1007/s40089-014-0133-4

Li J et al (2016) Biosynthesis of gold nanoparticles by the extreme bacterium Deinococcus radiodurans and an evaluation of their antibacterial properties. Int J Nanomedicine 11:5931–5944. https://doi.org/10.2147/IJN.S119618

Li GR, Xu H, Lu XF, Feng JX, Tong YX, Su CY (2013) Electrochemical synthesis of nanostructured materials for electrochemical energy conversion and storage. Nanoscale 5(10):4056–4069. https://doi.org/10.1039/c3nr00607g

Ma L et al (2017) Optimization for extracellular biosynthesis of silver nanoparticles by Penicillium aculeatum Su1 and their antimicrobial activity and cytotoxic effect compared with silver ions. Mater Sci Eng C 77:963–971. https://doi.org/10.1016/j.msec.2017.03.294

Mahmud KM et al (2022) Investigation of Antimicrobial Activity and Biocompatibility of Biogenic Silver Nanoparticles Synthesized using Syzigyum cymosum Extract. ACS Omega 7(31):27216–27229. https://doi.org/10.1021/acsomega.2c01922

Mythili R et al (2018) Biogenic synthesis, characterization and antibacterial activity of gold nanoparticles synthesised from vegetable waste. J Mol Liq 262:318–321. https://doi.org/10.1016/j.molliq.2018.04.087

Nakkala JR, Mata R, Gupta AK, Sadras SR (2014) European Journal of Medicinal Chemistry Biological activities of green silver nanoparticles synthesized with Acorous calamus rhizome extract. Eur J Med Chem 85:784–794. https://doi.org/10.1016/j.ejmech.2014.08.024

Nguyen THA et al (2022) Novel biogenic silver and gold nanoparticles for multifunctional applications: Green synthesis, catalytic and antibacterial activity, and colorimetric detection of Fe(III) ions. Chemosphere 287(P3):132271. https://doi.org/10.1016/j.chemosphere.2021.132271

Oves M et al (2018) Antimicrobial and anticancer activities of silver nanoparticles synthesized from the root hair extract of Phoenix dactylifera," *Mater. Sci. Eng. C*, vol. 89, no. August 2017, pp. 429–443. https://doi.org/10.1016/j.msec.2018.03.035.
P. Khandel, R. K. Yadaw, D. K. Soni, L. Kanwar, and S. K. Shahi, *Biogenesis of metal nanoparticles and their pharmacological applications: present status and application prospects*, vol. 8, no. 3. Springer Berlin Heidelberg, 2018.
Patil MP, Kim J, Seo YB, Kang M, Kim G (2021) Biogenic Synthesis of Metallic Nanoparticles and Their Antibacterial Applications Biogenic Synthesis of Metallic Nanoparticles and Their Antibacterial Applications. J Life Sci 31(9):862–872
Phanjom P, Zoremi E, Mazumder J, Saha M, Baruah SB (2012) Green Synthesis of Silver Nanoparticles using Leaf Extract of Myrica esculenta. Int. J. Nanosci. Nanotechnol 3(2), pp. 73–79. http://www.irphouse.com.
Pileni MP (2000) Fabrication and physical properties of self-organized silver nanocrystals. Pure Appl Chem 72(1–2):53–65. https://doi.org/10.1351/pac200072010053
Rai M, Yadav A, Gade A (2009) Silver nanoparticles as a new generation of antimicrobials. Biotechnol Adv 27(1):76–83. https://doi.org/10.1016/j.biotechadv.2008.09.002
Raja S, Ramesh V, Thivaharan V (2017) Green biosynthesis of silver nanoparticles using Calliandra haematocephala leaf extract, their antibacterial activity and hydrogen peroxide sensing capability. Arab J Chem 10(2):253–261. https://doi.org/10.1016/j.arabjc.2015.06.023
Rajkuberan C, Prabukumar S, Sathishkumar G, Wilson A, Ravindran K, Sivaramakrishnan S (2017) Facile synthesis of silver nanoparticles using Euphorbia antiquorum L. latex extract and evaluation of their biomedical perspectives as anticancer agents. J Saudi Chem Soc 21(8):911–919. https://doi.org/10.1016/j.jscs.2016.01.002
Rolim WR et al (2019) Applied Surface Science Green tea extract mediated biogenic synthesis of silver nanoparticles : Characterization , cytotoxicity evaluation and antibacterial activity. Appl. Surf. Sci., vol. 463, no. May 2018, pp. 66–74. https://doi.org/10.1016/j.apsusc.2018.08.203
S. Gwo, H. Chen, M. Lin, and X. Li, "Chem Soc Rev for plasmonics," pp. 13–19, 2016, doi: https://doi.org/10.1039/c6cs00450d.
Sakamoto M, Fujistuka M, Majima T (2009) Light as a construction tool of metal nanoparticles: Synthesis and mechanism. J Photochem Photobiol C Photochem Rev 10(1):33–56. https://doi.org/10.1016/j.jphotochemrev.2008.11.002
Salari Z, Danafar F, Dabaghi S, Ataei SA (2016) Sustainable synthesis of silver nanoparticles using macroalgae Spirogyra varians and analysis of their antibacterial activity Sustainable synthesis of silver nanoparticles using macroalgae Spirogyra varians. J Saudi Chem Soc 20(4):459–464. https://doi.org/10.1016/j.jscs.2014.10.004
Shaheen I, Ahmad KS (2020) Chromatographic identification of 'green capping agents' extracted from Nasturtium officinale (Brassicaceae) leaves for the synthesis of MoO3 nanoparticles. J Sep Sci 43(3):598–605. https://doi.org/10.1002/jssc.201900840
Sharma R, Garg R, Kumari A (2020) A review on biogenic synthesis, applications and toxicity aspects of zinc oxide nanoparticles. EXCLI J 19:1325–1340. https://doi.org/10.17179/excli2020-2842
Sharma D, Kanchi S, Bisetty K (2019) Biogenic synthesis of nanoparticles : A review. Arab J Chem 12(8):3576–3600. https://doi.org/10.1016/j.arabjc.2015.11.002
Singh P, Garg A, Pandit S, Mokkapati VRSS, Mijakovic I (2018) Antimicrobial effects of biogenic nanoparticles. Nanomaterials 8(12):1–19. https://doi.org/10.3390/nano8121009
Suganthy N, Sri Ramkumar V, Pugazhendhi A, Benelli G, Archunan G (2018) Biogenic synthesis of gold nanoparticles from Terminalia arjuna bark extract: assessment of safety aspects and neuroprotective potential via antioxidant, anticholinesterase, and antiamyloidogenic effects. Environ. Sci. Pollut. Res., vol. 25, no. 11, pp. 10418–10433. https://doi.org/10.1007/s11356-017-9789-4

Sunderam V, Thiyagarajan D, Lawrence AV, Mohammed SSS, Selvaraj A (2019) In-vitro antimicrobial and anticancer properties of green synthesized gold nanoparticles using Anacardium occidentale leaves extract. Saudi J. Biol. Sci. 26(3):455–459. https://doi.org/10.1016/j.sjbs.2018.12.001

Suresh J, Pradheesh G, Alexramani V, Sundrarajan M, Hong SI (2018) Green synthesis and characterization of zinc oxide nanoparticle using insulin plant (Costus pictus D. Don) and investigation of its antimicrobial as well as anticancer activities. Adv. Nat. Sci. Nanosci. Nanotechnol., vol. 9, no. 1. https://doi.org/10.1088/2043-6254/aaa6f1

Thakkar KN, Mhatre SS, Parikh RY (2010) "Biological synthesis of metallic nanoparticles", *Nanomedicine Nanotechnology*. Biol Med 6(2):257–262. https://doi.org/10.1016/j.nano.2009.07.002

Vinosha M et al (2019) Biogenic synthesis of gold nanoparticles from Halymenia dilatata for pharmaceutical applications: Antioxidant, anti-cancer and antibacterial activities. Process Biochem 85(April):219–229. https://doi.org/10.1016/j.procbio.2019.07.013

X. Li, H. Xu, Z. S. Chen, and G. Chen, "Biosynthesis of nanoparticles by microorganisms and their applications," *J. Nanomater.*, vol. 2011, 2011, doi: https://doi.org/10.1155/2011/270974.

X. H. Vu, T. T. T. Duong, T. T. H. Pham, D. K. Trinh, X. H. Nguyen, and V. S. Dang, "Synthesis and study of silver nanoparticles for antibacterial activity against Escherichia coli and Staphylococcus aureus," *Adv. Nat. Sci. Nanosci. Nanotechnol.*, vol. 9, no. 2, 2018, doi: https://doi.org/10.1088/2043-6254/aac58f.

Yang N et al (2017) Biogenic synthesis of silver nanoparticles using ginger (Zingiber officinale) extract and their antibacterial properties against aquatic pathogens. Acta Oceanol Sin 36(12):95–100. https://doi.org/10.1007/s13131-017-1099-7

Yu C, Yao Z, Hu B (2009) Preparation of polydimethylsiloxane/β-cyclodextrin/divinylbenzene coated 'dumbbell-shaped' stir bar and its application to the analysis of polycyclic aromatic hydrocarbons and polycyclic aromatic sulfur heterocycles compounds in lake water and soil by hig. Anal Chim Acta 641(1–2):75–82. https://doi.org/10.1016/j.aca.2009.03.031

Zhang Q, Kusada K, Kitagawa H (2021) Phase Control of Noble Monometallic and Alloy Nanomaterials by Chemical Reduction Methods. ChemPlusChem 86(3):504–519. https://doi.org/10.1002/cplu.202000782

Chapter 18
Biogenic Nanomaterials as Adsorbents for Mercury Remediation

Mxolisi Junior Kiwanuka, Philiswa N. Nomngongo, and Nomvano Mketo

Abstract Mercury has been in the environment since the beginning of time. Natural and anthropogenic activities are the main sources of mercury. The latter is in the top list of most toxic contaminants worldwide. The elimination of mercury from environmental matrices remains a burning issue in research. The traditional physiochemical treatments are considered expensive, and biogenic nanomaterials have gained an increasing attention. Plant extracts have been widely used as reducing and capping agents for the green synthesis of magnetic and metal oxide nanoparticles for the adsorptive removal of mercury in aqueous matrices. The synthesized metal oxide nanoparticles can be reused for at least five consecutive cycles without significant loss of removal efficiency. Additionally, microorganisms such as bacterial biofilms are used for the synthesis of nanoparticles such as bio selenium nanoparticles (SeNPs) for the immobilization of mercury. The synthesized SeNPs significantly reduces Hg^0 concentration in groundwater due to the reaction between S^0 and Hg^0 resulting in the formation of HgSe. Therefore, this book chapter will critically investigate the current and future developments of biologically synthesised nanoparticles for the remediation of mercury from environmental samples.

Keywords Biogenic nanoparticles · Mercury remediation · Plant extract · Microorganism

M. J. Kiwanuka · N. Mketo (✉)
Department of Chemistry, College of Science and Engineering and Technology, Florida Science Campus, University of South Africa, Roodepoort, Johannesburg 1710, South Africa
e-mail: nomvano.mketo@gmail.com; mketon@unisa.ac.za

P. N. Nomngongo
Department of Chemical Sciences, University of Johannesburg, Doornfontein 2028, 17011, Johannesburg, South Africa

M. P. Shah et al. (eds.), *Biogenic Nanomaterials for Environmental Sustainability: Principles, Practices, and Opportunities*, Environmental Science and Engineering,
https://doi.org/10.1007/978-3-031-45956-6_18

18.1 Introduction

18.1.1 Background Information

Mercury (Hg) is a naturally occurring element (Kulikova et al. 2019). The former is mostly present in cinnabar and is trace-level present in all environmental media. Since 1500, cinnabar and other ores have yielded about a million tons of metallic mercury (Chen and Yang 2012). There are three distinct species of mercury: elemental mercury, inorganic mercury, and organic mercury. Every mercury species has a unique set of behaviours and toxicity (Kulikova et al. 2019; Wilcox et al. 2012). Hg occurs as a liquid metal in its elemental state and can be transformed into a vapor at ambient temperature and pressure. Dental amalgams, batteries, barometers, thermometers, sphygmomanometers, and other devices mostly utilise this form of mercury. Mercuric and mercurous are the two different forms of inorganic mercury. Similar to oxidized elemental mercury, mercury salts are more lethal and water soluble than elemental mercury (Rice et al. 2014). Numerous goods, including pharmaceutical and cosmetic ones, use inorganic mercury. Alkyl mercury and phenyl mercury are two examples of the organic mercury compounds that are thought to be extremely hazardous. In contrast, alkyl mercury compounds are primarily found in the environment while the former is utilized as a preservative in medicine (Saturday 2018). Nevertheless, the presence of this element in the environment is not favourable due to its toxicity. Hitherto, a lot of research has been done for the accurate quantification of this toxic analyte in various environmental media.

18.1.2 Toxicity of Mercury

A wide range of negative consequences on human health, plant, and animal life, as well as the ecosystem, result from the presence of mercury in the environment. The chemical form of mercury affects its toxicity. Mercury toxicity in people is affected by a number of variables, including chemical form, food, age, concentration levels, state of health, and length of exposure (Bernhoft 2012). In addition, mercury plays no physiological role in people, and the body has no efficient way to eliminate it (Saturday 2018; Ferreira et al. 2015). The kidneys are where inorganic mercury builds up the most, followed by the liver. After acute mercuric mercury poisoning, the intestine and kidneys are the organs that are most damaged. Although renal failure may occur within 24 h due to necrosis of the tubular epithelium in the kidneys, the corrosive effects will mostly affect the colon. On the other hand, prolonged exposure to elemental mercury can result in neurological issues such tremors, delusions, memory loss, and anomalies of the central nervous system's neuro-cognitive processes (CNS). Finally, the most harmful kind of organic mercury is methyl mercury. The former has numerous adverse effects such as neurotoxicity, genotoxicity, and endocrine disruption (Azevedo and Rodriguez 2012; Zulaikhah et al. 2020; Annibaldi et al. 2019).

Hg mostly impacts the roots, seeds, flowers, and leaves of plants. The former has a strong affinity to react with sulfhydryl (SH) groups, an affinity to react with phosphate groups, and the capacity to substitute critical ions. It can also alter the permeability of cell membranes (Lima et al. 2019).

18.1.3 Occurrence and Fate of Hg in the Environment

Both anthropogenic and natural processes are responsible for the genesis of mercury in environmental matrices (Geneti et al. 2022). Volcanoes, forest fires, cinnabar (ore), and weathering of geologic reservoirs into the biosphere are examples of natural sources of mercury (Liu et al. 2021; Suvarapu and Baek 2017a). Since the beginning of industry, human activities have been a major cause of mercury pollution in the environment. Some of these activities include expanded mining, medical waste incinerators, municipal waste combustion, high rates of burning fossil fuels, and extensive use of raw materials containing mercury. Additionally, the production of metals, alkali, and cement contributes to the release of mercury into the environment (Suvarapu and Baek 2017b).

Mercury is easily converted into organomercury molecules like methylmercury (MeHg) in aquatic medium by chemical and biological mechanisms, which has a significant impact on its solubility, volatility, bioavailability, and toxicity. The most hazardous form of mercury is methylmercury, which can bioaccumulate more quickly than other trace metals along the food chain (Martins et al. 2013; Cegolon et al. 2023). To sum up, the biogeochemical cycle allows for the recycling of mercury in the environment. First, in the former, mercury is transformed into inorganic mercury by the evaporation of mercury from natural and anthropogenic sources. Rain aids in the spread of this chemical. Bacteria can transform mercury into organic compounds once it has entered the soil and water bodies (Ma et al. 2021).

18.1.4 Mercury Remediation Strategies

Due to its long-term tenacity, danger, and non-biodegradability, this poisonous element's build-up in the environment is a global problem. To date, numerous techniques and remediation plans have been created to clean up soil, sediments, and water that contain mercury. The former comprises roasting, solvent extraction, photocatalysis, ion exchange flotation, membrane separation, precipitation, coagulation, distillation, ultrafiltration, and adsorption (Geneti et al. 2022; Liu et al. 2021). Adsorbent materials, however, are receiving a lot more and more attention. Now, a variety of adsorbent materials are used, including activated carbons, compounds supported on silica gel, nanocellulose materials, and metal oxides. The unique properties of magnetite nanoparticles (Fe_3O_4), such as their high surface area to volume ratio and magnetite composition, have made them stand out among the adsorbents listed above. Numerous conventional physical and chemical methods for the synthesis of nanoparticles have been developed as a result of the research carried out by academic and industry experts over the years. However, most technologies are unsustainable because to the need for hazardous or expensive chemicals, high industrial temperatures, and pressures, growing risks to the environment and public health, excessive energy use, and waste creation. Alternative green methods that use eco-friendly precursors and processes under low-stress circumstances include biotechnological pathways including the utilization of bacteria, fungus, and plant extracts for nanoparticle synthesis (Boedicker et al. 2021; Kim et al. 2018).

18.1.5 Conventional Adsorbents Used for Mercury Remediation

Adsorption techniques have been and continue to be the most widely used in industry, and as a result, the most extensively researched methods for mercury clean up. Different conventional adsorbents have been used for mercury clean up. These consist of carbon nanotubes, biochar, mesoporous carbons, graphene oxide, activated carbon fibres, and zeolites, among others. There are various restrictions, including high cost and environmental impact, despite the fact that standard physical and chemical procedures were created for the large-scale manufacture of these nanomaterials (Kim et al. 2018).

18.2 Biogenic Nanomaterials

18.2.1 Synthesis

Nanomaterials that are produced by biological processes are known as "biogenic" materials. These nanoparticles have a wide range of applications in the biomedical, agricultural, electronics, and environmental remediation industries because of their exceptional biocompatibility and sensitivity (Forruque et al. 2022). Metal ions are reduced by organisms or their metabolites to stable nanomaterials in biological or "green" synthesis through redox processes. They accomplish this through the electron shuttle enzymatic metal reduction method and their Nicotinamide adenine dinucleotide phosphate hydrogen (NADPH) dependent reductase enzymes (El-Seedi et al. 2019). Although the production of nanomaterials using plant extracts has also been investigated, the utilization of microbes like bacteria, viruses, and fungus has received the most attention. Bacteria are thoroughly investigated for metal ion reduction to nanoparticles (Boedicker et al. 2021). By using their defense and protection mechanism against soluble metal ions, bacterial cells can produce metallic nanoparticles (Gautam et al. 2018). Bacteria can use either an external or an intracellular method to produce nanoparticles. The metal ions on the exterior of the microbial cells are declined extracellularly. However, when metal ions are contained within microbial cells, varied pH, oxygenation incubation times, and temperature state are inhibited. The fundamental benefit of employing bacteria for nanoparticle biosynthesis is that they are heavy metal resistant. In addition, bacteria are abundant in nature and can adapt to harsh environments.

According to Johnston et al. (2018), the bacteria *Delftia acidovorans* produces pure gold nanoparticles (AuNPs). A little amount of non-ribosomal delfibactin was used in their approach to monitor the resistance to harmful gold ion concentrations. By transforming gold ions into inert AuNPs that are harmless to bacterial organisms, delftibactin removes gold ions from solutions. Sintubin et al. (2021) reported on another intriguing work in which they looked at how lactic acid bacteria produce silver nanoparticles (AgNPs). Only four of the several examined bacterial species produced AgNPs: *Pediococcus pentosaceus, Lactococcus garvieae, Enterococcus faecium,* and other species of *Lactobacillus.* NPs acquired through bacteria are brought about within the cells after covering the produced supernatant with Ag^{+} or Au^{3+}.Ag ions are first biosorbed onto the cell wall during the biosynthesis of AgNPs, and then they are reduced to produce metallic nanoparicles. They went on to say that the cell wall may react as a capping agent for the nanoparticles, stregthning them and stop aggregation.

On the other hand, nanoparticles are also biosynthesized using viruses. A nucleic acid molecule enclosed in a protein sheath makes up a virus, an infectious bacterium. The most typical icosahedron and rod shapes for viruses are among several distinctive shapes. The highly dense and reactive surface made up of capsid proteins on viruses is the primary cause of their efficiency in interacting with metallic ions to produce nanoparticles. Recently, various projects have been carried out to create nanomaterials with the topological structures of viruses, which were inspired by the distinctive surface morphology of viruses (Niu et al. 2013). Tobacco mosaic virus, a plant virus, was utilized to examine how the presence of the virus affected the size and quantity of manufactured silver and gold nanoparticles. It was discovered that low viral concentration in combination with silver and gold salts and plant extracts produced smaller and more of them compared to the solution without the virus. On the other hand, less free nanomaterial was produced when there were more viruses present in the combination (Zhang et al. 2018).

18.2.1.1 Synthesis of Biogenic Nanoparticles by Fungi

Fungi have the ability to produce nanoparticles. In the former, metal ions are converted into nanoparticles by enzymes found in the cytoplasm and cell wall. The positive charge of the metal ions attracts the fungus and starts the biosynthetic process.

Additionally, several proteins that are stimulated by metal ions hydrolyze the ions. Because they can release large amounts of protein, fungi are able to produce more nanoparticles (Mughal et al. 2021). The former is able to produce both intracellularly and extracellularly. The cell wall is used in intracellular synthesis to transport metal ions, where the positive-charged ions interact with the negative-charged wall. These ions are converted to metal NPs in the cells by enzymes. These ions are converted to metal NPs in the cells by enzymes. It is anticipated that niacinamide adenine dinucleotide (NADH)-dependent enzymes, primarily nitrate reductases, will be involved in the extracellular process. These enzymes are released into the reaction vessel along with electron transporters like hydroxyquinoline (Basavaraja et al. 2008). In order to produce NADPH, which is converted into $NADP^+$ when the metal ions are reduced, nanoparticles (El-Seedi et al. 2019).

When it comes to creating nanoparticles, fungi offer a number of advantages over bacteria, which is why their application in the production of metallic nanoparticles has attracted a great deal of attention. Considerable advantages include the simplicity of scaling up and downstream processing, the economic viability, and the existence of mycelia, which provides a larger surface area. It was demonstrated that nanoparticles produced by fungi were more stable and monodisperse than those produced by other microorganisms (Pantidos 2014). The production of silver nanoparticles by several strains of *Fusarium oxysporum* was also examined, and the results demonstrated that a shuttle quinine and a nitrate-dependent reductase are in charge of the metal ions' extracellular reduction. In 2008, Basavaraja et al. (2001) investigated how the fungus *Fusarium semitectum* may be used to produce silver nanoparticles. Highly stable and crystalline silver nanoparticles are produced in solution by treating the fungal filtrate with an aqueous silver nitrate solution. The UV–vis and X-ray diffraction investigations might be used to infer how nanoparticles were formed. The size of the silver particles, which ranged from 10 to 60 nm, was typically spherical and was seen using transmission electron microscopy. In a study conducted by Mukherjee et al. (2009), the biological reduction of AuCl resulted in the formation of intracellular AuNPs that were localized on the surface of the mycelia of the fungus *Verticillium sp.*

18.2.1.2 Synthesis of Biogenic Nanoparticles Using Plant Extracts and Algae

Plants may be the finest potential candidates for the biosynthesis of nanoparticlesdue to their abundance, diversity, and accessibility. Plant extracts have been effectively used to reduce metallic salts and create Fe, Zn, Au, Ag, Pd, and Cu, Mn NPs in a range of shapes and sizes. Metal nanoparticles are bio-reduced in three primary ways using plant extracts. Metal ions are reduced and crystallized during the first step of activation. The second phase's characteristic is the enlargement of the local nanoparticles, commonly referred to as the "growth phase." The thermodynamic strength of the nanoparticles then increases. Not the least of which is that the termination phase gives the nanoparticles their ultimate structure (Mukherjee et al. 2001).

A wide variety of secondary metabolites are present in the cultures and extracts used to generate metallic nanoparticles, helping to reduce metal ions and stabilize thenanoparticles. Polysaccharides, enzymes, amino acids, vitamins, proteins, and organic acids are a few examples of the bioactive compounds.

18.2.1.3 Flavonoids and Terpenoids

Anthocyanins, isoflavonoids, flavonols, chalcones, flavones, and flavanones are only a few of the countless polyphenolic chemical substances known as flavonoids. These chemicals have the power to aggressively reduce metal ions to nanoparticles by chelating them. Nanoparticles can be created by flavonoids with different functional groups. One hypothesis states that the enol to keto tautomeric conversion of flavonoids produces a reactive hydrogen atom that can decrease metal ions and produce nanoparticles. For instance, it is thought that the production of silver nanoparticles from extracts of *Ocimum basilicum* (sweet basil) depends on the conversion of the flavonoids luteolin and rosmarinic acid from the enol- to the keto-form (Balaji et al. 2009). The ability of certain flavonoids to chelate metal ions with carbonyl groups or -electrons is an intriguing observation. Terpenoids and biosynthetic metallic nanoparticles frequently coexist, according to FTIR studies. It is believed that the hydroxyl group in eugenol undergoes deprotonation to produce an anion, which can then undergo further oxidation by metal ions to produce reduction and NPs. Additionally demonstrated to function as a capping and reducing agent when AgNPs are generated is the steroidal saponin diosgenin (El-Seedi et al. 2019).

18.2.1.4 Proteins and Sugars

Proteins can aid in the creation and reinforcement of metallic nanoparticles. A protein is a large biomolecule or macromolecule made composed of several long chains of amino acids. Numerous carbonyl groups on various nanoparticles may be clearly recognized in FTIR spectra, and carbonyl groups present in amino acid residues frequently serve as supportive ligands for nanoparticles, preventing them from aggregating and bolstering them in liquid environments. Studies similar to those on the synthesis of nanoparticles show that proteins can act as reducing agents and protective surface coats. However, free amino groups in proteins and peptides can also attach to silver nanoparticles. Over polymers and surfactants, capping proteins offer significant advantages in terms of cost efficiency, safety, and environmental friendliness for the generation of nanoparticles. Furthermore, they don't require certain circumstances for reactions (Gole et al. 2001; Balaji et al. 2009).

Plant sugar extracts may have an impact on the production and durability of metallic nanoparticles. Recent research has shown that non-soluble carbs, such as starch, affect how well sugar removes cap. For instance, soluble starch and b-Dglucose were used as structure-directing and strengthening agents in the production of silver and gold nanowires and spherical nanoparticles. It was discovered that the unusual stability of the silver nanoparticles generated in a plant extract (Rumex dentatus) was probably caused by the extract's high starch content. This prevented the nanoparticles from aggregating (Shervani and Yamamoto 2011).

18.2.1.5 Enzymes

Numerous investigations have shown that various enzymes are necessary for the production of distinctive metallic nanoparticles. Proteins that are naturally occurring and generated in great quantities by living things are known as enzymes. Since enzymes have a wide range of catalytic activity and may make NPs in a wide range of shapes and sizes, it is crucial to carefully choose the appropriate enzyme for a specific synthesis. For instance, the production of nanoparticles was carried out by the sulfate reductase enzymes produced by *Fusarium oxysporum* during the synthesis of CdS nanoparticles from an aqueous $CdSO_4$ solution. These enzymes converted sulfates from sulfate ions to very stable CdS nanoparticleswhen combined with water Cd^{2+} ions. The well-known coenzyme NADH could function as a reducing agent during the formation of nanoparticles (Ahmad et al. 2002; Palomo and Filice 2016).

18.2.1.6 Algae

Algae are single- or multicellular autotrophic aquatic photosynthetic creatures and constitute a subclass of the kingdom Protista. They lack the plant's multicellular structures yet are nevertheless eukaryotic creatures. The size of algae can vary widely, from tiny microalgae to enormous macroalgae. The majority of aquatic life relies on them as a food supply and they are recognized for creating oxygen. The presence of active chemicals in algae's cell walls is thought to be the cause of their capacity to synthesize nanoparticles. A common procedure for creating algae-assisted nanoparticles entails the following steps: (1) making algae extracts in solution at a high temperature; (2) making reagents; and (3) gestating the algae and reagent solutions before stirring for a certain amount of time. The metal precursor is combined with the algal extract to start the reaction. A shift in color indicates the start of a process that depicts nucleations, followed by the formation of nanoparticles where the nearby nucleonic particles combine to produce thermodynamically stable NPs with different geometries. The driving parameters at play include pH, temperature, concentration, and time, and the extract biocompounds enhance the production of nanoparticles (Mughal et al. 2021; Guilger-Casagrande and Lima 2019).

18.2.2 Characterization

Nanoparticle characterization is done specifically to evaluate the surface area and porosity, pore size, solubility, particle size distribution, aggregation, hydrated surface analysis, zeta potential, wettability, adsorption potential and shape, size of the interactive surface, crystallinity, fractal dimensions, orientation, and intercalation and dispersion of nanoparticles and nanotubes in nanocomposite materials. Indeed, characterisation has a major role in how nanoparticles behave (Kumar and Yadav 2009). Atomic force microscopy (AFM), transmission electron microscopy (TEM), scanning electron microscopy (SEM), dynamic light scattering (DLS), X-ray photoelectron spectroscopy (XPS), thermogravimetric analysis (TGA), and powder X-ray diffraction are some of the methods that can be used to determine nanoparticle parameters. Furthermore, (XRD),Fourier transform infrared spectroscopy (FT-IR), matrix-assisted laser desorption/ionization time-of-flight mass spectrometry (MALDI-TOF), dual polarization interferometry, nuclear magnetic resonance (NMR), nanoparticle tracking analysis (NTA) for evaluation of Brownian motion, and particle size analysis (Baer et al. 2013).

18.2.3 Mechanism of Adsorption

Adsorption is a surface phenomena that only happens on the sorbent's surface. In this case, the material that is adsorbed onto the surface is known as the sorbent, and the substance that is adsorbed is known as the sorbate. The sorbate molecules bind to the surface of the sorbents by molecular interactions, and they diffuse from the surface into the core of the sorbent materials, either in a monolayer or multilayer fashion. Although the primary documented mechanism for trace elements adsorption is the direct chemical binding of the metal to the mineral surface, reduction of metal ions to form immobilized secondary minerals is mostly described with biogenic nanoparticles (Villalobos et al. 2005). In 2012, Jiang et al. (2012), revealed that mercury may be immobilized by forming Hg-Se utilizing Se nanoparticles that were biogenerated. In their research, they reduced both Se (IV) and Hg (II) to create stable Hg-Se using the Hg-resistant bacterium *Shewanella putrefaciens 200*. Moreover, biogenic nanoparticles can serve as electron donors for the breakdown of trace elements by providing hydrogen gas or formate (Ali et al. 2018). Wiatrowski et al. (2006) demonstrated how *Shewanella oneidensis* MR-1 reduced mercury (II) to mercury (Hg^0), which required the presence of both electron donors and acceptors. Fumerate caused reduction, but the rate was at its maximum when ferric oxyhydroxide was utilized as the terminal electron acceptor..

18.2.4 Factors Affecting Synthesis and Adsorption

The synthesis, characterisation, and use of nanoparticles are all impacted by a number of variables. The pH of the solution, time, temperature, the concentration of the extracts utilized, the concentration of the raw material used, size, surface shape, and precoursor are some of the variables that impact the synthesis and adsorption of biogenic nanoparticles (Patra and Baek 2014; Pennycook et al. 2012). Hence, the final size and form will be produced by optimizing these characteristics. The impact of each component on the production and adsorption of biogenic nanoparticles is discussed below.

Particle size and shape: The characteristics of nanoparticles are significantly influenced by the particle size. For instance, it has been shown that as nanoparticle size increased, the melting point of the particles decreased. Similar energy properties across nanoparticles of various configurations make it simple to change their shape. The chemical characteristics of produced nanoparticles are significantly influenced by their dynamic nature and form (Baer et al. 2013; Yacamán et al. 1091).

pH: Both the size and the surface shape of the particle are impacted by pH's function in the production of biogenic NPs. According to Yang and Li, the form was less regular and tended to agglomerate at lower pH levels. The size of particles may be evenly created while synthesizing nanoparticles under various pH settings with the appropriate size and shape (Baker et al. 2013).

Time: The length of time at which the reaction medium is incubated also determines the quality of biogenic nanoparticles to be produced. The variations in the time may occur in many ways such as aggregation of particles due to long time storage; particles may shrink or grow during long storage; they may have shelf life, and so forth, that affects their potential (Baer 2011).

Temperature: One of the most important details regarding nanoparticle biosynthesis is that it happens at room temperature or at temperatures lower than 100 °C. Yet, a key element that affects the type of nanoparticles generated is the reaction medium's temperature. The percentage of gold nanotriangles compared to spherical particles was considerably reduced when Cymbopogon flexuosus was tested to manufacture gold nanoparticles at higher temperatures, whereas low temperature primarily encouraged nano triangle production (Rai et al. 2006).

Concentration: The size of the generated nanoparticles is also influenced by the concentration of reactants like precursors and reducing agents. This phenomenon could be brought on by the surface of prepared nuclei being bound by an excessive number of reducing agents, which accelerates the secondary reduction of metal ions. Because of this, nanoparticles are growing faster and becoming bigger. On the other hand, using an excessive amount of reducing agents may increase the bridging effect between the nanoparticles that are generated, leading to the aggregation of nanoparticles. This could be because too many metal ions were taken up on the surface of the produced nuclei, where a secondary reduction process took place and bigger nanoparticles were created. Apart from size, nanoparticles form will also be impacted.

18.2.5 Application of Biogenic Nanoparticles for Mercury Remediation

As we mentioned in our opening, mercury contamination is a worldwide issue because of its ecotoxicity and health hazards. The first stage in examining the possible effects of mercury on the environment is precise determination. However, mercury may be found in a wide variety of forms in the soil, water, and air, including elemental mercury, inorganic mercury, and methyl mercury (MeHg). These elements make determining mercury levels and determining environmental hazards extremely difficult (Chandrasoma et al. 2012). There are various analytical techniques used to detect mercury inductively coupled plasma mass spectrometry (ICP-MS), inductively coupled plasma-optical emission spectroscopy (ICP-OES), cold vapor technique with atomic absorption spectrometry (CV-AAS), and cold vapor technique with atomic fluorescence spectrometry (CV-AFS) (Jiang et al. 2012) (Table 18.1).

The apparent downsides, limits, or limitations of these approaches in terms of determining mercury, however, include the necessity for large sample quantities, high prices, labor-intensive testing processes, protracted testing timeframes, and specifications for bulky gear. As a result, their usefulness for immediate on-the-spot and in-the-moment evaluations is severely limited. Therefore, it is crucial to develop new analytical techniques and find out how to boost the sensitivity of low mercury detection. It has been demonstrated that biogenic nanoparticles can effectively remove mercury from a range of environmental matrices. In Hg cleaning, biological nanoparticles are frequently utilized as adsorbents. As was already said, biogenic nanoparticles are produced using biological components such as plant extracts and microorganisms. The production of nanoparticles by biological processes has been suggested as a potential replacement, providing a method that is energy-efficient, favorable economically, and safe for the environment (Khan et al. 2021). Several studies have reported the removal of mercury from various matrices using biogenic nanoparticles. The application of biogenic nanoparticles for mercury remediation are summarised in Table 18.2.

18.2.6 Challenges and Opportunities of Biogenic Nanomaterials

Although biogenic nanoparticles offer several advantages over conventional nanomaterials, there have also been some documented difficulties with these substances. Biogenic production of nanomaterials is a sluggish process compared to chemical and physical techniques. The use of biogenic nanoparticles to remove environmental contaminants has also been shown to result in the development of harmful intermediate complexes. Thus, it is crucial to keep track of how these complexes form and develop during the course of the treatment (Kumari et al. 2019). Moreover, the existence of uniformity, homogeneity, and mono nanoparticles in biologically based

Table 18.1 Advantages and disadvantages of conventional adsorbents for mercury remediation

Adsorbent	Principle	Advantage	Disadvantage	Refs.
Activated carbon	• Porous carbonaceous materials • Functional groups like carbonyl, carboxyl, phenol, lactone, and quinone are found in carbon structures • Primarily made from cattle, agricultural waste, and industrial waste	• Low cost • Increased accessible porosity relative to Green Analytical Chemistry (GAC)	• Slow kinetics • Bad selectivity • Limited design flexibility • Working capacity < 100%	Heidarinejad et al. (2020)
Activated carbon fibres	• The highly porous material known as "activated carbon fibers" (ACF) has a limited pore size distribution, a well-defined porous structure, and an aspect ratio of at least 10 μm	• Higher rates of adsorption • High availability of porosity • Elevated surface area • Increased media interaction effectiveness • Sufficient design flexibility • Simple regeneration, especially thermal)	• Expensive. • Extra processing steps • Difficult regeneration • When regenerated, a secondary pollution source	Hassan et al. (2019)
Carbon nanotubes	• Carbon nanotubes are one-nanometer-diameter cylinders made of graphene sheets that have been wrapped up • Carbon atoms are arranged in a hexagonal network	• Outstanding mechanical strength	• Very little and challenging to deal with • Extremely pricey to make	Ibrahim (2013)
Graphene oxides	• It is a single-atomic-layered material that is formed by the oxidation of graphite	• Low production cost • Large-scale production • Easy processing	• Generation of harmful gases such as ClO_2, NO_2 and N_2O_4	Yu et al. (2020)

(continued)

Table 18.1 (continued)

Adsorbent	Principle	Advantage	Disadvantage	Refs.
Biochar	• Solid carbon-rich residue produced from pyrolysis of plant derived biomass	• Cost-effective • Easy to apply • Efficient	• Requires modification using chemicals • Extremely high temperature is required and can lead to activation of other reactive compounds	Sohi et al. (2010)
Zeolites	• Natural substances with set pore widths and active sites in the crystal lattice that enable them to exchange ions, absorb gases and vapours, operate as molecular-scale sieves, and catalyze processes	• Flexible frameworks and compositions • Hydrothermal and physical stability • Non-toxicity • Exchangeable cations with large surface areas • Excellent cost–benefit ratio	• Zeolites that occur in nature are rarely pure and can vary in their contamination levels with other minerals, metals, and quartz	Koohsaryan and Anbia (2016)

nanoparticles is still debatable. To build green nanoparticles employing plant entities, there is a dearth of comprehensive understanding. The fundamental problem to be solved is the conversion of salt to ions, as well. The highest possible salt to ion conversion should be attained during plant-mediated synthesis. The final step is to adopt technology transfer techniques to manufacture nanoparticles at the industrial level (Patil and Chandrasekaran 2020). For the production of biogenic nanomaterials for many applications, particularly for heavy metal cleanup, there are several research oportunities. We will thus have greater room for discoveries and applications provided we adopt the right strategy.

18.3 Conclusion and Future Recommendation

The use of microorganisms, plant, and fungi for the synthesis of nanomaterials for the removal of mercury can successfully solve the challenges of water, soil and air pollution. Furthermore, the challenges of secondary wastes are eliminated when synthesising and applying these biogenic nanomaterials on environmental matrices. Nevertheless, it can be observed that there are more studies that are reported for the removal of mercury from environmental matrices using plant extracts than the other

Table 18.2 Application of biogenic nanoparticles for Hg remediation

Nanomaterial	Synthesis	Matrix	Analytical technique	Detection limit	Accuracy	Precision	Refs.
Magnetite (Fe_3O_4) NPs	Leaf extract	Aqueous solutions	Colorimetry	ND	90	ND	Geneti et al. (2022)
Gold (Au)-NPs	Leaf extract	Aqueous solutions	Sensor	0.49	100	3	Zohora et al. (2017)
Ag-NPs	Seeds extracts		SPR fluorescence	0.24	ND	ND	Jiang et al. (2012)
Ag-NPs	Bean extracts	Aqueous solutions	Colorimetry	0.13	ND	ND	Choudhary et al. (2019)
Se-NPs	Bacteria	Aqueous solutions	CVAFS	ND	ND	ND	Jiang et al. (2012)
Se-NPs	Bacteria	Soil	LC-HGAFS	ND	ND	ND	Wang et al. (2017)
Ag@AgCl-NPs	Algae	Water	ND	4.19	97–102.52	< 3.12	Karimi and Samimi (2019)
AuNPs	Enzyme	Fruits	ICP-MS	1.6–1.9	ND	ND	Hasan et al. (2019)
Se-NPs	Enzymes	Aqueous solution	Colorimetry	0.12	ND	ND	Cao et al. (2019)
DNA-Cu oxide NPs	Enzymes	Fish	ICP-MS	3.3	91–108	5.6–6.1	Lien et al. (2019)
Se-NPs	Fungi	Aqueous solutions	CVAFS/ ICP-MS	0.1	ND	ND	Jiang et al. (2012)
Se-NPs	Sponge	Aqueous solutions	CV-AAS	0.2	ND	ND	Ahmed et al. (2017)
Ag-NPs	Green sunlight	Aqueous solutions	Colorimetry	ND	ND	ND	Ahmed et al. (2015)

biological materials. Selenium nanoparticles were the most synthesised nanoparticles for mercury removal than the other metal-based nanoparticles. Aqueous matrices were the most examined matrices as compared to the other matrices. Therefore, there is a need for more research to be carried out for the removal of mercury using biogenic materials made from fungi, algae, bacteria and enzymes. Additionally, more investigations need to be conducted for the removal of mercury from other matrices such as food, soil, sediments, and fossil fuels using biogenic nanomaterials.

Acknowledgements This book chapter was compiled on the full support received from NRF-THUTHUKA (TTK170418227444) and University of South Africa, College of Science, Engineering and Technology (CSET) Chemistry Department (Florida Science campus).

References

Ahmad A et al (2002) Enzyme mediated extracellular synthesis of CdS nanoparticles by the fungus, Fusarium oxysporum. J Am Chem Soc 124(41):12108–12109. https://doi.org/10.1021/ja027296o

Ahmed KBA, Senthilnathan R, Megarajan S, Anbazhagan V (2015) Sunlight mediated synthesis of silver nanoparticles using redox phytoprotein and their application in catalysis and colorimetric mercury sensing. J Photochem Photobiol B Biol 151:39–45. https://doi.org/10.1016/j.jphotobiol.2015.07.003

Ahmed S, Brockgreitens J, Xu K, Abbas A (2017) A Nanoselenium Sponge for Instantaneous Mercury Removal to Undetectable Levels. Adv. Funct. Mater., vol. 27, no. 17. https://doi.org/10.1002/adfm.201606572

Ali T et al (2019) Overview of microbes based fabricated biogenic nanoparticles for water and wastewater treatment. J. Environ. Manage., vol. 230, no. September 2018, pp. 128–150. https://doi.org/10.1016/j.jenvman.2018.09.073.

Annibaldi A et al (2019) Determination of Hg in farmed and wild atlantic bluefin tuna (Thunnus thynnus L.) muscle. Molecules 24(7):1–16. https://doi.org/10.3390/molecules24071273

Azevedo R, Rodriguez E (2012) Phytotoxicity of Mercury in Plants: A Review. J Bot 2012:1–6. https://doi.org/10.1155/2012/848614

Baer DR (2011) Surface Characterization of Nanoparticles: critical needs and significant challenges. J. Surf. Anal. 17(3):163–169. https://doi.org/10.1384/jsa.17.163

Baer DR et al (2013) Surface characterization of nanomaterials and nanoparticles: Important needs and challenging opportunities. J. Vac. Sci. Technol. A Vacuum, Surfaces, Film vol. 31, no. 5, p. 050820. https://doi.org/10.1116/1.4818423

Baker S et al (2013) Plants: Emerging as nanofactories towards facile route in synthesis of nanoparticles. BioImpacts 3(3):111–117. https://doi.org/10.5681/bi.2013.012

Balaji DS, Basavaraja S, Deshpande R, Mahesh DB, Prabhakar BK, Venkataraman A (2009) Extracellular biosynthesis of functionalized silver nanoparticles by strains of Cladosporium cladosporioides fungus. Colloids Surfaces B Biointerfaces 68(1):88–92. https://doi.org/10.1016/j.colsurfb.2008.09.022

Basavaraja S, Balaji SD, Lagashetty A, Rajasab AH, Venkataraman A (2008) Extracellular biosynthesis of silver nanoparticles using the fungus Fusarium semitectum. Mater Res Bull 43(5):1164–1170. https://doi.org/10.1016/j.materresbull.2007.06.020

Bernhoft RA (2012) Mercury toxicity and treatment: A review of the literature. J. Environ. Public Health. https://doi.org/10.1155/2012/460508.

Boedicker JQ, Gangan M, Naughton K, Zhao F, Gralnick JA, El-Naggar MY (2021) Engineering Biological Electron Transfer and Redox Pathways for Nanoparticle Synthesis. Bioelectricity 3(2):126–135. https://doi.org/10.1089/bioe.2021.0010

Cao H, Xiao J, Liu L (2019) Enhanced oxidase-like activity of selenium nanoparticles stabilized by chitosan and application in a facile colorimetric assay for mercury (II). Biochem. Eng. J., vol. 152, no. September, p. 107384. https://doi.org/10.1016/j.bej.2019.107384

Cegolon L, Petranich E, Pavoni E (2023) Concentration of mercury in human hair and its associated factors in residents of the Gulf of Trieste (North-Eastern. Env. Sci. and Pollut. Res. 30:21425–21437

Chandrasoma A, Hamid AAA, Bruce AE, Bruce MRM, Tripp CP (2012) An infrared spectroscopic based method for mercury(II) detection in aqueous solutions. Anal Chim Acta 728:57–63. https://doi.org/10.1016/j.aca.2012.03.047

Choudhary MK, Garg S, Kaur A, Kataria J, Sharma (2020) Green biomimetic silver nanoparticles as invigorated colorimetric probe for Hg2+ ions: A cleaner approach towards recognition of heavy metal ions in aqueous media. Mater. Chem. Phys., vol. 240, no. September 2019, p. 122164. https://doi.org/10.1016/j.matchemphys.2019.122164

El-Seedi HR et al (2019) Metal nanoparticles fabricated by green chemistry using natural extracts: Biosynthesis, mechanisms, and applications. RSC Adv 9(42):24539–24559. https://doi.org/10.1039/c9ra02225b

Ferreira SLC et al (2015) Analytical strategies of sample preparation for the determination of mercury in food matrices - A review. Microchem J 121:227–236. https://doi.org/10.1016/j.microc.2015.02.012

Forruque S et al (2022) Green approaches in synthesising nanomaterials for environmental nanobioremediation : Technological advancements , applications , benefits and challenges. Environ. Res., vol. 204, no. PA, p. 111967. https://doi.org/10.1016/j.envres.2021.111967.

Gautam PK, Singh A, Misra K, Sahoo AK, Samanta SK (2019) Synthesis and applications of biogenic nanomaterials in drinking and wastewater treatment. J. Environ. Manage., vol. 231, no. June 2018, pp. 734–748. https://doi.org/10.1016/j.jenvman.2018.10.104.

Geneti ST et al (2022) Biogenic Synthesis of Magnetite Nanoparticles Using Leaf Extract of Thymus schimperi and Their Application for Monocomponent Removal of Chromium and Mercury Ions from Aqueous Solution. J Nanomater vol 2022, no. Ii. https://doi.org/10.1155/2022/5798824

Gole A et al (2001) Pepsin-gold colloid conjugates: Preparation, characterization, and enzymatic activity. Langmuir 17(5):1674–1679. https://doi.org/10.1021/la001164w

Guilger-Casagrande M, de Lima R (2019) Synthesis of Silver Nanoparticles Mediated by Fungi: A Review. Front. Bioeng. Biotechnol. 7(October):1–16. https://doi.org/10.3389/fbioe.2019.00287

Hasan A et al (2020) Nanozyme-based sensing platforms for detection of toxic mercury ions: An alternative approach to conventional methods. Talanta, vol. 215, no. December 2019, p. 120939. https://doi.org/10.1016/j.talanta.2020.120939

Hassan MF, Sabri MA, Fazal H, Hafeez A, Shezad N, Hussain M (2020) Recent trends in activated carbon fibers production from various precursors and applications—A comparative review. J. Anal. Appl. Pyrolysis, vol. 145, no. October 2019, p. 104715. https://doi.org/10.1016/j.jaap.2019.104715

Heidarinejad Z, Dehghani MH, Heidari M, Javedan G, Ali I, Sillanpää M (2020) Methods for preparation and activation of activated carbon: a review. Environ Chem Lett 18(2):393–415. https://doi.org/10.1007/s10311-019-00955-0

Ibrahim KS (2013) Carbon nanotubes-properties and applications: a review. Carbon Lett. 14(3):131–144. https://doi.org/10.5714/cl.2013.14.3.131

Jiang S, Ho CT, Lee JH, Van Duong H, Han S, Hur HG (2012) Mercury capture into biogenic amorphous selenium nanospheres produced by mercury resistant Shewanella putrefaciens 200. Chemosphere 87(6):621–624. https://doi.org/10.1016/j.chemosphere.2011.12.083

Karimi S, Samimi T (2019) Green and simple synthesis route of Ag@AgCl nanomaterial using green marine crude extract and its application for sensitive and selective determination of mercury," Spectrochim. Acta - Part A Mol. Biomol. Spectrosc., vol. 222, p. 117216. https://doi.org/10.1016/j.saa.2019.117216

Khan F et al (2022) Prospects of algae-based green synthesis of nanoparticles for environmental applications. Chemosphere, vol. 293, no. December 2021, p. 133571. https://doi.org/10.1016/j.chemosphere.2022.133571.

Kim TY, Kim MG, Lee JH, Hur HG (2018) Biosynthesis of nanomaterials by shewanella species for application in lithium ion batteries. Front. Microbiol vol. 9, no. NOV, pp. 1–6. https://doi.org/10.3389/fmicb.2018.02817

Koohsaryan E, Anbia M (2016) Nanosized and hierarchical zeolites: A short review. Cuihua Xuebao/chinese J. Catal. 37(4):447–467. https://doi.org/10.1016/S1872-2067(15)61038-5

Kulikova T, Hiller E, Jurkovič L, Filová F, Šottník P, Lacina P (2019) Total mercury, chromium, nickel and other trace chemical element contents in soils at an old cinnabar mine site (Merník, Slovakia): anthropogenic versus natural sources of soil contamination. Environ. Monit. Assess. 191, no. 5. https://doi.org/10.1007/s10661-019-7391-6.

Kumar V, Yadav SK (2009) Plant-mediated synthesis of silver and gold nanoparticles and their applications. J Chem Technol Biotechnol 84(2):151–157. https://doi.org/10.1002/jctb.2023

Kumari S, Tyagi M, Jagadevan S (2019) Mechanistic removal of environmental contaminants using biogenic nano-materials. Int J Environ Sci Technol 16(11):7591–7606. https://doi.org/10.1007/s13762-019-02468-3

Lien CW et al (2019) DNA engineered copper oxide-based nanocomposites with multiple enzyme-like activities for specific detection of mercury species in environmental and biological samples. Anal Chim Acta 1084:106–115. https://doi.org/10.1016/j.aca.2019.08.009

Lima FRD et al (2019) Critical mercury concentration in tropical soils: Impact on plants and soil biological attributes. Sci Total Environ 666:472–479. https://doi.org/10.1016/j.scitotenv.2019.02.216

Liu Y et al (2022) Detection and remediation of mercury contaminated environment by nanotechnology: Progress and challenges. Environ. Pollut., vol. 293, no. November 2021, p. 118557. https://doi.org/10.1016/j.envpol.2021.118557

Ma H et al (2021) Distribution of mercury in foliage, litter and soil profiles in forests of the Qinling Mountains, China. Environ. Res. vol. 211, no. November 2021, p. 113017. https://doi.org/10.1016/j.envres.2022.113017.

Martins C, Vasco E, Paixão E, Alvito P (2013) Total mercury in infant food, occurrence and exposure assessment in Portugal. Food Addit Contam Part B Surveill 6(3):151–157. https://doi.org/10.1080/19393210.2013.775603

Mughal B, Zaidi SZJ, Zhang X, Hassan SU (2021) Biogenic nanoparticles: Synthesis, characterisation and applications. Appl. Sci., vol. 11, no. 6. https://doi.org/10.3390/app11062598

Mukherjee P et al (2001) Bioreduction of AuCl4- ions by the fungus, Verticillium sp. and surface trapping of the gold nanoparticles formed. Angew. Chemie - Int. Ed. 40(19):3585–3588. https://doi.org/10.1002/1521-3773(20011001)40:19%3c3585::AID-ANIE3585%3e3.0.CO;2-K

Niu Y et al (2013) Nanoparticles mimicking viral surface topography for enhanced cellular delivery. Adv Mater 25(43):6233–6237. https://doi.org/10.1002/adma.201302737

Palomo JM, Filice M (2016) Biosynthesis of metal nanoparticles: Novel efficient heterogeneous nanocatalysts. Nanomaterials, vol. 6, no. 5. https://doi.org/10.3390/nano6050084

Pantidos N (2014) Biological Synthesis of Metallic Nanoparticles by Bacteria, Fungi and Plants. J. Nanomed. Nanotechnol., vol. 05, no. 05. https://doi.org/10.4172/2157-7439.1000233

Patil S, Chandrasekaran R (2020) Biogenic nanoparticles: a comprehensive perspective in synthesis, characterization, application and its challenges. J. Genet. Eng. Biotechnol vol. 18, no. 1. https://doi.org/10.1186/s43141-020-00081-3

Patra JK, Baek KH (2014) Green Nanobiotechnology: Factors Affecting Synthesis and Characterization Techniques. J. Nanomater., vol. 2014. https://doi.org/10.1155/2014/417305.

Pennycook TJ, McBride JR, Rosenthal SJ, Pennycook SJ, Pantelides ST (2012) Dynamic fluctuations in ultrasmall nanocrystals induce white light emission. Nano Lett 12(6):3038–3042. https://doi.org/10.1021/nl3008727

Rai A, Singh A, Ahmad A, Sastry M (2006) Role of halide ions and temperature on the morphology of biologically synthesized gold nanotriangles. Langmuir 22(2):736–741. https://doi.org/10.1021/la052055q

Rice KM, Walker EM, Wu M, Gillette C, Blough ER (2014) Environmental mercury and its toxic effects. J Prev Med Public Heal 47(2):74–83. https://doi.org/10.3961/jpmph.2014.47.2.74

Saturday A (2018) Mercury and its associated impacts on environment and human health: A review. J. Environ. Heal. Sci. 4(September):37–43. https://doi.org/10.15436/2378-6841.18.1906

Shervani Z, Yamamoto Y (2011) Carbohydrate-directed synthesis of silver and gold nanoparticles: Effect of the structure of carbohydrates and reducing agents on the size and morphology of the composites. Carbohydr Res 346(5):651–658. https://doi.org/10.1016/j.carres.2011.01.020

Sohi SP, Krull E, Lopez-Capel E, Bol R (2010) A review of biochar and its use and function in soil. Adv Agron 105(1):47–82. https://doi.org/10.1016/S0065-2113(10)05002-9

Suvarapu LN, Baek S (2017) Recent Studies on the Speciation and Determination of Mercury in Different Environmental Matrices Using Various Analytical Techniques_fDsowl. Int. J. Analy Chem., Vol. 2017, Article ID 3624015, 28 pages

Suvarapu LN, Baek SO (2017) Recent Studies on the Speciation and Determination of Mercury in Different Environmental Matrices Using Various Analytical Techniques," *Int. J. Anal. Chem.*, vol. 2017. https://doi.org/10.1155/2017/3624015

Villalobos M, Bargar J, Sposito G (2005) Trace Metal Retention on Biogenic Manganese Oxide Nanoparticles. Elements 1(4):223–226. https://doi.org/10.2113/gselements.1.4.223

Wang X et al (2017) Aerobic and anaerobic biosynthesis of nano-selenium for remediation of mercury contaminated soil. Chemosphere 170:266–273. https://doi.org/10.1016/j.chemosphere.2016.12.020

Wiatrowski HA, Ward PM, Barkay T (2006) Novel reduction of mercury(II) by mercury-sensitive dissimilatory metal reducing bacteria. Environ Sci Technol 40(21):6690–6696. https://doi.org/10.1021/es061046g

Wilcox J et al (2012) Mercury adsorption and oxidation in coal combustion and gasification processes. Int J Coal Geol 90–91:4–20. https://doi.org/10.1016/j.coal.2011.12.003

Yacamán MJ, Ascencio JA, Liu HB, Gardea-Torresdey J (2001) Structure shape and stability of nanometric sized particles. J. Vac. Sci. Technol. B Microelectron. Nanom. Struct., vol. 19, no. 4, p. 1091. https://doi.org/10.1116/1.1387089

Yu W, Sisi L, Haiyan Y, Jie L (2020) Progress in the functional modification of graphene/graphene oxide: A review. RSC Adv 10(26):15328–15345. https://doi.org/10.1039/d0ra01068e

Zhang Y, Dong Y, Zhou J, Li X, Wang F (2018) Application of plant viruses as a biotemplate for nanomaterial fabrication. Molecules, vol. 23, no. 9. https://doi.org/10.3390/molecules23092311

Zohora N, Kumar D, Yazdani M, Rotello VM, Ramanathan R, Bansal V (2017) Rapid colorimetric detection of mercury using biosynthesized gold nanoparticles. Colloids Surfaces A Physicochem. Eng. Asp. 532(April):451–457. https://doi.org/10.1016/j.colsurfa.2017.04.036

Zulaikhah ST, Wahyuwibowo J, Pratama AA (2020) Mercury and its effect on human health: A review of the literature. Int. J. Public Heal. Sci. 9(2):103–114. https://doi.org/10.11591/ijphs.v9i2.20416

Chapter 19
The Occurrence, Effect and Biodegradation of Antibiotics Using Metallic Biogenic Nanomaterials in Water

Njabulo S. Mdluli, Philiswa N. Nomngongo, and Nomvano Mketo

Abstract Antibiotic resistance is one of the most serious risks to human health today, requiring the use of more toxic, expensive, and ineffective medications. Antibiotic resistance can be caused by several factors, including overuse of antibiotics in humans and cattle. In this scenario, experts are looking for innovative ways to combat this alarming condition. Nanoparticles have emerged as new tools that can be used to combat deadly bacterial infections directly or indirectly to overcome antibiotic resistance. Despite the fact that nanoparticles are used in the pharmaceutical sector, there is ongoing concern regarding their potential damage to human health. This book chapter therefore addresses the use of more environmentally friendly, cheap, biodegradable and harmless biogenic nanoparticles in biodegradation of antibiotics in waste water. This book chapter will also address the limitations that were associated with the old nanoparticles that were not biogenic. Proper recommendations on how to improve the already present biogenic nanomaterial will be proposed.

Keywords Biogenic · Antibiotic · Biodegradable · Water · Nanoparticles and resistance

N. S. Mdluli · N. Mketo (✉)
Department of Chemistry, College of Science, Engineering and Technology (CSET), University of South Africa, Florida Science Campus, Johannesburg 1709, South Africa
e-mail: nomvano.mketo@gmail.com; mketon@unisa.ac.za

P. N. Nomngongo
Department of Chemical Sciences, University of Johannesburg, Doornfontein 2028, PO Box 17011, Johannesburg, South Africa

M. P. Shah et al. (eds.), *Biogenic Nanomaterials for Environmental Sustainability: Principles, Practices, and Opportunities*, Environmental Science and Engineering,
https://doi.org/10.1007/978-3-031-45956-6_19

19.1 Introduction

Antibiotics can either kill or inhibit microorganisms, since ancient times, antibiotics have been used to cure and prevent infections as well as support a wide range of medical interventions, from chemical therapy to organ donation. The well-known antimicrobial routes of antibiotics include inhibition of enzymes, DNA interference, RNA and protein production, and degradation of membrane structure (Wang et al. 2021). Antibiotics can be categorized based on their chemical makeup, range of action, or manner of action. For instance, there are two types of antibiotics: bacteriostatic and bactericidal (causing growth inhibition of bacteria) (Duarte et al. 2022). Narrow spectrum antibiotics only target one of the two types of bacteria, target both Gram-negative and Gram-positive bacteria (Wang et al. 2021). Antibiotics can be categorized into β-lactams, macrolides, tetracyclines, quinolones, aminoglycosides, sulphonamides, glycopeptides, and oxazolidinones based on their molecular structures. By adhering to penicillin binding protein, the β-lactam antibiotics prevent bacteria from synthesizing their cell walls (Japareng et al. 2015). The peptide units in the peptidoglycan layer are cross-linked by penicillin binding proteins. When β-lactams bind to penicillin binding proteins, the latter are inhibited, which results in cell lysis (Mubeen et al. 2021). There are several classes of antibiotics that have been reported in literature which include; sulfomanides, tetracyclines, fluoroquinolones, macrolides, trimethoprim, lincomycin, and others.

(a) Sulfonamides—these are a significant class of broad-spectrum antibiotics that have been widely utilized (Li et al. 2020). Sulfonamides structurally resemble para-aminobenzoic acid (PABA), a substrate used in bacterial cells to synthesize folic acid. Folic acid is required for the production of nucleic acids (DNA), hence sulfonamides prevent cell division and slow bacterial development by competing with PABA and preventing folic acid synthesis (Zhang and Li 2011). Unfortunately, it has been discovered that some bacteria are resistant to these common antibiotics, making it challenging to treat infections brought on by these germs (Yang et al. 2020). In wastewater treatment facilities in Europe, North America, East Asia, and Australia, sulfonamides have been discovered (Li et al. 2021). Additionally, under sulfonamides, the sulfamethoxazole antibiotic was the one that was found the most frequently followed by sulfamethazine, sulfapyridine, and sulfadiazine (Yang et al. 2020) (Fig. 19.1).

(b) Quinolones and fluoroquinolones—A family of broad-spectrum antibiotics known as quinolones has quinolone moiety as its fundamental structural component (Xiao et al. 2008). A subgroup of quinolones known as fluoroquinolones has a core ring that has been fluorinated (Jia et al. 2012). Quinolones prevent the growth of bacteria by preventing DNA helicases from doing their job, which is to unwind the double helical helix of DNA in preparation for replication or repair (Zhang and Li 2011). Additionally, topoisomerase II and topoisomerase IV in bacteria are affected negatively by quinolones, which has a deleterious impact on RNA polymerase and prevents RNA synthesis (Turiel et al. 2005).

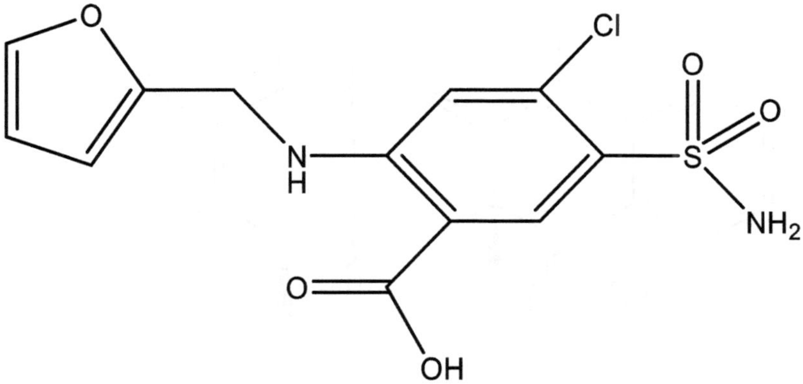

Fig. 19.1 The structure of Sulfonamide

Since the discovery of nalidixic acid in the 1960s, four generations of antibiotics have been created in this family (Turiel et al. 2005) (Fig. 19.2).

(c) Tetracyclines—These are a class of organic and semi-organic antibiotics that prevent the production of bacterial proteins (Saadati et al. 2016). The 30S inhibitors, which include tetracyclines and aminoglycosides, bind to the 30S ribosomal subunit, preventing aminoacyl -tRNAs from accessing the ribosome and so reducing protein synthesis. Tetracyclines and macrolides are frequently bacteriostatic, whereas aminoglycoside generally have a bactericidal effect. The production of nucleic acid (DNA and RNA) is essential for a cell to survive. Tetracyclines have a basic hydronaphthacene backbone structure with four fused rings and different replacements at the fifth, sixth, or seventh positions (Balakrishnan et al. 2023). Eight tetracycline antibiotics in all have been created for use by humans, animals, and the poultry industry. Tetracyclines were found to

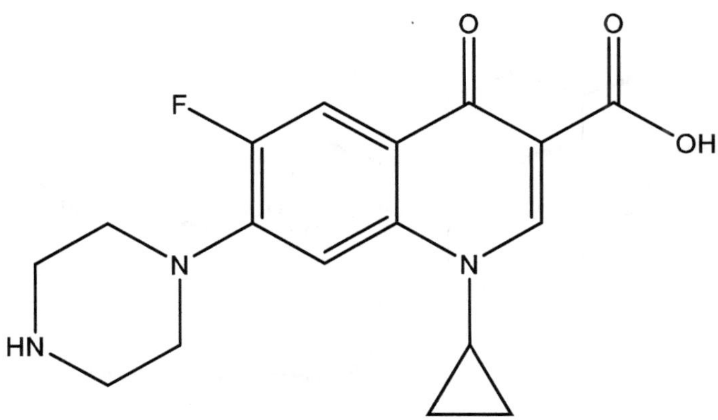

Fig. 19.2 The general structure of quinolone

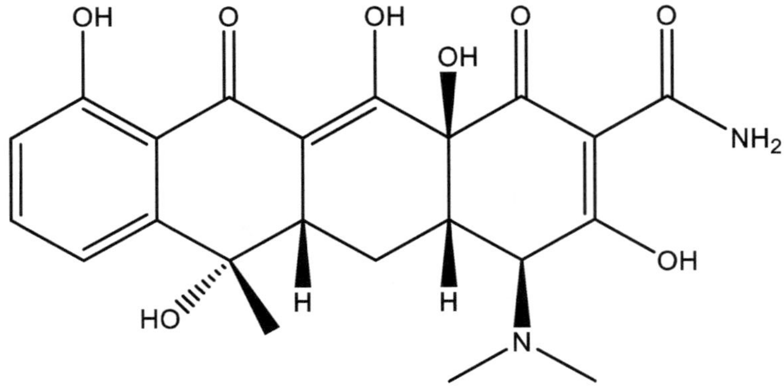

Fig. 19.3 The structure of tetracyclines

be used for veterinary purposes in more than 2500 tons and 1800 tons of tetracycline per year, respectively, in Europe (1998) and the United States (2001) (Kim et al. 2013) (Fig. 19.3).

(d) Macrolides—these are typically employed as penicillin substitutes because they impede bacterial protein synthesis by having a lactone ring that has been substituted with hydroxyl (or neutral or amino sugars), alkyl, and ketone groups (Xiao et al. 2008; Göbel et al. 2007). Protein synthesis is stopped when macrolides attach to the 50S ribosomal subunit and prevent the elongation of mRNA during translation. In wastewater effluent, six macrolides antibiotics and one metabolite, erythromycin-H_2O, have been found worldwide (Shearer et al. 2022). The primary metabolite of erythromycin, erythromycin- H_2O, is extremely unstable in conditions of severe acidity (Zhang and Li 2011) (Fig. 19.4).

(e) Trimethoprim—The synthetic antibacterial drug trimethoprim is a member of the diaminopyrimidines chemical class (Kraupner et al. 2020). Most often,

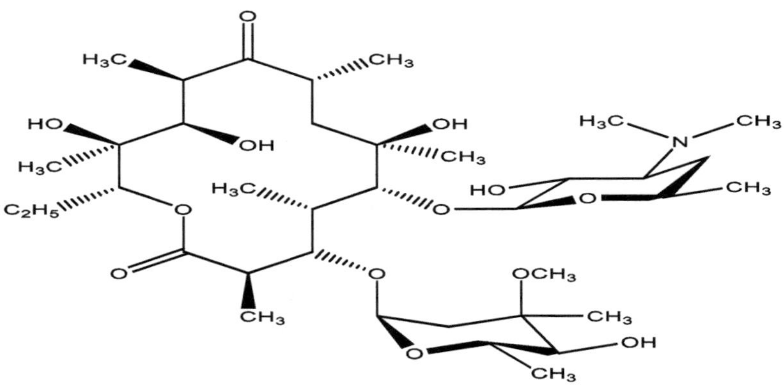

Fig. 19.4 The general structure of macrolides

urinary tract infections are treated with trimethoprim (Huovinen 1987). These chemicals prevent dihydrofolate reductase from working. In microbial and eucaryotic cells, an enzyme called 1.5.1.3 catalyzes the conversion of dihydrofolate to tetrahydrofolate. Unlike todihydrofolate reductase inhibitors aminopterin and methotrexate, which are structurally comparable to folate, the diaminopyrimidines are smaller than and structurally distinct from folate (Kraupner et al. 2020). Aminopterin and methotrexate inhibit both mammalian and microbialdihydrofolate reductase, whereas trimethoprim and other diaminopyrimidines are more effective at inhibiting microbial dihydrofolate reductase than mammalian dihydrofolate reductase (Kovalakova et al. 2020). Studies using X-ray crystallography have been unable to determine the cause of this discrepancy in potency. The substrate-binding site of Escherichia coli dihydrofolate reductase was found to match well with trimethoprim, but not mammalian dihydrofolate reductase. According to a global consumption survey conducted between 2000 and 2010, trimethoprim is one of the top five antibiotics used globally, which is what is causing the rise in trimethoprim resistance (Kraupner et al. 2020). Additionally, since sulfamethoxazole and trimethoprim, a dihydrofolate reductase inhibitor, are usually often prescribed together at a fixed ratio of 1:5, the detection data in raw influent samples is normally 1:5.4 which resembles the common prescription pattern (Zhang and Li 2011) (Fig. 19.5).

(f) Others—several other types of antibiotics have been reported in literature and these are lincomycin, clindamycin, chloramphenicol and thiamphenicol. All these antibiotics are normally reported in small concentration in water due to them being rarely used in human and animals (Japareng et al. 2015).

However, the overuse of antibiotics, high dosages that result in increased toxicity, prolonged hospitalization, and rising fatality, their widespread use in agriculture, and the absence of new antibiotic development, all directly contribute to the emergence of drug resistance in microorganisms (Mubeen et al. 2021; Nasir et al. 2020). In addition to the negative socioeconomic effects, antibiotic resistance to infectious pandemic

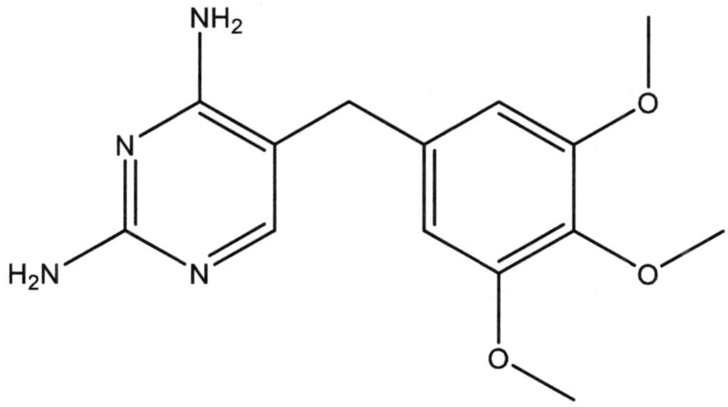

Fig. 19.5 The general structure of trimethoprim

diseases poses a severe public health concern (Anand 2022; Singh et al. 2018). There are several known feasible methods for preventing antibiotic resistance in microbes, including reducing the use of antimicrobial drugs, improving drug release, changing antibiotic targets, creating drugs to break down or alter microorganism enzymes, creating a biofilm coating that contains the bacteria, and avoiding antibiotic exposure (Li et al. 2021). Most importantly, no brand-new antibiotic varieties have been created in recent years (Jia et al. 2012). Additionally, the creation and promotion of novel antibiotics is a pricey and drawn-out process that calls for new chemicals, several clinical trials, and licensing (Tahrani et al. 2016). Resistance has emerged against a number of commonly used antibiotic types. The likelihood that bacteria may quickly acquire resistance to brand-new antibiotics would result in decreased antibiotic usage and sales, further deteriorating the public health and economic conditions. As a result, an increased risk of infection-related death would eventually result from the failure of antibiotic progress (Zhang and Li 2011).

With the advent of nanotechnology, the most important development in recent years, medicine has been updated. The market for nanotechnology products is steadily expanding. The revolutionary science of nanotechnology will have an impact on our efforts to enhance human health (Anand 2022). The longevity, effectiveness, toughness, adaptability, and unique physicochemical properties of nanoparticles have all been explored by the medical sector (Yang et al. 2020). They are used in a variety of therapeutic methods, including the focused administration of drugs, prognosis visual monitoring of therapy, and even the detection of tumors (Mubeen et al. 2021). Green techniques are used in the manufacture of biogenic nanoparticles. The creation of stable nanoparticles, the use of a biomass-based surface coating that increases the amount of active surface areas for biological interaction, the avoidance of hazardous byproduct formation, and additional stabilizing or reducing factors that eventually make the process economical are all benefits of green synthesis (Li et al. 2021; Vena et al. 2016; Gunatilake 2015).

19.2 Effects of Antibiotics on Aquatic Organisms

The presence of antibiotics in the water system is a result of anthropogenic stress brought on by the production of pharmaceuticals, the manufacture of personal care products, medical facilities, and agricultural areas (Costanzo et al. 2005). Additionally, the overuse of bactericides and bacteriostatic medications as well as agricultural runoff from animal feeding practices encourage the penetration of antibiotics into groundwater systems via surface (Kovalakova et al. 2020). Population growth, excessive antibiotic usage, poor medication prescribing, self-medication, excessive use of agricultural areas, and faulty wastewater treatment by pharmaceutical companies can all exacerbate the discharge of contaminants (Göbel et al. 2007). Drugs (antibiotics) have been discovered in pharmaceutical treatment facilities, hospital sewage treatment facilities, sewage treatment facilities, municipal sewage treatment facilities, and urban wastewater treatment facilities (UWWTP) (Li et al. 2019). However, these antibiotics can build up and undergo biotransformation in water systems because of misuse, limits in treatment facilities, and abuse (Costanzo et al. 2005). Due to the formation of multi-resistant bacterial genes at micro level, which in turn increases the chances of bacterial survival, these bio-accumulated antibiotics may have greater harmful consequences (Kümmerer 2009a). Since the twentieth century, antibiotics have gained a lot of attention due to the connection between their use and the emergence of resistance (Costanzo et al. 2005). Antibiotics are posing a major threat to the ecosystem despite the many benefits they provide for humans (Kümmerer 2009b). It has been observed at higher quantities than the disposal requirements, which raises the possibility of a concern down the road (Costanzo et al. 2005). Small doses of antibiotics in their active state, along with their parent components, have the potential to gradually alter the microbial colonies in water bodies (Kim et al. 2013). As basic antibiotics lose their effectiveness and functionality owing to microbial resistance, this therefore can negatively impact human health due to accumulating negative consequences of antibiotics (Liu et al. 2018).

There are different ways on which microorganisms can build resistance to medication, the common one is developing a physical habitat to resist the effects of better and newer drugs (Carvalho and Santos 2016). These microorganisms are either naturally existing at the site of origin or have the potential to change over time as a result of recurrent antibiotic exposure. Various tactics are used at an organizational and experimental level to combat the impacts of antibiotics on the environment (Yang et al. 2020). The most dependable of these processes, known as the Advanced Oxidation Process, is photocatalysis (Morsi et al. 2020; Amiralian et al. 2020). It is an efficient invention for the treatment of wastewaters containing organic pollutants that cannot be removed properly, such as antibiotics. The photon threshold energy affects the effectiveness of photocatalytic reactions (Liao et al. 2021). Superoxide, a significant reactive oxygen species, is produced during photocatalysis via a reductive process, which involves an interaction between electrons and molecular oxygen (Liao et al. 2021). Nanoparticles, metal oxides, and semiconductors have all been investigated as catalysts to increase superoxide production for pollutant degradation (Pan et al. 2020). A sustainable, environmentally acceptable remedial alternative to chemical-based remediation is the photocatalytic reaction.

Chemical-based methods can encourage the creation of halogenated species from some contaminants, which may be carcinogenic, by using hypochlorite and chlorine for the breakdown of antibiotics (Göbel et al. 2007). The Advanced Oxidation Process have surpassed chemically motivated oxidation approaches in popularity. They can oxidize and degrade antibiotic medications based on the concentration of the generated hydroxyl radicals and other non-radical species such singlet oxygen. The given irradiation settings have an impact on the photocatalysis response of the suspended photocatalysts (Xu et al. 2012). Similar to how electrons are excited from the valence band to the conduction band, the ground state electrons are stimulated to a higher energy level. This less hazardous, highly stable, economical, simple to make, and highly photoactive characteristics are desirable in a photocatalyst (Xu et al. 2012).

The widespread use of antibiotics has made their presence in both natural and manmade environments omnipresent, and almost everyone in the world is aware of this fact (Carvalho and Santos 2016). Surface and groundwater as well as other natural water sources have been shown to be contaminated with antimicrobials. Because they degrade more slowly, are improperly disposed of, are misused, are produced in large quantities, and are wasted, new antibiotics continue to survive in the ecosystem (Costanzo et al. 2005). New avenues of research have been opened up to address these problems as a result of the rising amounts of emerging contaminants. Due to slower rates of degradation, these are still extant in the environment. Human excretion, which is dependent on an individual's consumption and metabolism, is how they are released into water sources. To prevent their accumulating in the discharge, these composite materials must be removed promptly. New pollutants are becoming more prevalent, necessitating the modification of traditional scientific methods to address them because they have unidentified adverse effects on living things and the environment (Liu et al. 2018). The allowable threshold of a new pollutant, which is less harmful to the current environment when combined with other bio-chemicals and manmade chemicals, is unknown (Carvalho and Santos 2016). This class of pollutants affects human life in addition to disrupting the environment's delicate balance. When a medication created to treat a certain condition stops working, other medications are created to treat the same ailment, but they may also have additional side effects or impact other organs (Kümmerer 2009b). Emergence of persisters, antibiotic microbial resistance AMR, and antibiotic resistant genes ARGcan be foreseen when a chain of new medications is built up using outdated, unused medications that are dumped into the environment. Persisters are more common than emerging resistant types, according to reports (Costanzo et al. 2005). It takes time to identify new resistance, and during that time, the microbe may experience a number of additional changes (Liao et al. 2021). The last areas of sewage treatment facilities, or waste water treatment plant WWTPs, before entering aquatic systems, are where antibiotics are handled (Pan et al. 2020). Here, they can mix both horizontally and vertically and be transported farther down by weather changes and diffusion mechanisms (Sukri et al. 2019). Due to worries over water quality, it is currently rare to find pure, natural, safe drinking water. Antibiotic pollution has affected even tap water, which was once believed to be a safe supply of drinking water. The presence of macrolides, erythromycin, and clarithromycin in drinking tap water was confirmed by a study carried out in Madrid, Spain (Wang et al. 2021). During infusion and transportation through a waste water treatment plant WWTP, the drugs may be adsorbed to dispersed particles or may build on residue before returning via resuspension into the column (pipelines) (Wang et al. 2018). The remainder will be transferred to the sediments in the water body, while antibiotics with an affinity for adsorption can stay in the column. The aquatic systems' equilibrium is further hampered by the percolation of antibiotics, having a detrimental effect that is challenging to reverse (Liao et al. 2021). The inherent bio-composition of soil can have antimicrobial properties, so using sewage water as an alternate removal approach on agricultural regions. However, it is currently being questioned if natural pollutants are still present in these water sources despite these techniques of removal because human activities have boosted antimicrobial resistance and made urban aquatic systems more susceptible to antimicrobial pollution (Göbel et al. 2007). Although soil prevents contaminants from growing in auxiliary surface water, once they do, it is difficult to remediate the water. Natural infiltration and water outlet pipelines are two sources of groundwater contamination (Li et al. 2021). Sludge and fertilizer have repeatedly been released and accumulated, creating some significant hotspots that may facilitate the spread of antibiotics into the soil. The most common ways that contaminants are transferred to plants from sludge soils are by grazing by animals, uptake by root surfaces, leaf absorption, and root absorption (Ping Tai et al. 2020).

19.3 Occurrence of Antibiotics in Water

The destiny of antibiotics and antibiotic resistance genes (ARGs) in natural or manmade water environments is causing increasing ecological and health hazards (Wang et al. 2018). It has been noted that wastewater treatment plants' effluent is an important environmental source of antibiotics and ARGs (Chen et al. 2014). WWTP waste either recycled or released into surface water bodies, both of which carry potential risks of exposure. Rarely were the effects of WWTP effluents on the concentrations of antibiotics and ARGs in water bodies that received effluent as well as the effectiveness of antibiotic and ARG removal in reclaimed wastewater treatment plants (RWTPs) jointly examined (Chen et al. 2014).

Resistance to antibiotics in human pathogenic microorganisms has become more prevalent worldwide in recent decades as a result of the misuse of antibiotics in healthcare, agriculture, and animal husbandry (Kümmerer 2009b). This has posed a serious threat to ecological security and public health. ARGs, or antibiotic resistance genes, that are generally durable and accessible to other bacteria, can be released or disseminated by antibiotic-resistant bacteria in aquatic environments, increasing the population of these organisms. ARGs are also more prevalent in aquatic environments due to the presence of antibiotics (Wang et al. 2018). The expansion and co-pollution of antibiotics and ARGs have been recognized as posing a serious hazard to people's health. In municipal wastewater treatment facilities (WWTPs), antibiotics and ARGs are frequently found. In general, conventional WWTPs are capable of efficiently removing chemical oxygen demand, phosphate, nitrogen, and nitrogen from wastewater, but not ARGs and antibiotics. Antibiotic and ARG removal rates in WWTPs ranged between 79 and 88% and 1.0 and 2.0 log (Chen et al. 2014).

In urban secondary effluents, antibiotics have regularly been found in concentrations between 10 and 3000 ng L^{-1}. The commonly used antibiotics trimethoprim, quinolones, macrolides, tetracyclines, sulfonamides, and beta-lactams were present in the effluents from WWTPs (ECDC Annual epidemiological report for 2016, 2018). Quinolone resistance (e.g., qnrD, gyrA, qnrC, qnrS, andparC), macrolide resistance (e.g., ereA2, ermA,ereB, ermB,, and mefA), and tetracycline resistance are examples of ARGs (e.g., tetA). Furthermore, the abundances of ARGs in WWTP effluents ranged from 103 to 106 copies mL^{-1} (Tahrani et al. 2016). WWTP effluents, which are valuable water sources, are frequently used to replenish water to adjacent bodies of water or are reclaimed in reclaimed wastewater treatment plants after going through some advanced treatment. In surface water bodies around the world, antibiotics and ARGs have been found in a variety of species and abundances. We anticipate that WWTPs will serve as significant barriers restricting the discharge of antibiotics and ARGs for aquatic systems that receive wastewater (Tahrani et al. 2016). However, some evidence indicates that aquatic entities receiving wastewater from WWTPs may be significantly exposed to antibiotics and ARGs (Tahrani et al. 2016). Researchers discovered that the predominant species of antibiotics in effluents and coastal waterways, as well as the distance from WWTP discharge ports, both affected the amounts of antibiotics (Wang et al. 2018).

Different WWTP effluent applications, such as reclaiming or refilling local water bodies, could result in various exposures (Bai et al. 2022). ARGs and antibiotic hazards Rarely were the effects of WWTP effluents on the concentrations of antibiotics and ARGs in water bodies that receive effluent as well as the effectiveness of antibiotic and ARG removal in reclaimed wastewater treatment plants (RWTPs) jointly examined. A study was conducted investigating how secondary effluents from four WWTPs impacted the concentrations of antibiotics and ARGs in their downstream effluent-receiving lakes or rivers and RWTPs in three northern Chinese towns (Chen and Wang 2021).

When the ARB and ARG released from the human body return to their natural environments (such as food, water, and soil), they may raise the risk of environmental harm (Wang et al. 2018). In order to comprehend how clinical antibiotic resistance affects the environmental ARB, it is helpful to investigate antibiotic resistance in untreated hospital wastewater. Due to various antibiotic administration patterns, hospital wastewater may include varying levels of antibiotic resistance than other aquatic ecosystems. For instance, vancomycin, piperacillin, and cefotiam are antibiotics that are primarily utilized in hospitals (Li et al. 2021). Therefore, effluent from hospitals is where these antibiotic-resistant bacteria are most likely to be found (Kim et al. 2013). In order to determine the possible dangers associated with the release of clinical ARB and ARGs into the environment, this can be investigated by assessing the pollution levels of antibiotics, ARB, ARGs, and mobile genetic elements (MGEs) in untreated hospital wastewater. In Xinxiang city's wastewater, for instance, a wide variety of antibiotics from various families was selected, and the presence of sulfonamides, quinolones, tetracyclines, macrolides, lincomycin, cephalexin, and trimethoprim was analyzed (Kovalakova et al. 2020). This investigation involved keeping an eye on the amount of antibiotics present in swine farm wastewater. This is due to the fact that swine production utilized the most antibiotics when compared to other species. Because of their effectiveness, broad spectrum, and low cost, sulfonamides, tetracyclines, fluoroquinolones, macrolides, and other antibiotics including lincomycin, trimethoprim, bacitracin, and ormetoprim were frequently employed in the swine business (Kovalakova et al. 2020). However, after administration, up to 30% to 90% of the parent components of antibiotics are eliminated through the urine or feces (Tahrani et al. 2016). Wastewater, sludge, and manure were all shown to contain significant amounts of antibiotic residues.

There are eleven groups of antibiotics that are mostly reported in literature and they include aminoglycosides, macrolides, polymyxins, lactams, oxazolidinones, glycopeptides, quinolones(fluoroquinolones), streptogramins, sulfonamides, tetracyclines, and other substances (a class of drugs that includes chloramphenicol, trimethoprim, lincomycin, thiamphenicol, and clindamycin) (Göbel et al. 2007).

19.4 Degradation of Antibiotics in Water

Public concern has been raised by the widespread use of antibiotics, particularly their misuse or abuse (Kong et al. 2015). A substantial amount of wastewater containing antibiotics is produced during the creation and usage of antibiotics and dumped into the environment, severely contaminating it (Wang and Zhuan 2020). The remaining antibiotics are persistent and challenging for traditional biological treatment techniques to break down. Long-term antibiotic usage in the environment may lead to the emergence of bacteria and genes resistant to antibiotics, hastening the spread of antibiotic resistance and jeopardizing ecological and human health systems (Kong et al. 2015). Numerous techniques, such as coagulation, membrane separation, adsorption, and biodegradation, have been used to remove antibiotics from water and wastewater. For biodegradation, the advanced oxidation process (AOPs) has been used (Wu and Hu 2020). The AOPs have the ability to break down antibiotics or transform them into tiny molecules, which could lessen the inhibitory effects of antibiotics on microorganisms and increase the rate of biodegradation and elimination. In advanced oxidation techniques, strong oxidation agents including ozone (O_3), superoxide radical (O_2), and hydroxyl radical (OH) are employed to break down organic pollutants (Chen and Wang 2021). The AOPs can be categorized as Fenton or Fenton like oxidation, photocatalytic oxidation, ozonation/catalytic ozonation, ionizing radiation, non-thermal plasma (NTP) and electrochemical oxidation. The following are the different types of AOPs, their advantages and disadvantages (Chen and Wang 2021).

19.4.1 Ozonation or Catalytic Ozonation

Catalytic ozonation, often known as ozonation, is an environmentally favorable wastewater treatment method (Wang et al. 2021). Numerous refractory organic contaminants can be oxidized by ozone, which has an oxidation potential of 2.07 V (Baker et al. 2017). Organic contaminants can be directly degraded by the ozone molecule. Additionally, ozone and water can interact through the use of catalyst to produce hydroxyl radicals (OH), which have a greater propensity to oxidize (Wang and Zhuan 2020). As a result, both homogeneous and heterogeneous catalytic ozonation can be employed to improve the breakdown efficiency of organic contaminants. Liquid catalysts, particularly transition metal ions like Cr^{3+}, Cd^{2+}, Co^{2+}, Cu^{2+}, Ag^{+},

Fe^{2+}, Mn^{2+}, Ni^{2+}, and Zn^{2+} in reaction solution, are utilized in the homogeneous catalytic ozonation process. (Chen and Wang 2021). These catalysts can stimulate ozone to produce hydroxyl radicals and increase the effectiveness of degradation (Chen et al. 2014). Solid catalysts such metal oxide, activated carbon, porous materials, and their composite materials are added to the reaction fluid in the heterogeneous catalytic ozonation process. The degradation of antibiotics by ozonation or catalytic ozonation is mostly affected by several factors which include ozone concentration and pH value of the matrix (Wang et al. 2021).

19.4.2 Electrochemical Oxidation

In the process known as electrochemical oxidation, organic materials are oxidized and transformed into non-toxic and harmless substances or degraded. Direct oxidation and indirect oxidation are both forms of electrochemical oxidation, and they typically occur simultaneously. During the direct oxidation process, organic compounds in water may directly interact with the anode, losing electrons in the process to create tiny molecular molecules (Chen and Wang 2021). In the case of indirect oxidation, anions in the water react with the anode to create intermediate products that have a significant oxidizing power, and these intermediate products oxidize and breakdown organic materials further. Electrolytes are involved in this process. Differing electrolytes will result in varied strong oxidative products and degradation efficiencies (Chen and Wang 2021). The hydroxyl radical is a kind of intermediate oxidant generated by indirect electrochemical oxidation and deposited on the anode surface (OH). Additionally, hydroxyl radicals (OH) have the ability to oxidize organic materials, producing carbon dioxide and tiny molecular molecules (Chen et al. 2014).

19.4.3 Fenton Oxidation

Fenton reagent is the name given to the mixture of ferrous salt and hydrogen peroxide (Anjum et al. 2019). Methods of Fenton oxidation are frequently employed in wastewater treatment. The Fenton reagent (and H_2O_2) are added to the wastewater in the Fenton oxidation Fe^{2+} technique, which can react to produce hydroxyl radicals (OH). These radicals have the potential to oxidize or degrade antibiotics (Wu and Hu 2020). Fenton oxidation has advantages such as improved degradation effectiveness and ease of application. Numerous operational parameters, such as pH level, temperature, H_2O_2 concentration, and Fe^{2+} concentration, have an impact on the treatment efficiency (Kong et al. 2015). Fenton oxidation, which is limited to acidic conditions, has a number of drawbacks, including the production of a significant volume of difficult-to-dispose-of iron-containing sludge. Other catalysts are utilized to substitute Fe^{2+} in the Fenton-like oxidation process in order to get around these drawbacks (Wu and Hu 2020).

19.4.4 Ionizing Radiation

Ionizing radiation, such as gamma rays and electron beams, is a newly developed method for the direct or indirect destruction of organic contaminants (Wang and Zhuan 2020). Various active species are produced during the radiolytic process of water. Organic contaminants can be reduced by solvated electrons and can be oxidized by hydroxyl radicals (OH) (Bai et al. 2022). Ionizing radiation can degrade antibiotics in different ways depending on the absorbed dose, baseline pH, organic matter, and water matrix (Wang et al. 2018).

19.4.5 Photocatalytic Oxidation

It has been extensively researched how to degrade organic contaminants via photocatalytic oxidation. Photocatalysts are made of semi-conductor materials including TiO_2, SnO_2, WO_3, and ZnS (Balakrishnan et al. 2023). Photocatalysts take in energy and then excite to create electrons with high reducing power and holes with high oxidizing power. The energized electrons can decrease O_2 to produce superoxide radical (O_2) (Balakrishnan et al. 2023). While holes travel to the photocatalysts' surface, H_2O will be oxidized to produce hydroxyl radicals. The superoxide radical (O_2) or the hydroxyl radical may then degrade the organic pollutants (OH) (Bai et al. 2022). Due to its high catalytic effectiveness, stability, and lack of secondary pollutants, TiO_2 is the most often used catalyst (Wang and Zhuan 2020).

19.4.6 Non-Thermal Plasma (NTP)

Non-thermal plasma (NTP), which is often produced using electrical energy, is thought to be a new AOP technology for contaminant elimination that complements the pollutant abatement strategy, particularly with regard to complicated, hazardous, and/or persistent organic chemicals (Mirón et al. 2020). It is well known that gas-phase plasmas and plasmas that ignite in liquids or come into contact with liquids can produce chemically varied and reactive species with high oxidation levels, such as OH, HO_2, O, and H_2O_2. without secondary contamination, redox potential (ORP) (Zhang et al. 2021). Over the past ten years, successful tests have been conducted on the use of a number of NTP discharges, including dielectric barrier discharge, pulsed corona discharge, contact glow discharge, and gliding arc discharge, for the degradation of antibiotics in wastewater (Anjum et al. 2019). However, there are significant challenges to industry's use of the technology.

19.5 Biogenic Nanoparticles and Wastewater Treatment

Due to the use of natural products from plants, bacteria, algae, fungi, or yeast, such as polyphenols, vitamins, amino acids, carbohydrates, biopolymers, and natural surfactants, biogenic synthesis requires less time and resources than conventional processes (Kong et al. 2015). Rare research have used top-down techniques in biological synthesis, which is typically bottom-up and capable of synthesizing medium to tiny nanomaterials (NMs) (Wang and Zhuan 2020). NMs have been synthesized from a wide range of biogenic sources, many of which offer interesting advantages over traditional chemical procedures, such as creating a reduced faulty surface, having a greater production rate, and requiring less energy (Zhang et al. 2021). The ability of biogenic nanomaterials, such as metallic nanomaterials and semiconductors, to function as antioxidants and antibacterial, as well as their activity, stability, affordability, and low toxicity, have garnered a lot of interest (Chen and Wang 2021). However, much less is known about the photocatalytic activity of biogenic nanomaterials. The use of biogenic nanomaterials as photocatalysts, photo-electrocatalysts, and photo-antimicrobials to assist advanced oxidation processes (AOPs) has been the subject of recent research (Wu and Hu 2020). Since UV radiation is the only source of activity for semiconductors made on biogenic metal oxide (such as, ZnO, SnO_2 and TiO_2), their application has been severely constrained (Wang and Zhuan 2020). But by utilizing electrochemically active biofilms, which result in a narrowing of the semiconductor band gap, the problem has been resolved (Kong et al. 2015).

High surface-to-volume ratios in biological nanomaterials provide a lot of reactive sites and reduced NM loads (Mubeen et al. 2021). As a result, less nanomaterial is required for water treatment, less waste is produced, and less environmental pollution occurs (Mubeen et al. 2021). Syntheses that are green, eco-friendly, and based on living things are preferred to traditional ones that frequently involve poisonous organic substances and solutions that can be cytotoxic, carcinogenic, and damaging to the environment (Abdullahi Ari et al. 2021). The biological methods utilized to create nanoparticles are readily accessible. Additionally, when using biogenic nanomaterials for environmental applications like water treatment, there is also a large knowledge gap about the necessity to address the toxicity of biogenic nanomaterials with photocatalytic activity. The composition of the molecules covering biogenic materials is yet unclear, as was previously stated, and the range of biogenic sources that can be employed to create biogenic nanomaterial serves to emphasize the need for more research(Rozhin et al. 2021).

Because of the chemical makeup of biogenic NM stabilizing agents, biogenic NMs tend to be larger and more complicated than conventional NMs. The surface of biogenic NMs has been found to include a variety of stabilizers, from large biopolymers (such proteins and cellulose) to tiny chemical compounds. (Xiao et al. 2008). Once stabilized, synthetic NMs frequently only require one polymer or a few additional compounds and exhibit more consistency; but, when production costs and environmental benefits are taken into account, biogenic NMs appear to be by far the superior choice(Jia et al. 2012). However, we have discovered that in order to choose the optimal alternative and identify the circumstances in which one material should be preferred over another, a better, more accurate comparison of the effectiveness of traditional and biogenic NMs is necessary. Only a few research, in particular, have described the use of biogenic NMs with photocatalytic properties to eliminate pathogens in water or decompose organic pollutants (Yang et al. 2020).

Because it greatly influences their level of environmental friendliness and increases their sustainability, production, regulation, control, and remediation, the green-synthesis technique is wanted in the "development and research of smart NMs (Carvalho and Santos 2016). It's interesting to synthesize biogenic NMs in a non-toxic, environmentally friendly manner, particularly when applications for extensive water disinfection and/or potential human exposure are concerned (Kümmerer 2009b).

The two primary strategies for NM synthesis are top down and bottom up. Molecular self-assembly, atomic layer deposition, and electrodeposition are examples of bottom-up procedures. Top-down techniques include lithography, thermal or laser ablation, chemical etching, milling or attrition, thermal or laser ablation, and repetitive quenching (Kong et al. 2015). To the best of our knowledge, top-down methods for biogenic NM synthesis have not been used in any recent investigations(Mubeen et al. 2021). The majority of biogenic NM synthesis procedures in the literature employ bottom-up strategies. The bottom-up approaches examine the development of particles from the form metal atoms that are produced from ionic or molecular precursors in a manner comparable to chemical processes as opposed to physical processes, which depend on the division of bulk metals (Wu and Hu 2020). The most important phase in the extracellular preparative process is managing atom aggregation to guarantee the size and consistency of the MNPs, which is challenging to accomplish utilizing biogenic synthetic techniques (Mirón et al. 2020). The intracellular preparative method, on the other hand, is more suitable for producing homogenous, tiny nanoparticles. A synthetic biology strategy that results in the intracellular creation of several kinds of metallic nanoparticles can also be used to produce biogenic NMs. For instance, magnetotactic bacteria create ferromagnetic particles to enable their direction and movement within a specific geomagnetic field (Mirón et al. 2020).

When NMs are formed, the bottom-up synthesis process begins, made from an ion of the target substance. The biogenic NMs are synthesized by dissolving a raw balk or metal in a proper solvent, and after that, a reducing agent is added to begin the reduction process (Wu and Hu 2020). The nanoparticles are stabilized to maintain the desired size by surfactants, polymers, or surface stabilizers that are readily available (Rozhin et al. 2021). For the production of nanoparticles, bottom-up procedures are far more common, and numerous techniques have been developed (Abdullahi Ari et al. 2021). For example, spray pyrolysis, laser pyrolysis, aerosol techniques, chemical and electrochemical precipitation, vapor deposition, sol–gel processes, homogeneous nucleation from liquid or vapor, or heterogeneous nucleation on substrates, can all be used to create nanoparticles (Ahmed et al. 2017). Additionally, phase segregation under increased temperatures can be used to prepare nanoparticles.

19.5.1 Application of Biogenic Nanoparticles in Wastewater

Due to their antibacterial properties, metallic nanoparticles like silver, iron, copper, zinc, and titanium can be employed to treat multidrug resistant (MDR) microorganisms (Wu and Hu 2020). It's important to note that biogenic nanoparticles are mostly used for due to their biocompatibility and long-term stability, antibacterial uses (Ahmed et al. 2017). The combined oxidative stress, metal ion release, and non-oxidative stress that these nanoparticles experience are what make them antimicrobial. In some instances, green metallic nanoparticles produced by bacteria have been investigated for use as antimicrobial agents against a variety of harmful pathogens (Rozhin et al. 2021). For instance, the pathogens *Salmonella enterica, Candida albicans, Bacillus cereus, Escherichia coli, Bacillus anthracis,*and *Vibrio parahaemolyticus*, were all sensitive to biogenic AgNPs isolated from the bacteria *Bhargavaea indica* DC1, *Sporosarcinakoreensis*DC4, *and Brevibacteriumfrigoritolerans* DC2. *Escherichia coli* copper nanoparticles were capable of killing *Escherichia coli, Proteus vulgaris,* and *Staphylococcus aureus* (CuNPs) (Ahmed et al. 2017). These nanoparticles showed enhanced antibacterial action when coupled with various antibiotics, including oleandomycin, lincomycin, vancomycin, penicillin G, novobiocin, and rifampicin (Rozhin et al. 2021). Additionally, studies on zinc oxide have shown that it has antibacterial properties against *S. aureus, P. aeruginosa*, and*E. Coli* (Mubeen et al. 2021). Therefore, it is of paramount importance to state the green metallic nanoparticles that are key in control of the development of resistant antibiotic microbes (Satapathy et al. 2017).

19.5.2 Metallic Nanoparticles Produced Using Environmentally Friendly Methods and Their Impact on Harmful Microbes

The resources needed for biogenic synthesis are continuously replenished in nature, which acts as a biolaboratory (Satapathy et al. 2017; Asadzadeh et al. 2021). Important biomaterials are available to use the difficult and remarkable biocompatible yield of nanoparticles, in contrast to the conventional chemical technique (Batterjee et al. 2022). Comparing green synthesis to conventional methods, there are various benefits which includes nonlinear photosensitive, chemical stability, and thermal characteristics as well as a higher surface area to volume ratio, which improves their capacity to react with catalysts (Sharma et al. 2019). The greatest significant medical advances in human history have been credited to the use of nanoparticles in biomedicine (Phoon et al. 2020). Metallic nanoparticles differ physically and chemically from their bulk counterparts due to their high surface-to-volume ratio. It can therefore be noted that the use of the traditional metallic nanoparticles was limited as it was much costly to synthesize the nanoparticles, the use of high volumes of chemicals which results to production of secondary waste and mostly the nanoparticles that are not biogenic can be much selective compared to the biosynthesized (Satapathy et al. 2017). The different functional groups on the biogenic nanoparticle can improve removal of wide range of different analytes, depending of the chemical composition of the target analyte. Researchers in academia and industry are currently favoring the biocompatible, affordable, and secure method of creating nanoparticles (NPs) from biomass, such as plant extracts, bacteria, fungi, and algae, using green technologies (Tahrani et al. 2016). In addition, the development of environmentally friendly and green NP synthesis has found extensive use in the food, space, consumer healthcare, pharmaceutical, and cosmetic industries. Such biogenic NPs have recently been identified as important functional agents that contribute to disease prevention and infection control in the healthcare setting" (Singh et al. 2018).

In recent studies, inorganic metal oxides like TiO2, CuO, and ZnO have been synthesized and used (Nguyen et al. 2022). The most intriguing of these metal oxides are ZnO NPs since they are easy to make in adequate yield using environmentally friendly techniques. Zinc oxide (ZnO), an n-type semiconducting metal oxide, retains its natural reducing nature and has a wide range of applications in electronics, optics, and medical systems (Phoon et al. 2020). The US has designated ZnO as a" generally recognized as safe" metal oxide (El-Shahat et al. 2019). The following are some of thebiogenic metallic nanoparticles mostly used in prevention of antibiotic resistance.

(A) **Silver nanoparticles (AgNPs)**—are used in the pharmaceutical, paint, ointment, food, textile, and packaging sectors for their exceptional bactericidal and fungicidal capabilities (Abdullahi Ari et al. 2021). Numerous studies have been conducted on the large-scale green synthesis of AgNPs in various sizes and forms from yeast, bacteria, fungus, and plants (Azhar et al. 2016). It has been established that the main antibacterial effect of AgNPs is either caused by the release of silver ions or by the deposition of nanoparticles inside cells (Noor et al. 2022). Damage to cell membranes, interference with energy metabolism, production of oxidative stress from the development of ROS, and transcriptional suppression are the primary components of the detailed mechanism (Bai et al. 2022). It has been shown that the protein groups in bacterial cell walls and plasma membranes that contain sulfur and phosphorus interact with the silver ions generated by AgNPs (Noor et al. 2022). Furthermore, this leads to an electrochemical imbalance in the cells and allows silver ions to flow through the plasma membrane of the bacterial cell and into its cytoplasm, where they interact with the internal components and permanently injure the cell. Additionally, it has been demonstrated that silver ions damage DNA and RNA, disrupt and destabilize the outer membrane, inhibit respiratory enzymes that produce ROS, and inhibit proteins and enzymes required for the generation of ATP (Wang and Zhuan 2020). Due to their tiny size and great surface area, nanoparticles have a high chance of passing through the cell membrane and peptidoglycan. The increased sensitivity of Gram-negative bacteria to nanoparticles than Gram-positive bacteria has been explained in terms of this phenomena (Zhang et al. 2021).

(B)**Zinc Oxide Nanoparticles (ZnO-NPs)**—Various biological resources are utilized in the synthesis of zinc oxide nanoparticles (ZnO-NPs), including agents that reduce. They are transparent, semiconducting, non-toxic materials with effective photocatalysis. Plant materials such as leaves, roots, and other components are used to create ZnO-NPs (Ezeuko et al. 2021). fruits, flowers, rhizomes, and bark. ZnONPs have uses in drug delivery and anticancer therapy and exhibit possible antibacterial activity and excellent photo degradation.Metallic nanoparticles ZnO-NPs are also extensively studied for their antibacterial properties. *Salmonella, E. coli, S. aureus* and *Listeria* monocytogenes, are only a few of the many Gram-positive and Gram-negative bacteria that have shown sensitivity to ZnO-NPs (Abdullahi Ari et al. 2021). When bacterial cells are treated with ZnO-NPs, ROS are produced, lipids are peroxidized, proteins, DNA, and reducing sugars seep through the membrane, and cell viability increases the ZnO-NPs (Japareng et al. 2015).

(C) **Copper Nanoparticles (CuO-NPs)**—or cupric oxide nanoparticles, have become extremely important because of their use in agriculture, pharmaceuticals, cosmetics, cosmetic surgery, transportation, and power (Durán-Álvarez et al. 2016). It is CuO-NPs are extremely simple to make chemically, but have various drawbacks, such as poor potency, high toxicity, unfriendliness to the environment, and expensive cost (Phoon et al. 2020). CuO-NPs are created from a variety of biogenic sources, including bacteria, pectin, chitosan, alginate, leaf extracts, and other polysaccharides. It has proved difficult to create, in contrast to gold, silver, and other nanoparticles (Xu et al. 2012). CuO-NPs are unstable because of their propensity to oxidize when exposed to an aqueous medium. While there are a few papers describing the manufacture of metallic CuO-NPs from copper salts under non inert conditions, there are relatively few reports describing the synthesis of metallic CuO-NPs under inert conditions (Su et al. 2019). Biological production of CuO-NPs is relatively recent compared to other nanoparticles, and strategies are being investigated to make it simple and environmentally friendly. The mechanism behind the antibacterial activity is assumed to be electrostatic contact between Cu^{+2} and the plasma membrane of CuO-NPs, disrupting the membrane and killing bacteria (Anand 2022).

(D) **Gold nanoparticles**—are one of the most extensively investigated biogenic nanoparticles (AuNPs) (Ahmed et al. 2017). The AuNPs are mainly spherical, triangular, and hexagonal in shape, while rod-shaped nanoparticles have also been described in a number of investigations (Satapathy et al. 2017). AuNPs are produced using the entire plant or by combining a number of different elements that serve as reducing agents (Kaur et al. 2022). It's interesting to note that the size and form of produced nanoparticles are determined by the type of extracts employed as bioreductants. One significant instance of how AuNPs were produced with a variety of sizes (4–77 nm) and shapes (hexagonal, triangular spherical, and rod) is the Galaxaura elongate-derived AuNPs (El-Shahat et al. 2019). The impact of pH on the size of AuNPs is yet another significant finding. Mango peel extract is pH-sensitive, varying from 9 to 2 andwas said to produce nanoparticles with core sizes of 6 nm and 18 nm. The biocompatibility of AuNPs to microbial cells is well established, and they lack bacteriostatic or bactericidal activity (Singh et al. 2018).

(E) **Titanium Dioxide Nanoparticles (TiO_2-NPs)**—Titanium dioxide nanoparticles (TiO_2- NPs) have intriguing optical, dielectric, antibacterial, and catalytic properties that make them appealing for use in a range of catalyst industries, sensors, biosensors, solar cells, and as image-contrast agents in medical diagnostic procedures

(Kaur et al. 2022). It is usual practice to produce TiO2 -NPs with various morphologies, such as nanorods and nanotubes, employing various reducing and stabilizing chemicals (Bandala et al. 2020). Another method is hydrothermal processing, which is simple and inexpensive. However, green routes must be found to have a consistent supply in enough quantities without having any negative environmental consequences (Singh et al. 2018). TiO_2-NPs also display antibacterial action through a variety of pathways, indicating that the likelihood of microbial cells developing resistance to these nanoparticles is quite low. TiO_2-NPs' ability to kill bacteria including *S. aureus, P. aeruginosa, E. faecalis*and *E. coli*has been clearly proven.TiO_2-NPs generate reactive oxygen species (ROS) when exposed to ultraviolet light, which is one of the ways that it kills bacteria (Kong et al. 2015). By interfering with oxidative phosphorylation and disrupting the cell membrane, the ROS created cause cell death. In a recent study, it was found that exposure to TiO_2 photocatalysis significantly lowers the coenzyme-independent respiratory chains, the capacity for biosynthesis, the ability to transport and take up iron and phosphorus, and the rate at which regulatory signaling is inactivated in cells (Bai et al. 2022).

Additionally, for future purposes, researchers will be able to build safer nanomaterials with the aid of the biosynthesis of NPs utilizing waste materials. This will also help researches to advance knowledge of NPs' health and safety considerations. Because the biomaterial-based approaches do not require the use of hazardous chemicals, useful materials can be manufactured readily even at acceptable scale. Diverse capping agents, including biomolecules and polysaccharides, which can function as chelating/reducing and capping agents for the production of NPs, are the subject of considerable research. The resultant particles are thereby shielded from further reactions and aggregation, so enhancing their stability and lifetime. Single-pot reactions without the need of extra surfactants, capping agents, or templates are the norm for greener NP synthesis techniques.

19.6 Conclusion

Biogenic metallic nanoparticles, either alone or in combination with antibiotics, have demonstrated great activity against a variety of MDR pathogens in order to solve these issues. It's critical to consider these nanoparticles' dispersion, absorption, active targeting, and excretion from the body when employing them as medication carriers to treat infections at specific places. Self-cleaning coatings for mobile phones, washing machines, computer keyboards, water treatment, textiles, food packaging, cosmetics, agriculture (nanopesticides and nanofertilizers), and self-cleaning coatings for textiles and food products are just a few examples of applications outside of the biomedical field. The biogenic nanoparticles, though, have not yet been made available for sale for these uses. The real difficulty in using biogenic nanoparticles is striking the correct balance between the cost of manufacture, scalability, and usefulness.

Acknowledgements The authors would like to thank South African National Research Foundation-THUTHUKA (113951) and University of South Africa for their financial support.

References

Abdullahi Ari H, Adewole AO, Ugya AY, Asipita OH, Musa MA, Feng W (2021) Biogenic fabrication and enhanced photocatalytic degradation of tetracycline by bio structured ZnO nanoparticles," Environ. Technol. (United Kingdom), vol. 0, no. 0, pp. 1–16

Ahmed S, Chaudhry SA, Ikram S (2017) A review on biogenic synthesis of ZnO nanoparticles using plant extracts and microbes: A prospect towards green chemistry. J. Photochem. Photobiol. B Biol., vol. 166, pp. 272–284

Amiralian N et al (2020) Magnetic nanocellulose : A potential material for removal of dye from water. J Hazard. Mater vol. 394, no. February, p. 122571

Anand U et al (2022) Safer plant-based nanoparticles for combating antibiotic resistance in bacteria: a comprehensive review on its potential applications, recent advances, and future perspective," Sci. Total Environ vol 821

Anjum M, Miandad R, Waqas M, Gehany F, Barakat MA (2019) Remediation of wastewater using various nano-materials. Arab J Chem 12(8):4897–4919

Asadzadeh SN, Malakootian M, Mehdipoor M, Neyestanaki DK (2021) The removal of tetracycline with biogenic CeO2 nanoparticles in combination with US/PMS process from aqueous solutions: Kinetics and mechanism. Water Sci Technol 83(6):1470–1482

Azhar MR, Abid HR, Sun H, Periasamy V, Tadé MO, Wang S (2016) Excellent performance of copper based metal organic framework in adsorptive removal of toxic sulfonamide antibiotics from wastewater. J Colloid Interface Sci 478:344–352

Bai X, Chen W, Wang B, Sun T, Wu B, Wang Y (2022) Photocatalytic Degradation of Some Typical Antibiotics: Recent Advances and Future Outlooks. Int J Mol Sci vol. 23, no. 15

Baker S, Pasha A, Satish S (2017) Biogenic nanoparticles bearing antibacterial activity and their synergistic effect with broad spectrum antibiotics: Emerging strategy to combat drug resistant pathogens. Saudi Pharm. J. 25(1):44–51

Balakrishnan A, Chinthala M, Polagani RK, Vo D-VN (2023) Removal of tetracycline from wastewater using g-C3N4 based photocatalysts: A review. Environ Res 216(P3):114660

Bandala ER, Stanisic D, Tasic L (2020) Biogenic nanomaterials for photocatalytic degradation and water disinfection: A review. Environ Sci Water Res Technol 6(12):3195–3213

Batterjee MG, Nabi A, Kamli MR, Alzahrani KA, Danish EY, Malik MA (2022) Green Hydrothermal Synthesis of Zinc Oxide Nanoparticles for UV-Light-Induced Photocatalytic Degradation of Ciprofloxacin Antibiotic in an Aqueous Environment. Catalysts 12(11):1–17

Carvalho IT, Santos L (2016) Antibiotics in the aquatic environments: A review of the European scenario. Environ Int 94:736–757

Chen X, Wang J (2021) Degradation of antibiotic Cephalosporin C in different water matrices by ionizing radiation: Degradation kinetics, pathways, and toxicity. Sci Total Environ 791:148253

Chen C, Li J, Chen P, Ding R, Zhang P, Li X (2014) Occurrence of antibiotics and antibiotic resistances in soils from wastewater irrigation areas in Beijing and Tianjin, China. Environ Pollut 193:94–101

Costanzo SD, Murby J, Bates J (2005) Ecosystem response to antibiotics entering the aquatic environment. Mar Pollut Bull 51(1–4):218–223

Dan A, Li L, Tai YP, Zhang XM, Tao R, Yang Y (2020) Behavior assessment of sulfonamides and N4-acetyl sulfonamides from wastewater effluent in subsurface constructed wetlands: Removal, distribution, and biotransformation. Chem. Eng. J vol. 396

Duarte AC, Rodrigues S, Afonso A, Nogueira A, Coutinho P (2022) Antibiotic Resistance in the Drinking Water: Old and New Strategies to Remove Antibiotics, Resistant Bacteria, and Resistance Genes. Pharmaceuticals 15(4):1–22
Durán-Álvarez JC, Avella E, Ramírez-Zamora RM, Zanella R (2016) Photocatalytic degradation of ciprofloxacin using mono- (Au, Ag and Cu) and bi- (Au-Ag and Au-Cu) metallic nanoparticles supported on TiO2 under UV-C and simulated sunlight. Catal Today 266:175–187
El-Shahat M, Abdelhamid AE, Abdelhameed RM (2020) Capture of iodide from wastewater by effective adsorptive membrane synthesized from MIL-125-NH2 and cross-linked chitosan. Carbohydr Polym vol 231, p 115742
Ezeuko AS, Ojemaye MO, Okoh OO, Okoh AI (2021) Potentials of metallic nanoparticles for the removal of antibiotic resistant bacteria and antibiotic resistance genes from wastewater: A critical review. J Water Process Eng vol 41, no. March, p. 102041
Göbel A, McArdell CS, Joss A, Siegrist H, Giger W (2007) Fate of sulfonamides, macrolides, and trimethoprim in different wastewater treatment technologies. Sci Total Environ 372(2–3):361–371
Gunatilake SK (2015) Methods of Removing Heavy Metals from Industrial Wastewater. Journal of multidsciplinary Engineering Science Studies. no. 2912–1309. Vol. 1
Huovinen P (1987) MINIREVIEW Trimethoprim Resistance. Antimicrob Agents Ch 31(10):1451–1456
Japareng AASA, Efaq L, Noman A, Norli JDBI (2015) Removal of heavy metals and antibiotics from treated sewage effluent by bacteria. Clean Technol Environ Policy 17(8):2101–2123
Jia A, Wan Y, Xiao Y, Hu J (2012) Occurrence and fate of quinolone and fluoroquinolone antibiotics in a municipal sewage treatment plant. Water Res 46(2):387–394
Kaur M, Gautam A, Guleria P, Singh K, Kumar V (2022) Green synthesis of metal nanoparticles and their environmental applications. Curr. Opin. Environ. Sci. Heal. 29:100390
Kim H, Hong Y, Park JE (2013) Sulfonamides and tetracyclines in livestock wastewater," Chemosphere, vol. 91, no. 7, pp 888–894
Kong D et al (2015) Cathodic degradation of antibiotics: Characterization and pathway analysis. Water Res 72:281–292
Kovalakova P, Cizmas L, McDonald TJ, Marsalek B, Feng M, Sharma VK (2020) Occurrence and toxicity of antibiotics in the aquatic environment: A review. Chemosphere 251:126351
Kraupner N, Ebmeyer S, Hutinel M, Fick J, Flach CF, Larsson DJ (2020) Selective concentrations for trimethoprim resistance in aquatic environments. Environ. Int., vol. 144, no. September, p. 106083
Kümmerer K (2009a) Antibiotics in the aquatic environment - A review - Part I. Chemosphere 75(4):417–434
Kümmerer K (2009b) Antibiotics in the aquatic environment - A review - Part II. Chemosphere 75(4):435–441
Li Z, Li M, Zhang Z, Li P, Zang Y, Liu X (2020) Antibiotics in aquatic environments of China: A review and meta-analysis. Ecotoxicol. Environ. Saf., vol. 199, no. December 2019, p. 110668
Li Z, Dai R, Yang B, Chen M, Wang X, Wang Z (2021) An electrochemical membrane biofilm reactor for removing sulfonamides from wastewater and suppressing antibiotic resistance development: Performance and mechanisms. J. Hazard. Mater., vol. 404, no. PB, p. 124198
Li L et al (2021) Complete degradation and detoxification of ciprofloxacin by a micro-/nanostructured biogenic mn oxide composite from a highly active mn2+-oxidizing pseudomonas strain," Nanomaterials, vol. 11, no. 7
Liao Q, Rong H, Zhao M, Luo H, Chu Z, Wang R (2021) Interaction between tetracycline and microorganisms during wastewater treatment: A review. Sci Total Environ 757:143981
Liu X, Lu S, Guo W, Xi B, Wang W (2018) Antibiotics in the aquatic environments: A review of lakes, China. Sci Total Environ 627:1195–1208
Mirón C, Sánchez R, Prats S, Todolí JL (2020) Total polyphenol content and metals determination in Spanish virgin olive oils by means of a dispersive liquid-liquid aerosol phase extraction method and ICP-MS. Anal Chim Acta 1094:34–46

Morsi R, Bilal M, Iqbal HM, Ashraf SS (2020) Laccases and peroxidases: The smart, greener and futuristic biocatalytic tools to mitigate recalcitrant emerging pollutants. Sci. Total Environ., vol. 714
Mubeen B et al (2021) Nanotechnology as a Novel Approach in Combating Microbes Providing an Alternative to Antibiotics. Antibiotics, vol. 10, no. 12
Noor NN, Kamaruzaman NH, Al-Gheethi A, Mohamed RM (2022) Degradation of antibiotics in aquaculture wastewater by bio-nanoparticles: A critical review. Ain Shams Eng. J., no. xxxx, p. 101981
Nasir AM et al (2020) Recent progress on fabrication and application of electrospun nanofibrous photocatalytic membranes for wastewater treatment: A review. J. Water Process Eng., vol. 40, no. November 2020, p. 101878
Nguyen NTT, Nguyen LM, Nguyen TTT, Liew RK, Nguyen DTC, Van Tran T (2022) Recent advances on botanical biosynthesis of nanoparticles for catalytic, water treatment and agricultural applications: A review. Sci Total Environ 827:154160
Pan T et al (2020) A Sm-doped Egeria-densa-like ZnO nanowires@PVDF nanofiber membrane for high-efficiency water clean. Sci Total Environ 737:139818
Phoon BL et al (2020) Conventional and emerging technologies for removal of antibiotics from wastewater. J. Hazard. Mater vol. 400, no. April, p. 122961
Rozhin A, Batasheva S, Kruychkova M, Cherednichenko Y, Rozhina E, Fakhrullin R (2021) Biogenic silver nanoparticles: Synthesis and application as antibacterial and antifungal agents. Micromachines, 12, no. 12
Saadati F, Keramati N, Ghazi MM (2016) Influence of parameters on the photocatalytic degradation of tetracycline in wastewater: A review. Crit Rev Environ Sci Technol 46(8):757–782
Satapathy S, Kumar S, Sukhdane KS, Shukla SP (2017) Biogenic synthesis and characterization of silver nanoparticles and their effects against bloom-forming algae and synergistic effect with antibiotics against fish pathogenic bacteria. J Appl Phycol 29(4):1865–1875
Sharma D, Kanchi S, Bisetty K (2019) Biogenic synthesis of nanoparticles: A review. Arab J Chem 12(8):3576–3600
Shearer L, Pap S, Gibb SW (2022) Removal of pharmaceuticals from wastewater: A review of adsorptive approaches, modelling and mechanisms for metformin and macrolides. J Environ Chem Eng 10(4):108106
Singh P, Garg A, Pandit S, Mokkapati VR, Mijakovic I (2018) Antimicrobial effects of biogenic nanoparticles. Nanomaterials, vol. 8, no. 12
Su Y et al (2019) Metallic nanoparticles induced antibiotic resistance genes attenuation of leachate culturable microbiota: The combined roles of growth inhibition, ion dissolution and oxidative stress. Environ Int 128(April):407–416
Sukri M et al (2020) Arsenic adsorption mechanism on palm oil fuel ash (POFA) powder suspension. J. Hazard. Mater., vol. 383, no. September 2019, p. 121214
Tahrani L, Van Loco J, Ben Mansour H, Reyns T (2016) Occurrence of antibiotics in pharmaceutical industrial wastewater, wastewater treatment plant and sea waters in Tunisia. J. Water Health, vol. 14, no. 2, pp. 208–213
Turiel E, Bordin G, Rodríguez AR (2005) Determination of quinolones and fluoroquinolones in hospital sewage water by off-line and on-line solid-phase extraction procedures coupled to HPLC-UV. J Sep Sci 28(3):257–267
Vena MP, Jobbágy M, Bilmes SA (2016) Science of the Total Environment Microorganism mediated biosynthesis of metal chalcogenides; a powerful tool to transform toxic ef fluents into functional nanomaterials. Sci Total Environ 565:804–810
Wang J, Zhuan R (2020) Degradation of antibiotics by advanced oxidation processes: An overview. Sci Total Environ 701:135023
Wang Q, Wang P, Yang Q (2018) Occurrence and diversity of antibiotic resistance in untreated hospital wastewater. Sci Total Environ 621:990–999

Wang R, Ji M, Zhai H, Guo Y, Liu Y (2021) Occurrence of antibiotics and antibiotic resistance genes in WWTP effluent-receiving water bodies and reclaimed wastewater treatment plants. Sci Total Environ 796:148919

Wu S, Hu YH (2021) A comprehensive review on catalysts for electrocatalytic and photoelectrocatalytic degradation of antibiotics. Chem. Eng. J., vol. 409, no. November 2020, p. 127739

Xiao Y, Chang H, Jia A, Hu J (2008) Trace analysis of quinolone and fluoroquinolone antibiotics from wastewaters by liquid chromatography-electrospray tandem mass spectrometry. J Chromatogr A 1214(1–2):100–108

Xu P et al (2012) Use of iron oxide nanomaterials in wastewater treatment: A review. Sci Total Environ 424:1–10

Yang CW, Liu C, Chang BV (2020) Biodegradation of amoxicillin, tetracyclines and sulfonamides in wastewater sludge. Water (Switzerland), vol. 12, no. 8

Zhang T, Li B (2011) Occurrence, transformation, and fate of antibiotics in municipal wastewater treatment plants. Crit Rev Environ Sci Technol 41(11):951–998

Zhang T et al (2021) Degradation of cefixime antibiotic in water by atmospheric plasma bubbles: Performance, degradation pathways and toxicity evaluation. Chem Eng J 421(P2):127730

Chapter 20
Biogenic Silver Nanoparticle and Their Applications

Lalit Mohan, Raksha Anand, Lakhan Kumar, and Navneeta Bharadvaja

Abstract Nanoparticles ranges in size between 1 and 100 nm. Nanoparticles can be synthesized using physical and chemical processes but these methods pose various harmful impact on environment. Now the focus has shifted towards biogenic synthesis as these methods are highly sustainable. The characteristics of nanoparticles depends upon various factors such as incubation conditions, molar concentration of metal precursor solution, pH and biogenic extract employed for synthesis. Various biogenic sources for the production of nanoparticles are bacteria, fungi, plant and its parts, and algae. Silver (Ag) has many inherent properties due to which it has been used over period of time. Biogenic silver nanoparticles (Ag-NPs) have various applications in wide research areas including nanomedicines, biosensing and others. Biogenic Ag-NPs have been extensively studied for its antimicrobial action and it has been found that Ag-NPs are highly effective against gram negative bacterial species as compared to gram positive ones. Ag-NPs have been studied their dye degrading and heavy metal removing capabilities and it has been found that biogenic Ag-NPs have high potency as an environment remediating agent.

Keywords Plant-based NPs · Microbial-based NPs · Antimicrobials · Environmental remediation

20.1 Introduction

The science of nanotechnology includes materials in nano-scale range usually between 1 and 100 nm. Nanotechnology works at nano-scale range and thus, plays a major role in various fields like bioengineering, dentistry and medical, pharmaceuticals and others (Chaudhary et al. 2020a).There are various physical and chemical methods that are being employed for the synthesis of nanomaterials such as

L. Mohan · R. Anand · L. Kumar · N. Bharadvaja (✉)
Department of Biotechnology, Delhi Technological University, Delhi 110042, India
e-mail: navneetab@dce.ac.in

M. P. Shah et al. (eds.), *Biogenic Nanomaterials for Environmental Sustainability: Principles, Practices, and Opportunities*, Environmental Science and Engineering,
https://doi.org/10.1007/978-3-031-45956-6_20

microwave mediated, electrochemical method, photochemical methods, laser evaporation methods, thermolysis, co-precipitation, sol–gel method and sonochemical method (Kolahalam et al. 2019).These methods involve high input of energy, expensive and toxic chemicals such as sodium borohydride or hydrazine as reducing agents for synthesis and consequently produces harmful by-products which poses negative impact on the environment thus, these methods are non-sustainable and cannot be employed for large-scale production basis (Arif and Uddin 2021a; Rauwel et al. 2015). Thus, to overcome the drawbacks of physical and chemical methods, green synthesis approach has gained momentum in the past decades. Green synthesis or the biological methods of synthesis has gained attention owing to their economical and environment friendly nature and high thermal stability of the produced nanoparticles (Arif and Uddin 2021a).The various solvents and reducing agents employed for the synthesis of nanoparticles (NPs) affects the morphology (size, shape), physicochemical properties and the use of synthesized NPs (Gour and Jain 2019). For green synthesis, microorganisms such as bacteria, fungi, yeast and algae and higher organisms such as plants have been used for the synthesis of metal NPs. Employing these organisms reduces the overall production cost, leaves no harmful by-products andmakes the overall synthesisenvironment friendly and sustainable (Abdelghany et al. 2018).

Silver (Ag) possesses many fascinating properties due to which it has been extensively used in many fields such as medical, pharmacy, remediation. Green synthesis of Ag-NPs has been highly studied over the past decade due to its wide variety of applications. The morphology and stability of Ag-NPs can be regulated according to their application by regulating the biogenic synthesis parameters such as pH, light intensity, incubation temperature, precursor metal salt ($AgNO_3$) concentration, concentration of biogenic extract employed for synthesis (Siddiqi et al. 2018). For example, Ag-NPs employed for drug delivery are greater than 100 nm in size whereas those used as antimicrobial agents have a size ranging between 15 and 80 nm (Liu et al. 2021). These unique properties of Ag-NPs enable them to be used in the fields of biosensing, nanomedicines, pharmacy and for enhancing fuel efficiency. Some of the practical applications of Ag-NPs includes their use as antimicrobial agents, anti-cancerous agents, bioremediation candidates, optoelectronics and as a water purifying agent (Rout et al. 2018; Kumar et al. 2013; Banerjee et al. 2014; Wang et al. 2021). Furthermore, Ag-NPs are cost effective, highly abundant in nature, and possess high potential as compared to gold NPs or any other metallic NPs (Abdelghany et al. 2018; Qamer et al. 2020). The current review aims to provide a brief account and advancements in the field of green synthesis of Ag-NPs employing plants and microbes (bacteria, fungi and algae) and their various applications in diverse fields such as healthcare and environmental remediation.

20.2 Biogenic Ag-Nps Synthesis

Synthesis of Ag-NPs can take place employing a number of methods including physical, chemical and biological methods as depicted in Fig. 20.1. Amongst all, biological synthesis methods are the most utilized ones for NPs synthesis due to enhanced stability and safety of NPs. Biogenic Ag-NPs are highly stable, environment friendly, easy to scale up and sustainable as compared to Ag-NPs produced from any other method (Gour and Jain 2019). In the recent times, NPs have been synthesized using microorganisms, extract of plant parts (leaves, seeds, bark), enzymes and metabolites of arthropods and agricultural wastes. Biogenic Ag-NPs due to their low production cost, stability and safety find its uses in various fields such as medicine, pharmaceuticals, environment remediation and in energy enhancement. The compounds like alkaloids, flavonoids, phenolic acids, proteins, enzymes, RNA and DNA present in extract of plants and microbes acts as capping and reducing agent responsible for the conversion of Ag^+ ions into Ag-NPs (Arif and Uddin 2021b). Production of biogenic Ag-NPs involves some basic steps: preparation of extract of biological material; mixing and incubation of extract with $AgNO_3$ solution of desired or optimized concentration; and finally confirming synthesis via visual color change of the mixture to brown color or via UV-spectrophotometer. The produced Ag-NPs are then characterized using various methods such as scanning electron microscopy (SEM), transmission electron microscopy (TEM), X-ray diffraction (XRD), Fourier-transform infrared spectroscopy (FTIR) (Dotto et al. 2012; Khaleelullah et al. 2012; Ramakrishna et al. 2016; Sharma et al. 2016; Selvam and Sivakumar 2012).

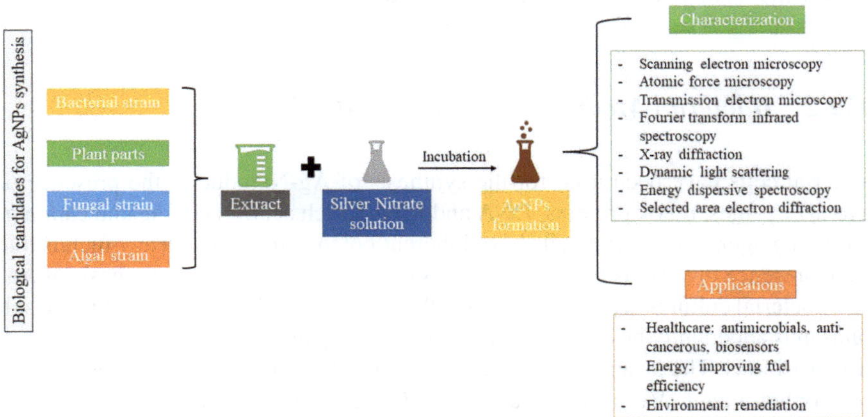

Fig. 20.1 Various physical, chemical and biological methods for synthesis of Ag-NPs

20.2.1 Plant Mediated Ag-NPs Synthesis

Plants and their parts such as leaves, stems, fruits, roots and flowers have been used for green synthesis of Ag-NPs. Plant based production is highly rapid, non-pathogenic, ecofriendly, economical and a one step process. The various phyto-compounds involved in reduction, capping and stabilization of nanoparticles are amino acids, alkaloids, alcohols, carbohydrates, vitamins, polyphenols, glycosides and flavonoids (Abdelghany et al. 2018). For Ag-NPs synthesis, plant extract obtained from different plant parts is incubated with precursor metal salt solution ($AgNO_3$) for a defined amount of time. The time duration of incubation depends upon nature of plant extract (phytocompounds), metal salt concentration, extract concentration, pH of extract, contact time, exposure to light and stirring (Arif and Uddin 2021a; Abdelghany et al. 2018). Pirtarighat et al., reported the synthesis of Ag-NPs using *Salvia spinosa* extract with particle size ranging from 19 to 125 nm (Pirtarighat et al. 2019). Leaf extract of *Ocimum sanctum* have been used by Jain and Mehta for the synthesis of spherical Ag-NPs whose size ranged from about 12 to 20 nm (Jain and Mehata 2017a). Studies conducted by Banerjee et al. reported the synthesis of Ag-NPs (size up to 200 nm) using 1 mM $AgNO_3$ solution and leaf extract of *Musa balbisiana, Azadirachta indica* and *Ocimumtenuiflorum.* They also reported synthesis of Ag-NPs using seeds of *Vigan radiata* and *Cicer arietinum* (Banerjee et al. 2014). Garibo et al., reported green synthesis of quasi spherical AgNPs of average size 5 nm using extracts obtained from *Lysilomaacapulcensis* (Garibo et al. 2020). *Momoridicacharantia* have been reported to synthesize spherical AgNPs of size range 11–16 nm (Vanlalveni et al. 2021). Pahal et al. reported the synthesis of Ag-NPs employing leaflet extracts of *Triticum aestivum* and *Oryza sativa* (Pahal et al. 2022).

20.2.2 Bacterial Mediated Ag-NPs Synthesis

Bacterial cells have been used for the synthesis of Ag-NPs due to the presence of several enzymes, redox proteins, DNA and RNA which act as reducing and capping/stabilizing agents. Bacterial cells have inherent potency to reduce heavy metals due to the presence of various functional groups in the cell wall. For the synthesis of Ag-NPs, bacterial cultures are first grown as cell suspension and then precursor metal salt solution is added into the culture medium followed by either stirring or incubation at still conditions. The progress in formation of Ag-NPs can be monitored by UV–Vis spectrophotometer (Vanlalveni et al. 2021). *Pseudomonas aeruginosa, Escherichia coli, Bacillus cereus, Klebsiella* sp., *Arthrobacter gangotriensis* and *Lactobacillus* sp., are some of the most commonly employed bacterial species for the synthesis of Ag-NPs. Ag-NPs have also been synthesized using actinomycetes. Various enzymes secreted by cell wall and membrane aids in reduction of Ag^+ to Ag^0 (Arif and Uddin 2021a; Gour and Jain 2019; Sachin and Karn 2021). Yusof et al., synthesized spherical Ag-NPs of size range 14 ± 4.7 nm employing *Lactobacillus plantarum* TA4 and

2 mM solution of $AgNO_3$ (Mohd Yusof et al. 2020). *Paenarthrobacter nicotinovorans* has been used for the synthesis of spherical Ag-NPs within the size range of 13-27 nm (Huq and Akter 2021). Saleh and Alwan synthesized spherical Ag-NPs using *Klebsiella pneumoniae* culture supernatant of the size range between 26.48 to 44.42 nm (Saleh and Khoman 2020). Ibrahim et al., reported the synthesis of Ag-NPs (size range 5–7.06 nm) by using *Bacillus cereus* and 1 mM $AgNO_3$ (Ibrahim et al. 2021). Saeed et al. reported the synthesis of Ag-NPs of size range 5–50 nm using the secondary metabolites of *Brevundimonas diminuta, Escherichia coli* and *Exiguobacterium aurantiacumm* (Saeed et al. 2020). Kathwate et al., demonstrated the use of *Pseudomonas aeruginosa* for the the production of Ag-NPs (Deshmukh et al. 2022).

20.2.3 Fungal Mediated Ag-NPs Synthesis

Fungi harbors various proteins due to which they are commonly used as reducing and stabilizing agents for the synthesis of Ag-NPs. Fungal agents are easy to handle, have high yields and produces very low toxic residues. Fungal biomolecules enhance the biological activity of nanoparticles and also improves the stability of nanoparticles. Fungi and yeasts have high metal tolerance and are easy to manipulate and handle. Fungi harbors more than 6,400 bioactive compounds (Firdhouse and Lalitha 2015). Fungi secretes large number of extracellular proteins which provides stability of nanoparticles (Guilger-Casagrande and Lima 2019). Fungi mediated NP synthesis can take place either intracellularly or extracellularly. For intracellular synthesis, metal precursor salt solution is added along the mycelial cultures which is then internalized by mycelia. In case of intracellular synthesis, extraction of NPs is required employing chemical cell lysis, centrifugation followed by filtration. For extracellular synthesis, precursor metal solution is incubated with aqueous fungal filtrate containing biomolecules. Extracellular synthesis results in free NPs formation (Abdelghany et al. 2018). *Fusarium*sp have been reported to be one of the best candidates for the production of Ag-NPs (Rai et al. 2021). Feroze et al. reported the synthesis of spherical Ag-NPs from *Penicillium oxalicum* of size ranging from 60 to 80 nm (Feroze et al. 2020). Ag-NPs have been synthesized extracellularly in size range of 1-24 nm by using *Aspergillus sydowii* (Wang et al. 2021). Velhal et al. synthesized Ag-NPs ranging in size from 2 to 13–80 nm employing *Aspergillus terrerus* (Velhal et al. 2016). Liu et al. reported the synthesis of Ag-NPs falling in size range of 50-500 nm using *Metschnikowia* (Liu et al. 2021).

20.2.4 Algal Mediated Ag-NPs Synthesis

Algae are the most abundant and easily available organism, this property of algae makes them the best candidates for the synthesis of nanoparticles (Agarwal et al.

2021). Algal mediated Ag-NPs synthesis involves three major steps, (i) algal extract preparation in distilled water or in organic solvents by boiling for predetermined duration, (ii) preparation of molar solution of precursor metal salt and (iii) incubation of algal extract along with molar solution of precursor metal solution under controlled conditions (Agarwal et al. 2021; Wang et al. 2017). Various biomolecules present in algae such as polysaccharides, proteins, cytochromes and other pigments causes the reduction and stabilization of metal ions (Sánchez-López et al. 2020). Algal based NP synthesis takes short time as compared to any other biogenic synthesis method (Meroni et al. 2021). Algal NP synthesis takes place both extracellularly and intracellularly (Chaudhary et al. 2020a). Intracellular synthesis of NPs require additional step of recovery of synthesized NPs from within the cells (Peiris et al. 2018). Edison et al., synthesized spherical Ag-NPs of size ~ 25 nm employing *Caulerpa racemose* (Edison et al. 2016). Ag-NPs of size range 8–12 nm have been synthesized using algal extract of *Chlorella pyrenoidosa* (Aziz et al. 2015). Spherical NPs have been reported to be synthesized from *Janiarubins* (El-Rafie et al. 2013). Azizi et al. reported the synthesis of spherical Ag-NPs in size range 5–15 nm using the algal extract of *Sargassum muticum* (Azizi et al. 2021; Li et al. 2015).

20.3 Applications of Ag-Nps

20.3.1 Ag-NPs as Antimicrobial Agents

Biogenic Ag-NPs are emerging as new age antimicrobials and the research in this area has accelerated over the period of time due to the emergence of antimicrobial resistance strains (LewisOscar et al. 2016). Nanoparticles are considered as an effective antibacterial agents because they interact directly with bacterial cells and consequently overcome the antibiotic resistance mechanisms adopted by bacterial species (Wang et al. 2017). Biogenic Ag-NPs are advantageous over conventional antibacterial agents as they prevent the emergence of antimicrobial resistance strains, they are cheap to synthesize, rapid synthesis occurs and environment friendly (Amin 2020).These NPs are reported to use one of the three mechanisms or a combination of mechanisms for causing bactericidal effect (i) increased production of reactive oxygen species, (ii) penetration within membrane, and (iii) interaction with cellular components like DNA, RNA and cell organelles (Yeh et al. 2020). Dhavale et al. demonstrated the synthesis of Ag-NPs using *Amphiroa fragilissima* and its application as an antibacterial agent targeting *Bacillus subtilis*, *Escherichia coli* and *Staphylococcus sp* (Dhavale et al. 2020). *Chlorella vulgaris* extract was used for the biosynthesis of Ag-NPs and was used as antibacterial agents against *E. coli* and *Pseudomonas aeruginosa* (Annamalai and Nallamuthu 2016). *Ocimum sanctum* was used for the synthesis of Ag-NPs and was found effective against *E. coli* (Jain and Mehata 2017b). Yusuf et al. synthesized Ag-NPs employing *Lactobacillus plantarum* and the synthesized particles were found to be effective against both gram positive (*S. aureus* and *S. epidermidis*) and negative (*E. coli* and *Salmonella* sp) (Mohd Yusof et al. 2020). *Aspergillus sydowii* was used for synthesis of Ag-NPs and was found effective against various fungal strains such as *Candida albicans, C. glabrata, Fusarium solani* (Wang et al. 2021).

20.3.2 Ag-NPs Mediated Heavy Metal Remediation

Recently, bio-nanobiotechnology has gained a lot of interest in research especially in the area of environment. Nanoparticles are considered to be the best material for heavy metal remediation as they have high surface activity, large surface-to-volume ratio and unique physical as well as chemical characteristics. Type of nanoparticles and their physical, chemical and magnetic properties plays a major role in the abatement of heavy metal (Thekkudan et al. 2017). Metal-based nanoparticles like Ag, Au, Fe and metal-oxide nanoparticle are widely used for remediation of heavy metals like Cd, Cu, Cr, Zn, Pb and Hg, etc. Ag-NPs can remediate mercury, cadmium, chromium, cobalt, lead, etc. (Al-Qahtani 2017). It is observed that remediation ability of Ag-NPs is dependent on the reduction potential of heavy metals (Yang et al. 2023). Thus, it can be concluded that for every different heavy metal, a different type of nanomaterial is required. Attasi and Nsiah reported that 20 nm Ag-NPs is able to remediate 92.92% lead and 53.34% cobalt within 14 days (Attatsi and Nsiah 2020). On the other hand, El-Tawil et al. observed that Ag-quartz nanocomposite enhances the removal efficiency of mercury to 96% within 1 h (El-Tawil et al. 2019). Many algal species like *Sargassum muticum, Turbinaria ornate, Sargassum polycystum, Turbinaria conoides, Gilidiellaa cerosa, Sargassum wightiigrevilli, Padina pavonica, Colpmenia sinusa*are able to synthesize different size of Ag-NPs (Chaudhary et al. 2020b). Ficus benjamina leaf extract has been utilized for the synthesis of Ag-NPs which further has been employed for the removal of Cd^{2+} from the contaminated sources (Al-Qahtani 2017).

20.3.3 Ag-NPs Mediated Dye Degradation

Biogenic NPs have been studied for their potential use in dyes degradation present in contaminated water bodies. NPs help in effective adsorption of these dyes owing to large surface area of NPs (Patel et al. 2015). NPs are also known to degrade dyes into simpler non-toxic forms. Ag-NPs have been used for the degradation of dyes such as Congo Red, Coomassie blue, Malachite green and methyl orange (Husain et al. 2022). Ag-NPs derived from *Microchaete*sp have been reported to demonstrate 84.6% removal efficiency of Methyl red in contaminated water sources (Husain et al. 2021). Aziz et al. reported the synthesis of Ag-NPs using extracts of *Chlorella* *pyrenoidosa* and demonstrated the degradation of methylene blue from waste water (Aziz et al. 2015). Ag-NPs have been synthesized using algal extracts of *Ulva lactuca* and was subsequently used for the degradation of methyl orange from contaminated water sources (Kumar et al. 2013). *Viburnum opulus* fruit extract has been used for the synthesis of Ag-NPs which was further used for the degradation of tartrazine, brilliant green and carmoisine (David and Moldovan 2020). Raj et al. reported the removal or degradation efficiency of 86.68% for methyl orange, 93.60% for methylene blue, 88.80% for 4- nitrophenol and 92.20% in case of congo red by Ag-NPs synthesized using *Terminalia arjuna* (Raj et al. 2020). Ag-NPs synthesized using *Phaseolus vulgaris* extract was used for removal of reactive red-141 dye with 97% efficiency (Rani et al. 2020). *Pestalotipsis versicolor* has been used for the synthesis of Ag-NPs and has been used for the degradation of azo dyes such as Rhodamine B, Congo red and Orange G (Rajput et al. 2017).

20.4 Conclusion

Research in the field of nanoscience has accelerated over the period of years due to various advantages offered by nanoparticles such as large surface area to volume ratio, small size and ease of production. There are various methods of synthesis of nanoparticles including physical, chemical and biological methods. Biogenic methods are widely used due non-toxic and sustainable nature of synthesis. The various biological agents used for the synthesis of NPs are plants and parts, fungi, bacteria and algae. Biogenic Ag-NPs have wide range of applications such as antimicrobial properties, anti-cancerous activities, environmental remediation. Researchers have conducted a large number of studies employing Ag-NPs for their antimicrobial properties and found out that Ag-NPs are one of the most prominent antimicrobial activities as compared to any other metallic and metallic oxide NPs. Ag-NPs have been found to be highly effective for environmental remediation strategies such as heavy metal removal and dye degradation from waste water. Thus, it can be concluded that biogenic Ag-NPs can be a sustainable alternative to the traditionally used antimicrobial agents which can also prevent the emergence of antimicrobial resistance. Biogenic Ag-NPs are also a cheap and highly effective alternative for dye degradation and removal of heavy metals form waste water.

Acknowledgements The authors would like to thank institution for constant support.

References

Abdelghany TM, Al-Rajhi AMH, Al Abboud MA, Alawlaqi MM, Ganash Magdah A, Helmy EAM et al (2018) Recent Advances in Green Synthesis of Silver Nanoparticles and Their Applications: About Future Directions. A Review. Bionanoscience 8:5–16. https://doi.org/10.1007/s12668-017-0413-3

Agarwal P, Gupta R, Agarwal N. Advances in Synthesis and Applications of Microalgal Nanoparticles for Wastewater Treatment. J Nanotechnol 2019;2019. https://doi.org/10.1155/2019/7392713.

Al-Qahtani KM (2017) Cadmium removal from aqueous solution by green synthesis zero valent silver nanoparticles with Benjamina leaves extract. Egypt J Aquat Res 43:269–274. https://doi.org/10.1016/j.ejar.2017.10.003

Amin HH (2020) Biosynthesized silver nanoparticles using Ulva lactuca as a safe synthetic pesticide (in vitro). Open Agric 5:291–299. https://doi.org/10.1515/opag-2020-0032

Annamalai J, Nallamuthu T (2016) Green synthesis of silver nanoparticles: characterization and determination of antibacterial potency. Appl Nanosci 6:259–265. https://doi.org/10.1007/s13204-015-0426-6

Arif R, Uddin R (2021a) A review on recent developments in the biosynthesis of silver nanoparticles and its biomedical applications. Med Devices Sensors 4:1–20. https://doi.org/10.1002/mds3.10158

Arif R, Uddin R (2021b) A review on recent developments in the biosynthesis of silver nanoparticles and its biomedical applications. Med DEVICES SENSORS 4:7–9. https://doi.org/10.1002/mds3.10158

Attatsi IK, Nsiah F (2020) Application of silver nanoparticles toward Co(II) and Pb(II) ions contaminant removal in groundwater. Appl Water Sci 10:1–13. https://doi.org/10.1007/s13201-020-01240-0

Aziz N, Faraz M, Pandey R, Shakir M, Fatma T, Varma A et al (2015) Facile Algae-Derived Route to Biogenic Silver Nanoparticles: Synthesis, Antibacterial, and Photocatalytic Properties. Langmuir 31:11605–11612. https://doi.org/10.1021/acs.langmuir.5b03081

Azizi S, Namvar F, Mahdavi M, Ahmad M Bin, Mohamad R. Biosynthesis of silver nanoparticles using brown marine macroalga, Sargassum muticum aqueous extract. Materials (Basel) 2013;6:5942–50. https://doi.org/10.3390/ma6125942.

Banerjee P, Satapathy M, Mukhopahayay A, Das P (2014) Leaf extract mediated green synthesis of silver nanoparticles from widely available Indian plants: synthesis, characterization, antimicrobial property and toxicity analysis. Bioresour Bioprocess 1:3. https://doi.org/10.1186/s40643-014-0003-y

Chaudhary R, Nawaz K, Khan AK, Hano C, Abbasi BH, Anjum S (2020a) An Overview of the Algae-Mediated Biosynthesis of Nanoparticles and Their Biomedical Applications. Biomolecules 10:1498. https://doi.org/10.3390/biom10111498

Chaudhary R, Nawaz K, Khan AK, Hano C, Abbasi BH, Anjum S (2020b) An overview of the algae-mediated biosynthesis of nanoparticles and their biomedical applications. Biomolecules 10:1–36. https://doi.org/10.3390/biom10111498

David L, Moldovan B. Green Synthesis of Biogenic Silver Nanoparticles for E ffi cient Catalytic Removal of Harmful Organic Dyes 2020.

Deshmukh N, Kadam D, Deshmukh P, Kathwate G (2022) Utilization of biosurfactant produced by Pseudomonas aeruginosa strain LTR1 for the synthesis of the silver nanoparticles and their application as antimicrobial agents. J Appl Biol Biotechnol 10:35–41. https://doi.org/10.7324/JABB.2022.100205

Dhavale R, Jadhav S, Sibi G (2020) Microalgae mediated Silver nanoparticles (AG-NPS) synthesis and their biological activities. J Crit Rev 7:15–20. https://doi.org/10.31838/jcr.07.02.04

Dotto GL, Cadaval TRS, Pinto LAA (2012) Preparation of bionanoparticles derived from Spirulina platensis and its application for Cr (VI) removal from aqueous solutions. J Ind Eng Chem 18:1925–1930. https://doi.org/10.1016/j.jiec.2012.05.005

Edison TNJI, Atchudan R, Kamal C, Lee YR (2016) Caulerpa racemosa: a marine green alga for eco-friendly synthesis of silver nanoparticles and its catalytic degradation of methylene blue. Bioprocess Biosyst Eng 39:1401–1408. https://doi.org/10.1007/s00449-016-1616-7

El-Rafie HM, El-Rafie MH, Zahran MK (2013) Green synthesis of silver nanoparticles using polysaccharides extracted from marine macro algae. Carbohydr Polym 96:403–410. https://doi.org/10.1016/j.carbpol.2013.03.071

El-Tawil RS, El-Wakeel ST, Abdel-Ghany AE, Abuzeid HAM, Selim KA, Hashem AM (2019) An Overview of the Algae-Mediated Biosynthesis of Nanoparticle. Heliyon 5:1–9. https://doi.org/10.1016/j.heliyon.2019.e02415

Feroze N, Arshad B, Younas M, Afridi MI, Saqib S, Ayaz A (2020) Fungal mediated synthesis of silver nanoparticles and evaluation of antibacterial activity. Microsc Res Tech 83:72–80. https://doi.org/10.1002/jemt.23390

Firdhouse MJ, Lalitha P (2015) Biosynthesis of Silver Nanoparticles and Its Applications. J Nanotechnol 2015:1–18. https://doi.org/10.1155/2015/829526

Ganapathy Selvam G, Sivakumar K. Phycosynthesis of silver nanoparticles and photocatalytic degradation of methyl orange dye using silver (Ag) nanoparticles synthesized from Hypnea musciformis (Wulfen) J.V. Lamouroux. Appl Nanosci 2015;5:617–22. https://doi.org/10.1007/s13204-014-0356-8.

Garibo D, Borbón-Nuñez HA, de León JND, García Mendoza E, Estrada I, Toledano-Magaña Y et al (2020) Green synthesis of silver nanoparticles using Lysiloma acapulcensis exhibit high-antimicrobial activity. Sci Rep 10:12805. https://doi.org/10.1038/s41598-020-69606-7

Gour A, Jain NK (2019) Advances in green synthesis of nanoparticles. Artif Cells, Nanomedicine Biotechnol 47:844–851. https://doi.org/10.1080/21691401.2019.1577878

Guilger-Casagrande M, de Lima R (2019) Synthesis of Silver Nanoparticles Mediated by Fungi: A Review. Front Bioeng Biotechnol 7:1–16. https://doi.org/10.3389/fbioe.2019.00287

Huq MA, Akter S (2021) Bacterial Mediated Rapid and Facile Synthesis of Silver Nanoparticles and Their Antimicrobial Efficacy against Pathogenic Microorganisms. Materials (basel) 14:2615. https://doi.org/10.3390/ma14102615

Husain S, Afreen S, Hemlata, Yasin D, Afzal B, Fatma T. Cyanobacteria as a bioreactor for synthesis of silver nanoparticles-an effect of different reaction conditions on the size of nanoparticles and their dye decolorization ability. J Microbiol Methods 2019;162:77–82. https://doi.org/10.1016/j.mimet.2019.05.011.

Ibrahim S, Ahmad Z, Manzoor MZ, Mujahid M, Faheem Z, Adnan A (2021) Optimization for biogenic microbial synthesis of silver nanoparticles through response surface methodology, characterization, their antimicrobial, antioxidant, and catalytic potential. Sci Rep 11:770. https://doi.org/10.1038/s41598-020-80805-0

Jain S, Mehata MS. Medicinal Plant Leaf Extract and Pure Flavonoid Mediated Green Synthesis of Silver Nanoparticles and their Enhanced Antibacterial Property. Sci Rep 2017:1–13. https://doi.org/10.1038/s41598-017-15724-8.

Jain S, Mehata MS (2017a) Medicinal Plant Leaf Extract and Pure Flavonoid Mediated Green Synthesis of Silver Nanoparticles and their Enhanced Antibacterial Property. Sci Rep 7:15867. https://doi.org/10.1038/s41598-017-15724-8

Khaleelullah MMSI, Murugan M, Radha K V., Thiyagarajan D, Shimura Y, Hayakawa Y. Synthesis of super-paramagnetic iron oxide nanoparticles assisted by brown seaweed Turbinaria decurrens for removal of reactive navy blue dye. Mater Res Express 2017;4. https://doi.org/10.1088/2053-1591/aa9131.

Kolahalam LA, Kasi Viswanath IV, Diwakar BS, Govindh B, Reddy V, Murthy YLN (2019) Review on nanomaterials: Synthesis and applications. Mater Today Proc 18:2182–2190. https://doi.org/10.1016/j.matpr.2019.07.371

Kumar P, Govindaraju M, Senthamilselvi S, Premkumar K (2013) Photocatalytic degradation of methyl orange dye using silver (Ag) nanoparticles synthesized from Ulva lactuca. Colloids Surfaces B Biointerfaces 103:658–661. https://doi.org/10.1016/j.colsurfb.2012.11.022

LewisOscar F, Vismaya S, Arunkumar M, Thajuddin N, Dhanasekaran D, Nithya C. Algal Nanoparticles: Synthesis and Biotechnological Potentials. Algae - Org. Imminent Biotechnol., 2016, p. 13. https://doi.org/10.5772/62909.

Li X, Xu H, Chen ZS, Chen G. Biosynthesis of nanoparticles by microorganisms and their applications. J Nanomater 2011;2011. https://doi.org/10.1155/2011/270974.

Liu X, Chen J-L, Yang W-Y, Qian Y-C, Pan J-Y, Zhu C-N et al (2021) Biosynthesis of silver nanoparticles with antimicrobial and anticancer properties using two novel yeasts. Sci Rep 11:15795. https://doi.org/10.1038/s41598-021-95262-6

Meroni G, Filipe JFS, Martino PA. In vitro antibacterial activity of biological-derived silver nanoparticles: Preliminary data. Vet Sci 2020;7. https://doi.org/10.3390/vetsci7010012.

Mohd Yusof H, Abdul Rahman N, Mohamad R, Zaidan UH (2020) Microbial Mediated Synthesis of Silver Nanoparticles by Lactobacillus Plantarum TA4 and its Antibacterial and Antioxidant Activity. Appl Sci 10:6973. https://doi.org/10.3390/app10196973

Pahal V, Kumar P, Kumar P, Kumar V (2022) Antibacterial activity and hormetic response of silver nanoparticles synthesized using leaflet extract of wheat (Triticum aestivum) and rice (Oryza sativa) crop plants. J Appl Biol Biotechnol 10:154–167. https://doi.org/10.7324/JABB.2022.100219

Patel A, Enman J, Gulkova A, Guntoro PI, Dutkiewicz A, Ghorbani Y, et al. Integrating biometallurgical recovery of metals with biogenic synthesis of nanoparticles. Chemosphere 2021;263. https://doi.org/10.1016/j.chemosphere.2020.128306.

Peiris MMK, Fernando SSN, Jayaweera PM, Arachchi NDH, Guansekara TDCP (2018) Comparison of Antimicrobial Properties of Silver Nanoparticles Synthesized from Selected Bacteria. Indian J Microbiol 58:301–311. https://doi.org/10.1007/s12088-018-0723-3

Pirtarighat S, Ghannadnia M, Baghshahi S (2019) Green synthesis of silver nanoparticles using the plant extract of Salvia spinosa grown in vitro and their antibacterial activity assessment. J Nanostructure Chem 9:1–9. https://doi.org/10.1007/s40097-018-0291-4

Qamer S, Romli MH, Che-Hamzah F, Misni N, Joseph NMS, AL-Haj NA, et al. Systematic Review on Biosynthesis of Silver Nanoparticles and Antibacterial Activities: Application and Theoretical Perspectives. Molecules 2021;26:5057. https://doi.org/10.3390/molecules26165057.

Rai M, Bonde S, Golinska P, Trzcińska-Wencel J, Gade A, Abd-Elsalam KA et al (2021) Fusarium as a Novel Fungus for the Synthesis of Nanoparticles: Mechanism and Applications. J Fungi 7:139. https://doi.org/10.3390/jof7020139

Raj S, Singh H, Trivedi R, Soni V. Biogenic synthesis of AgNPs employing Terminalia arjuna leaf extract and its efficacy towards catalytic degradation of organic dyes 2020:1–10. https://doi.org/10.1038/s41598-020-66851-8.

Rajput K, Sharma J, Agrawal PK (2017) Mycosynthesis of silver nanoparticles using endophytic fungus Pestalotiopsis versicolor and investigation of its antibacterial and azo dye degradation efficacy. Kavaka 71:65–71

Ramakrishna M, Rajesh Babu D, Gengan RM, Chandra S, Nageswara RG (2016) Green synthesis of gold nanoparticles using marine algae and evaluation of their catalytic activity. J Nanostructure Chem 6:1–13. https://doi.org/10.1007/s40097-015-0173-y

Rani P, Kumar V, Singh PP, Matharu AS, Zhang W, Kim K-H et al (2020) Highly stable AgNPs prepared via a novel green approach for catalytic and photocatalytic removal of biological and non-biological pollutants. Environ Int 143:105924. https://doi.org/10.1016/j.envint.2020.105924

Rauwel P, Küünal S, Ferdov S, Rauwel E (2015) A Review on the Green Synthesis of Silver Nanoparticles and Their Morphologies Studied via TEM. Adv Mater Sci Eng 2015:1–9. https://doi.org/10.1155/2015/682749

Rout GK, Shin HS, Gouda S, Sahoo S, Das G, Fraceto LF et al (2018) Current advances in nanocarriers for biomedical research and their applications. Artif Cells, Nanomedicine Biotechnol 46:1053–1062. https://doi.org/10.1080/21691401.2018.1478843

Sachin K, Karn SK (2021) Microbial Fabricated Nanosystems: Applications in Drug Delivery and Targeting. Front Chem 9:1–15. https://doi.org/10.3389/fchem.2021.617353

Saeed S, Iqbal A, Ashraf MA (2020) Bacterial-mediated synthesis of silver nanoparticles and their significant effect against pathogens. Environ Sci Pollut Res 27:37347–37356. https://doi.org/10.1007/s11356-020-07610-0

Saleh MN, Khoman AS (2020) Bio-synthesis of silver nanoparticles from bacteria Klebsiella pneumonia: Their characterization and antibacterial studies. J Phys Conf Ser 1664:012115. https://doi.org/10.1088/1742-6596/1664/1/012115

Sánchez-López E, Gomes D, Esteruelas G, Bonilla L, Lopez-Machado AL, Galindo R et al (2020) Metal-based nanoparticles as antimicrobial agents: An overview. Nanomaterials 10:1–39. https://doi.org/10.3390/nano10020292

Sharma A, Sharma S, Sharma K, Chetri SPK, Vashishtha A, Singh P et al (2016) Algae as crucial organisms in advancing nanotechnology: a systematic review. J Appl Phycol 28:1759–1774. https://doi.org/10.1007/s10811-015-0715-1

Siddiqi KS, Husen A, Rao RAK (2018) A review on biosynthesis of silver nanoparticles and their biocidal properties. J Nanobiotechnology 16:14. https://doi.org/10.1186/s12951-018-0334-5

Thekkudan VN, Vaidyanathan VK, Ponnusamy SK, Charles C, Sundar SL, Vishnu D et al (2017) Review on nanoadsorbents: A solution for heavy metal removal from wastewater. IET Nanobiotechnol 11:213–224. https://doi.org/10.1049/iet-nbt.2015.0114

Vanlalveni C, Lallianrawna S, Biswas A, Selvaraj M, Changmai B, Rokhum SL (2021) Green synthesis of silver nanoparticles using plant extracts and their antimicrobial activities: a review of recent literature. RSC Adv 11:2804–2837. https://doi.org/10.1039/D0RA09941D

Velhal SG, Kulkarni SD, Latpate RV (2016) Fungal mediated silver nanoparticle synthesis using robust experimental design and its application in cotton fabric. Int Nano Lett 6:257–264. https://doi.org/10.1007/s40089-016-0192-9

Wang L, Hu C, Shao L (2017) The antimicrobial activity of nanoparticles: present situation and prospects for the future. Int J Nanomedicine 12:1227–1249. https://doi.org/10.2147/IJN.S121956

Wang D, Xue B, Wang L, Zhang Y, Liu L, Zhou Y (2021) Fungus-mediated green synthesis of nano-silver using Aspergillus sydowii and its antifungal/antiproliferative activities. Sci Rep 11:10356. https://doi.org/10.1038/s41598-021-89854-5

Yang J, Hou B, Wang J, Tian B, Bi J, Wang N, et al. Nanomaterials for the removal of heavy metals from wastewater. Nanomaterials 2019;9. https://doi.org/10.3390/nano9030424.

Yeh YC, Huang TH, Yang SC, Chen CC, Fang JY (2020) Nano-Based Drug Delivery or Targeting to Eradicate Bacteria for Infection Mitigation: A Review of Recent Advances. Front Chem 8:1–22. https://doi.org/10.3389/fchem.2020.00286

MIX
Papier aus verantwortungsvollen Quellen
Paper from responsible sources
FSC® C105338

FSC
www.fsc.org

If you have any concerns about our products,
you can contact us on
ProductSafety@springernature.com

In case Publisher is established outside the EU,
the EU authorized representative is:
Springer Nature Customer Service Center GmbH
Europaplatz 3, 69115 Heidelberg, Germany

Printed by Libri Plureos GmbH
in Hamburg, Germany